ELEMENTS OF PETROCHEMICAL ENGINEERING

PROF. SAIKAT MAITRA

Vice-Chancellor, Maulana Abdul Kalam Azad University of Technology, West Bengal (Formerly West Bengal University of Technology, West Bengal)

OM PRAKASH GUPTA

Ex DGM, HRD
Bokaro Steel Plant
Steel Authority of India Limited (SAIL)

KHANNA BOOK PUBLISHING CO. (P) LTD.

Publisher of Science, Technology and Engineering Books

4C/4344, Ansari Road, Darya Ganj, New Delhi - 110002
Phone: 011-23244447-48, **Mobile:** +91-99109 09320
E-mail: contact@khannabooks.com
Website: www.khannabooks.com

CONTAINS MORE THAN 200 MULTICHOICE QUESTIONS & ANSWERS AND ABOUT 125 SHORT QUESTIONS & ANSWERS DRAWN FROM EXAMINATION PAPERS OF VARIOUS UNIVERSITIES & TECHNICAL INSTITUTES

Price : ₹ 425.00

Elements of Petrochemical Engineering
Om Prakash Gupta

Copyright © Khanna Book Publishing Co. (P) Ltd.

This book is sold subject to the condition that it shall not, by way of trade or otherwise, be lent, resold, hired out, or otherwise circulated without the publisher's prior consent in any form of binding or cover other than that in which it is published and without a similar condition including this condition being imposed on the subsequent purchaser and without limiting the rights under copyright reserved above, no part of this publication may be reproduced, stored in or introduced into retrieval system, or transmitted any form or by any means (electronic, mechanical, photocopying, recording or otherwise), without the prior written permission of both the copyright owner and the above mentioned publisher of this book.

ISBN : 978-93-86173-36-2

Edition: 2017
Reprint: 2020

Published by:

KHANNA BOOK PUBLISHING CO. (P) LTD.
4C/4344, Ansari Road, Darya Ganj, New Delhi-110002
Phone: 011-23244447/48 **Mobile:** 91-9910909320
E-Mail: contact@khannabooks.com

Printed in India by:
Vaibhav Products

PREFACE TO THE FIRST EDITION

Petrochemicals are compounds & polymers derived from petroleum and used in the chemical market. Among the major petrochemical products are plastics, synthetic fibers, synthetic rubber, detergents, & nitrogeneous fertilizers. Many other important chemical industries such as paints, adhesives, aerosols, insecticides, & pharmaceuticals may involve one or more petrochemical products within their manufacturing steps.

It gives me immense pleasure to present this book titled Elements of Petrochemical Engineering. This book is meant for the students, teachers and practicing engineers.

To prepare a book on this subject in its true sense would be very exhaustive requiring many volumes to do real justice to the subject. However, attempt has been made to assimilate the essentials of the subject in lucid manner in a single volume.

This book contains the manufacture, properties & applications of important petrochemicals in 34 chapters. Important informations about feedstocks & applications of petrochemical derived products, Status of Indian petrochemical industry and environmental standards for the petrochemical plant are given in the appendices. It also contains short questions & answers and multiple choice questions & answers drawn from the examination papers of various engineering colleges for the benefits of the students.

This book is targetted to benefit the following :

- Diploma in engineering students
- Degree in engineering students
- AMIE, AMIIM, AMIICHE Students
- Faculty members and teaching staff
- Practising engineers/professionals

The scope and coverage of the subject matter presented can be gauged from glancing the contents portion of the book to judge its suitability and relevance for a particular target group.

Extensive use of internet resources have been made in gathering the relevant informations and the subject matter has been presented in a simple, lucid and easy to understand manner for easy grasping of the subject.

Latest and updated informations/data/statistics pertaining to the subject matter has been included in this edition for the benefit of the readers.

Constructive criticisms and suggestion for further improvement of the book are solicited.

BRIEF TABLE OF CONTENTS

DETAILED TABLE OF CONTENTS

Chapter 1

INTRODUCTION TO PETROCHEMICALS

Petrochemicals are chemical compounds elements recovered derived partially entirely from petroleum or natural gas hydrocorbons intended for chemical markets.

The petrochemical industry produces various kinds of chemical products such as polymers, fibers or rubber, from such raw materials as petroleum, LPG, natural gas and other hydrocarbons through many different production processes.

Hydrocarbons, the source material, are used to produce a variety of components including ethylene, propylene, butadiene and pyrolysis gasoline through non-catalytic thermal decomposition reaction with steam (steam cracking).

The feedstock to ethylene process varies depending on the availability of resources in each country. For example, in Japan, naphtha (crude gasoline) produced by refining crude oil is often used. In recent years, many ethylene plants that use light gas as a feedstock which comes from refinery and natural gas plant are built in Middle East. When light gas is used as a feedstock, products other than ethylene are produced in smaller quantities, which leads to plants that produce propylene by dehydrating propane'.

The petrochemical industry is wide-ranging, creating a variety of chemical products such as shown below:

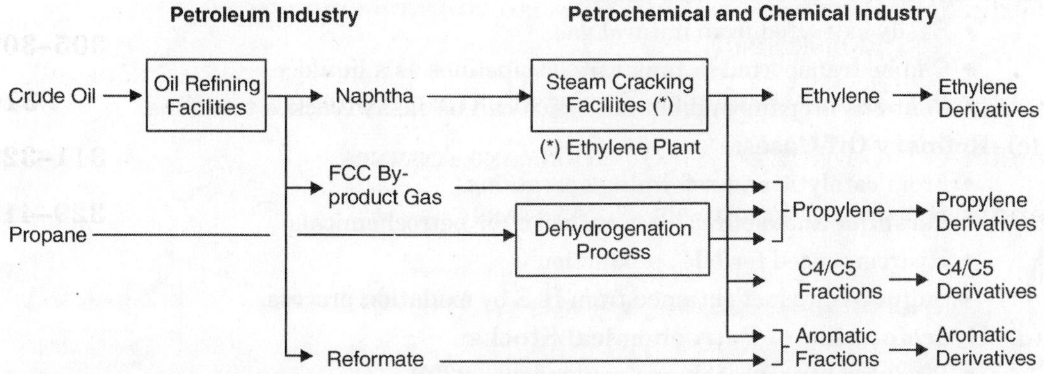

Fig. 1.1

About 5 % of the oil and natural gas consumed each year is needed to make all the petrochemical products. Petrochemicals play an important role in our food, clothing, shelter and leisure. Because of low cost and easy availability, oil & natural gas are the main sources of raw materials for most petrochemicals.

CLASSIFICATION OF PETROCHEMICALS

Petrochemicals are broadly classified into three categories:

(a) **Light Petrochemicals:** These are mainly used as bottled fuel and raw materials for other organic chemicals. The lightest of these—methane, ethane and ethylene—are gaseous at room temperature. The next lightest fractions comprise petroleum ether and light naphtha with boiling points between 27 and 88ºC.

(b) **Medium Petrochemicals:** Hydrocarbons with 6–12 carbon atoms are called "gasoline", which are mainly used as automobile fuels. Octane, with eight carbons, is a particularly good automobile fuel, and is considered to be of high quality. Kerosene contains 12 to 15 carbons and is used in aviation fuels, and also as solvents for heating & lighting.

(c) **Heavy Petrochemicals:** These can be generally categorized as diesel oil, heating oil and lubricating oil for engines & machinery. They contain around 15 and 18 carbon atoms with boiling points between 300 and 400ºC. The heaviest fractions of all are called "bitumens" and are used to surface roads or for waterproofing. Bitumens can also be broken down into lighter hydrocarbons using a process called "Cracking".

Principal Raw Materials for petrochemicals are as follows:

(a) **Natural Gas (largely methane):**

- Major raw material in U.S., now available in large quantity in India also.
- Ethane (~4% in natural gas) is the most valuable product since ethylene can be produced by thermal cracking:

  Ethane <u>cracking</u> Ethylene (75% yield)

  However, most of the ethylene comes from refinery gases.

(b) **Liquefied Petroleum Gas (LPG):**

- Easily extracted from natural gas.
- Can be transported in tank cars or pipelines as a liquid.
- Contains propane and butane which can be easily cracked to olefins.

(c) **Refinery Off-Gases:**

- From catalytic and reforming operations.
- Paraffinic and olefinic fractions used for petrochemicals.
- Hydrogen used for NH_3 production.
- Sulfur by-product obtained from H_2S by oxidation process.

(d) **Hydroforming of Petrochemical Stocks:**

- Produces benzene, toluene, and xylene (BTX)
- Gives large quantities of hydrogen.

(e) **Naphtha and Fuel Oil:**

- Accumulate in large quantities in Indian refineries due to lack of end use.

- Used to make (a) synthesis gas by catalytic reforming or partial combustion, (b) petrochemical starting materials such as ethylene, acetylene, and propylene by steam cracking.

(f) Petroleum Coke:

- Source of carbon for calcium carbide and acetylene or electrodes for aluminum industry.

These raw materials contain one or several predominant hydrocarbons which can be separated so that compounds are synthesized as follows:

- Chemicals From C_1 Compounds (Methane and Synthesis Gas).
- Chemicals From C_2 Compounds (Ethylene and Acetylene).
- Chemicals From C_3 Compounds (Propylene).
- Chemicals From C_4 Compounds (Butanes and Butenes)
- Chemicals From Aromatics (Benzene, Toluene and Xylene)
- Pesticides.
- Pharmaceutical Industry.
- Polymers.

PETROCHEMICALS

There are three basic group of raw materials that make up major part of Petrochemical industry:

- Synthesis Gas from Coal or Natural Gas Hydrogen, Ammonia, Urea, Methanol, GTL/ CTL and derivatives
- **Aromatics:**
  Benzene, Toluene and Xylenes (BTX) and its derivatives
- **Olefines:**
  Ethylene, Propylene and its derivatives

Basic Feed Stocks for Petrochemicals are:

- Ethylene
- Propylene
- Para-xylene
- Butadiene

Basic Building Blocks for Petrochemicals are:

- Olefin
- Aromatics
- Synthesis Gas

Major Source of Olefins are:

- Naphtha Cracker
- Gas Cracker
- FCC (Fluid Catalytic Cracker)

Major Sources of Aromatics are:

©
- Naphtha Catalytic Reforming
- Pyrolysis Gasoline from Cracker

Primary Feed Stock-naphtha

Route	Secondary Feed Stock/Products
Aromatization	BTX (Benzene, Toluene, Xylene)
Cracking	C_2, C_3, C_4 & Higher Olefins

Feed Stock From Gas & LPG of FCC (Fluid Catalytic Cracker):

- C_2 Oleffins (Ethylene and Acetylene)
- C_3 Oleffin (Propylene)
- C_4 Olefin (Butene)

The most important petrochemicals are:

- Lower alkenes (olefins): e.g. ethylene, propylene and butadiene.
- Aromatics: benzene, toluene, xylene (BTX)
- Ammonia, methanol, synthesis gas (H_2 + CO)
  - Most final products are produced directly or indirectly from the previous compounds (building blocks).
  - The feedstock chosen for the production of the chemical products depends on availability, prices & production units.
  - For the production of light alkenes, there is a difference, for example, between the USA and the rest of the world (e.g., Europe and Japan). In the USA, the production of light alkenes is from alkanes such as ethane and propane. In Europe and Japan, the production of light alkenes is from naphtha. In Saudi Arabia, ethane is the main source for the production of lower alkenes. Figures below shows the production of lower alkenes from crude oil and natural gas, respectively.

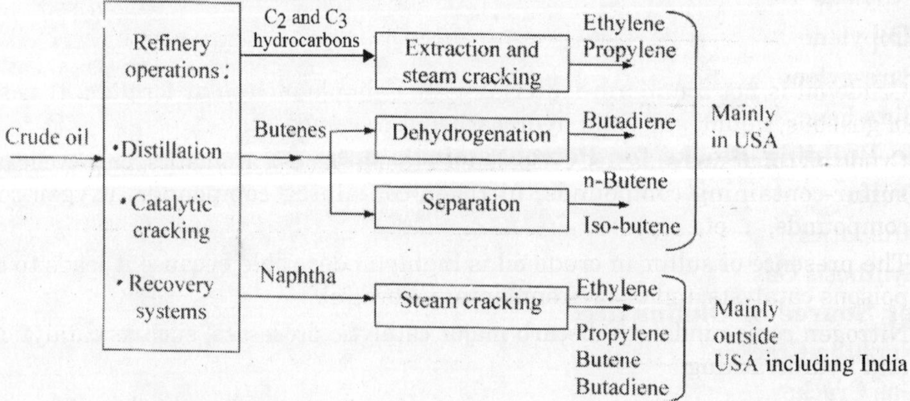

Figure 1.2: Lower alkenes (olefins) production from oil.

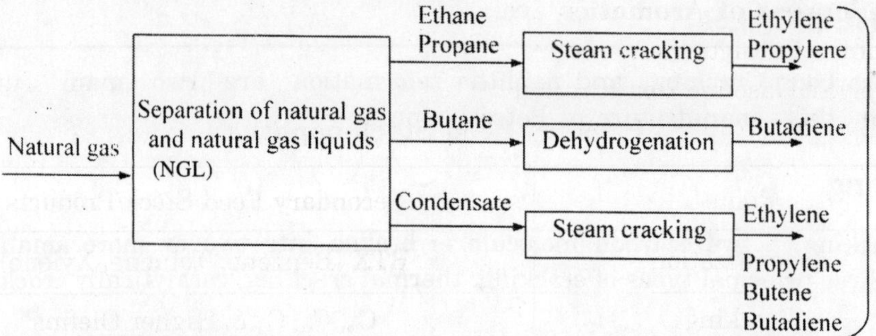

Figure 1.3: Lower alkenes (olefins) from natural gas.

Natural Gas

- Natural gas is a mixture of hydrocarbons with methane as the main constituent (upto 90%). It can be found in porous reservoirs, either associated with crude oil (associated gas) or in reservoirs in which no oil is present (non-associated gas).
- Natural gas is important as a source of energy and also as a raw material for the petrochemical industry.
- Natural gas usually contains small amounts of non-hydrocarbon gases such as carbon dioxide, nitrogen, and hydrogen sulfide.
- Natural gas classifies as 'dry' or 'wet' natural gas. Dry natural gas contains only small amounts of condensable hydrocarbons (at ambient temperature.). Non-associated gas is usually dry.
- Natural gas contains substantial amounts of ethane, propane, butane, and C5 + hydrocarbons, which liquefy on compression at ambient temperature called natural gas liquids (NGL).
- The terms 'sweet' and 'sour' natural gas denote the absence or presence of H_2S and mercaptans (RSH).
- Condensable hydrocarbons are removed from a wet natural gas and part can be sold as liquefied petroleum gas (LPG).

Crude Oil

- Crude oil is not a uniform material with a simple molecular formula. It is a mixture of gaseous, liquid, and solid hydrocarbon compounds.
- Crude oil conatains alkanes, cycloalkanes (naphthenes), aromatics, poly-cyclic aromatics, sulfur-containing compounds, nitrogen-containing compounds, oxygen-containing compounds, ... etc.
- The presence of sulfur in crude oil is highly undesirable because it leads to corrosion, poisons catalysts and is environmentally harmful.
- Nitrogen compounds can disturb major catalytic processes, such as catalytic cracking and hydrocracking.

UNIT OPERATIONS IN MANUFACTURING PROCESSES OF PETROCHEMICALS

Hydrocarbons Cracking and naphtha reformation are two main unit operations involved in them manufacture of Petrochemicals:

(a) Cracking

In cracking, a hydrocarbon molecule is broken into two or more smaller fragments. There are three principal types of cracking: thermal cracking, catalytically cracking and hydro cracking.

Thermal cracking for fuel production is performed by subjecting a feedstock to temperature usually in excess of 455°C and at above atmospheric pressures with the objective of converting a residual crude fraction or a heavy distillate into gasoline and light distillate.

Catalytic cracking is performed in presence of a catalyst at temperature between 455°-540°C and at above atmospheric pressure. The process 'converts a distillate feedstock into gasoline as the primary product with production of light hydrocarbons.

Hydro cracking process operates at elevated pressure in the presence of hydrogen and catalyst at temperature generally less than 430°C.

(b) Reforming

The purpose of reforming naphtha is to rearrange or reform the molecular structure of hydrocarbon, particularly with the objective of producing aromatics. The chemical processes involved in reformation are as follows:

- Dehydrogenation of naphthenes to aromatics.
- Dehydrocyclisation of paraffins to form aromatics.
- Isomerisation of paraffins to more highly branched isomers

As the reaction proceeds, the reformed products will contain increasing concentration of aromatics and decreasing concentration of heavy paraffins. These reforming reactions are regulated by metal catalysts in an environment of hydrogen under moderate pressure.

Petrochemical Process Requirements

- **Reactors:** Reactors are the most important units in petrochemical processes. Petrochemicals are manufactured by simple reactions using relatively purer feedstocks. Therefore, reaction chemistry for petrochemicals manufacture is very well established All petrochemical processes heavily depend upon chemical transformation to first product and the purification.
- **Separation:** With distillation being the most important unit operation to separate the unreacted feed and generated petrochemical product, the separation processes also play a major role in the process. Where multiple series parallel reactions are involved, the separation process assumes a distillation sequence to separate all products from the feed. A characteristic feed recycle will be also existent in the process topology. Apart from this, other separation technologies used in petrochemical processing units include phase separators, gravity settling units and absorption columns. All these separation technologies is well exploited to achieve the desired separation.

- **Dependence on Reaction pathway:** A petrochemical can be produced in several ways from the same feedstock. For instance, phenol can be produced using the following pathways.
  - Peroxidation of Cumene followed by hydrolysis of the peroxide.
  - Two stage oxidation of Toluene.
  - Chlorination of Benzene and hydrolysis of chloro-benzene.
  - Direct oxidation of Benzene.

  In the above reaction schemes, there are two reaction pathways for phenol from benzene i.e., either chlorination of benzene or oxidation of benzene. Therefore, choosing the most appropriate technology for production is a trivial task.
- **Complexity in pathway:** In the above Cumene example case, it is to be noted that toluene hydrodealkylation produces benzene which can be used to produce phenol. Therefore, fundamentally toluene is required for the generation of various petrochemicals such as benzene and phenol. There is no hard and fast rule to say that a petrochemical is manufactured using a suggested route or a suggested intermediate petrochemical. Intermediate petrochemicals play a greater role in consolidating the manufacture of other downstream petrochemicals.

Petrochemicals Product Catagories is shown below:

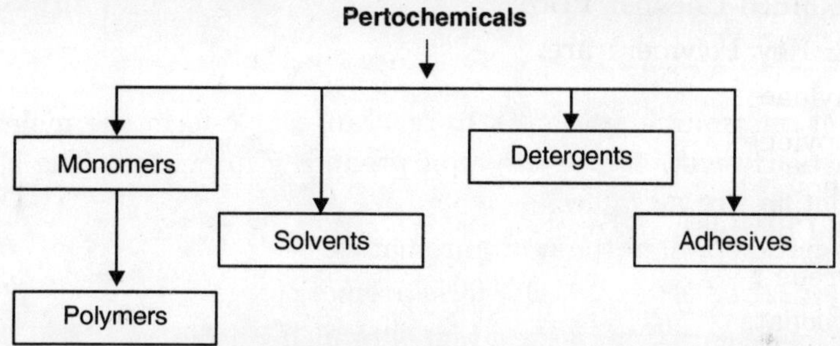

Types of Polymers Produced from Petrochemicals is Shown Below:

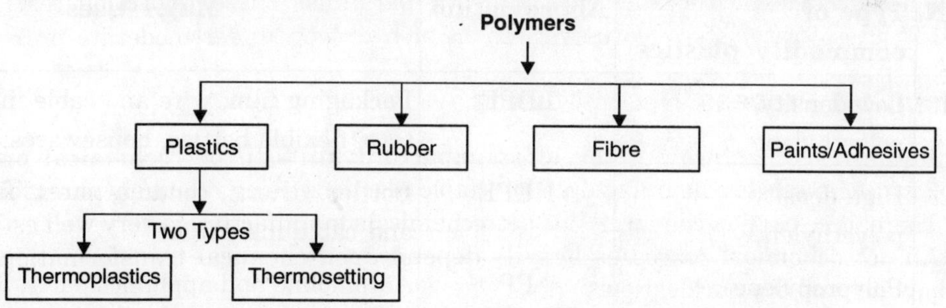

Polymeric materials with specific properties and having different application areas are:

- **Plastics:** polymer characterized by rigidity & stiffness having low recoverable deformation
- **Rubber:** polymer characterized by flexibility & high amount of recoverable deformation.

- **Fiber:** polymer having high melting point with high amount of crystallinity when stretched.
- **Paints & adhesives:** polymer with tackiness enabling good bond strength.

Advatages of Plastics are as follows:

- Properties can be altered to suit the needs
- Good strength/weight ratio
- Tough & durable
- Not toxic
- Easy processability
- Chemical/Moisture Resistant
- Fire retardant
- Do not rust/corrode.
- Resistant to microbes
- Attactive colours
- Good weatherability
- Practically maintenance free
- Reusable/Recyclable

Available At Much Cheaper Price

Important & Key Polymers are:

- Poly Ethylene
- Poly Propylene
- Poly Esters
- Polyvinyl Chloride
- Poly Styrene
- Poly Carbonate

Important Types of Commodity plastics are tabalated below:

S.N.	Type of commodity plastics	Abbreviation	Major Uses
1.	Low-density polyethylene	LDPE	Packaging film, wire and cable insulation, toys, flexible bottles, housewares, coatings
2.	High-density polyethylene	HDPE	Bottles, drums, conduit, sheet, film, wire and cable insulation
3.	Polypropylene	PP	Automobile and appliance parts, furniture, cordage, webbing, carpeting, film packaging
4.	Poly (vinyl chloride)	PVC	Construction, rigid pipe flooring, wire and cable insulation, film and sheet
5.	Polystyrene	PS	Packaging (foam and film), foam insulation, appliances, Housewares, toys

Principal Types of Engineering Plastics are listed below:

S.N.	Type of engineering plastics	Abbreviation	
1.	Acetal[a]	POM	
2.	Polymide[b]	—	
3.	Polyamideimide	PAI	
4.	Polyarylate	—	
5.	Polybenzimidazole	PBI	
6.	Polycarbonate	PC	
7.	Polyester[c]	—	
8.	Polyether ether ok ketone	PEEK	
9.	Polyetherimide	PEI	
10.	Polyimide	PI	
11.	Poly(phenylene oxide)	PPO	
12.	Poly(phenylene sulfide)	PPS	
13.	Polysulfone[d]	—	

[a] Common name for polyform aldehyde, Abbreviation refers to" poly (oxynethylene).

[b] Principally nylons 6 and 66.

[c] Principally polyethylene terephthalate (PET) and poly (buty/lene terephthalate) (PBT).

[d] Several types marketed.

Principal Types of Thermosetting Plastics are tabulated below:

S.N.	Type of Thermo-setting plastics	Abbreviation	Typical Uses
1.	Phenol–formaldehyde	PF	Electrical and electronic equipment, automobile parts, utensil handles, plywood adhesives Particle board binder
2.	Urea-formaldehyde	UF	Similar to PF polymers; also treatment of textiles, coatings
3.	Unsaturated polyester	UP	Construction, automobile parts, boat hulls, marine accessories, corrosion-resistant ducting, pipe, tanks, etc., business equipment
4.	Epoxy	—	Protective coatings, adhesives, electrical and electronics applications, industrial flooring, highway paving materials, composites
5.	Melamine-formaldehyde	MF	Similar to UF polymers; decorative panels, counter and table tops, dinnerware

Principal Types of Synthetic Fibres are tabulated below:

S.N.	Type of Synthetic fibres	Description
	Cellulosic:	
1.	Acetate rayon	Cellulose acetate
2.	Viscose rayon	Regenerated cellulose
	Noncellulosic:	
3.	Polyester	Principally poly(ethylene terephthalate)
4.	Nylon	Includes nyton 66, nylon 6, and a variety of other aliphatic and aromatic polyamides
5.	Olefin	Includes polypropylene and copolymers of vinyl chloride, with lesser amounts of acrylonitrile, vinyl acetate, or vinylidene chloride (copolymers ccnsisting of more than 85% vinyl chloride are called *vinyon* fibers)
6.	Acrylic	Contain at least 80% acrylonitrile; included are *modacrylic* fibers comprising acrylonitrile and about 20% vinyl chloride or vinylidene chloride

Principal Types of Synthetic Rubbers are tabulated below:

S.N.	Type	Description
1.	Styrene-butadiene	Copolymer of the two monomers in various proportions depending on properties desired; called SBR for styrene-butadiene rubber.
2.	Poly butadiene	Consists almost entirely of the cis-1,4 polymer
3.	Ethylene-propylene	Often abbreviated EPDM for ethylene-propylene-diene monomer; made up principally of ethylene and propylene units with small amounts of a diene to provide unsaturation.
4.	Polychloroprene	Principally the trans-1,4 polymer, but also some cis-1,4 and 1,2 polymer; also known as *neoprene* rubber.
5.	Polyisoprene	Mainly the cis-1,4 polymer; sometimes called "synthetic natural rubber".
6.	Nitrile	Copolymer of acrylonitrile and butadiene, mainly the latter.
7.	Butyl	Copolymer of isobutylens and isoprene, with only small amounts of the latter.
8.	Silicone	Contains inorganic backbone of alternating oxygen and methylated silicon atoms; also called polysiloxane
9.	Urethane	Elastomers prepared by linking polyethers through urethane groups.

Chapter **2**

PRODUCTION OF METHANOL (METHYL ALCOHAL)

INTRODUCTION

When 'synthesis gas' (H_2 + CO) is subjected to high pressure and moderate temperature conditions, it converts to methanol.

Followed by this, the methanol is separated using a series of phase separators and distillation columns.

Chemical Reactions involved in methanol production from synthesis gas (CO + H_2):

Desired/reaction: $CO + 2H_2 \rightarrow CH_3OH$

Side reactions: $CO + 3H_2 \rightarrow CH_4 + H_2O$

$2CO + 2H_2 \rightarrow CH_4 + CO_2$

All above reactions are exothermic.

Undesired reaction: $zCO + aH_2 \rightarrow$ alcohols + hydrocarbons

A flowsheet of Process Technology involed in methanol production from synthesis gas is shown below:

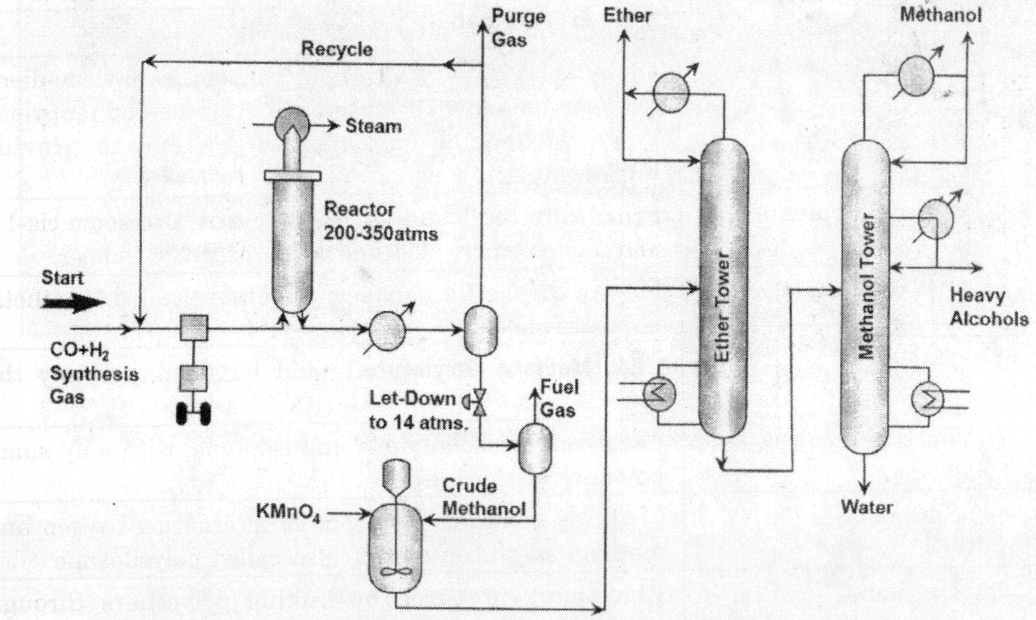

Figure 2.1: Flow sheet of manufacture of methanol from Synthsis Gas

PROCESS DESCRIPTION OF METHANOL PRODUCTION

Hydrogen and carbonmonoxide in a mole ratio of 2.25 is compressed to 200-350 atm mixed with recycle gas, and fed to a high pressure converter. Internal preheat is usually employed. The reactor is copper-lined steel and contains a mixed catalyst of zinc, chromium, manganese, or aluminum oxides. The temperature is maintained at 300-375°C.

The exit gases are cooled by heat exchange with reactants, then with water. The methanol condenses under full operating pressure to maximize yields (~50% conversion per pass). The liquid methanol is depressurized, purified by permanganate (KMnO4) to remove traces of ketones, aldehydes, & other such impurities, sent to a stripper to remove light ends such as dimethyl ether, and to a fractionator to separate methanol from the higher mol. wt. compounds.

ICI has developed a methanol process which operates at only 50 atmospheres pressure using a copper catalyst.

Methane Derivatives are shown below.

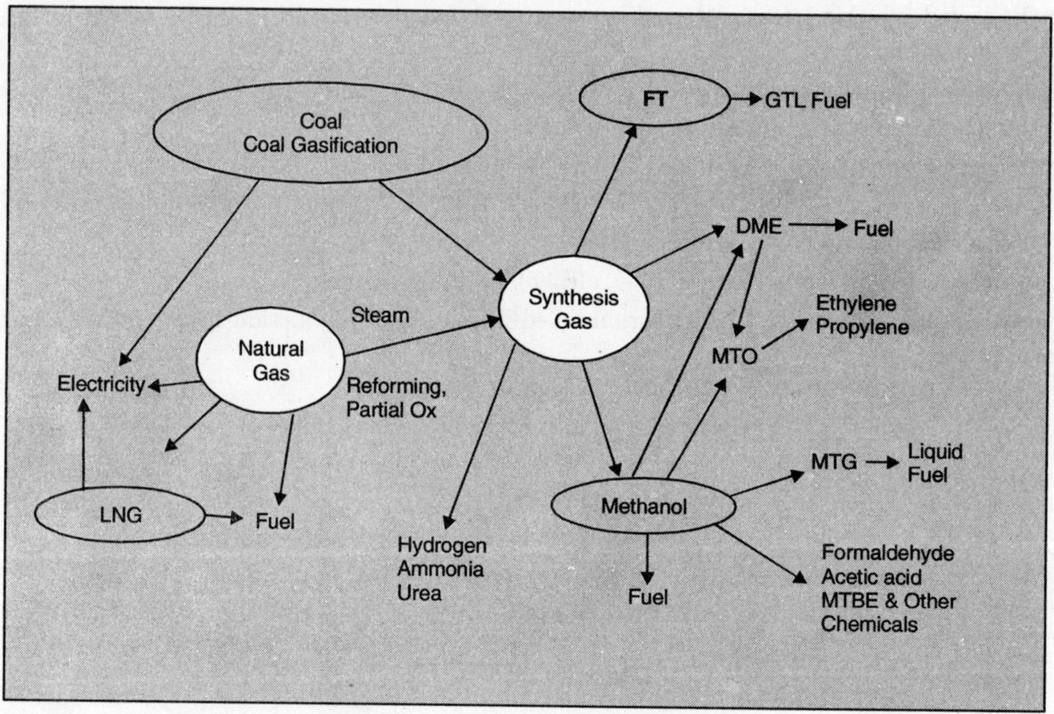

Figure 2.2: Methane Derivatives

Chemicals produced from methane by various unit operations/processes are shown below:

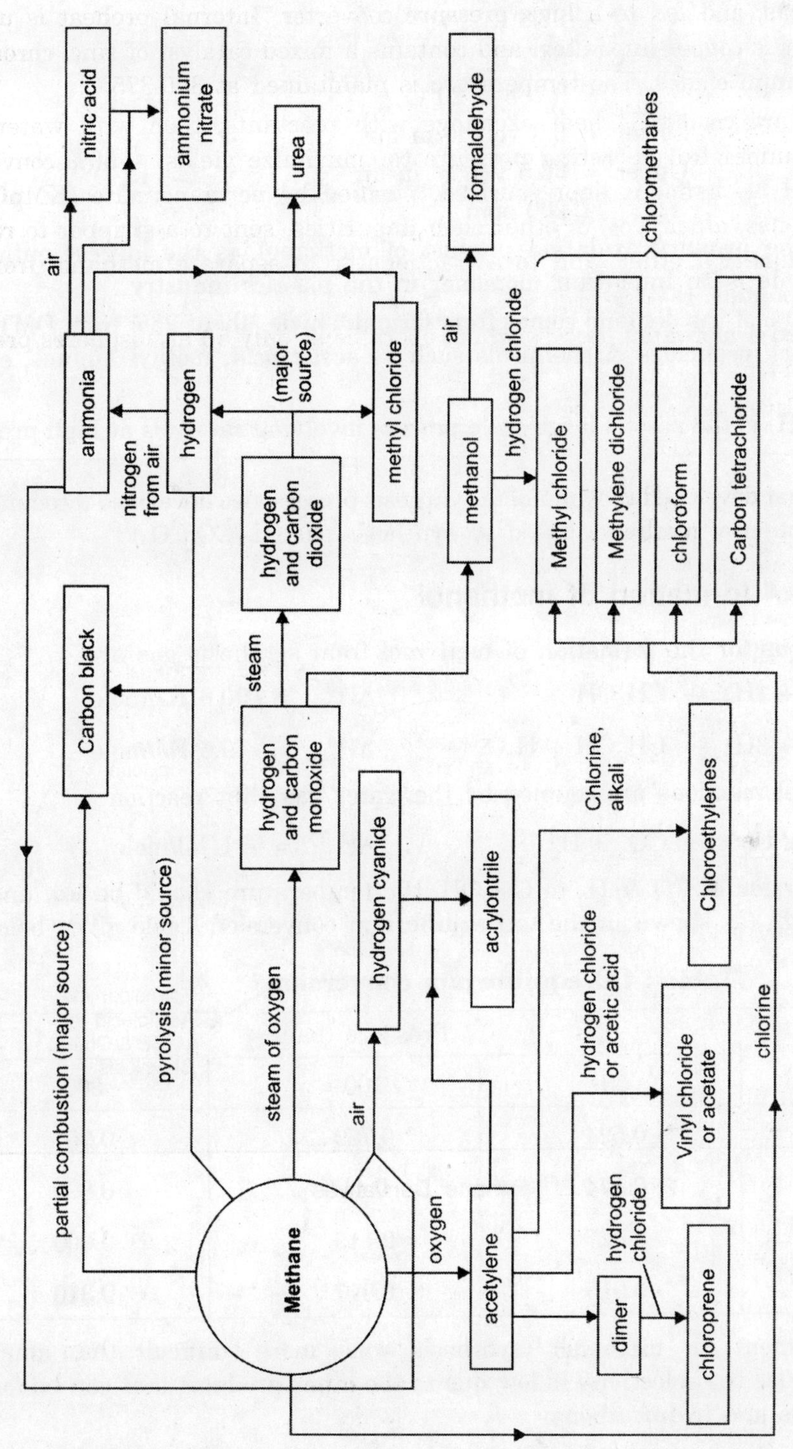

Figure 2.3: Chemicals produced from methane

Properties of Methanol (CH_3OH)

Mol. wt.:		32.04
M.P.:		— 97.8°C
B.P.;		64.7°C
Sp.gr.:		0.788 @ 20°C
Explosive limits:	Lower =	6.0 vol. % in air
	Upper =	36.5 vol. % in air
Toxicity limits:		200 ppm

Formaldehyde, the primary oxidation product of methanol, is the largest outlet for methanol. Formaldehyde is an important monomer in the plastics industry.

Approximately 50% of the demand comes from formaldehyde, about 25% from DMT, and the balance from drugs, pesticides & chemicals such as acetic acid, methyl amines, esters, solvents, etc.

Methanol (CH_3OH) is the second large-scale process involving catalysis at high pressure & temperature.

The same team that developed the ammonia synthesis process also developed a commercial process for the production of methanol based on synthesis gas ($H_2/CO/CO_2$)

Thermodynamics of formation of methanol

The main reactions for the formation of methanol from synthesis gas are:

$$CO + 2H_2 \rightarrow CH_3OH \qquad \Delta H^0_{298} = -90.8 \text{ KJ/mole}$$

$$CO_2 + 3H_2 \rightarrow CH_3OH + H_2O \qquad \Delta H^0_{298} = -49.6 \text{ KJ/mole}$$

The two methanol reactions are coupled by the water gas shift reaction:

$$CO + H_2O \rightarrow CO_2 + H_2 \qquad \Delta H^0_{298} = -41 \text{ KJ/mole}$$

For higher conversion of CO & H_2 to CH_3OH, the temperature should be low and the pressure should be high (as shown in the CO equilibrium conversion Table given below.)

Table : CO equilibrium conversion

Temp. (K)	Pressure (bar)		
	50	100	300
525	0.524	0.769	0.951
575	0.174	0.44	0.825
625	0.027	0.145	0.600
675	0.015	0.017	0.310

Catalyst development for methanol synthesis was more difficult than ammonia synthesis. This is because the selectivity is low due to the other products that can be formed such as higher alcohols and hydrocarbons.

The original catalysts (ZnO-Cr_2O_3) were only active at high temperature. Therefore, the pressure has to be high (250-350 bar) to reach acceptable conversions. The catalysts that are active at low temperature were not resistance to impurities.

The catalysts ($Cu/ZnO/Al_2O_3$) is very selective. In modern plants, the catalysts are active at low temperature which led to "low-pressure plants".

The reaction temperature is critical. A low temperature is favorable from thermodynamic point of view, but the rate of reaction is also low at lower temperature. The pressure for modern plant is usually in the range of 50-100 bar.

The ideal synthesis gas for CH_3OH production has a H_2/CO ratio of about 2. A small amount of CO_2 (about 5%) increases the catalyst activity. When H_2/CO ratio is lower than 2, it leads to increased by product formation (higher alcohols, ...etc.). A higher ratio results in a less efficient plant due to the excess hydrogen present in syngas which has to be purged.

Methanol synthesis

- The first industrial plants were based on a catalyst that was resistant to impurities but not very active and selective. Therefore, to achieve a reasonable conversion to methanol, these plants were operated at high pressure.
- All modern processes are low pressure processes with plant capacities ranging from 150 to 6000 t/d.

Process steps:

- In a modern plant, an adiabatic reactor is used with a single recycle gas.
- The reaction is quenched by adding cold reactant gas at different heights in the catalyst bed.
- The product is quenched and the methanol is separated from the syngas in the flash, drum. The syngas is recycled back to the reactor.
- The crude methanol is separated in two columns. The first column removes gases and other light impurities. The second column separates methanol from heavy alcohols (side stream) and water.

Note: PRODUCTION OF SYNTHESIS GAS

Synthesis gas (or syngas) is a general term used to describe mixtures of hydrogen (H_2) and carbon monoxide (CO) in various ratios.

Mixtures	Main uses	Example
H_2	Refinery hydrotreating and hydrocracking	$C_6H_6OH + H_2 \rightarrow C_6H_6 + H_20$
$3H_2{:}1N_2$	Ammonia	$3H_2 + N_2 \rightarrow + 2NH_3$
$2H_2{:}1CO$	Methanol	$2H_2 + CO \rightarrow CH_3OH$
$1H_2{:}1CO$	Aldehydes (Hydro formulation)	$CH_3CH = CH_2 + CO + \rightarrow$ O CH_3 C HCH CH_3
CO	Acids (formic and acidic)	O $CH_3OH + CO$ -> CH_3 C —OH

Synthesis gas is used for the production of different materials and is also a source of pure hydrogen and pure carbon monoxide.

Syngas may be produced from a variety of raw materials ranging from natural gas to coal. The choice for a particular raw material depends on cost and availability of feedstock.

Syngas can be produced by any one of the following three processes:

- **Steam reforming** of natural gas or light hydrocarbons (eg. naphtha) optionally in the presence of oxygen or carbon dioxide.
- Partial oxidation of (heavy) hydrocarbons with steam & oxygen.
- Partial oxidation of coal (gasification) with steam & oxygen.

Steam reforming is the reaction of hydrocarbons with steam in the presence of a catalyst.

e.g. $CH_4 + H_2O \rightarrow CO + 3H_2$ $\Delta H^0_{298} = -206$ KJ/mole

Reforming (in gas industry) is commonly used for the conversion of a hydrocarbon by reacting it with O - containing molecules, usually H_2O, CO_2 and/or O_2

Partial oxidation (also called steam/oxygen reforming) is a non-catalytic reaction of hydrocarbons with oxygen and usually also steam.

e.g. $CH_4 + 1/2O_2 \rightarrow CO + 2H_2$ $\Delta H^0_{298} = -36$ KJ/mole

Coal gasification is more common term to describe partial oxidation of coal. A combination of steam reforming (endothermic reaction) & partial oxidation (exothermic reaction) is referred to as (autothermic reforming).

A General Flowsheet for the manufacture of synthetic gas by different routes is given below:

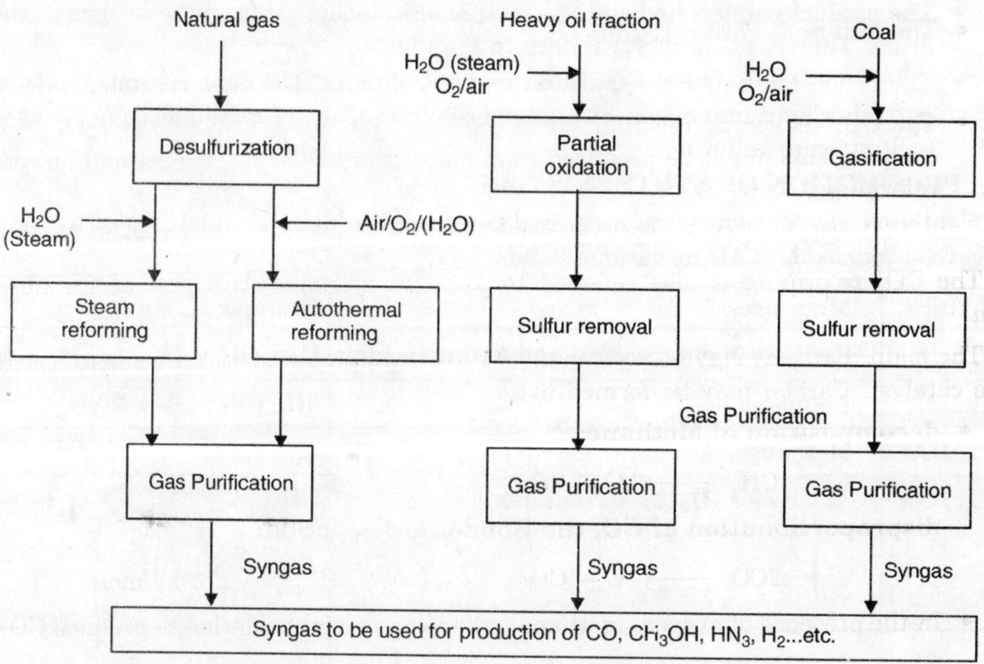

Figure 2.4: General flow sheet for the production of syngas

Process description

- The steam reforming feed usually has to be desulfurized. Sulfur is a poison of metal catalysts because it can block active sites by the formation of sulfides.
- In steam reforming and in downstream, many reactors based catalysts are used, therefore sulfur compounds must be removed to less than 1 ppm.
- When sulfur is present as FeS, it can be removed by:
  - Adsorption (e.g. activated carbon).
  - Reaction with an oxide (for example, ZnO).
  - Scrubbing with a solvent (e.g. NaOH).
- Feed purification of coal and heavy hydrocarbon is not possible, therefore partial oxidation process is used for such feeds. Sulfur compounds are then removed after the formation of syngas.
- Depending on the use of syngas, the syngas may be treated in several ways to adjust the ratio of H_2 to CO. Examples: (1) water-gas shift reaction may be used to reduce the contents of CO and increase the ratio of hydrogen: $CO + H_2O = CO_2 + H_2$. (2) Separation of H_2 from CO can be applied using distillation process.

Synthesis Gas from Natural Gas

Natural gas consists mainly of methane and higher alkanes. For methane in presence of steam, the most important reactions are:

- The steam reforming reaction:

$$CH_4 + H_2O \rightarrow CO + 3H_2, \qquad \Delta H^0_{298} = 206 \text{ kJ/mole}$$

- The water-gas shift reaction:

$$CO + H_2O \rightarrow CO_2 + H_2, \qquad \Delta H^0_{298} = -41 \text{ kJ/mole}$$

- The CO_2 reforming reaction: some processes require synthesis gas with a high CO content. This might be produced from methane and CO_2 in a reaction known as CO_2 reforming:

$$CH_4 + CO_2 \rightarrow 2CO + 2H_2, \qquad \Delta H^0_{298} = 247 \text{ kJ/mole}$$

The CO_2 reforming is also referred to as "dry reforming", because of the absence of steam.

The main reactions may be accompanied by carbon formation which leads to deactivation of the catalyst. Carbon may be formed by:

- **decomposition of Methane:**

$$CH_4 \rightleftharpoons C + 2H_2, \qquad \Delta H^0_{298} = 75 \text{ kJ/mole}$$

disproportionation of CO, the Boudouard reaction:

$$2CO \rightleftharpoons C + CO_2, \qquad \Delta H^0_{298} = -173 \text{ kJ/mole}$$

- In the presence of oxygen, methane undergoes partial oxidation to produce CO & H_2:

$$CH_4 + \frac{1}{2} O_2 \rightleftharpoons CO + 2H_2, \qquad \Delta H^0_{298} = -36 \text{ kJ/mole}$$

- Side reactions such as the complete oxidation of methane to CO_2 and F_2O may occur:

$$CH_4 + 2O_2 \rightleftharpoons CO_2 + 2H_2O , \qquad \Delta H^0_{298} = -803 \text{ kJ/mole}$$

- Oxidation of the formed CO & H_2 may also occur:

$$CO + \frac{1}{2} O_2 \rightleftharpoons CO_2 , \qquad \Delta H^0_{298} = -284 \text{ kJ/mole}$$

$$H_2 + \frac{1}{2} O_2 \rightleftharpoons H_2O , \qquad \Delta H^0_{298} = -242 \text{ kJ/mole}$$

Steam Reforming Process

Steam reforming is carried out at high temperature (> 1000 K). A catalyst (supported nickel) is required to accelerate the reaction due to the very high stability of methane. The catalyst is contained in tubes, which are placed inside a furnace that is heated by combustion of fuel. The steam reforming consists of two sections: convection section & Radiation section. In the convection section, heat recovered from the hot flue gases is used for preheating of the natural feed process, the process steam & for generation of superheated steam. In the radiant section of the furnacc, the reforming reactions take place.

PRODUCTION OF FORMALDEHYDE FROM METHANOL (METHYL ALCOHOL)

Chapter 3

Formaldehyde is produced from methanol by catalytic oxidation-dehydrogenation of methanol.

Chemical Reaction involved in production of formaldehyde (HCHO) from CH_3OH(methonol):

(a) Oxidation: $CH_3OH + 0.5\ O_2 \rightarrow HCHO + H_2O$

(b) Pyrolysis: $CH_3OH \rightarrow HCHO + H_2$

(c) Undesired reaction: $CH_3OH + 1.5\ O_2 \rightarrow 2H_2O + CO_2$

In the above reactions, the first and third are exothermic reactions but the second reaction is endothermic. The reactions are carried out in vapour phase.

Catalyst: Silver or zinc oxide catalysts on wire gauge are used.

Operating temperature & pressure: Near about atmospheric pressure & 500-600 °C.

Proces description

- Air is sent for pre-heating using reactor outlet product and heat integration concept.
- Eventually heated air and methanol are fed to a methanol evaporator unit which enables the evaporation of methanol as well as mixing with air. The reactor inlet temperature is 54°C.
- The feed ratio is about 30-50 % for CH_3OH: O_2
- After reaction, the product is a vapour mixture with temperature 450 - 600°C.
- After reaction, the product gas is cooled with the heat integration concept and then eventually fed to the absorption tower.
- The absorbent in the absorption tower is water as well as formaldehyde rich water.
- Since formaldehyde rich water is produced in the absorption, a portion of the rich water absorbent solution from the absorber is partially recycled at a specific section of the absorber.
- From the absorber formaldehyde + methanol rich water stream is obtained as the bottom product.
- The stream is sent to a light end stripper eventually to remove any light end compounds that got absorbed in the stream. The vapors from the light end unit consisting of light end compounds can be fed at the absorption unit at specific location that matches with the composition of the vapors in the absorption column.

- Eventually, the light end stripper bottom product is fed to a distillation tower that produces methanol vapour as the top product and the bottom formaldehyde + water product (37 % formaldehyde concentration).

Flowsheet of formaldehyde production from methanol is shown below:

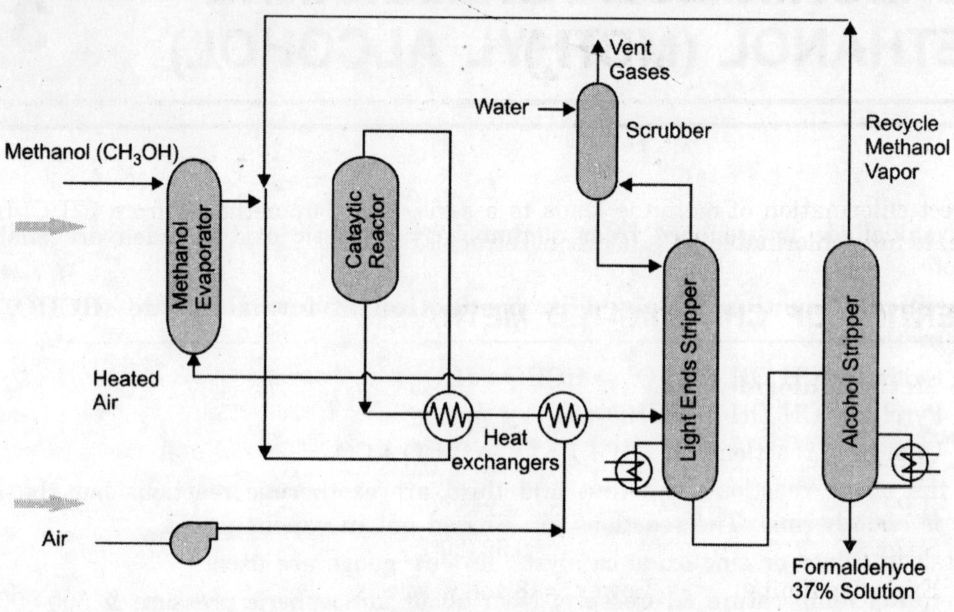

Figure 3.1: Flow sheet of Formaldehyde production from methanol

Properties of Formaldehyde (HCHO):

Mol. wt.:		50.49
M.P.:		— 118.°C
B.P.:		— 19°C
Sp.gr.:		0.815 (at 20°C)
Explosive limits:	Lower =	7 vol. % in air
	Upper =	73 vol. % in air
Toxicity limits:		10 ppm

Solubility: Soluble in water, alcohol & polar solvents, slightly soluble in hydrocarbons, chloroform, ether.

The principal use of formaldehyde is in the manufacture of phenolic, urea and melemine resins.

Raw material requirements:

Basis: 1 ton of formaldehyde (90% yield)	
Methanol (99 + % purity)	1.04 tons
Air	1,760 Nm^3

PRODUCTION OF CHLOROMETHANES BY DIRECT CHLORINATION OF METHANE (CH_4)

Chapter **4**

The direct chlorination of methane leads to a series of chloromethenes from CH_3Cl (methyl chloride) to fully chlorinated CCl_4 (Carbon Tetrachloride).

PROPERTIES OF CHLORINATED METHANES

- **Methyl Chloride (CH_3Cl):**

Mol. wt.:		50.49
M.P.:		— 97.7°C
B.P.:		— 24.1°C
Sp.gr.:		0.92 (20°C)
Explosive limits:	Lower =	8 vol. % in air
	Upper =	20 vol. % in air
Toxicity limits:		50 ppm

Solubility: Soluble in water, alcohol, acetone, chloroform, and ether

- **Methylene Chloride (CH_2Cl_2):**

Mol. wt.:	84.95
M.P.:	— 97.6°C
B.P.:	40.1°C
Sp.gr.:	1.33

Solubility: Slightly soluble in water (1.8% @ 20°C); completely miscible with alcohols and ether

- **Chloroform ($CHCl_3$):**

Mol. wt.:	119.4
M.P.:	— 63.5°C
B.P.:	61.2°C
Sp.gr.:	1.49
Toxicity limits:	100 ppm

Solubility: Slightly soluble in water (0.8% @ 20°C); completely miscible with alcohols, aromatics, ethers

- **Carbon Tetrachloride (CCl_4):**

Mol. wt.:	153.84
M.P.:	—22.6°C
B.P.:	76.8°C
Sp.gr.:	1.59
Toxicity limits:	25 ppm

Solubility: Slightly soluble in water (0.08% @ 20°C); completely miscible with alcohols, aromatics, and ethers

The uses of chlorinated methanes are as follows:

- Solvents
- Intermediate products, e.g.,

  CH_3C1 for methylations

  + Si (Roschow process) → $(CH_3)_2SiCl_2$ (silicones)

  + Pb → $(CH_3)_4Pb$

  (important anti knocking agent) (next to $(C_2H_5)_4Pb$)

  + Na cellulose → methylcellulose
- $CC1_4$ and $CHC1_3$ for synthesis of hydrocarbons containing F and Cl

Classification of Processes of production of chloromethanes:

- Direct thermal methane chlorination—a direct substitution process for making all four chlorinated derivatives depending on mole ratio of CH_4/Cl and reaction conditions.
- CH_3C1 by reaction of CH_3OH + HCl—competitive; can be coupled with methane chlorination to use by-product HCl.
- $CC1_4$ by reaction of CS_2 + $C1_2$—older process, no longer economical.

Quantitative requirements:

- Basis: Depends on mole ratio CH_4/Cl_2

  For typical feed ratio of 1:8, the product compositions are:

  CH_3Cl—60 wt. %, CH_2Cl_2—28%; $CHCl_3$—9%; CCl4—3%; yields are 99% based on Cl_2 and 85-90% based on CH_4.

  For 1 ton production of products listed above:

  Chlorine (99% yield):1.53) tons; Methane (85% yield): 0.305 ton or 385 Nm^3

  Chloromethanes namely methyl chloride (CH_3Cl), methylene chloride (CH_3Cl_2), Chloroform ($CHCl_3$) and Carbon Tetrachloride (CCl_4) are produced by direct chlorination of CH_2 in a gas phase reaction without any catalyst.

Chemical Reactions involved in production of chloromethanes:

$$CH_4 + Cl_2 \rightarrow CH_3Cl + H_2Cl$$
$$CH_3Cl_2 \rightarrow CH_3Cl + H_2Cl$$
$$CH_2Cl_2 \rightarrow CHCl_3 + H_2Cl$$

$$CHCl_3 + Cl_2 \rightarrow CCl_4 + H_2Cl$$

- The reactions are highly exothermic.
- The feed molar ratio affects the product distribution. When $CH_4/C1_2$ is about 1.8, then more CH_3C1 is produced. On the other hand, when CH_4 is chosen as a limiting reactant, more of $CC1_4$ is produced. Therefore, depending upon the product demand, the feed ratio is adjusted.

Process description:

- Methane and $C1_2$ are mixed and sent to a furnace.
- The furnace has a jacket or shell and tube system to accommodate feed pre-heating to desired furnace inlet temperature (about 280 - 300 °C).
- To control temperature, N_2 is used as a diluent at times.
- Depending on the product distribution desired, the $CH_4/C1_2$ ratio is chosen.
- The product gases enter an integrated heat exchanger that receives separated CH_4 (or a mixture of CH_4 + N_2) and gets cooled from the furnace exit temperature (about 400 °C).
- Eventually, the mixture enters an absorber where water is used as an absorbent and water absorbs the HC1 to produce 32 % HC1.
- The trace amounts of HC1 in the vapour phase are removed in a neutralizer fed with NaOH.

Flow sheet for manufacture of chloromethane by direct chlorination of methane is shown below.

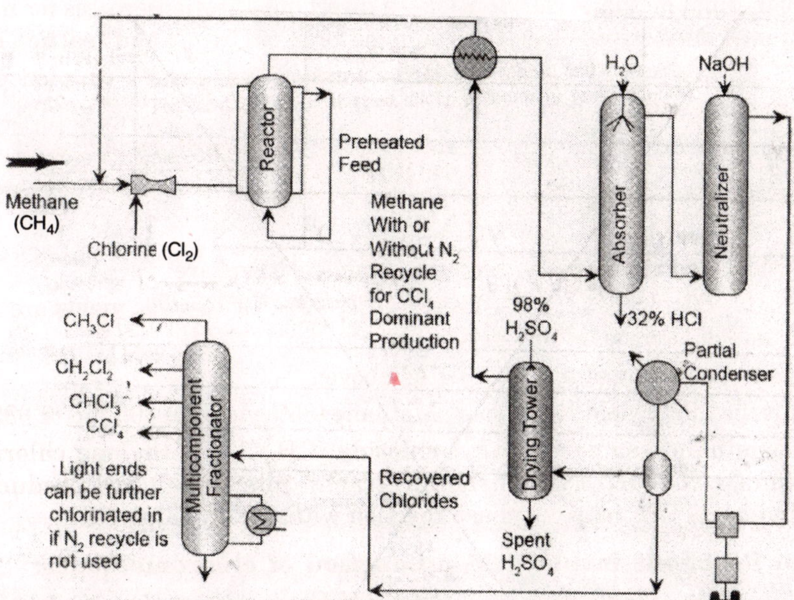

Figure 4.1: Flowsheet for Chloromethane production by direct chlorination of methane

- The gas eventually is compressed and sent to a partial condenser followed with a phase separator. The phase separator produces two streams namely a liquid stream consisting of the chlorides and the unreacted CH_4/N_2.

- The gaseous product enters a dryer to remove H_2O from the vapour stream using 98% H_2SO_4 as the absorbent for water from the vapour.
- The chloromethanes enter a distillation sequence. The distillation sequence consists of columns that sequentially separate CH_3Cl, CH_2Cl_2, $CHCl_3$ and CCl_4.

MANUFACTURE OF TRICHLOROETHYLENE ($ClHC = CCl_2$)

Chapter 5

Trichloroethylene represents one of the major chemicals produced from acetylene, of which a substantial fraction comes from natural gas. Other chemicals derived from acetylene are acrylonitrile, vinyl chloride and acetate monomers.

Properties of Trichloroethylene ($ClHC = CCl_2$):

Mol. wt.:	131.40
M.P.:	— 73 °C
B.P.:	87.2°C
Sp.gr.:	1.47
Toxicity limits:	200 ppm

Solubility: Slightly soluble in water (0. 1 % @ 20 °C), miscible with alcohols and ethers

Vapor degreasing of metal parts before electroplating or painting is the major end use.

As an intermediate, a pyrolysis reaction to perchloroethylene yields an end chemical in high demand as a dry cleaning agent.

Classification of Production processes

- Vapor phase dehydrochlorination of tetrachloroethane, using $BaCl_2$ catalyst.
- Liquid phase dehydrochlorination of tetrachloroethane, using milk of lime.

Chemical reactions involved in vapor phase dehydrochlorination of tetrachloroethane:

(a) $C_2H_2 + 2Cl_2 \xrightarrow{FeCl_3} C_2H_2Cl_4$ (tetrachloroethane)

(b) $C_2H_2Cl_4 \xrightarrow[250-300°C]{BeCl_3} C_2HCl_3 + HCl$

Raw materials requirements:

Basis: 1 ton of trichloroethylene (95% ultimate yield, 90% conversion)

Chlorine:	1.15 tons
Acetylene:	0.21 ton

A flowsheet of trichloroethylene by dehydro chlorination of tetrachloroethane is shown below:

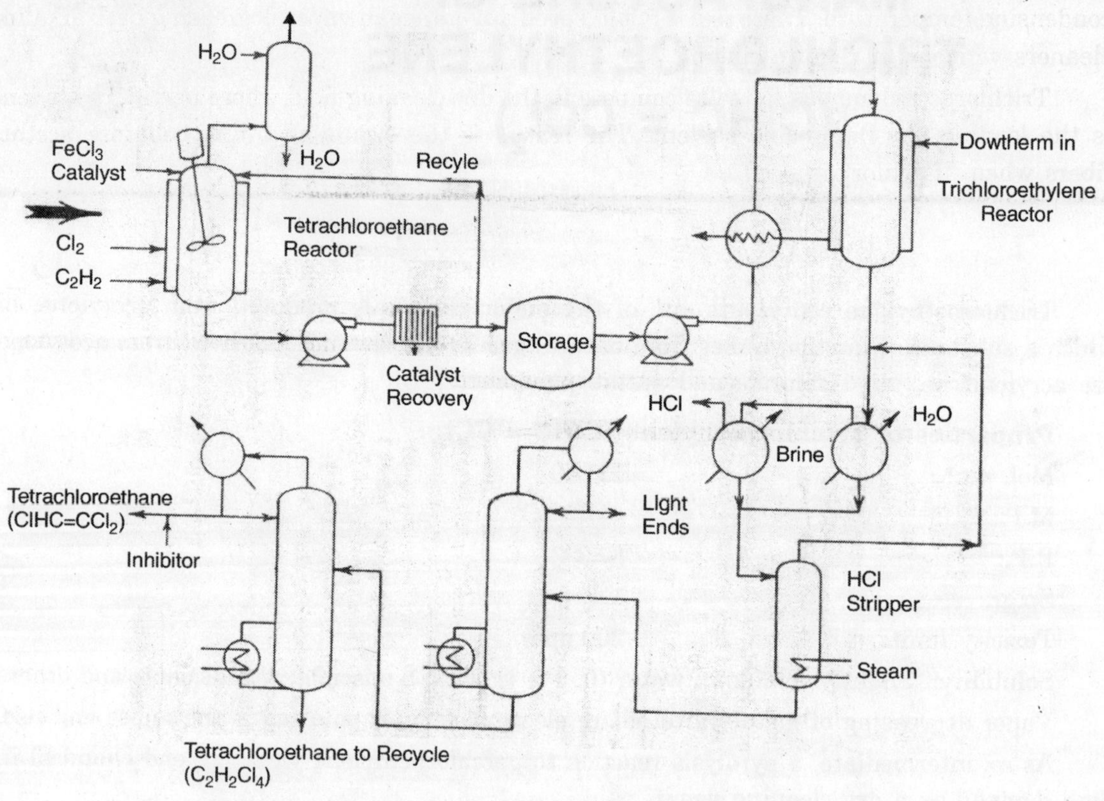

Figure 5.1: Trichloroethylene by vapor phase dehydrochlorination of tetrachloroethane via acetylene-Cl_2 route.

Process technology

Tetrachloroethane is produced by feeding chlorine & acetylene into a reactor containing recycle tetrachloroethane. Ferric or antimony chloride is the catalyst used in low concentration. The reaction is exothermic and is held at 80-100°C by external cooling. Product is removed and purified by distillation, or by settling-filtration to remove catalyst.

The vapor phase dehydrochlorination is accomplished by preheat and reaction at 250-300°C over $BaCl_2$ catalyst deposited on porous carbon. The conversion to trichloroethylene is 90% per pass with little side reaction. Degassing of HCl with steam and a dual column distillation produces acid-free grade "Trichlor."

The vapor phase process is preferable to the liquid phase-lime neutralization process for dehydrochlorination if the HCl by-product can be utilized at a reasonable value.

Trichloroethylene has replaced carbon tetrachloride in vapor degreasing of light metal parts. Its lower boiling point compared to perchloroethylene allows more solvent condensation & cleaning during the effective cleaning period before the piece heats up to an equilibrium condensing temperature. There is a 4-5 fold speed advantage in vapor degreasing over alkaline cleaners which also require rinsing.

Trichloroethylene has failed to compete in the dry cleaning field where perchloroethylene is the leading non-flammable solvent. The reason is the loss of dye from cellulose acetate fibers when "Trichlor" is used.

MANUFACTURE OF PERECHLOROETHYLENE ($Cl_2C = CCl_2$)

Chapter 6

This is the last of the three major chlorinated solvents, the other two being $CC1_4$ and $CHCl = CCl_2$.

Properties of Perchloroethylene:

Mol. wt.:	165.8
M.P.:	–22.4°C
B.P.:	121.2°C
Sp.gr.:	1.61
Toxicity limits:	200 ppm

Solubility: Slightly soluble in water (0.1% @ 20° C), miscible with alcohols and ethers.

Dry cleaning solvent is its principal use, with markets expanding in the coin machine, dry cleaning business.

Classification of Production Processes:

- Pyrolysis of $CC1_4$.
- Propane chlorination to yield CCl, co-product.
- Chlorination of trichloroethylene to pentachloroethylene followed by lime dehydrochlorination.

Pyrolysis of $CC1_4$ is the preferred method:

Chemical reactions involved in pyrolysis of CCl_4:

(a) $2CCl_4 \rightarrow \underset{\text{hexachloroethane}}{Cl_3C \bullet CCl_3} + Cl_2$

(b) $Cl_3C \bullet CCl_3 \rightarrow Cl_2C = CCl_2 + 2Cl_2$

(c) $2CCl_4 \rightarrow Cl_2C = CCl_2 + 2Cl_2$

Raw material requirements:

Basis: 1 ton perchloroethylene (95% yield); 0.86 ton of $C1_2$;
Carbon tetrachloride:2.0 tons

Process Technology

$CC1_4$ is vaporized, passed into resistively heated carbon electrode furnace at 800-900°C, and pyrolyzed to perchloroethylene, hexachloroethane, & chlorine. The effluent goes to a partial condenser, held at 80-85°C where Cl_2 plus unreacted CCl_4 are separated from heavy ends. The CCl_4 is stripped of chlorine in a condenser and recycled.

The heavy ends are distilled to separate perchloroethylene from hexachloroethane. The latter can be recycled for further pyrolysis or further purified via crystallization.

In a process modification, propane can be chlorinated to yield $CC1_4$ which subsequently pyrolyses to "Perchlor."

A flowsheet for production of perchloroethylene by pyrolysis of carbon tetrachloride (CCl_4) is shown below.

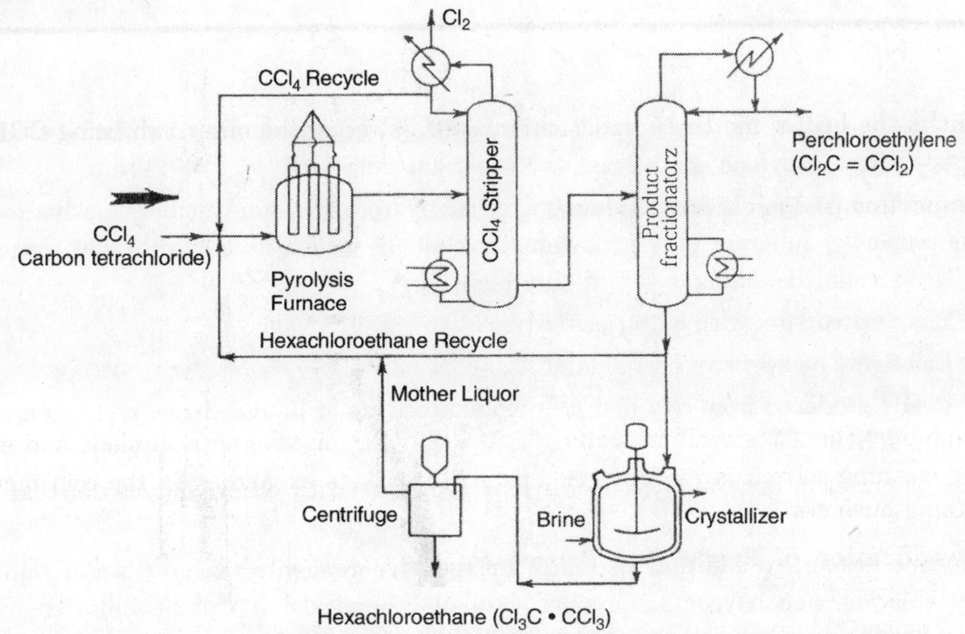

Figure 6.1: Perchloroethylene by pyrolysis of carbon tetrachloride

PRODUCTION OF ETHYLENE ($CH_2 = CH_2$)

Chapter

7

Ethylene is known as the "king of petrochemicals" because more commercial chemicals are produced from ethylene than from any other intermediate.

This unique position of ethylene among other hydrocarbon intermediates is'due to some favorable properties inherent in the ethylene molecule as well as to technical and economical factors. These could be summarized in the following:

- Simple structure with high reactivity.
- Relatively inexpensive compound.
- Easily produced from any hydrocarbon source (gas or liquid) through steam cracking and in high yields.

Less by-products generated from ethylene reactions with other compounds than from other olefins.

Ethylene reacts by addition to many inexpensive reagents such as water, chlorine, hydrogen chloride, and oxygen to produce valuable chemicals. It can be initiated by free radicals or by coordination catalysts to produce polyethylene, the largest-volume thermoplastic polymer. It can also be copolymerized with other olefins producing polymers with improved properties. For example, when ethylene is polymerized with propylene, a thermoplastic elastomer is obtained.

Ethylene and its derivatives

Ethylene is one of the most important fundamental chemicals in the petrochemical industry as it is the source material for a variety of products such as polyethylene resin, ethylene glycol, vinyl chloride resin, acetic acid, styrene, and alpha olefin which are produced by polymerization, oxidation, alkylation, hydration, or the addition of halogen.

The most important chemicals based on ethylene is shown below.

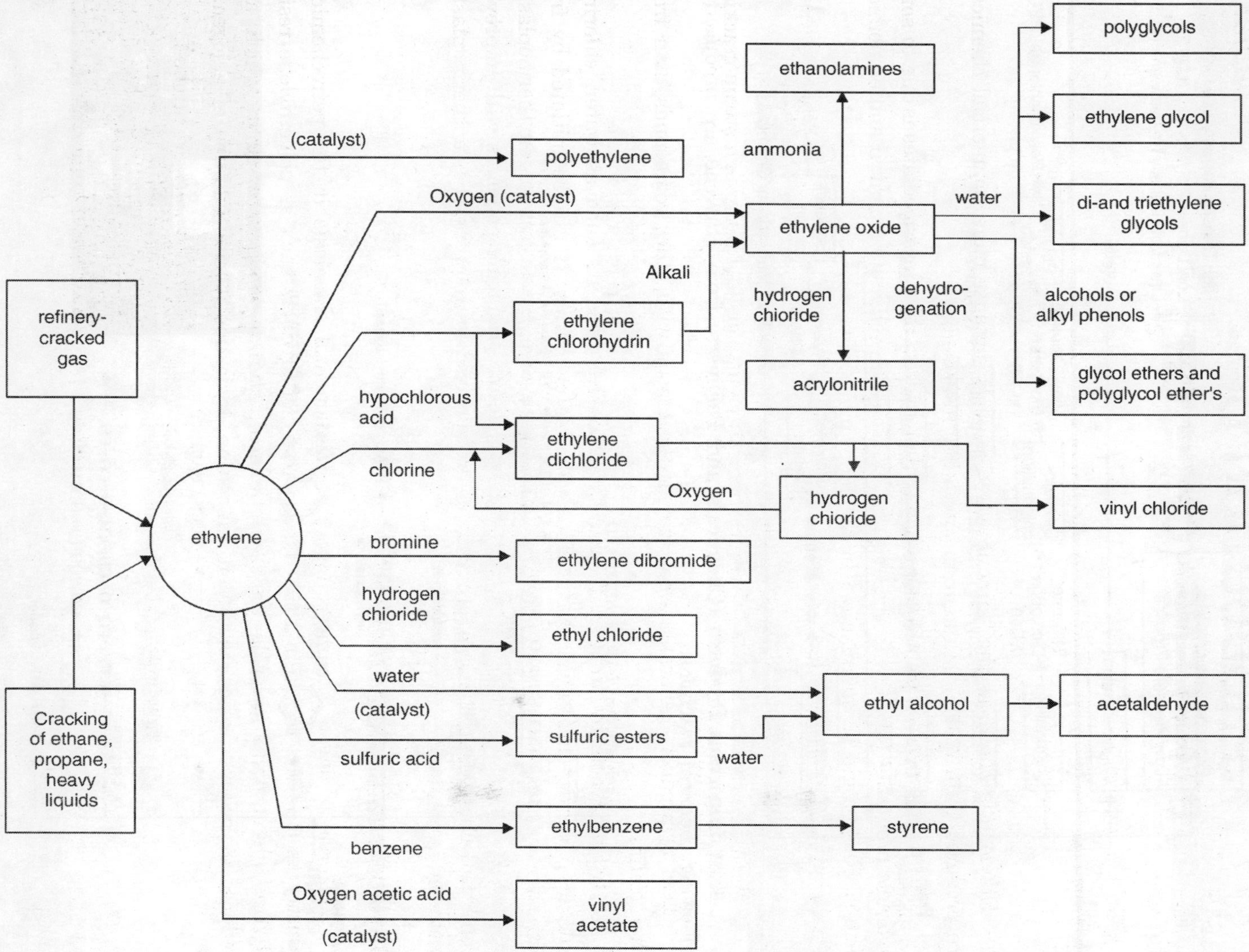

Ethylene and its derivatives are shown below.

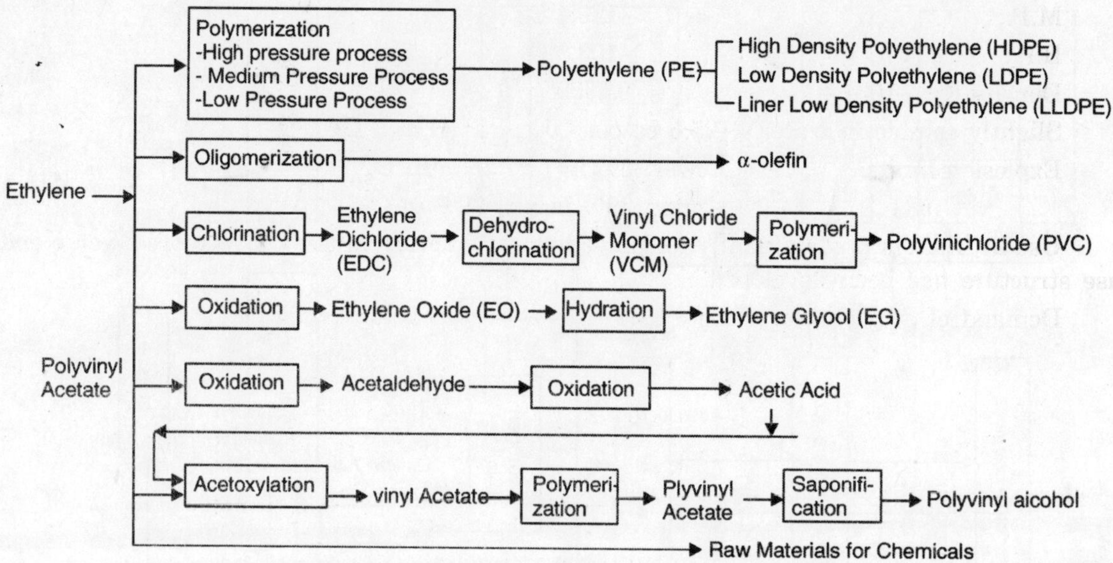

Most Important Products (Chemicals and Polymers) from ethylene are produced by following chemical reactions:

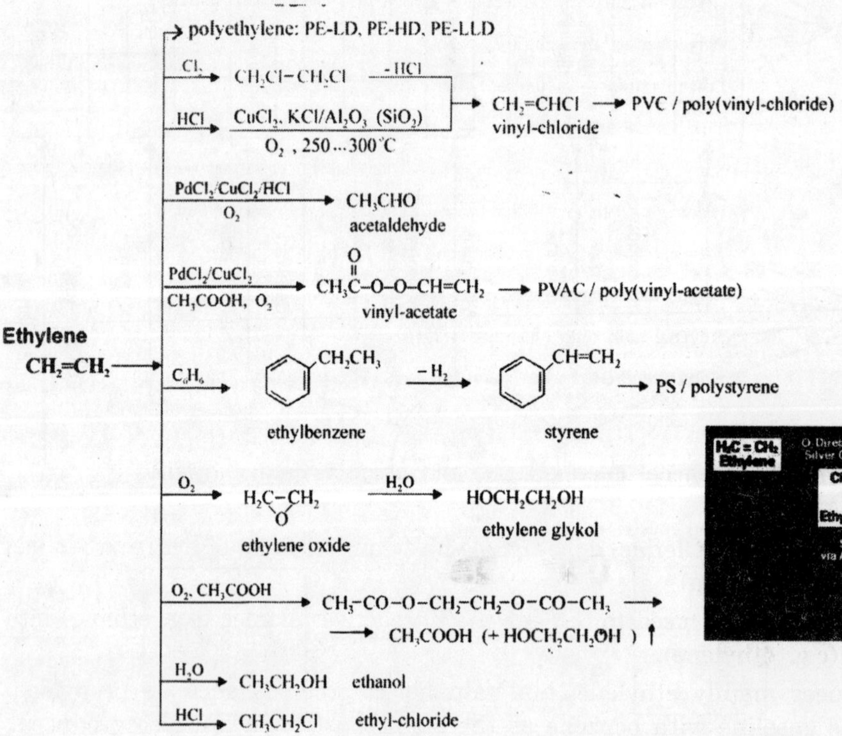

Ethylene

$CH_2{=}CH_2$ →

→ polyethylene: PE-LD, PE-HD, PE-LLD

$\xrightarrow{Cl_2}$ CH_2Cl-CH_2Cl $\xrightarrow{-HCl}$ $CH_2{=}CHCl$ (vinyl-chloride) → PVC / poly(vinyl-chloride)

$\xrightarrow{HCl}$ $CuCl_2, KCl/Al_2O_3\ (SiO_2)$, O_2, 250...300°C → $CH_2{=}CHCl$ (vinyl-chloride)

$\xrightarrow[O_2]{PdCl_2/CuCl_2/HCl}$ CH_3CHO acetaldehyde

$\xrightarrow[CH_3COOH,\ O_2]{PdCl_2/CuCl_2}$ $CH_3\overset{O}{\overset{\|}{C}}-O-O-CH{=}CH_2$ vinyl-acetate → PVAC / poly(vinyl-acetate)

$\xrightarrow{C_6H_6}$ ethylbenzene ($C_6H_5CH_2CH_3$) $\xrightarrow{-H_2}$ styrene ($C_6H_5CH{=}CH_2$) → PS / polystyrene

$\xrightarrow{O_2}$ H_2C-CH_2 (O bridge) ethylene oxide $\xrightarrow{H_2O}$ $HOCH_2CH_2OH$ ethylene glykol

$\xrightarrow{O_2,\ CH_3COOH}$ $CH_3-CO-O-CH_2-CH_2-O-CO-CH_3$ →

→ CH_3COOH (+ $HOCH_2CH_2OH$) ↑

$\xrightarrow{H_2O}$ CH_3CH_2OH ethanol

$\xrightarrow{HCl}$ CH_3CH_2Cl ethyl-chloride

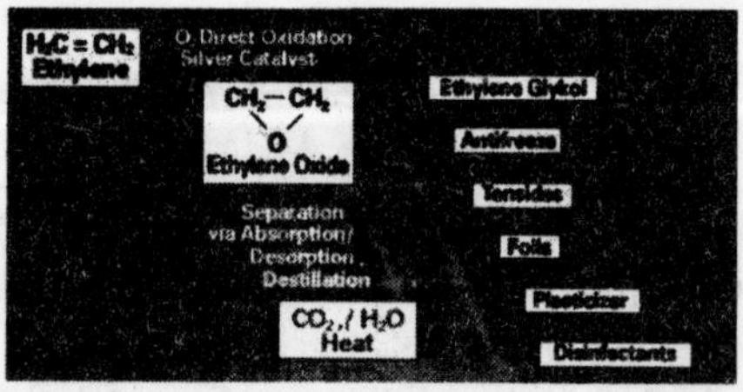

$H_2C=CH_2$ Ethylene
O Direct Oxidation Silver Catalyst
CH_2-CH_2 (O) Ethylene Oxide
Separation via Absorption/ Desorption, Destillation
CO_2 / H_2O Heat
Ethylene Glykol
Antifreeze
Tensides
Foils
Plasticizer
Disinfectants

Properties of Ethylene ($CH_2 = CH_2$) also called ethene (an olefin):

Mol. wt.:	28.03
M.P.:	–169.4°C
B.P.:	–103.8°C
Density @ : –103.8°C	0.5699
Slightly soluble in water	(25.6 ce per 100 gmH_2O at 0°C
Explosion limits:	Lower: 3% in air, 2,9% in O_2
	Upper: 29% in air, 79% in O_2

Use in India is predominantly for polyethylene, whereas in the U.S. a more diverse end use structure has been developed.

Demand of ethylene for various end uses is shown below in figure.

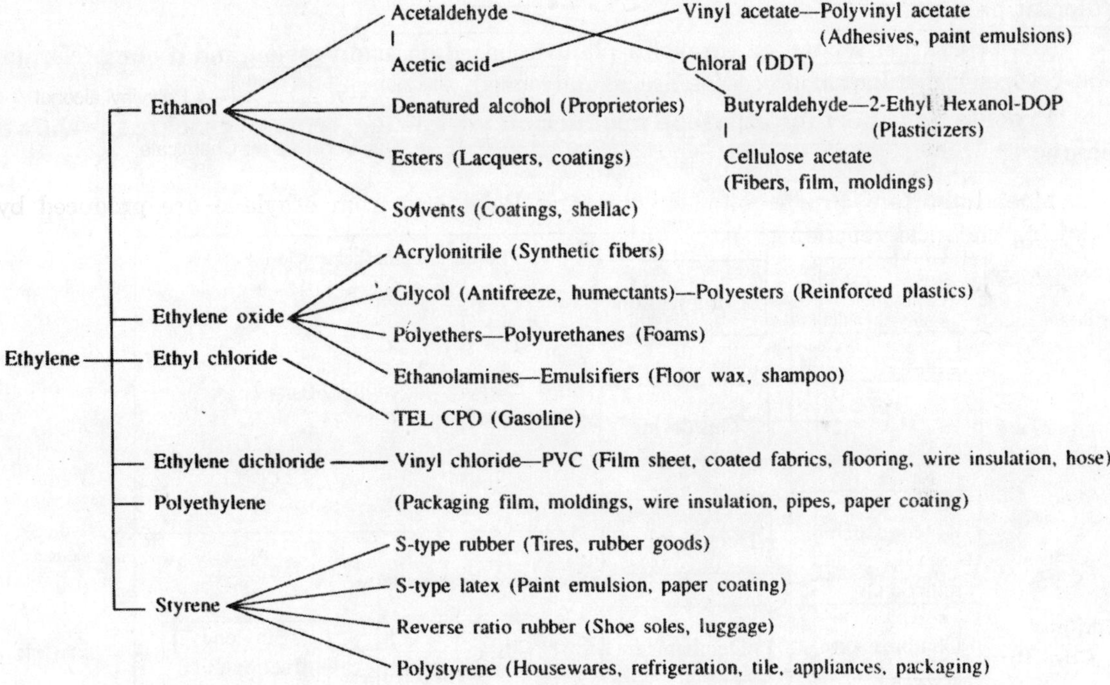

Figure 7.1: Synthetic chemical intermediates and products from ethylene.

Production of Lower Alkenes (Olefins) eg; ethylene by steam cracking of hydrocarbon gas or liquid hydrocarbons (eg, naphtha)

Steam cracking is the process used to convert the unreactive alkanes (e.g. ethane) into much reactive alkenes (e.g. ethylene).

This process produces mainly ethylene, but valuable co-products such as propylene, butadiene, and pyrolysis gasoline with benzene as the main constituent are also produced.

In steam cracking (also called thermal cracking), a hydrocarbon feed is thermally. cracked in the presence of steam, yielding a complex product mixture. Steam is not cracked but it functions primarily as a diluent, allowing higher conversion.

Feedstocks range from light saturated hydrocarbons gases such as ethane & propane to liquid naphtha and light & heavy gas oil.

In USA, ethane (from natural gas) is the primary feedstock for the production of ethylene. In Europe India and Japan, naphtha is the major feedstock.

In Saudi Arabia, ethane is the primary feedstock for the ethylene production.

Process Description

Cracked gas is based on either gas or liquid feedstocks (see Figures)

The conventional process flow sheet for the production of olefins varies from one process to another and depends on the type of feedstock & the degree of recovery desired for the different products.

The separation section on an olefin plant is based on compression and cooling. Normal and cryogenic distillation processes are usually used.

Process flowsheet for ethylene production via gas (eg, ethane) cracking is shown below:

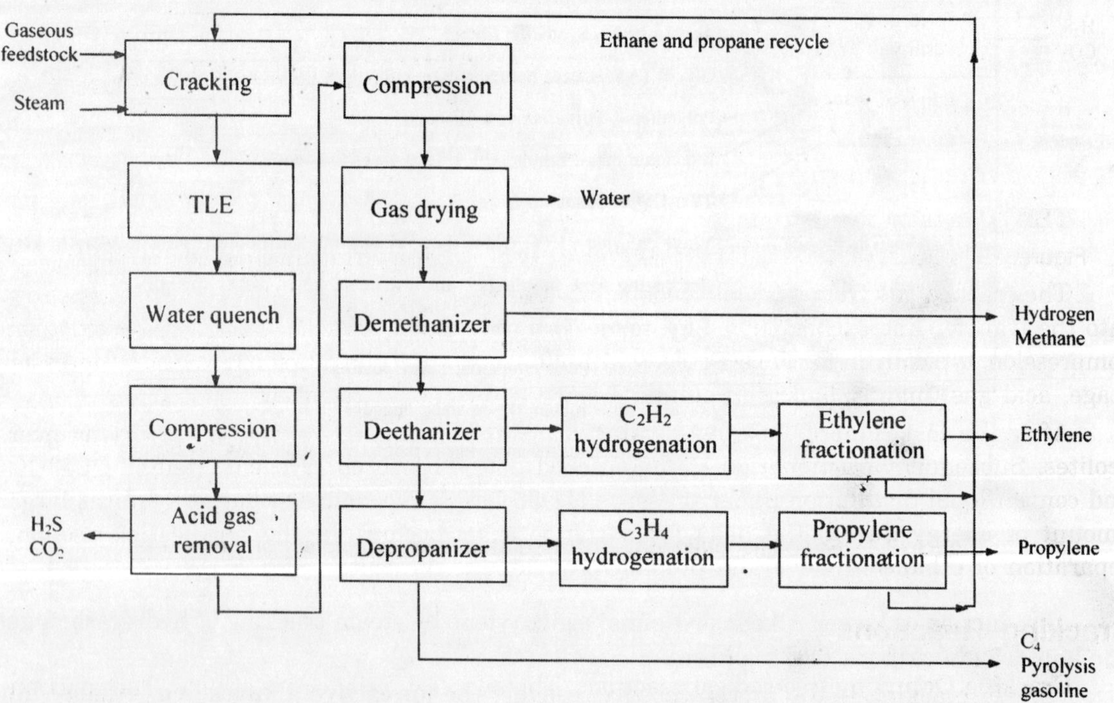

TLE– Transfer line exchanger

Figure:7.2 process flow sheet for producing ethylene (ethene) via gas cracking.

Process flowsheet for ethylene production via liquid (eg, naphtha) cracking is shown below:

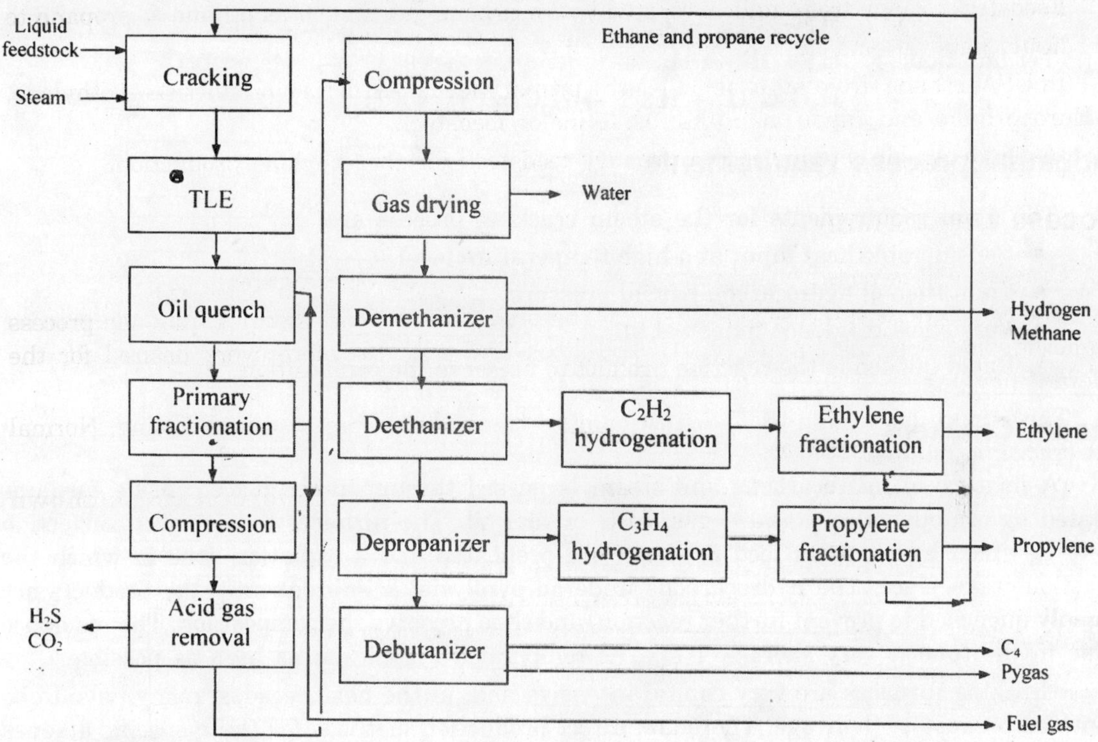

TLE– Transfer line exchanger

Figure:7.3 process flowsheet for producting ethylene (ethene) via liquid naphtha cracking.

The cracked gas from gaseous feedstocks (ethane, propane, butane) begins with its entry into the transfer-line exchanger (TLE), followed by direct quench with water, and multistage compression, typically in four to six stages with intermediate cooling. Before the last compressor stage, acid gas (mainly hydrogen sulfide & carbon dioxide) is removed.

After the last compressor stage, water removal takes place by chilling & drying over zeolites. Subsequent fractionation of the cracked gas is based on cryogenic (temp. < 273 K) and conventional distillation under pressure (15-35 bar). Cryogenic distillation requires large amount of energy due to the need for a refrigeration system. In the purification section, separation of ethane/ethylene and propane/propylene is difficult.

Cracking Reactions

Cracking Occurs by free-radical reactions. The simplest feedstock is ethane. The reaction is initiated by cleavage of the C-C in an ethane molecule resulting in the formation of two methyl radicals.

Initiation: $H_3C–CH_3 \rightarrow H_3C^\bullet + H_3C^\bullet$

Propagation: $H_3C^{\bullet} + H_3C\text{–}CH_3 \rightarrow CH_4 + H_3C\text{–}C^{\bullet}H_2$

$H_3C\text{-}C^{\bullet}H_2 \rightarrow H_2C = CH_2 + H^{\bullet}$

$H^{\bullet} + H_3C\text{–}CH_3 \rightarrow H_2 + H_3\,C\text{–}C^{\bullet}H_2$

$H_3C\text{–}C^{\bullet}H_2 \rightarrow$ etc.

Termination: $H^{\bullet} + H^{\bullet} \rightarrow H_2$

$H_3C\text{–}C^{\bullet}H_2 + H_3C^{\bullet} \rightarrow H_2C = CH_2 + CH_4$

Industrial process requirements

The main requirements for the steam cracking process are:

- Considerable heat input at a high temperature level.
- Limitation of hydrocarbon partial pressure.
- Very short residence times (<1 s).
- Rapid quench of the reaction product to preserve the composition.

Steam Crackers

A mixture of hydrocarbons and steam is passed through tubes placed inside furnaces heated by combustion of natural gas, LPG or fuel oil. The furnaces consist of a convection zone in which hydrocarbon feed & steam are preheated and a radiation zone in which the reactions take place. The hydrocarbons undergo pyrolysis, & subsequently the products are rapidly quenched to prevent further reaction, and thus preserve the composition. The residence time in the tube is very short (« 1 s). The temperature is chosen as high as possible.

Cracking furnaces are very capital intensive and, in the next decades, many have to be replaced because of their age. Therefore, direct production methods for these specific alkenes receive increasingly more attention. A logical choice of process is the Catalytic dehydrogenation of the corresponding alkanes.

Some characteristics of the catalytic dehydrogenation of alkanes are:

- The reaction is endothermic and high temperatures are required.
- At high temperatures, secondary reactions such as cracking and coke formation are possible.
- The thermodynamic equilibrium limits the conversion per pass so that a substantial recycle stream is required.
- Several catalytic dehydrogenation processes are available, which differ in the type of catalyst used, the reactor design, the method of heat supply, and the method used for catalyst regeneration.

Typical Operating conditions of Naphtha Cracker:

- Naphtha cracker constitutes:
  - Hot section : Heaters
  - Mid section: Quench Tower
  - Cold section: Recovery
- Operating temperature of heaters : 950°C
- Cold section:

– Minimum temperature for methane recovery: 180°C
Pressure : Around 35 Kg./cm²

Naphtha Cracker unit Block flow Diagram is shown below:

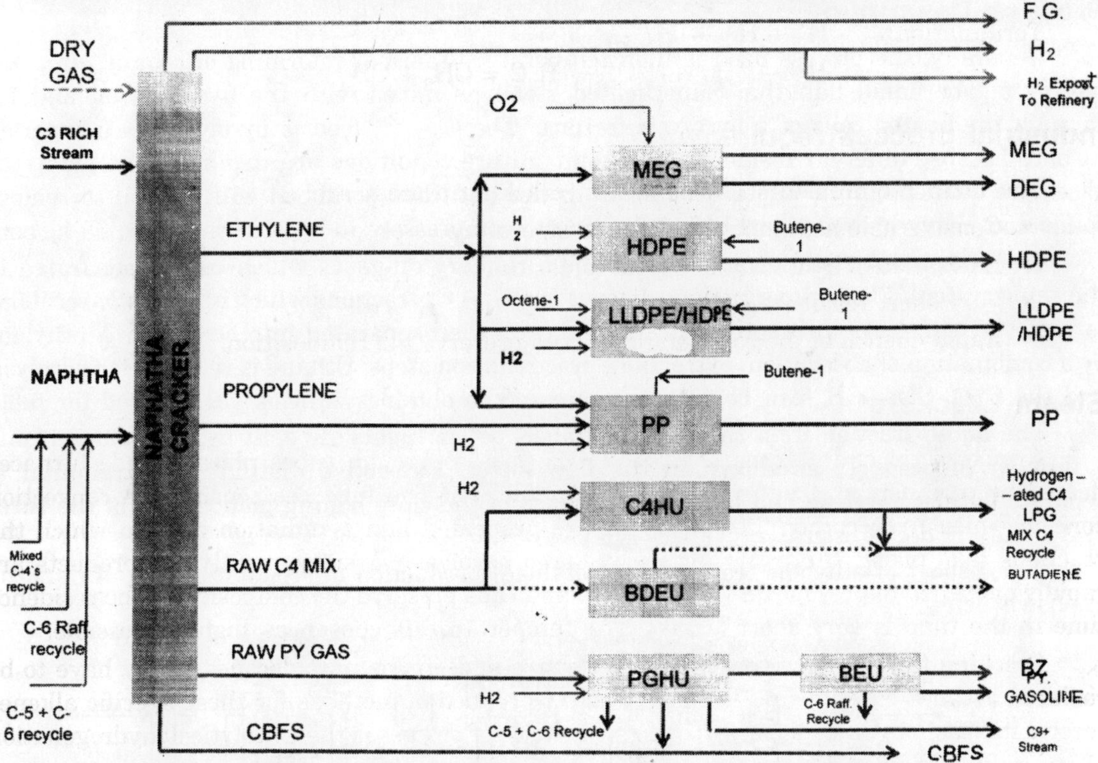

Possible Methods of Production of ethylene include:

- Steam pyrolysis (cracking) of petroleum from LPG and naphtha feedstocks.
- Thermal pyrolysis of ethane and/or propane—not flexible.
- Dehydration of ethanol—used in India, but not competitive for large-scale requirement in the long term.

Demand for ethylene will exceed supply in years to come and import of cryogenic ethylene by tankship from Saudi Arabia and Qatar may assume an important place in ethylene supply. Middle Eastern suppliers, Saudi Arabia and Qatar, are low cost producers and have considerable freight advantage to India.

Steam Cracking of Petroleum Hydrocarbons involves the following chemical reactions:

Chemical Reactions

$C_2H_{2r+2} + H_2O + O_2$ (optional) $\xrightarrow{700-800°C}$ C_2H_4 (4-15%) + C_2H_6 + C_2H_2 (7-13%) + H_2 (25-30%) + $CO + CO_2 + CH_4 + C_3H_6 + C_3H_8$

$+ C_4H_{10} + C_4H_8 + C_4H_6 + C +$ Heavy oil fraction

Co-products: acetylene, propylene, butylene, butadiene, aromatic such as benzene, toluene, xylene, heavy oil residues etc.

Process Description

Generally, the process used is high temperature thermal reforming of ethane, propane, butane and/or liquid naphtha. Superheated steam is mixed with the hydrocarbon and fed through the heated coils of a pyrolysis furnace. The $C_2 - C_4$ feed is pyrolyzed in a separate furnace because different residence time-temperature conditions are required. The pyrolyzed gases are quenched in a waste heat steam boiler and then scrubbed with gas oil to remove solids and heavy hydrocarbons before sending to compressors to boost pressure to 35 kg/cm^2.

The compression station may also handle refinery off-gases which can be separated in the same system. Flash vaporization removes the $C_1 - C_2$ fractions which are caustic-scrubbed to remove CO_2, dried with activated alumina, and then separated into acetylene & ethylene by a combination of absorption, extraction, fractionation steps. Ethane is recycled for pyrolysis and the CH_4, CO, & H_2 can be further processed to obtain synthesis gas or used for fuel.

The liquid fraction from the flash chamber. (C_3 & higher) is split by fractionation into a number of products as shown in the flow sheet. Extractive distillation is required to separate butane-butylene and butadiene because of the close boiling point range of the three compounds under pressure.

A flowsheet of ethylene (co-product-acetylene) production of steam creaking of naphtha is shown below:

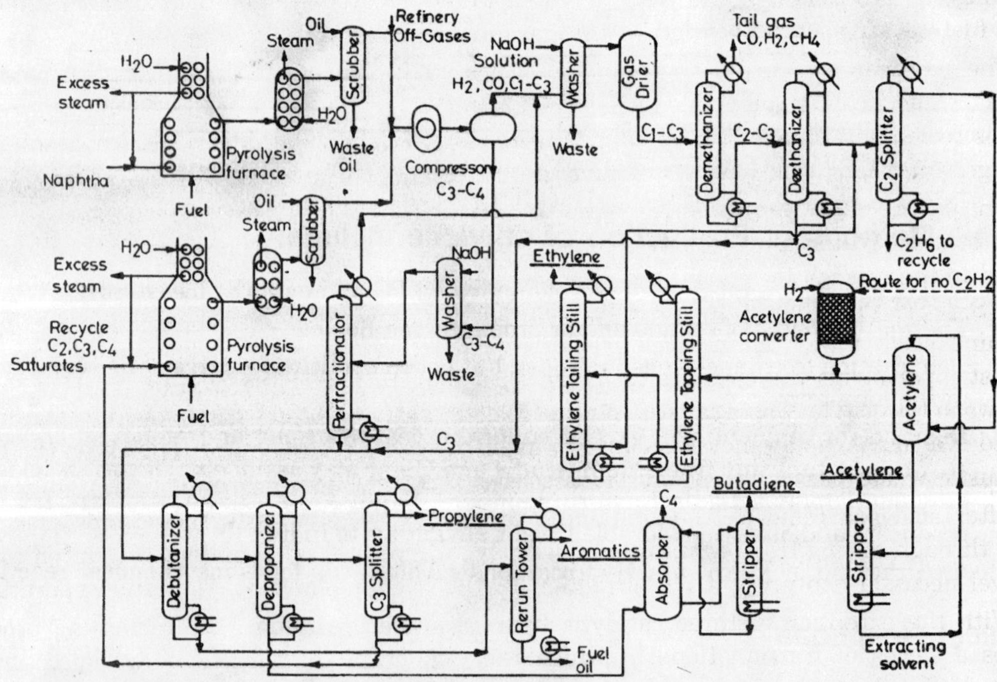

Figure 7.4: Ethylene & acetylene production by steam cracking of naphtha

ETHYLENE/PROPYLENE (OLEFINS) PRODUCTION SCHEME IN AN INDIAN PLANT

Ethylene and propylene are produced by thermal cracking of hydrocarbons. A process flow diagram for a plant producing ethylene from naphtha and or natural gas is shown in Figure 7.5. Naphtha and/or natural gas, diluted with steam, is fed in parallel to a number of gas or oil fired tubular pyrolysis furnaces. The fuel for these furnaces is supplied from gas & oil fractions recovered from the cracked gas in later separation stages i.e ethane & propane. Ethane & propane, which are present in the cracked gas, are separated in later distillation steps, are mixed and recycled through a separate cracking furnace. The flue gas from this pyrolysis furnace is emitted through furnace stack.

During this operation, coke accumulates on the inside walls of the reactor and each furnace are periodically taken out of service for removal of the accumulated coke. Present day, plants are de-coked on-line i.e. by passing steams air through the coil while the furnace is maintained at an elevated temperature, effectively burning the carbon out of the coil. While a furnace is being de-coked, the exhaust is diverted to an emission control device (Vent A in Fig.) whose main function is to reduce particulate emissions.

The cracked gas leaving the pyrolysis furnace is rapidly cooled (oil) to 250 to 300°C and steam is generated. The gaseous streams are then further quenched by the injection of recycled pyrolysis fuel oil from the gasoline fractionators. The quenched cracked gas passes to the gasoline fractionators where pyrolysis fuel oil is separated. Most of the fuel oil is cooled and recycled to the oil quenching operation. The surplus fuel oil passes to the stripper, where light fractions are removed, and then it is sent to fuel oil storage. The light fraction removed in the fuel oil stripper is recycled to the gasoline fractionators.

The overhead stream from the gasoline fractionators passes to the water quench tower, where C_5's and heavier compounds are separated. Most of the water separated in the quench tower, is cooled and recycled to the quench tower. Part of the water is passed through dilution steam generators to generate steam and bleed outside as effluent.

On leaving the water quench tower, the pyrolysis gas is compressed to about 3.5 KPa in five stages. Water and organic fractions condensed during compression & cooling are recycled to the quench tower and gasoline stripper, respectively. Following compression, acid gas (H_2S & CO_2) is removed by absorption in diethanolamine (DEA)/other similar solvents in the amine wash tower (amine treator) followed by a caustic wash (caustic scrubber). The amine stripper strips the acid gas from the saturated DEA and the DEA is recycled to the amine wash tower. The waste caustic solution & blow down from the DEA cycle are neutralized, stripped of acid gas, and removed as liquid waste streams. The acid gas stripped from the DEA and caustic waste passes to an emissions control device (Vent C) to control H_2S emissions.

After acid gas removal, the remaining process gas stream is further compressed and passed through drying traps containing a desicant, where the water content is reduced to the low level necessary to prevent ice formation in the low-temperature distillation operations.

With the exception of three catalytic hydrogenation operations, the remaining process involves a series of fractionations in which the various product fractions are successively separated. The de-methaniser separates a mixture of hydrogen and methane from the C_2 & heavier components of the process gas. The de-methaniser overhead stream (hydrogen & methane) is further separated into hydrogen rich & methane rich streams in the low temperature chilling section. The methane rich stream is used primarily for furnace fuel. Hydrogen is required in the catalytic hydrogenation operations.

The de-ethaniser separates the C_2 components (ethylene, ethane & acetylene) from the C_3, and heavier components. Following catalytic hydrogenation of acetylene to ethylene by the acetylene converter, the ethylene-ethane split is made by the ethylene fractionators. The overhead from the ethylene fractionators is removed as the purified ethylene product, and the ethane fraction is recycled to the ethane/propane cracking furnace.

The de-ethaniser bottoms (C_3 and heavier compounds) pass to the de-propanizer where a C_3–C_4 split is made. The de-propaniser overhead stream (primarily propylene & propane) passes to a catalytic hydrogenation reactor (C_3 converter), where traces of propadiene and methyl acetylene are hydrogenated. Following hydrogenation, the C_3 fraction passes to the propylene fractionators, where propylene is removed overhead as a purified product. The propane is recycled to the ethane/propane pyrolysis furnace.

The C_4 and heavier components from the de-propaniser pass to the de-butaniser, where a C_4–C_5 split is made. The overhead G_4 stream is removed & may be treated as follows:

- For recovery of butadiene (Procticed in IPCL Vadodara, HPL Haldia, NOCIL)
- Hydrogenation in Catalytic Hydrogenation Unit to form Butane with/without Butadiene extraction (HPL Haldia/RIL, Hazira)

A process flowsheet for ethylene & co-products manufacture in given below in figure:

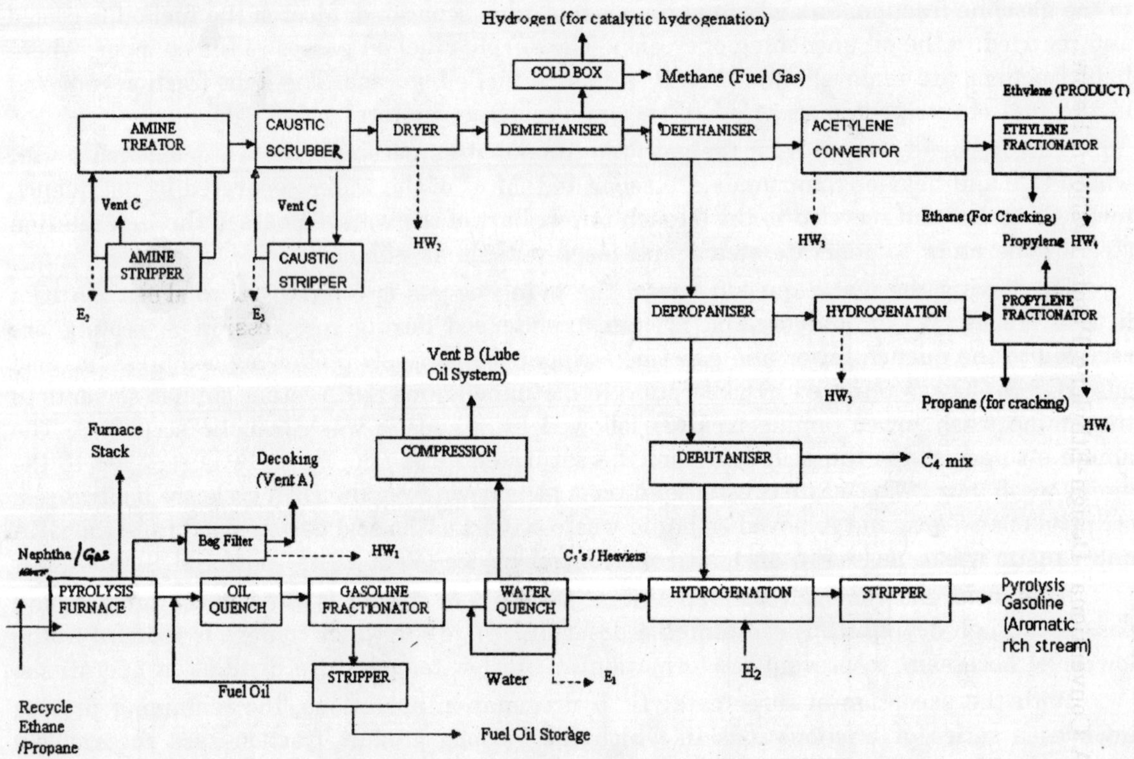

Figure 7.5: Flow diagram of ethylene manufacture

A Conventional Design Ethylene Plants is shown Below:

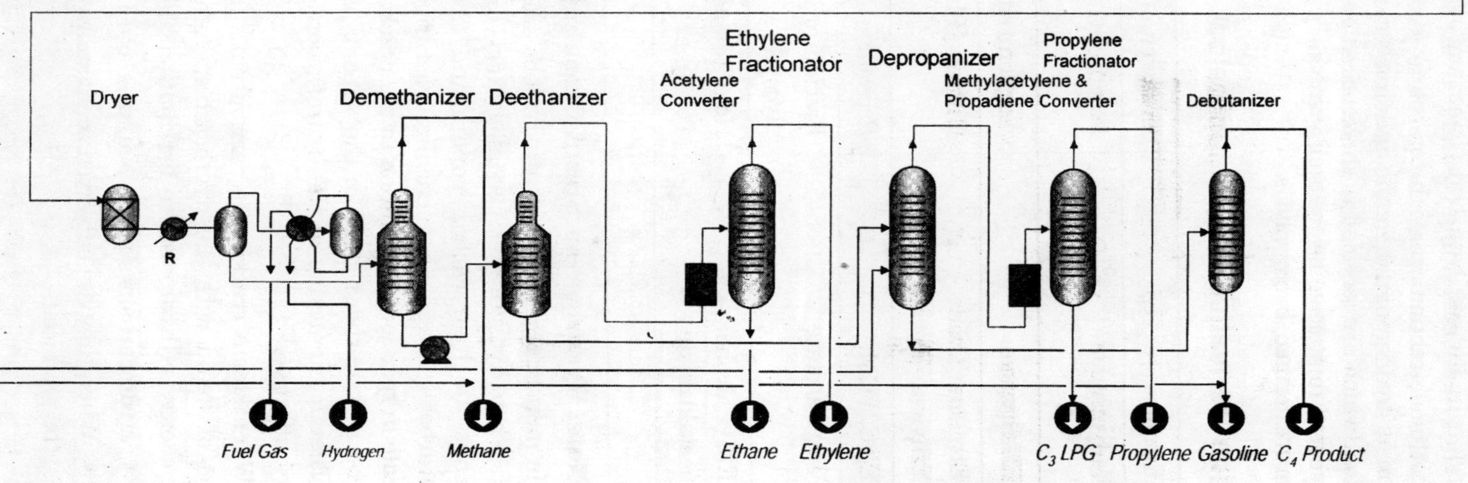

- Used as feed into the ethylene process (RIL Hazira)
- Use as fuel for furnaces in the ethylene process

The stream containing C_5 and heavier component from the debutaniser is combined with the bottoms fraction from the gasoline stripper as raw pyrolysis gasoline. The combined stream is hydrogenated in the gasoline treatment section. Following the stripping of light fraction, which are recycled to the cracked-gas compressor, the C_5 and heavier compounds are transferred to storage as treated pyrolysis gasoline. This stream contains benzene and other aromatics formed by pyrolysis.

Olefins Derivatives are listed below.

Ethylene (C_2H_4)	Propylene (C_3H_6)	Butadiene (C_4H_6)
Polyethylene	Polypropylene	Polymers and copolymers with styrene and ACN
EB-Styrene	Acrylonitrile	ABS
Ethylene Oxide, Ethylene glycols	Cumene-Phenol	n Butene-MEK
EDC, VCM	Propylene Oxide	Isobutene-TBA
Vinyl acetate	Epichlorohydrin, glycerol	Disobutene triisobutene & polyisobutene
Ethyl alcohol acetaldehyde	Nonylphenol Tetramer	Isobutene-MTBE
		Motor Fuel Alkylate

Note: Hydrocarbon Steam Cracking for Petrochemicals:

In industrial processes, hydrocarbons are contacted with H_2O, depending upon the desired effect. When hydrocarbon vapors at very high pressures are contacted with water, water quenches the hydrocarbon vapors and transforms into steam. The steam cracking of hydrocarbons is an anti-quenching operation, and will involve the participation of water molecule in reactions in addition to the cracking of the carbon bond on their own. Since steam and the hydrocarbons react in the vapour phase, tne reaction products can be formed very fast. Therefore cracking of the hydrocarbons on their own as well as by steam in principle is very effective.

When steam cracking is carried out, in addition to the energy supplied by the direct contact of steam with the hydrocarbons, steam also takes part in the reaction to produce wider choices of hydrocarbon distribution along with the generation of H_2 and CO.

- Hydrocarbons such as Naphtha and LPG have lighter compounds.
- When they are subjected to steam pyrolysis, then good number of petrochemicals can be produced.

- These include primarily ethyelene & acetylene along with other compounds such as propylene, butadiene, aromatics (benzene, toluene & xylene) and heavy oil residues.
- The reaction is pretty complex as we produce about 10 to 12 compounds in one go.
- The flowsheet will be reaction-separation-recycle system only but the separation system will be pretty complex.
- Important separation tasks include. elimination of CO & CO_2, purification of all products such as ethylene, acetylene etc.
- Typical feed stocks are Naphtha & LPG.
- Reaction temperature is about 700 - 800°C.

Process flowsheet is shown below in figure.

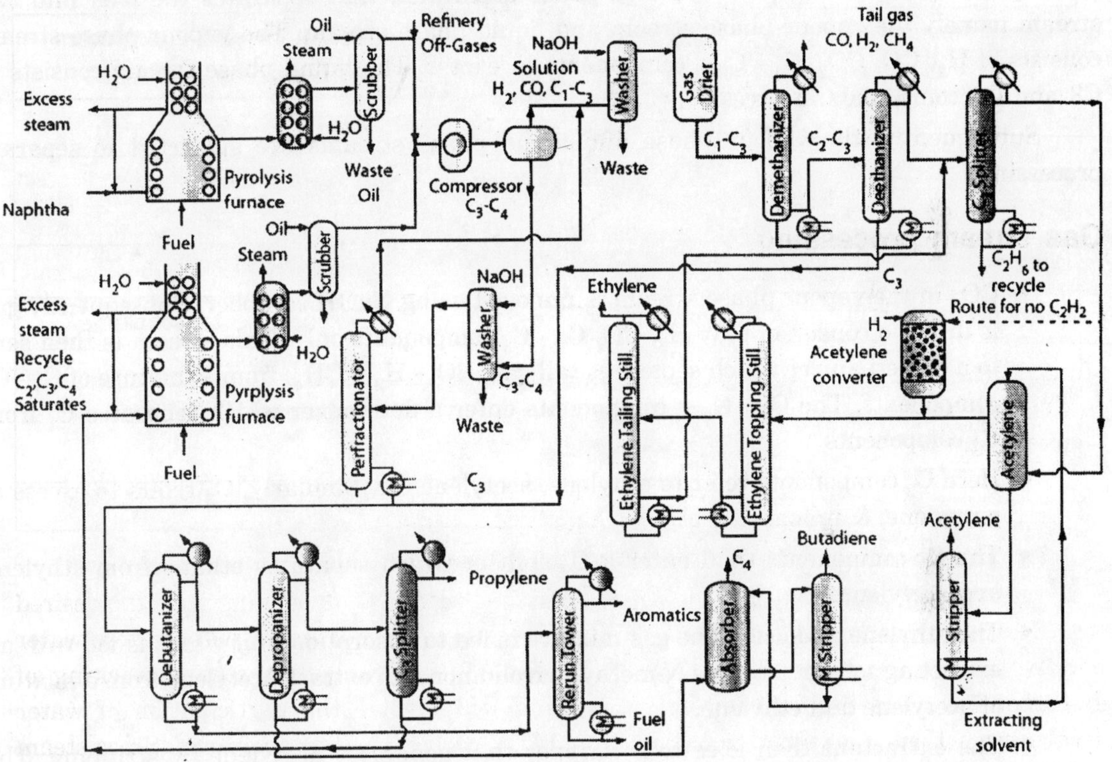

Figure. Flowsheet of Hydrocarbon Steam Cracking for producing Petrochemicals

Process description

Naphtha/LPG saturates is mixed with superheated steam and fed to a furnace fired with fuel gas + fuel oil as fuels to generate heat. The superheated steam is generated from the furnace itself using heat recovery boiler.

The C_2 - C_4 saturates are fed to a separate furnace fed with fuel gas + fuel oil as fuels to generate heat.

In the furnace, apart from the steam cracking, steam is also generated. This is by using waste heat recovery concept from the combustion gases in the furnace.

After pyrolysis reaction, the products from the furnace are sent to another heat recovery steam boiler to cool the product streams (from 700 - 800 °C) and generate steam from water.

After this operation, the product vapours enter a scrubber that is fed with gas oil as absorbent. The gas oil removes solids and heavy hydrocarbons.

Separate set of waste heat recovery boiler and scrubbers are used for the LPG furnace and Naphtha steam cracking furnaces.

After scrubbing, both product gases from the scrubbers are mixed and fed to a compressor. The compressor increases the system pressure to 35 Kg/cm^2.

The compressed vapour is fed to a phase separation that separates the feed into two stream namely the vapour phase stream and liquid phase stream. The vapour phase stream consists of H_2, CO, CO_2 C_1 - C_3 + components in excess. The liquid phase stream consists of C3 and C4 compounds in excess.

Subsequently, the vapour phase and liquid phase streams are subjected to separate processing.

Gas stream processing

- CO_2 in the vapour phase stream is removed using NaOH scrubber. Subsequently gas is dried to consist of only H_2, CO, C_1 - C_3 components only. This stream is then sent to a demethanizer which separates tail gas (CO + H_2 + CH_4) from a mixture of C_1 - C_3 components. The C_2 - C_3 + components enter a dethanizer which separates C_2 from C_3 components.
- Here C_2 components refer to ethylene, acetylene etc. Similarly, C_3 refers to excess of propylene, & propane.
- The C_2 components then enter a C_2 splitter which separates ethane from ethylene and acetylene.
- The ethylene and acetylene gas mixture is fed to absorption unit which is fed with an extracting solvent (such as N-methylpyrrolidinone) to extract Acetylene from a mixture of acetylene and ethylene.
- The extractant then goes to a stripper that generates acetylene by stripping. The regenerated solvent is fed back to the absorber.
- The ethylene stream is fed to a topping & tailing still to obtain high purity ethylene and a mixture of ethylene & acetylene as the top & bottom products. The mixture of ethylene and acetylene is sent back to the C_2 splitter unit as its composition matches to that of the C_2 splitter feed.

Liquid stream processing

- The liqiuid stream consists of C_3,C_4, aromatics and other heavy oil components is fed to a NaOH scrubber to remove CO_2.

- Eventually it is fed to a pre-fractionator. The pre-fractionator separates lighter components from the heavy components. The lighter components are mixed with the vapour phase stream and sent to the NaOH vapour phase scrubber unit.
- The pre-fractionator bottom product is mixed with the deethanizer bottom product.
- Eventually the liquid mixture enters a debutanizer that separates C_3, C_4 components from aromatics and fuel oil mixture. The bottom product eventually enters a distillation tower that separates aromatics and fuel oil as top & bottom products respectively.
- The top product then enters a depropanizer that separates C_3s from C_4 components.
- The C_4 components then enter an extractive distillation unit that separates butane + butylenes from butadiene. The extractive distillation unit consists of a distillation column coupled to a solvent stripper. The solvent stripper produces butadiene and pure solvent which is sent to the distillation column.
- The C_3 components enter a C_3 splitter that separates propylene from propane + butane mixture. The saturates mixture is recycled to the saturates cracking furnace as a feed stream.

MANUFACTURE OF ETHYLENE DICHLORIDE (EDC) AND VINYL CHLORIDE MONOMER (VCM)

Chapter 8

Ethylene dichloride (EDC) is an intermediate product used primarily in the manufacture of Vinyl chloride monomer. The process consists of a direct chlorination operation, a cracking/pyrolysis operation (for hydro-de-chlorination of EDC to VCM), an oxy-chlorination step (for utilising HCI removed in hydro-de-chlorination stage), product finishing and waste treatment operations. The raw materials for the direct chlorination process are chlorine and ethylene. Oxy-chlorination involves the treatment of ethylene with oxygen and HCI. Oxygen for oxy-chlorination generally is added by feeding air or purified oxygen as feed material.

Ethylene dichloride (EDC)

EDC is of interest as it is one of the intermediates for production of vinyl chloride monomer (VCM), which polymerizes to polyvinyl chloride (PVC).

Properties of Ethylene Dichloride ($ClCH_2 \bullet CH_2Cl$)

Mol. wt.:		98.97
M.P.:		–35.3°C
B.P.:		83.7°C
Sp.gr. @ 20 °C:		1.257
Flash point:		15.5°C
Ignition Temperature:		412°C
Explosive limits:	Lower =	6.2% in air
	Upper =	15.9% in air
Maximum toxicity limit:		75-100ppm

End Use of EDC include

- Vinyl chloride
- Antiknock agent
- Solvent

Methods of Production of EDC

- Reaction of chlorine with ethylene in liquid or vapor phase
- By-product of direct chlorination of ethane to ethyl chloride

- By-product of chlorinated hydrocarbons
  Ethylene + Chlorine Reaction
  $CH_2 = CH_2 + C1_2 \rightarrow ClCH_2 \bullet CH_2C1$ (EDC)

Raw material requirements

Basis: 1 ton of $C_2H_4Cl_2$ (95% yield)	
Ethylene:	0.30 ton
Chlorine:	0.75 ton
Ethylene dibromide catalyst—trace	
Cooling water	48 tons
Co-products: HCl, propylene chloride, polychloroethanes	

process description

Ethylene (with or without C_2H_6, CH_4 & H_2 diluent) is mixed with chlorine and bubbled through a liquid phase reactor. Ethylene dichloride product serves as the reacting medium. Heat of reaction is controlled by external heat exchange and recycle, or by coil or jacket heat transfer, to hold the reactor at 45-50°C with a pressure of 1.5-2 atms. Traces of ferric chloride or ethylene dibromide as catalyst gives 90-95% yield with little dimer formation.

The gaseous products are cooled to strip the acid gas of EDC. The liquid product is alkali washed and fractionated.

Salient features of EDC Production process is as follows:

- Undesired products are Propylene dichloride and Polychloroethanes.
- Reaction occurs in a liquid phase reactor with ethylene dichloride serving as the liquid medium and reactants reacting the liquid phase.
- Catalyst is $FeCl_3$ or Ethylene dibromide.
- C_2H_4 and Cl_2 are mixed and sent to the liquid phase reactor.
- Here, the feed mixture bubbles through the ethylene dichloride product medium.
- Reactor operating conditions are 50°C and 1.5-2 atms.
- The reaction is exothermic. Therefore, energy is removed using either cooling jacket or external heat exchanger.
- To facilitate better conversion, circulating reactor designs are used.
- $FeCl_3$ traces are also added to serve as catalyst.
- The vapour products are cooled to produce two products namely a vapour product and a liquid product. The liquid product is partially recycled back to the reactor to maintain the liquid medium concentration.
- The vapour product is sent to a refrigeration unit for further cooling which will further extract ethylene dichloride to liquid phase and makes the vapour phase bereft of the product.
- The liquid product is crude ethylene dichloride with traces of HCl. Therefore, acid wash is carried out first with dilute NaOH to obtain crude ethylene dichloride. A settling tank is allowed to separate the spent NaOH solution and crude $C_2H_4Cl_2$ (as well liquid).

- The crude ethylene dichloride eventually enters a distillation column that separates the ethylene dichloride from the other heavy end products.
- The vapour phase stream is sent to a dilute NaOH solution to remove HC1 and produce the spent NaOH solution. The off gases consist of H_2, CH_4, C_2H_4 and C_2H_6.

A flowsheet of EDC production by chlorination of ethylene is given below.

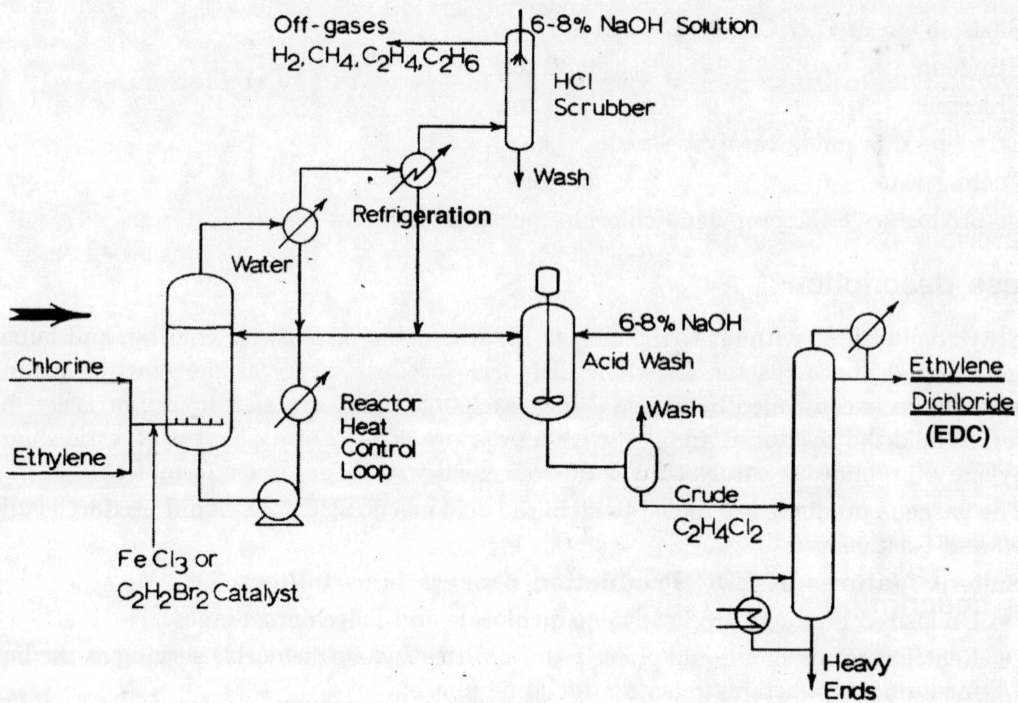

Figure 8.1: Flowsheet of Ethylene dichloride (EDC) production, by chlorination of ethylene

Vinyl Chloride Monomer (VCM)

Properties of VCM (CH_2CL)

Mol. wt.:		62.50
M.P.:		–153.8°C
B.P.:		–13.81 °C
Sp.gr. @ 20°C:		0.983

Soluble in $CC1_4$, $(C_2H_5)_2O$, and alcohols; slightly soluble in water

Flash point:		–77 °C
Explosive limits:	Lower =	5% in air
	Upper =	23% in air
Maximum toxicity limits:		500 ppm

Vinyl chloride monomer (VCM) is strictly an intermediate in polyvinyl chloride (PVC) production.

Methods of Production of VCM

- Ethylene dichloride thermal pyrolysis
- Acetylene-HCl reaction
- Ethylene dichloride-caustic reaction

The first two processes are described since they are often combined to keep a chlorine balance.

Ethylene Dichloride Thermal Pyrolysis Process of VCM Production:

Chemical reaction involved in VCM production:

$$CH_2C1 \bullet CH_2C1 \xrightarrow{500°C} CH_2 = CHC1 + HC1$$

Raw materials requirements

Basis: 1 ton of vinyl chloride (99.5% pure, 95% yield)

Ethylene dichloride:	1.65 tons
Heating steam:	2.0 tons
Cooling water	30 tons
Electricity	1.5KWH
Fuel gas	5.7 Nm^3

Co-product: 0.65 ton HC1

Process description

Ethylene dichloride (EDC) vapor at 4 kg/cm^2 is dried by silica gel and sent to a stainless steel tubular cracking furnace. This is externally flue gas fired and controlled at 480-520 °C. The contact surface catalyst within the tubes is pumice or charcoal. The conversion per pass is around 50% and the ultimate yield is 95%. Spray quenching with cold EDC prevents backward reaction. Uncondensed gases are sent to a surface heat exchanger to remove the balance of the EDC and vinyl chloride monomer (VCM).

The non-condensibles containing HCl are either sent to the acetylene-HCl process in an adjacent process area or water scrubbed to recover HCl.

The condensate is fractionated with the EDC bottoms returned for recycle and vinyl chloride monomer taken from overhead, stabilized & sent to storage.

Flowsheet of vinyl chloride monomer (VCM) production by pyrolysis of ethylene dichloride is given below.

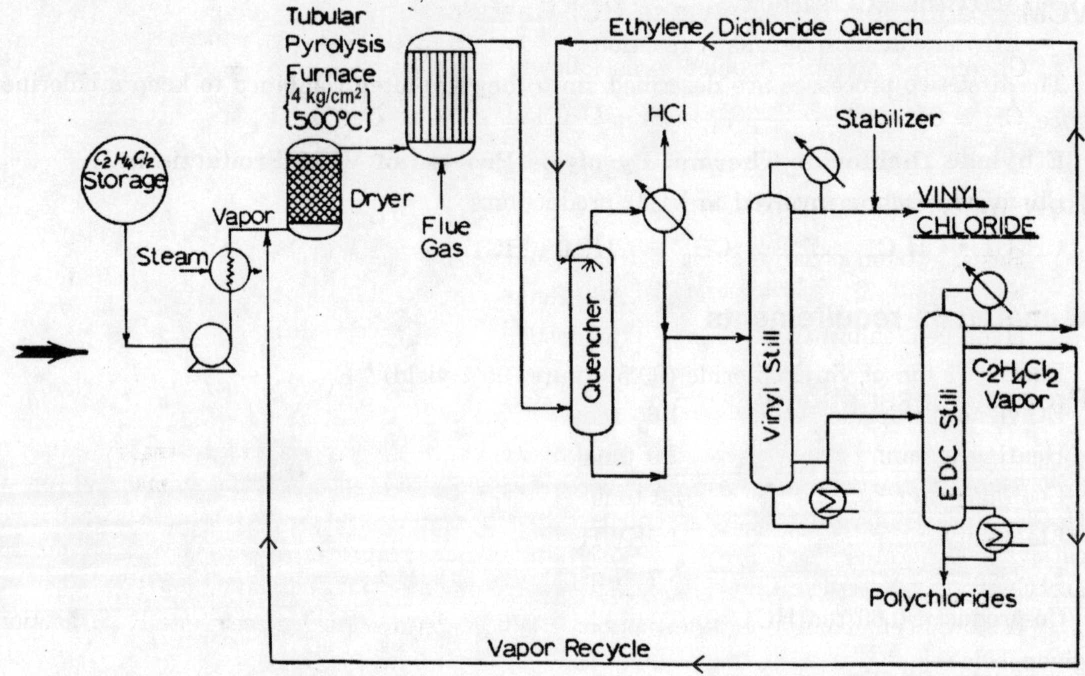

Figure 8.2: Flowsheet of Vinyl chloride production via ethylene dichloride pyrolysis

Salient features of VCM production by pyrolysis of EDC is as follows:

- Charcoal is used as the catalyst.
- The reaction is a reversible gas phase reaction.
- Ethylene dichloride is initially vaporized using a heat exchanger fed with process steam.
- Ethylene vapors then enter a dryer that removes traces of water molecules.
- After drying, the vapors enter a pyrolysis furnace operated at 4 kg/cm^2 and 500°C. The furnace is similar to a shell and tube arrangement with the gases entering the tube side and hot flue gas goes past the tubes in the shell side.
- The product vapors eventually enter a quenching tower in which cold ethylene dichloride is used to quench the product gases and cool them.
- The gases from the quench tower then enter a partial condenser which produces HCl as a gas and the liquid stream consisting of vinyl chloride, unreacted ethylene dichloride and polychlorides.
- The liquid stream from the quench tower as well as the condenser is fed to the vinyl still which produces the vinyl chloride product. The product is stabilized using a stabilizer as vinyl chloride is highly reactive without stabilizer.
- The bottom product from the vinyl still is fed to a distillation column which separates the ethylene dichloride from the polychlorides. The ethylene dichloride vapors are

recycled back to the cracking furnace and the ethylene dichloride liquid is sent to the quenching tower to serve as the quenching liquid.

VCM production by Acetylene-HCl Reaction:

Chemical reaction in volved is as follows:

$$\underset{\text{Acetylene}}{CH \equiv CH} + HCl \xrightarrow[\text{1atm}]{160\text{–}200°C} \underset{\text{VCM}}{CH_2 = CHCl}$$

Row materials requirements:

Basis: 1 ton of vinyl chloride (99.5% pure, 97% yield)

Acetylene:	0.462 ton
Hydrogen chloride:	0.60 ton

Process description:

Acetylene and dry HC1 in 5-10% molar excess are vapor blended by jet mixing in a pipe and passed through a tubular catalytic reactor containing carbon pellets impregnated with $HgCl_2$. The temperature is maintained at 160°C and is gradually raised to 200°C as the catalyst deteriorates. The effluent gases contain vinyl chloride which is separated from unreacted acetylene plus hydrogen chloride, these unreacted materials being recycled.

A flowsheet of vinyl choride monomer (VCM) production by acetylene-HCl reaction is given below:

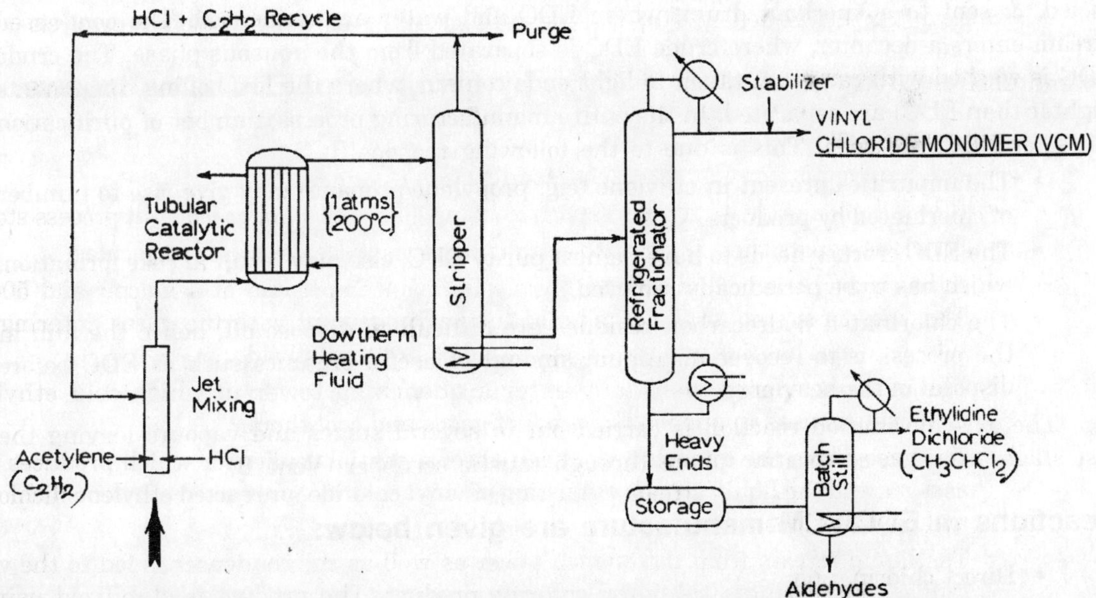

Figure 8.3: Flowsheet of vinyl chloride production via acetylene-HCl reaction

EDC/VCM Production in an Indian Petrochemical Plant

Basic operations that are used in a process using air for the oxy-chlorination steps as in IPCL, Vadodara are shown in figure.

In the direct-chlorination step of the balanced process, equimolar amounts of ethylene & chlorine are reacted in EDC, which acts as liquid medium in a high temperature chlorination reaction (HTC). Most commercial plants carry out the reaction in presence of ferric chloride catalyst. The reaction mass (in vapour form) is quenched in quench column and distilled in distillation column for EDC separation. The vent from the EDC column is treated in caustic scrubber before releasing any un-reacted gases/products to atmosphere (vent A). The heavy ends from distillation column are taken to incinerator. The dry pure EDC produced is sent to intermediate storage before cracking.

Pure EDC liquid is taken from intermediate storage, preheated & vaporized and sent to pyrolysis furnace where EDC is cracked at 490°C. The cracked gases containing VCM, HCI & un-cracked EDC are quenched rapidly in quench column. Bottoms are sent to tar still for recovery of EDC & heavy ends are sent to storage. The vapours leaving quench column are fed to HCL column for separation of HCL (as 100%) from column top by propylene-refrigerated condenser. The HCL column bottom containing EDC & VCM is next fed to VCM column to separate VCM product from top, which is dried & taken for storage. The VCM column bottoms are taken to flash distillation column to recover EDC.

In the oxy-chlorination reaction, the ethylene, anhydrous hydrogen chloride (generated in EDC cracking) and air are reacted in the vapour phase at 200 to 315°C in either fixed bed or fluid bed reactors. A mixture of copper chloride and other chlorides is used as catalyst.

The products of reaction from the oxy-chlorination reactors are quenched with water, cooled, & sent to a knockout drum, where EDC and water are condensed. The condensed stream enters a decanter, where crude EDC is separated from the aqueous phase. The crude EDC is washed with caustic & taken to light ends column, where the low boiling components (lighter than EDC) are separated. In the entire manufacturing process, number of purification steps for EDC, are seen. This is due to the following reasons.

- The impurities present in ethylene (e.g. propylene/propane) may give rise to number of chlorinated by-products.
- The EDC cracker needs to have highest purity EDC, else may result in coke formation, which has to be periodically removed.
- The chlorinated hydrocarbon residues are difficult to dispose off, hence the aim in the process is to recover maximum amount of useful organics such as EDC before disposal of the heavier.

The oxy-chlorination reaction is carried out in several stages and vapours leaving the last stage are released to atmosphere through caustic scrubber (Vent B).

Reactions in EDC/VCM manufacture are given below:

- Direct chlorination:

$$\underset{\text{Acetylene}}{CH_2 = CH_2} + \underset{\text{Chlorine}}{Cl_2} \longrightarrow \underset{\text{Ethylene Dichloride (EDC)}}{ClCH_2CH_2Cl}$$

- EDC pyrolysis:

$$\underset{\text{Ethylene dichloride (EDC)}}{2ClCH_2\ CH_2Cl} \longrightarrow \underset{\text{Vinyl chloride}}{2CH_2 = CHCl} + 2HCl$$

- Oxy-chlorination:

$$\underset{\text{Ethylene}}{CH_2 = CH_2} + \underset{\text{Hydrochloric acid}}{2HCl} + \underset{\text{Oxygen}}{\tfrac{1}{2}O_2} \longrightarrow \underset{\text{Ethylene dichloride}}{ClCH_2CH_2Cl} + \underset{\text{Water}}{H_2O}$$

- Overall reaction:

$$\underset{\text{Ethylene}}{2CH_2 = CH_2} + \underset{\text{Chlorine}}{Cl_2} + \underset{\text{Oxygen}}{\tfrac{1}{2}O_2} \longrightarrow \underset{\text{Vinyl chloride}}{2CH_2 = CHCl} + \underset{\text{Water}}{H_2O}$$

Several by-products like 1,1,2 tri-chloro-ethane, ethyl chloride, chloro-butanes, chloro-ethylene, ethyl chloride are produced.

A Flow diagram of EDC/VCM production by balance process as practiced at Indian Petrochimical Corporation Limited (IPCL), Vadodara (Gujrat) is shown below.

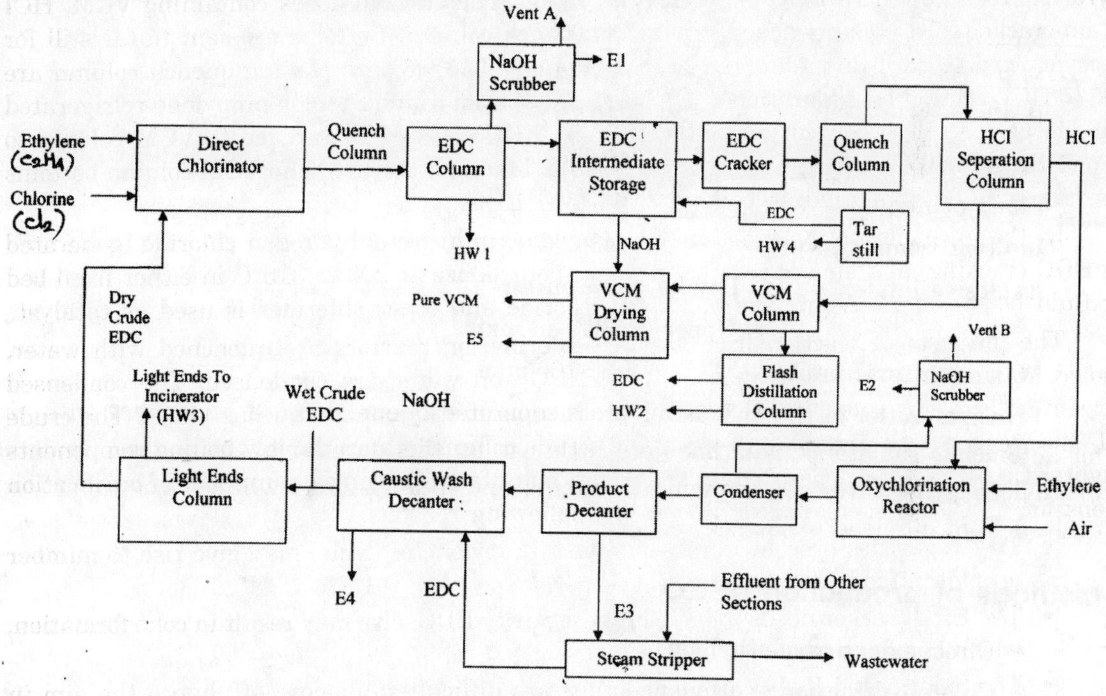

Figure 8.4: Flow diagram of EDC/VCM Manufacture at IPCL, Vadodara

PRODUCTION OF ETHYLENE OXIDE (EO)

Chapter 9

Properties of Ethylene Oxide ($CH_2 — CH_2$ with O bridging, epoxide ring):

Mol.:		44.05
M.P.:		—111.7°C
B.P.:		10.7°C
Sp.gr. @ 0°C:		0.896
Flash point:		—15°C
Ignition temperature:		430 °C
Explosive limits:	Lower =	3% in air
	Upper =	80% in air
Maximum toxic limit:		25-100 ppm

Principal outlet in the U.S. is for production of ethylene glycol, a permanent antifreeze for automobile radiators. India has very little use for this derivative, but instead finds use in ethylene glycol polyesters, non-ionic detergents, and ethanolamines which are useful products derived from ethylene oxide.

Methods of production of EO

- Direct oxidation of ethylene
- Chlorohydrination of ethylene

Direct Oxidation of Ethylene to produce EO:

Chemical reaction involved is as follows:

$$CH_2 = CH_2 + \frac{1}{2}O_2 \xrightarrow[\substack{250\text{-}300°C \\ 4\text{-}5 \text{ atms.}}]{\text{air or oxygen, AgO}} \underset{\diagdown O \diagup}{CH_2 — CH_2} + (CO_2 + H_2O); \Delta H = -29.2 \text{ Kcal}$$

Side reaction: $(CO_2 + H_2O)$

Raw materials requirements

Basis: 1 ton of ethylene oxide (99% purity, 70% yield)	
Ethylene:	0.92 ton
Air:	9.0 tons
Silver:	0.3 kg in fixed bed
	0.7 kg in fluidized bed
Ethylene dichloride suppressor:	10-15 kg
Electricity:	1,500 KWH
Steam:	0.1 ton
Water:	180 tons

Process description of EO production by oxidation of ethylene

Ethylene of 95-98% purity and air are compressed separately, mixed together giving 3-10% C_2H_4 volume concentration and passed over a catalyst of silver oxide on a porous inert carrier such as alumina. A side reaction suppressing agent such as ethylene dichloride is added to the feed to reduce the competitive oxidation reaction to CO_2 + H_2O.

The reaction is highly exothermic and is best carried out in a fixed bed tubular reactor in which heat transfer salt or Dowthenn is pumped around the tube within the shell to maintain a 250-300°C temperature. Heat is recovered in a waste-heat steam boiler. A short residence time of 1 sec in plug flow is the design with an ultimate yield of 60-70%.

A Flowsheet for production of EO by oxidation of ethylene is given below:

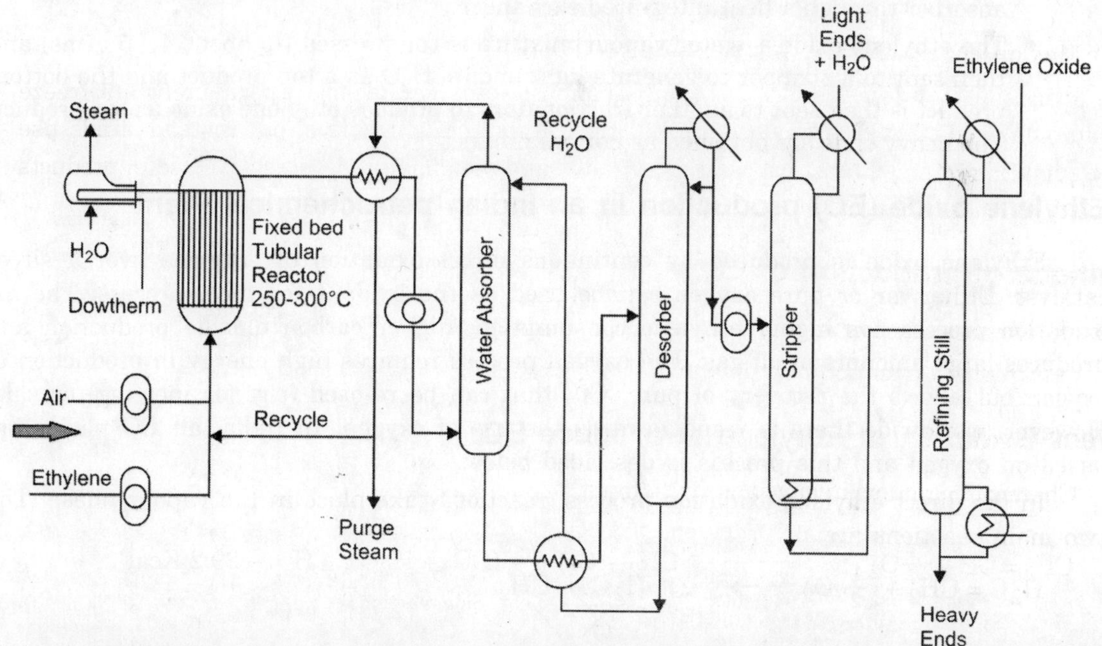

Figure 9.1: Flowsheet of production of ethylene oxide (EO) by oxidation of ethylene

The effluent gases from the reactor are water-washed under pressure. The absorbed ethylene oxide is sent to a packed bed desorber-fractionator tower and taken overhead. It still contains a large amount of water vapor plus some impurities. This stream is compressed to 4-5 atms., and fractionated twice to remove light ends, water, and high boiling polymers.

Salient features of the EO production process is as follows:

- Air and ethylene are separately compressed and along with recycle stream area sent to the shell and tube reactor.
- The reactor is fed on the shell side with Dowtherm fluid that serves to maintain the reaction temperature. A dowtherm fluid is a heat transfer fluid, which is a mixture of two very stable compounds, biphenyl & diphenyl oxide. The fluid is dyed clear to light yellow to aid in leak detection.
- The hot dowtherm fluid from the reactor is sent to a waste heat recovery boiler to generate steam.
- The vapour stream is cooled using a integrated heat exchanger using the unreacted vapour stream generated from an absorber.
- The vapour stream is then sent to the heat integrated exchanger and is then sent back to the reactor and a fraction of that is purged to eliminate the accumulation of inerts such as Nitrogen and Argon.
- The product vapors are compressed and sent to a water absorber which absorbs ethylene oxide from the feed vapors. Eventually, the ethylene oxide rich water stream is sent to a stripper which desorbs the ethylene oxide + water as vapour and generates the regenerated water as bottom product. The regenerated water reaches the absorber through a heat integrated exchanger.
- The ethylene oxide + water vapour mixture is compressed (to about 4 - 5 atms) and then sent to a stripper to generate light ends + H_2O as a top product and the bottom product is then sent to another fractionators to produce ethylene oxide as top product. The heavy ends are obtained as bottom product.

Ethylene oxide (EO) production in an Indian petrochemical Plant

Ethylene oxide is produced by continuous direct oxidation of ethylene over a silver catalyst. Either air or pure oxygen can be used as the oxidant for the . process. The air oxidation process has higher ethylene consumption, higher carbon dioxide production and produces large amounts of off-gas. The oxygen process requires high energy in production of oxygen but allows the recovery of pure. CO_2 that can be re-used (e.g. for inerting) or sale. However, worldwide there is trend towards the use of oxygen. In India, all EO plants are based on oxygen and this process is described below.

In the direct ethylene oxidation process, reactions take place in the vapour phase. The two main reactions are:

$$H_2C = CH_2 + \underset{\text{oxygen}}{\tfrac{1}{2}\, O_2} \longrightarrow \underset{\text{Ethylene Oxide}}{H_2C - CH_2 \text{ (bridged by O)}} \quad \ldots\ldots\ldots\ldots \text{I}$$

$$\underset{\text{Ethylene}}{H_2C = CH_2} + \underset{\text{Oxygen}}{3O_2} \longrightarrow \underset{\text{Carbon dioxide}}{2CO_2} + \underset{\text{Water}}{2H_2O} \quad \cdots\cdots\cdots \quad II$$

The formation of 25 to 30% of the ethylene to carbon dioxide & water, as given in Reaction (II) is major drawback of the oxidation process. Reaction (II) also releases thirteen times as much heat energy as does Reaction (I). Reaction (II) can be suppressed by replacing the catalyst regularly and by carefully controlling the temperature on the surface of the catalyst, thereby limiting the conversion of ethylene to CO_2 and H_2O on each catalyst pass to less than 30%.

In the Oxygen oxidation process, ethylene and oxygen enter the reactor. The reaction takes place over a silver catalyst packed in tubes, the heat from the reaction is dissipated by a jacket of heat transfer fluid. Reaction temperature and pressure are maintained at 220 to 280°C and 1 to 3 MPa. The activity of the catalyst can be enhanced by the addition of promoters such as alkali metals or alkali earth metals. Catalyst inhibitors such as Ethylene dichloride are added to suppress conversion of ethylene. to carbon dioxide while not interfering with the primary reaction. In addition to the main by-product, carbon dioxide, small amounts of acetaldehyde (less than 1%) and traces of formaldehyde are also produced.

Descriptions of Streams and Vents Illustrated in Fig. 9.2

Code Number	**Description in Fig 9.2**
Stream	
1	Ethylene feed, >98 mole %
2	Oxygen feed, >97-99 mole %
3	Recycle to primary reactor, 0.006 % EO
4	Primary reactor product gas, 2 % EO
5	CO_2 purge stream
6	CO_2-free recycle to primary reactor
7	CO_2-rich CO_2 absorbent ($KHCO_3$)
8	Reactivated CO_2 absorbent ($KHCO_3$)
9	Absorber bottoms, minor EO levels
10	Recycle water to absorbers
11*	Reabsorber Vent
12	Recycle Water to Reabsorber
Vent	
A	CO_2 desorber vent (CO_2 purge)
B	Inert Purge
Effluent	
El	CO_2 absorbent
E2	Water stripped during EO purification

The effluent from the reactor (stream 4) contains Ethylene oxide, Ethylene and carbon dioxide. It is cooled, compressed, and passed through the primary absorber. As it passes up the packed column absorber counter current to cold water, the ethylene oxide and some of the carbon dioxide, hydrocarbons and aldehydes dissolve in the water.

Most of the unabsorbed gas that exits the top of the absorber is cooled and becomes the recycle stream (stream 3). A smaller portion of the unabsorbed gas stream (stream 5) is purged to prevent the accumulation of inert gases such as CO_2 and Argon (which is present as an impurity in oxygen). The purge gas is passed through a CO_2 absorber, which uses potassium carbonate as an absorbent, then (as stream 6) joins the recycle to the reactor. The spent potassium carbonate (stream 7) is reactivated in the $C0_2$ stripper, then recycled to the CO_2 absorber (stream 8). The CO_2 is vented from the CO_2 stripper (Vent A). A vent stream (Vent B) is taken from the recycle gas in order to reduce the build-up of inerts like ethane, argon & nitrogen, impurities present in the ethylene & oxygen feedstock. The inerts vent is typically used as fuel gas & burned (e.g. in cracker furnace or stream boiler) & or flared (IPCL, Dahej).

The dilute aqueous solution of EO from the absorber are taken to a stripper, where EO and dissolved inert are distilled under reduced pressure. The stripper water virtually free of EO is re-circulated to the absorbers (stream 10). The crude EO from the desorber is then reabsorbed in water. The EO from the re-absorber is sent for refining/purification. The off-gas from the re-absorber (steam 11) contains large quantity of ethylene and is recycled back to the oxidation stage. The final product 99.5 mole percent EO is stored under nitrogen pressure.

Ethylene Glycol (EG) manufacture from Ethylene oxide (EO):

The crude EO from re-absorber/purified EO product is hydrolysed with water in a pipe reactor. The large amount of excess water is removed in a multiple effect evaporator. The water removed from the evaporator (stream 13) is used back in MEG reactor. The Ethylene glycol is then taken to a MEG Column where MEG is separated from higher glycols. The column bottoms (containing higher glycols) are further taken for separation of di-ethylene glycol and tri-ethylene glycol column residues are sold as heavier glycols/polyethylene glycols. Fig. 9.3 presents the flow diagram of Ethylene glycol manufacturer stream descriptions of vents and stream in Fig. 9.3 is given below:

Code Number	Description in Fig 9.3
Stream	
13	Recycle water to MEG Reactor
14	MEG with trace moisture
15	Higher glycols streams (containing DEG/TEG/Polymeric glycols)
Effluent	
E3	Water removed during evaporation & drying of EO
E4	Water from vacuum system (MEG/DEG)

Various chemical reactions taking place during EO/EG manufacture are listed below:

Main reactions

1. $\underset{\text{Ethylene}}{C_2H_4} + \underset{\text{Oxygen}}{\frac{1}{2}O_2} \longrightarrow \underset{\text{Ethylene oxide}}{C_2H_4O}$

2. $\underset{\text{Ethylene oxide}}{C_2H_4O} + H_2O \xrightarrow[10\text{–}15\,kg/cm^2]{150\text{–}175°C} \underset{\text{Mono Ethylene glycol (MEG)}}{HO\bullet CH_2\bullet CH_2OH}$

3. $C_2H_4O + OH\bullet CH_2 - CH_2\bullet OH \longrightarrow \underset{\text{Di-ethylene glycol (DEG)}}{CH_2\bullet OH\bullet CH_2 - O - CH_2\bullet CH\bullet CH_2}$

4. $\underset{\text{EO}}{C_2H_4O} + \underset{\text{(DEG)}}{CH_2\bullet OH\bullet CH_2 - O - CH_2\bullet OH\bullet CH_2} \longrightarrow \underset{\text{Tri-thylene glycol (TEG)}}{HO\,(C_2H_4O)_3\,H}$

Side reactions

1. $C_2H_4 + 3O_3 \longrightarrow 2CO_2 + 2H_2O)$

2. $2C_2H_4 + O_3 \longrightarrow \underset{\text{Acetaldehyde}}{2CH_3CHO)}$

Flow diagram of ethyene oxide (EO) Manufacture by direct oxideation of ethylene is given below in figure 9.2

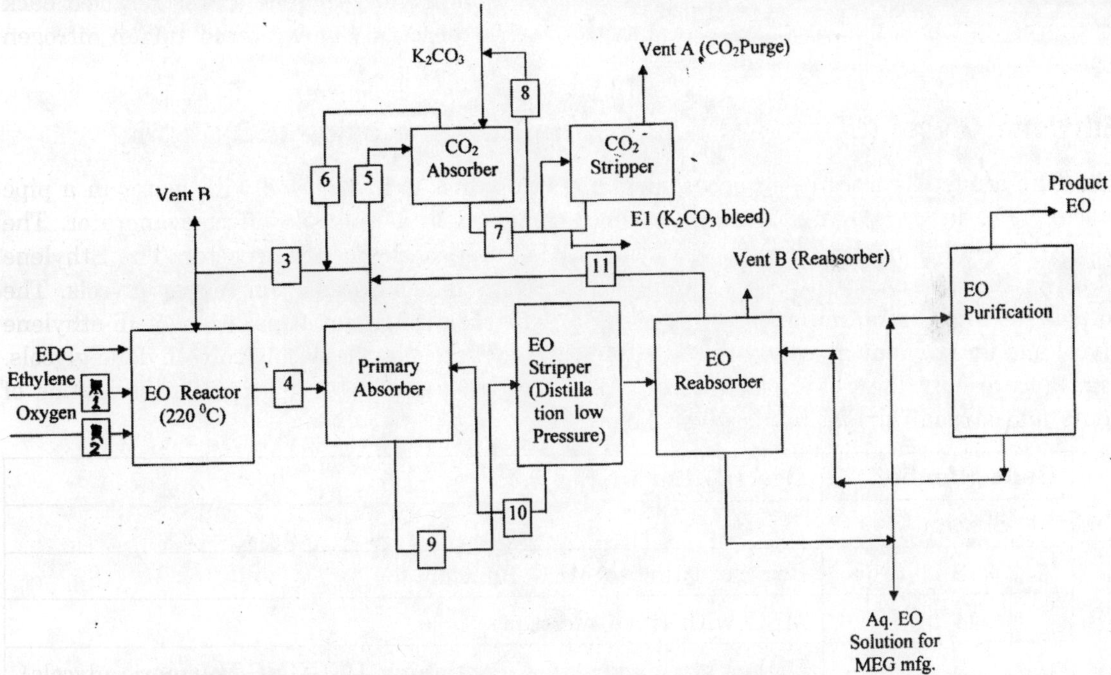

Figure 9.2: Flow Diagram of EO Manufacture by oxidation of ethylene

Flow diagram of ethylene glycol manufacture is given below in figure 9.3

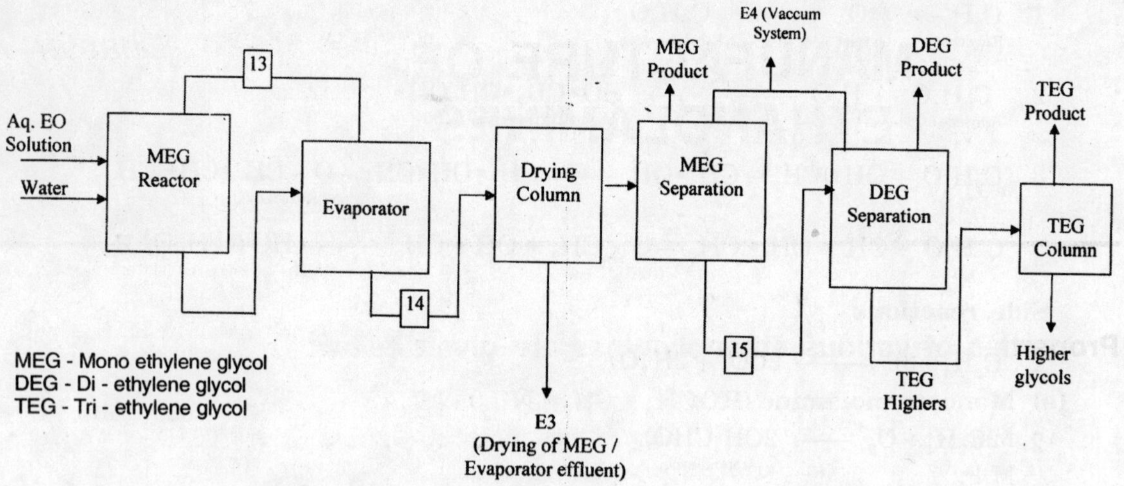

Figure 9.3: Flow Diagram of ethylene Glycol Manufacture

MANUFACTURE OF ETHANOLAMINES

Chapter 10

Properties of various ethanolamines are given below:

(a) Monoethanolamine ($HOCH_2 \bullet CH_2 \bullet NH_2$) (MEA)

Mol. wt.: 61.08

M.P.: 10.5°C

B.P.: 170.5 °C

Sp. gr. @ 20°C: 1.02

Appearance: Colorless, moderately viscous oil

Odor: Mild ammoniacal

Soluble in water, alcohol, chloroform, ether; slightly soluble in benzene

Flash point: 95 °C

Purity: 80-98%, containing higher amines and water; specified on B.P. range

(b) Diethanolamine [$(HO \bullet CH_2 \bullet CH_2)_2 \bullet NH$] (DEA):

Mol. wt.: 105.2

M.P.: 28 °C

B.P.: 270°C (decomp.)

Sp. gr. @ 30°C; 1.09

Appearance: Colorless crystals @ 20 °C

Soluble in water, alcohol; slightly soluble in ether; insoluble in benzene

Flashpoint: 138°C

Ignition temperature: 660°C

Purity: 94-95%, containing mono and triethanolamine

(c) Triethanolamine [$(HO \bullet CH_2 \bullet CH_2)_3N$] (TEA):

Mol. wt. 149.2

M.P.: 21.2°C

B.P.: 360°C (decomposes.)

Sp. gr. @ 20°C: 1.13

Appearance: Pale yellow oil

Odor: Slight ammoniacal

Soluble in water, alcohol, chloroform; slightly soluble in ether

Purity: 80% technical grade with approximately 2% mono-and 15% diethanolamine; 98% high purity grade.

Uses of ethonolemines

Ethanolamines are valuable products whose main use is in the manufacture of detergents by reaction with fatty acids. Ethanolamine can be used directly in gas purification to remove undesired acidic components. The fatty acid salts of ethanolamines can be used as components of soaps and cosmetic creams. Finally, the ethanolamines may be used as intermediates in the manufacture of morpholine, ethyleneimine and ethylene diamine.

Methods of Production

There is only one basic process, namely exothermic reaction of ethylene oxide with ammonia.

Chemical Reactions; (E is the limiting reagent)

A set of complex series reactions exists:

$E + A \rightarrow M$ E = ethylene oxide (in vapor phase)

$E + M \rightarrow D$ A = ammonia (in aqueous phase)

$E + D \rightarrow T$ M, D, T = mono-, di-, and triethanolamines respectively.

The exact chemical equations are:

$$\underset{\diagdown\ \text{O}\ \diagup}{CH_2 - CH_2} + NH_3 \longrightarrow NH_2CH_2CH_2OH$$

$$\underset{\diagdown\ \text{O}\ \diagup}{CH_2 - CH_2} + NH_2CH_2CH_2OH \longrightarrow NH(CH_2CH_2OH)_2$$

$$\underset{\diagdown\ \text{O}\ \diagup}{CH_2 - CH_2} + NH(CH_2CH_2OH)_2 \longrightarrow N(CH_2CH_2OH)_3$$

Product ratios depend on conditions used.

Raw materials Requirements:

Basis: 1 ton of mixed ethanolamines (e.g., 70% mono-, 25% di-, and 5% tri- with 95% ethylene oxide utilization)

Ethylene oxidel	0.81 ton
Ammonia:	0.24 ton

Process Description

Ethylene oxide and aqueous ammonia (25-30% NH_3) are fed to a stirred tank reactor which can be operated under a wide range of conditions depending on product distribution. Ammonolysis is exothermic and heat must be removed. The product stream is flashed to remove NH_3 for recycle. The bottoms contain a water solution of the amines which are separated in four fractionating columns, the last two under vacuum.

A Flowsheet of production of ethonolomines using ammonia and ethylene oxidie is given below.

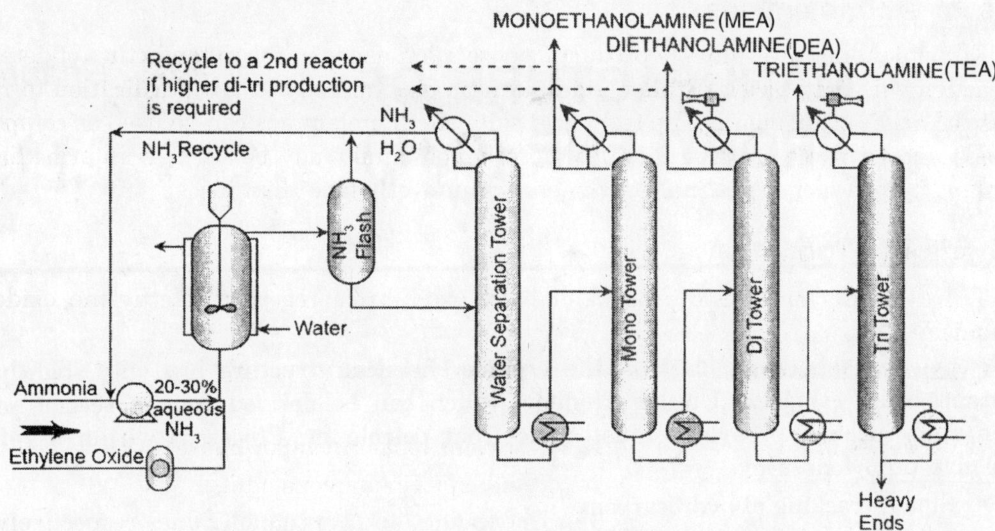

Figure 10.1: Flowsheet for production of ethanolamines

Salient features of ethanolamines production is listed below:

- Ammonia is mixed with ammonia recycle stream from the process and pumped to the CSTR where liquid phase ammonolysis takes place.
- Ethylene oxide is compressed and fed to the CSTR.
- The CSTR operating pressure will be such that the feed (and product) mixtures do not vaporize and good liquid phase reaction can occur.
- The reactor is cooled using water in the cooling jacket as the reactions are mildly exothermic
- The product stream is then sent to a flash unit that separates NH_3 + H_2O as a vapour stream and water + ethanolamines as a liquid stream.
- The ammonia + water stream is recycled to mix with the fresh ammonia and enter the reactor.
- The bottom product from ammonia flash unit is sent to a water separation tower that again removes dissolved ammonia in the ethanolamine rich solution. Once again ammonia + water are generated and this stream is also recycled to mix with fresh ammonia feed.
- The bottom product consists of crude mixture of ethanolamines and heavy ends.
- This mixture is fed to a monoethanolamine tower first to separate the monoethanol amine from the other two and heavy ends
- The bottom product from the first distillation tower then enters the second and third distillation towers which are operated under vacuum to produce diethanolamine and triethanolamine as top products. The bottom product from the last distillation tower is the heavy ends product.

MANUFACTURE OF PROPYLENE (C_3H_6)

Chapter

11

Propylene as building block in the organic chemical structure has only slightly less importance than ethylene. Unlike ethylene, which can be derived from a second source, fermentation alcohol, propylene must come from petroleum. Processes within a refinery which give propylene are:

- Steam cracking of hydrocarbons.
- Direct pyrolysis of propane.
- Refinery off-gases.

Following figure gives the chemical intermediates & Products for various uses made from propylene.

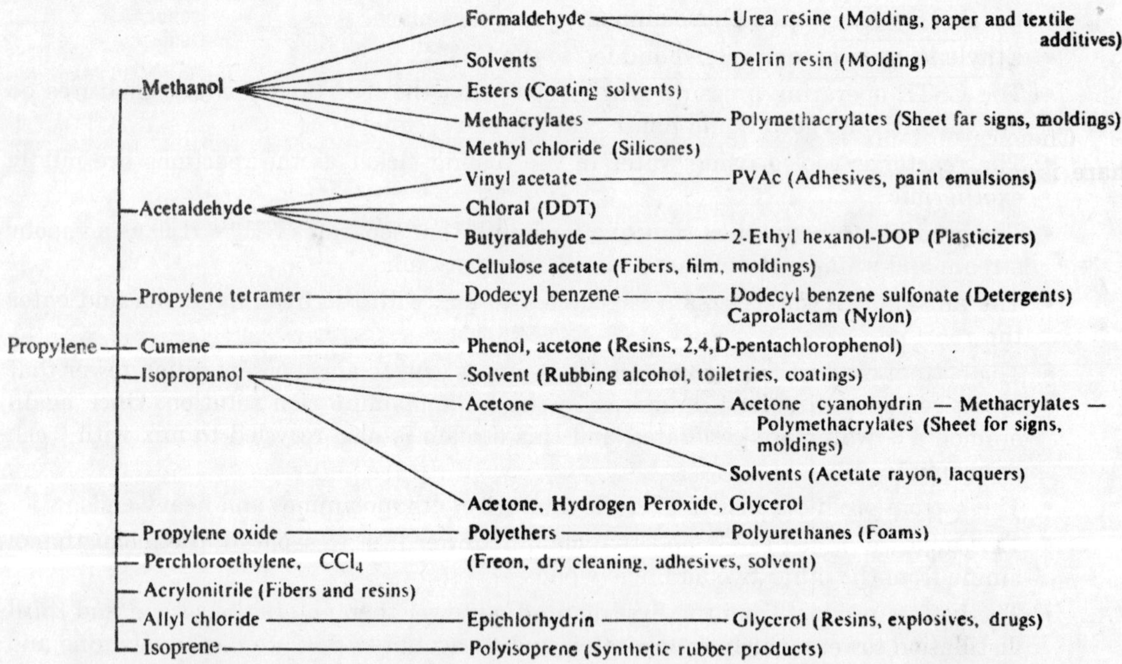

Figure 11.1: Chemical intermediate and products from propylene

Propylene and Its Derivatives produced by various unit operations/processes are shown below.

Propylene is used to produce polypropylene resin, acrylonitrile, acrylic acid, propylene oxide, isopropyl alcohol, and acetone through polymerization, oxidation, alkylation, hydration and the addition of halogen.

Propylene is as important a basic chemical in the petrochemical industry as ethylene.

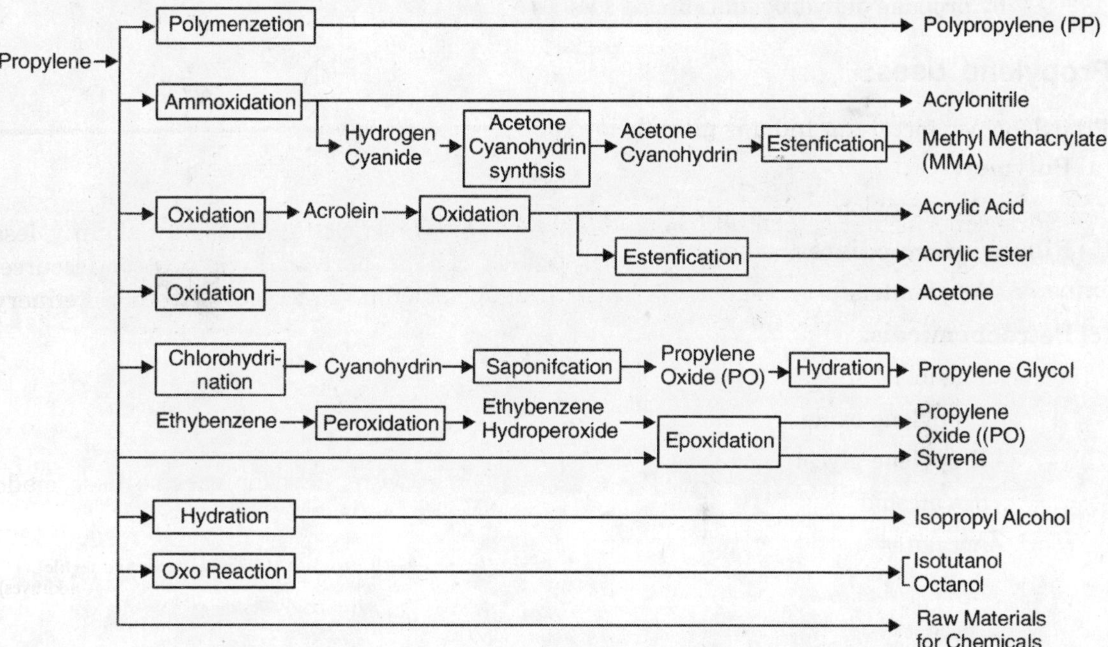

Chemical formula of Most important propylene derivatives/products & their production share is given below:

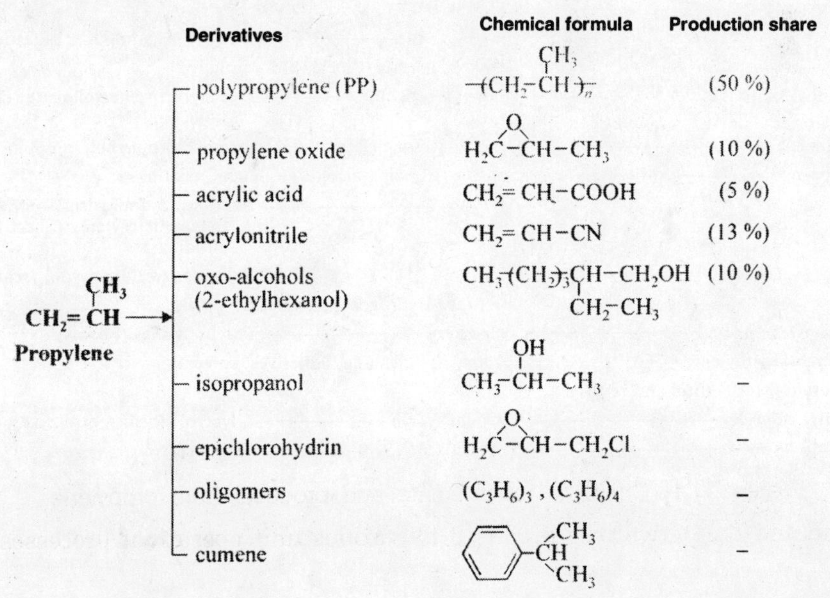

Propylene production

- by steam cracking/pyrolysis 65%
- from refinery gases (FCC) 30%
- by propane dehydrogenation 5%

Propylene uses:

Propylene use picture in India is given below.

(a) **Polymer:**

- Polypropylene

(b) **Fibres/intermediates:**

- Acrylonitrile

(c) **Petrochemicals:**

- Phenol
- Propylene oxide
- Propylene glycol
- oxo alcohal
- Acrylates

PRODUCTION OF ISOPROPANOL (ISO-PROPYL ALCOHOL) FROM PROPYLENE

Chapter 12

Properties of Isopropanol [$(CH_3)_2 \cdot CHOH$]

Mol. wt.	60.1
M.P.	–89.5°C
B.P.	82.5°C
Sp. gr. @ 20°C:	0.789
Miscible with water, alcohols, ether	
Flash point:	11.6°C
Ignition temperature:	400 °C
Explosive limits (in air):	2-12%
Toxic limit concentration:	200-400 ppm

Isopropanol (IPA) is used largely as an intermediate for acetone production with the next largest use as a multipurpose industrial solvent.

Methods of Production of iso-propanol

- Direct catalytic hydration of propylene.
- Hydration of propylene via sulfation and hydrolysis.
- Hydration of Propylene via Sulfation and Hydrolysis:

Chemical reactions involved

(a) **Sulfation:**

$$\underset{\text{Propylene}}{CH_3 = CH_2} + H_2SO_4 \longrightarrow \underset{\text{isopropyl acid sulfate}}{(CH_3)_2CH(OSO_3H)}$$

(b) **Hydrolysis:**

$$CH_3\underset{\displaystyle H}{\underset{|}{CH}} - \underset{\displaystyle O-SO_3}{\underset{|}{CH_3}} + H_2O \longrightarrow \underset{\text{iso-propanol}}{\begin{matrix} CH_3 \diagdown \\ \quad CHOH \\ CH_3 \diagup \end{matrix}} + H_2O_4$$

(c) **Ether formation — side reaction**

$$\underset{\text{diisopropyl sulfate}}{\begin{matrix} (CH_3)_2CH \diagdown \quad \diagup O \\ S - OH \\ (CH_3)_2CH \diagup \quad \diagdown O \end{matrix}} + H_2O \longrightarrow \underset{\text{diisopropyl ether}}{\begin{matrix} (CH_3)_2CH \diagdown \\ O \\ (CH_3)_2CH \diagup \end{matrix}} + H_2SO_4$$

Raw materials requirements

Basis: 1 ton of isopropanol (88 wt.% purity, 95% yield)
Propylene: 0.680 ton
Sulfuric acid (75%): 11 kg
Sodium hydroxide: 18 kg
By-products: Isopropyl ether (optional), polymers

Process description

Propylene, mixed with other C_2 – C_3 hydrocarbons, is absorbed in 70-75% H_2SO_4 in a packed or sieve tray absorption tower or in a series of agitated reactors operating at 20-25 atms. The strongly exothermic sulfation reaction is kept at 20-30°C by refrigerated brine circulation. The solution is hydrolyzed in a steam stripper which removes the organic phase as a vapor. The weak acid bottoms are sent to the H_2SO_4 concentration plant after neutralization in a caustic solution. The product vapor is fractionally condensed, with the non-condensible olefin fraction returned to the sulfation step.

The crude isopropanol is distilled to remove the more volatile isopropyl ether. A portion of the reflux is shunted to a decanter where the ether-rich top layer is pumped back to the sulfation step or dried for product sales. The water-alcohol bottom layer is returned to the reflux-column feed.

A flowsheet of iso-propanol production by hydration of propylene via sulfation and hydrolysis is given below.

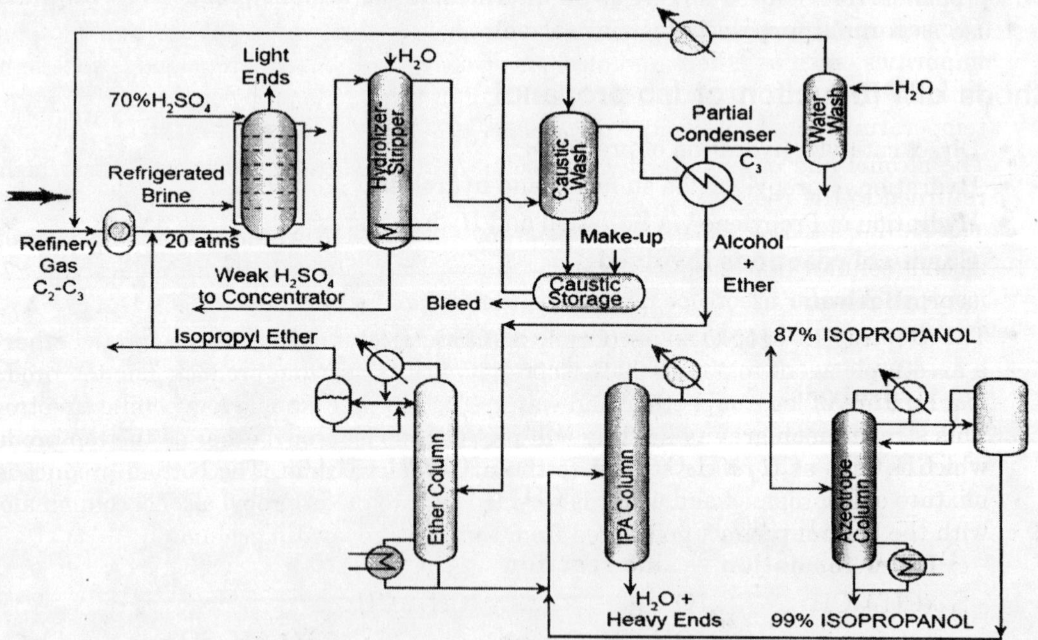

Figure 12.1: Flowsheet of Isopropanol manufacture

Ether column bottoms are pumped to the isopropanol column where an 87% IPA azeotrope is taken off the top. This can be sold as product or sent to a ternary azeotropic distillation column where isopropyl ether removes the water as overhead and the bottoms are 99% absolute isopropanol.

Salient features of the iso-propanol manufacture is listed below:

- Either pure propylene or a mixture of Propylene and other C_2, C_3 components can be fed to a reactor.
- The hydrocarbon feed is compressed and fed to the reactor at about 20 - 25 kg/cm^2.
- Sulphuric acid of 70% acid strength is fed in a countercurrent mode to the tray column where reactive absorption takes place. Here, sulfonation reaction takes place.
- The reaction is highly exothermic and therefore, refrigerated brine is used to control the temperature in the absorber. Jacketed arrangement is preferred for the tray absorption column to circulate the refrigerated brine in the cooling jacket.
- After reaction, the unreacted light ends such as saturated components will leave the unit as the gas stream.
- The sulfonated product rich stream is then sent to a hydrolyzer cum stripper where isopropanol is produced and is vaporized due to existing stripper temperatures.
- The hydrolyzer is fed with water to facilitate the conversion of the sulfonate product.
- The isopropanol rich vapors then enter a caustic wash unit to remove the acidic impurities.
- The isopropanol rich vapors then enter a partial condenser which separates the unreacted propylene from the alcohol + ether mixture. Here, propylene is separated as the vapour and alcohol + ether is separated as the liquid stream.
- The separated propylene gas is once again subjected to water wash to remove soluble impurities (such as ethers and alcohols). Subsequently, pure propylene is sent to mix with the fresh feed stream. Before sending to the unit, the propylene is cooled to room temperature so as to have identification conditions as the fresh feed stock.
- The alcohol and ether enter a ether column that separates isopropyl ether which is returned to the reactor.
- The bottom product consisting of isopropyl alcohol and water is sent to a isopropyl alcohol column that produces water + heavy ends as the bottom product and 87 % isoprpanol-water azeotrope mixture as the top product.
- The azeotrope is sent to an azeotropic distillation column that uses isopropyl ether as a azeotropic agent to obtain 99 % isopropanol as the bottom product. The top product is a mixture of isopropyl ether and water. The top product is a low boiling azeotrope. This stream upon gravity settling will produce the isopropyl ether as the top product which is sent as a reflux stream to the azeotropic column. The bottom product is a mixture of isopropanol and water is recycled back to the isopropyl alcohol column along with the bottom product generated from the ether separating column.

MANUFACTURE OF ACETONE BY DEHYDROGENATION OF ISO-PROPANOL

Chapter 13

Properties of Acetone ($CH_3{\bullet}CO{\bullet}CH_3$) :

Mol. wt.		508
M.P.:		— 95.1°C
B.P.:		56.5°C
Sp. gr. @ 20 °C:		0.79
Miscible with water, alcohol and ethyl ether		
Flash point:		—9.4°F
Ignition temperature:		592 °C
Explosive limits:	Lower =	2.2% in air
	Upper =	13.0% in air
Toxic limit concentration:		500 ppm

Acetone was primarily used as a solvent, but the emphasis has now shifted to acetone-derived chemicals.

Use of acetone in various sectors in India is as follows:

(a) Pharmaceuticals
(b) Pesticides
(c) Rubber Chemicals
(d) Surface coating/paints
(e) Explosives
(f) Di Acetone Alcohol
(g) MIBK
(h) Cellulose Acetate
(i) Acetylene gas cylinder
(j) Acrylics

Methods of Production of acetone

- Catalytic dehydrogenation of isopropanol
- Co-product of phenol process utilizing cumene
- Co-product of glycerin-H_2O_2 process

- Oxidation of butane
- Oxidation of propylene (Wacker process)
- Fermentation of molasses

Catalytic Dehydrogenation of Isopropanol to produce acetone

Chemical reactions involved:

(a) $(CH_3)_2CHOH \longrightarrow CH_3COCH_3 + H_2O$

(b) $(CH_3)_2CHOH + \frac{1}{2} O_2 \text{ (air)} \longrightarrow CH_3COCH_3 + H_2O$

Ra material requirements

Basis: 1 ton of acetone (99% pure, 90% yield)

Isopropanol (88% grade) : 1.30 tons

Process description

Isopropanol vapor compressed to 3 atms. is preheated by reactor effluent heat exchange and passed through a tubular catalytic reactor maintained at 500 °C with the catalyst consisting of copper or brass deposited on porous carrier. The hot reaction gases pass through a water-cooled condenser and thence into a water scrubber where final traces of isopropanol and acetone are removed from hydrogen. Condensate and water scrub liquor are fractionated to give product grade acetone overhead and dilute isopropanol bottoms. These are further fractionated to give 88% isopropanol recycle with the water recycled to the water scrubber.

A flowsheet of acetone production by catalytic dehydrogenation of iso-propanol is given below.

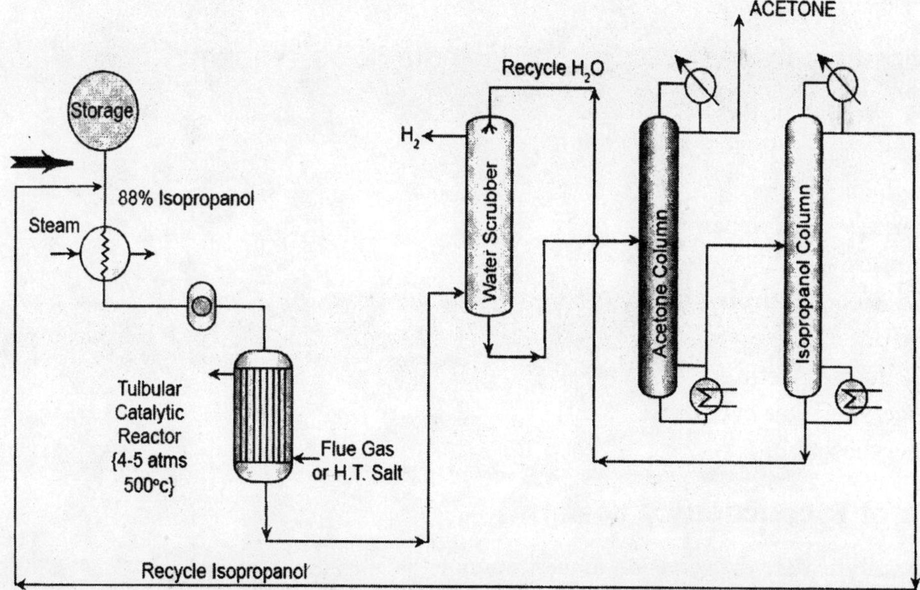

Figure 13.1: Flowsheet of acetone manufacture from isopropanol

Salient features of acetone production process is given below.

- First, Isopropanol is heated using steam to vaporize the same.
- Then, Isopropanol is compressed to desired reactor pressure i.e., 4 – 5 atms.
- The compressed Isopropanol then enters a catalytic shell and tube reactor in the tube side. The tube is packed with the porous copper catalyst.
- The reactor is operated at 400 – 500°C using flue gas for heating. The flue gas is passed in the shell side of the shell and tube reactor.
- After reaction, the gases are condensed using cooling water condenser. The condensed isopropanol and acetone are sent for fractionation.
- The gases consisting of the remaining quantities of isopropanol and acetone are absorbed into water using a water scrubber.
- The acetone + isopropanol obtained from the condenser and water + isopropanol + acetone are sent to an acetone fractionator that separates acetone as the top product and isopropanol + water as bottom product.
- The bottom product isopropanol + water from the acetone fractionators is sent to a isoprpopanol column.
- This column produces water as the bottom product and isopropanol as the top product.
- The water is cooled using a water condenser and sent to the water scrubber as fresh water solvent.
- Aceone is used 'as a polar, aprotic solvent in a variety of organic reactions. One important property for which it is used as laboratory solvent is because it does not form an azeotrope with water.
- Acetone is also used in various medical & cosmetic applications. It also forms an important component in food additives & food packaging.

Chapter 14

PRODUCTION OF CUMENE (ISOPROPYL BENZENE) FROM PROPYLENE

Properties of Cumene [$C_6H_5 \bullet CH \bullet (CH_3)_2$]

Mol. wt.:	120.19
M.P.:	— 96.9°C
B.P.:	152.5°C
Sp. gr. @ 20°C:	0.862
Miscible with alcohols, ether, insoluble in water	
Flash point:	39°C
Ignition temperature:	138°C
Toxic limit concentration:	250 ppm

Cumene was first developed as a high octane addition for engine fuels. This use has been overshadowed by the phenol process which uses cumene as the starting material and produces acetone as a co-product.

Cumene Production by Propylene Alkylation of Benzene

Chemical reactions

(a) **Main reaction**

$$\underset{\text{Benzene}}{C_6H_6} + \underset{\text{Propylene}}{CH_3 \bullet CH = CH_2} \longrightarrow \underset{\text{isopropyl benzene (Cumene)}}{C_6H_5 \bullet C_3H_7}$$

(b) **Side reaction.**

$$C_6H_6 + nCH_3CH = CH_2 \longrightarrow \underset{\text{polyisopropyl benzene}}{C_6H_{6-n} \bullet (C_3H_7)_n}$$

Raw materials Requirements

Basis: 1 ton of cumene (>99.5% purity, 94% yield based on propylene)

Benzene:	0.76 ton
Propylene:	0.41 ton
Co-products:	Polyisopropyl benzene, mainly diisopropyl benzene

Process Description

Propylene-propane feedstock from refinery off-gases from a naphtha-steam cracking plant is mixed with benzene and pumped at 25 atms. into the top of a reactor packed stage-wise with H_3PO_4 impregnated catalyst. The temperature is maintained at 250°C by adding cold propane at each stage to absorb the heat of reaction.

The reactor effluent is depropanized and the propane split into quench or product streams. The depropanized bottoms are separated into benzene, cumene, and polycumenes in the remaining two stills.

A flowsheet of cumene production by proplylene alkylation of benzene is shown below.

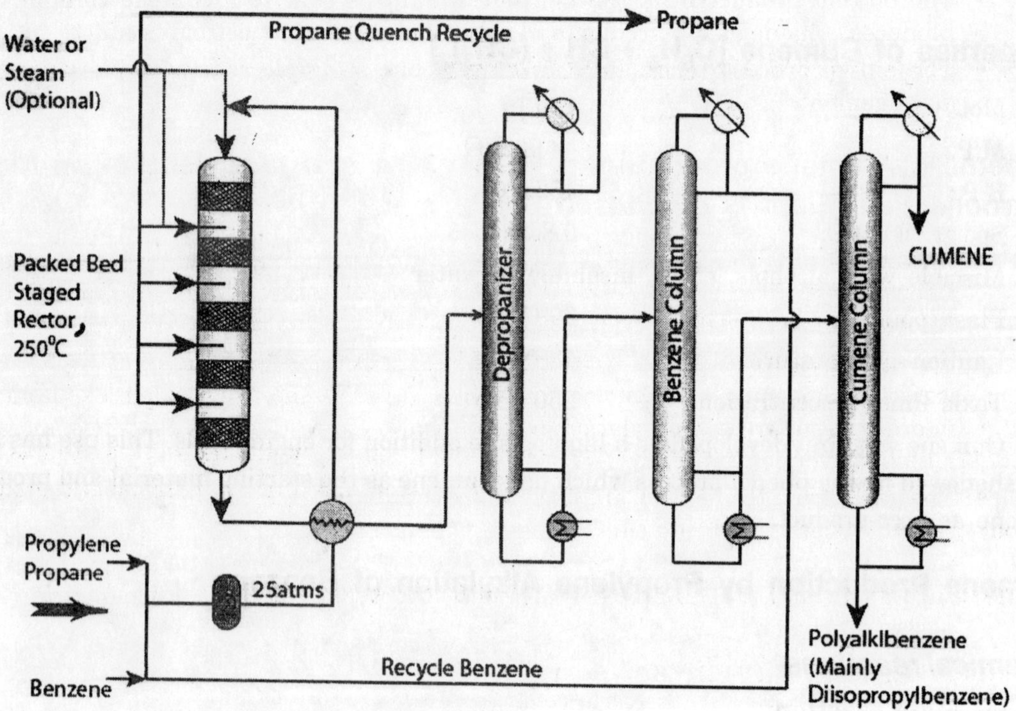

Figure 14.1: Flowsheet of cumene production

Salient features of cumene production is listed below:

- Propylene is obtained from refinery processes as a mixture of propylene and propane.
- The mixture along with benzene is compressed to 25 atms.
- Eventually, the mixture enters a heat integrated exchanger to heat the feed mixture.
- The feed mixture enters a packed bed reactor.
- The stream distribution in the packed bed reactor corresponds to cold shot arrangement i.e., cold propane from the distillation column in the process is added after every reactor with the product stream so that the temperature of the stream is controlled.
- Here, propylene is the limiting reactant and therefore, all propylene undergoes conversion.

- Here, propane does not react but is a diluent or inert in the system. In that way it controls the reaction temperature.
- The reactor units are maintained at about 250°C.
- The product vapors are cooled using the heat integrated exchanger.
- The vapors then pass to a depropanizer which separates propane from the product mixture.
- The bottom product consisting of benzene, cumene and polyalkyl benzenes enters another distillation column which separates benzene from the mixture of cumene and polyalkyl benzene. The benzene stream is recycled to enter the compressor.
- The bottom product from the benzene column is sent to a cumene column which produces cumene as top product and poly alkyl benzene as bottom product.
- The entire process technology is nothing but a simple reactor separator recycle arrangement.

Scheme of Cumene production from benzene and propylene in an Indian Petrochemical Plant is discussed below:

- Chemical grade Propylene or a Propylene cut containing up to 60% of Propane and Benzene are mixed and fed into a reactor, where they are brought into contact with a catalyst consisting of Phosphoric acid supported 'on Kieselghur or Pumice.
- The reactor temperature is kept at 200-250°C with a pressure range of 15-35 bar. An excess of Benzene (in the molecular ratio of 5:1 Benzene to Propylene)is maintained in order to suppress dealkylation, oligomerization and other side reactions and attain a high conversion rate.
- The gases from the reactor are used to heat incoming feed before entering the recycle column where any un-reacted benzene is recovered and recycled. The remainder of the liquid stream is fed into the cumene distillation column where heavy by-products such as di- and tri- Iso-propyl-benzene are recovered as bottoms while pure cumene passes overhead.

$$\underset{\text{Benzene}}{C_6H_6} + \underset{\text{Propylene}}{CH_3CH = CH_2} \longrightarrow \underset{\text{Cumene (isopropyl benzene)}}{C_6H_5CH(CH_3)_2}$$

PRODUCTION OF ACRYLONITRILE FROM PROPYLENE

Chapter 15

Properties of Acrylonitriie ($CH_2 = CHCN$)

Mol. wt.:		53.06
M.P.:		– 82°C
BP:		78.5°C
Sp. gr. @ 20°C:		0.807
Soluble in most organic solvents and water		
Flash point:		0°C
Ignition temperature:		480°C
Explosive limits:	Lower =	3.1% in air
	Upper =	17.0% in air
Toxic limit		15 ppm

Acrylonitrile is the cheapest of the acrylic monomer group and has good reactivity at $C \equiv N$ bond and the double bond. Its end uses are largely for polymers.

The worldwide pattern of acrylonitrile use are for the production of acrylic fibres, ABS & SAN plastics, nitrile rubber, adiponitrile, acrylamide etc.

Methods of Production of acrylonitrile

- Propylene-ammonia-air oxidation reaction.
- Acetylene-hydrogen cyanide reaction
- Ethylene cyanohydrin pyrolysis
- Acetaldehyde-hydrogen cyanide reaction

Propylene- Ammonia-Air Oxidation Process for production of acrylonitrile

Chemical reaction involved

$$\underset{\text{Propylene}}{CH_2 = CH - CH_3} + NH_3 + 3/2O_2 \longrightarrow \underset{\text{Acrylonitrile}}{CH_2 = CHCN} + 3H_2O$$

Raw materials requirements

Basis: 1 ton of acrylonitrile (99.0+ %, 80% yield)

Propylene: 0.996 ton

Ammonia:	0.45 ton
Air:	5.2 tons or 4,000 Nm^3

By-products:

Acetonitrile (CH_3CN):	0.09 ton
Hydrogen cyanide (HCN):	0.08 ton

Process description

A C_3 refinery stream is introduced along with ammonia and air via spargers into a fluidized bed catalytic reactor. Conditions within the reactor are 1.5-3 atms. pressure, a temperature of 400-500°C, and a contact time of several seconds.

The reactor effluent is scrubbed with water to remove the desired products in an aqueous solution which is further fractionated to give wet acrylonitrile and acetonitrile. Both are further purified by azeotropic and conventional distillation.

A flowsheet for production by acrylonitrile from propylene is shown below.

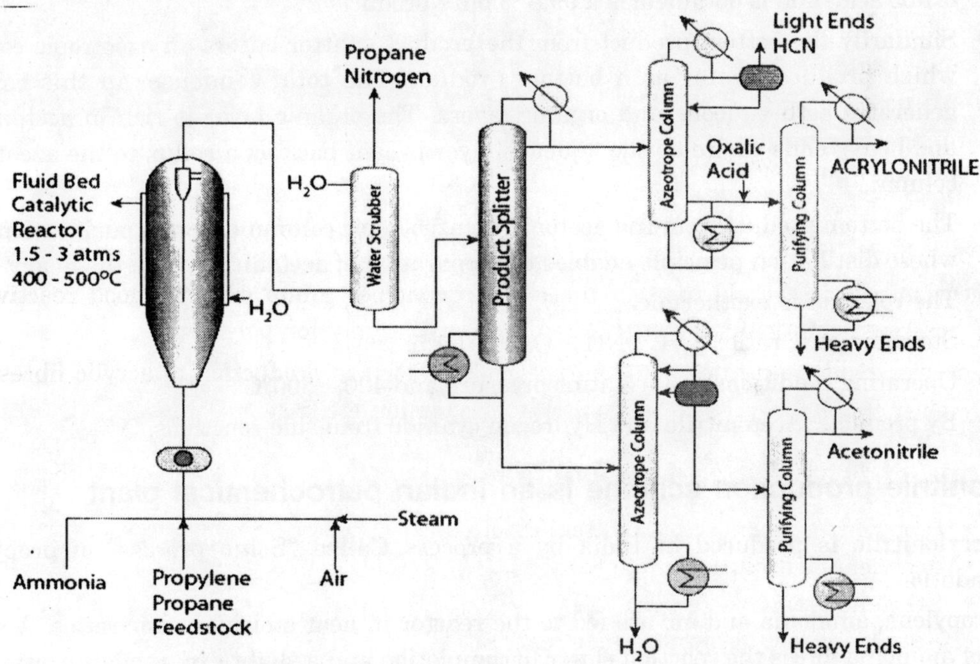

Figure 15.1: Flowsheet of acrylonitrile production

Salient features of acrylonitrile production is listed below

- Propylene + Propane, air Ammonia & Steam are compressed to required pressure and are sent to the fluidized catalytic reactor consisting of the Mo-Bi spherical catalyst. The reactor is maintained at 400 - 500°C.
- Cyclone separator is also kept in the fluidized bed reactor in which catalyst and product gases are separated after fludization. The contact time for fluidization is of the order of seconds.

- The product vapors then enter a water scrubber that does not absorb propane and nitrogen from the products. The products absorbed in the water include acrylonitrile, acetonitrile and other heavy ends.
- The very dilute acryolonitrile (about 3 %) solution in water is sent to a fractionator. The fractionators separates acrylonitrile + heavy ends + HCN + light ends as a top product stream and acetonitrile + water + heavy ends as a bottom product.
- The top product then enters an extractive distillation column with water as extractant. The azeotropic distillation column vapour is partially condensed to obtain a vapour, aqueous and organic layer. The vapour consists of Light ends and HCN and is let out. The organic layer consists of acrylonitrile and heavy ends is sent for further purification. The aqueous layer is sent as a reflux to the azeotropic column. Addition of water enables the formation of a heterogenous azeotropic mixture at the top.
- The bottom product from the azeotropic distillation column enters a product purification unit along with oxalic acid where acrylonitrile is further purified from heavy ends (+ oxalic acid) and is obtained as a 99.5 % pure product.
- Similarily the bottom product from the product splitter enters an azeotropic column which produces water as a bottom product. The total condenser in this column generates both aqueous and organic layers. The organic layer is rich in acetonitrile and heavy ends where as the aqueous layer is sent back as a reflux to the azeotropic column.
- The bottom product from the acetonitrile azeotropic column enters a purification unit where distillation principle enables the separation of acetonitrile from the heavy ends.
- The reaction is exothermic.
- Stoichiometric ratio: $C_3H_6 : NH_3 : O_2$ = 1:1:1.5.
- Operating conditions: 1.5 – 3 atms pressure and 400 – 500°C.
- By products: Acetonitrile and Hydrogen cyanide from side reactions.

Acrylonitrile production scheme is an Indian petrochemical plant

Acrylonitrile is produced in India by a process Called "Sohio process" of propylene ammoxidation.

Propylene, ammonia and air are fed to the reactor in near stoichiometric ratios. A slight excess of ammonia forces the reaction closer to completion and a slight excess of air continually regenerates the catalyst used in the reaction.

The vapour-phase reaction takes place in a fluidised bed reactor in presence of a catalyst. The conversion of propylene in the reactor is essentially complete. The yield of acrylonitrile monomer from this reaction is typically 79 weight percent with approximately 2 and y molar percent yields of each of the by-products acetonitrile and hydrogen cyanide (HCN) respectively. The stream exiting the reactor contains acryonitrile and by-products but also un-reacted oxygen, carbon monoxide, carbon dioxide, and nitrogen.

The propylene ammoxidation reaction generates substantial quantities of heat, which must be removed. Heat removal from the product stream is generally accomplished by utilizing excess heat to generate steam in a waste heat boiler. The cooled product stream leaving the waste heat boiler passes to a quench tower followed by scrubber, where unreacted ammonia is scrubbed by using sulphuric acid.

The quenched product stream is passed to a counter-current absorber, which removes acrylonitrile, acetonitrile, HCN in solution form. The unabsorbed gases are vented to atmosphere. Acrylonitrile, acetonitrile and HCN are separated in a series of distillation columns.

In the first recovery column, acrylonitrile & HCN are separated from acetonitrile and water. Water is then removed from the acetonitrile in the acetonitrile column and recycled to the absorber. The heavy bottoms from the acetonitrile column (HW1) are discharged to the effluent treatment plant after hot alkali digestion. The acetonitrile by-product may be recovered to 99 percent purity for commercial sale.

The crude acrylonitrile exiting the recovery column is passed to storage facilities and then to a light-ends column where HCN is recovered. The HCN by-product may be further purified to 99 + percent for sales, or maybe incinerated in case of lack of demand. In the final product column, heavy ends are removed (HW_2) and incinerated. The acrylonitrile product obtained from the "Sohio process" has a purity of 99 + percent.

Chemical Reactions involved:

$$\underset{\text{Propylene}}{C_3H_6} + \underset{\text{Oxygen}}{3O_2} + \underset{\text{Ammonia}}{3NH_3} \longrightarrow \underset{\text{Hydrogen cynide}}{3HCN} + \underset{\text{Water}}{6H_2O}$$

$$\underset{\text{Propylene}}{C_3H_6} + \underset{\text{Oxygen}}{1.5O_2} + \underset{\text{Ammonia}}{1.5NH_3} \longrightarrow \underset{\text{Acetonitrile}}{1.5CH_3CN} + \underset{\text{Water}}{3H_2O}$$

$$\underset{\text{Propylene}}{C_3H_6} + \underset{\text{Oxygen}}{1.5O_2} + \underset{\text{Ammonia}}{NH_3} \longrightarrow \underset{\text{Acrylonitrile}}{CH_2 = CHCN} + \underset{\text{Water}}{3H_2O}$$

Side Reactions

$$\underset{\text{Propylene}}{C_3H_6} + \underset{\text{Oxygen}}{O_2} \longrightarrow \underset{\text{Acrolin}}{CH_2 = CHCHO} + \underset{\text{Water}}{HO_2}$$

$$\underset{\text{Propylene}}{C_3H_6} + \underset{\text{Oxygen}}{3/2O_2} \longrightarrow \underset{\text{Acrylic Acid}}{CH_2 = CHCOOH} + \underset{\text{Water}}{H_2O}$$

$$\underset{\text{Propylene}}{C_3H_6} + \underset{\text{Oxygen}}{3/2O_2} \longrightarrow \underset{\text{(Acetic Acid)}}{CH_3COOH}$$

$$\underset{\text{Propylene}}{C_3H_6} + \underset{\text{Oxygen}}{9/2\,O_2} \longrightarrow \underset{\text{Carbon dioxide}}{3CO_2} + \underset{\text{Water}}{3H_2O}$$

$$\underset{\text{Propylene}}{C_3H_6} + \underset{\text{Oxygen}}{O_2} \longrightarrow \underset{\text{Carbon Mono oxide}}{3CO} + \underset{\text{Water}}{3H_2O}$$

$$\underset{\text{Ammonia}}{NH_3} + \underset{\text{Oxygen}}{\tfrac{3}{4}\,O_2} \longrightarrow \underset{\text{Nitrogen}}{1/2N_2} + \underset{\text{Water}}{3/2\,H_2O}$$

A flowsheet for acrylonitrile production by propylene ammoxidation (sohio process) is given below.

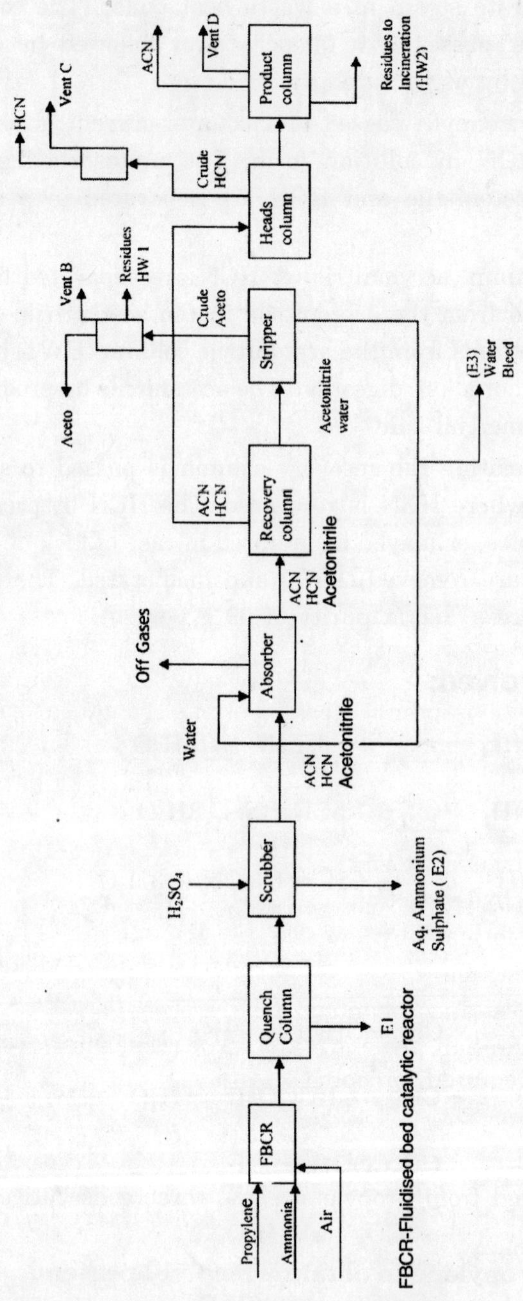

Figure 15.2: Flowsheet of acrylonitrile Manufacture by 'Sohil Process'

PRODUCTION OF ISOPRENE FROM PROPYLENE

Chapter 16

Properties of Isoprene ($CH_2 = CH - C(CH_3) = CH_2$)

Mol. wt.:	68.11
M.P.:	—120°C
B.P.:	34 °C
Sp. gr. @ 20°C:	0.681
Miscible with alcohols, ether.	
Flash point:	—53 °C
Ignition temperature:	220 °C
Toxic limit:	250 ppm

Polymerized isoprene (polyisoprene) closely resembles natural rubber chemically and adds flexibility to the synthetic rubber. This is the only market for the isoprene monomer.

Methods of Production of isoprene (C_5H_8)

- Isomerization of propylene dimer
- Isobutylene-formaldehyde reaction
- Methyl ethyl ketone-formaldehyde reaction
- Acetone-acetylene reaction
- By product from isopentane/isopentene dehyodrogenation

Propylene dimer isomerisation process is the most economical process which producess high purity grade isoprene required for polymerisation.

Chemical reactions

Isoprene is manufactured from propylene using three consecutive reactions.

These are:

(a) Dimerization of propylene to obtain 2-methyl-l-pentene

$$\underset{\text{(Propylene)}}{2.H_2C = CH - CH_3} \xrightarrow[\substack{(C_2H_7)_3\text{ Al Catalyst}\\ 150\text{–}250°C,\ 200\text{ atm}}]{\text{dimerization}} \underset{\text{(2-methyl-1-pentene)}}{H_2C = \overset{\overset{\large CH_3}{|}}{C} - CH_2 - CH_2 - CH_3}$$

(b) Dimerization of propylene to obtain 2-methyl-l-pentene:

$$\underset{\text{(2-methyl-1-pentene)}}{H_2C = \overset{\overset{\displaystyle CH_3}{|}}{C} - CH_2 - CH_2 - CH_3} \xrightarrow[\text{acid Catalyst}\ 150\text{–}300°C]{\text{isomerization}} \underset{\text{(2-methyl-2pentene)}}{H_3C - \overset{\overset{\displaystyle CH_3}{|}}{C} - CH - CH_2 - CH_3}$$

(c) Pyrolysis of 2-methyl-2-pentene to obtain isoprene along with methane as a side product:

$$\underset{\text{(2-methyl-2-pentene)}}{H_2C = \overset{\overset{\displaystyle CH_3}{|}}{C} - CH_2 - CH_2 - CH_3} \xrightarrow[\text{HBR}\ 650\text{–}800°C]{\text{pyrolysis}} \underset{\text{(isoprene)}}{CH_2 - \overset{\overset{\displaystyle CH_3}{|}}{C} - CH - CH_2} + \underset{\text{(inethane)}}{CH_4}$$

Raw material requirements

Basis:	1 ton of isoprene (99.0 + % purity, 85% yield)
Propylene:	1.46 tons

Minor quantities of tripropyl aluminum $(C_3H_7)_3Al$ and hydrogen bromide catalysts.

Process description

A compressed propylene refinery cut is mixed with tripropyl aluminum, catalyst plus carrier solvent at 200 atms. in a stirred reactor fitted with heat transfer surface to remove the exothermic heat of reaction. A conversion of 60-95% with a temperature range of 150-250°C is possible.

The reactor effluent passes directly to a flash tower which separates the catalyst and carrier solvent (saturated naphtha fraction). The top products, 2-methyl-1-pentene, unreacted propylene, and light hydrocarbons present in the feed are separated by successive distillations.

Isomerization occurs in a packed bed reactor containing an acid type catalyst. Process conditions range from 150 - 300°C. Fractionation and recycle of dimer is required to obtain a 98-99% ultimate yield.

The final step involves pyrolysis of 2-methyl-2-pentene to isoprene. This is done in a tubular cracking furnace with HBr as a catalyst and steam as the diluent. Reaction occurs in the range of 650-800°C and residence times of 0.3 to 0.5 second. Acetylenic compounds are avoided under these conditions.

Since three reactions are involved that have distinct operating conditions as well as catalyst requirements, the process involves a series of reactor-separator-recycle networks.

Features of Dimerization reaction

- Catalyst is tripropyl aluminium.
- A solvent carrier is used that enables the recovery of the catalyst from the product stream.
- Operating conditions are 150 - 200°C and 200 atms.

Features of Isomerization reaction

- The reaction requires an acid catalyst on porous carrier.
- Operating conditions are 150 - 300°C and normal pressure conditions.

Features of Pyrolysis reaction

- HBr catalyst is used.
- Operating conditions are 650 - 800°C and normal pressure.

A flowsheet of isoprene production from propylene is shown below in figure.

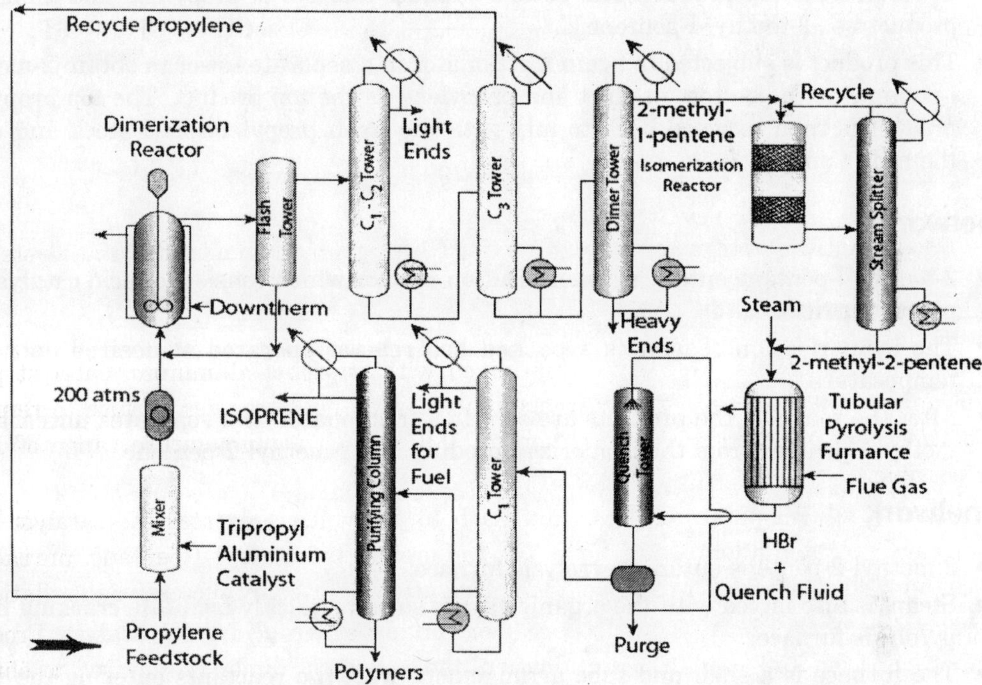

Figure 16.1: Flowsheet of isoprene production from propylene

The process technology consists of three major blocks namely:

- Dimerization-separator-recycle network (**DSR**)
- Isomerization-separator-recycle network (**ISR**)
- Pyrolysis-separator-recycle network (**PSR**)

DSR network

- Propylene feed is mixed along with unreacted propylene (recovered in a distillation column in the DSR) to enter a mixer.
- The mixer is fed with tripropyl aluminium catalyst (this catalyst is in liquid phase).
- The mixture is compressed to about 200 atms. At these conditions, the mixture is in the liquid phase.

- The liquid phase dimerization feed enters the dimerization reactor.
- The dimerization reactor is a CSTR with dowtherm fluid circulated in the cooling jacket to remove the heat generated from the exothermic reaction.
- After the reaction, the liquid stream is subjected to flash operation. From the flash tower a vapour and liquid stream are produced. The liquid stream consists of the catalyst + solvent where as the vapour stream consists of the unreacted propylene, light ends and the product formed.
- The liquid stream from the flash tower is recycled back to the dimerization CSTR.
- The vapour stream enters a fractionation tower that produces light ends such as C_1 - C_2 as the top product. The bottom product consists of propylene and dimerized product i.e., 2-methyl-l-pentene.
- This product is subjected to again fractionator in a separate tower to obtain 2-methyl-l-pentene as the bottom product and propylene as the top product. The top propylene product is then recycled back to mix with the fresh propylene feed stock and enter the mixer unit.

ISR network

- 2-methyl-l-pentene enters an isomerisation reactor which consists of acid catalyst (on porous carrier).
- The isomerisation reactor is a packed bed reactor operated at desired operating temperature.
- After the reaction, the products are sent to a fractionator that separates unreacted 2-methyl-l-pentene from the isomerized product i.e., 2-methyl-2-pentene.

PSR network

- 2-methyl-2-pentene enters a pyrolysis furnace.
- Steam is also mixed with the organic vapors so as to quickly facilitate cracking in the pyrolysis furnace.
- The furnace is a shell and tube arrangement with the reactants entering the tubes and the hot flue gases enter the shell.
- HBr catalyst also enters the tubes of the pyrolysis furnace.
- After reaction, the product gases (HBr + product + reactant) are taken out and are quenched with a solvent in a quench tower. The quenching then produces a gas liquid mixture which is sent to a phase separator unit. The liquid stream from the phase separator stream consists of HBr + solvent and this is sent for quenching.
- Fresh HBr can be generated from the HBr + solvent obtained from the phase separator.
- The gas fraction from the phase separator enters a C_1 tower that separates fuel gas and other light ends as a top product.
- The bottom product from this tower consists of polymeric compounds and isoprene is fed to a purification tower to fractionate isoprene from the polymeric compounds.

Chapter 17

MANUFACTURE OF EPICHLOROHYDRIN AND PROPYLENE OXIDE

EPICHLOROHYDRIN

(1-chloro-2,3-epoxypropang) is another propylene derivative of industrial importance. Its manufacture involves the following stages:

- High temperature chlorination of propylene to Allyl chloride.
- Conversion of allyl chloride to dichlorohydrin with hypochlorous acid in an aqueous medium.
- Selective dehydrochlorination of dichlorohydrins to epichlorohydrin with an alkali.

PROPYLENE OXIDE

Properties of Propylene Oxide ($CH_3 - CH - CH_2$, with O bridging CH and CH_2)

Mol. wt.:		58.08
M.P.:		—104.4°C
B.P.:		33.9°C
Sp. gr. @ 20°C:		0.831
Soluble in water, alcohols, ether		
Flashpoint:		—34°C
Explosive limits:	Lower =	2.1% in air
	Upper =	22% in air
Toxic limit:		250 ppm

Propylene oxide is an importent chemical to study because of its similarities and differences as compared with ethylene oxide. For example, the toxic character of ethylene glycol, the hydrate of ethylene oxide, has created markets for propylene glycol in cellophane, food, and drug industries. The largest use for the glycol is in polyester resins.

Use of Propylene Oxide include the following:

(a) Polypropylene glycol and propoxylated products for polyurcthanes
(b) Propylene glycol
(c) Miscellaneous: glycol ethers, glycerin, isopropanolamine, etc.

Methods of Production of propylene oxide

Classification of Processes:

- Chlorhydrin route
- Direct oxidation

CHLORHYDRIN PROCESS OF PROPYLENE OXIDE MANUFACTURE

Chemical reactions

(a) Chlorhydrination:

$$CH_3CH = CH_2 + HOCl \longrightarrow \underset{\text{Chlorhydrin}}{CH_3 \bullet CHCl \bullet CH_2OH}$$

(b) Dehydrochlorination:

$$2CH_3 \bullet CHCl \bullet OH + Ca(OH)_2 \longrightarrow 2\underset{\text{Propylene oxide}}{CH_3\,CH - CH_2} + CaCl_2 + 2H_2O$$

(with O bridging CH and CH_2 in the epoxide ring)

Raw material requirements

Basis: 1 ton of propylene oxide (99 + % purity, 80% yield)

Propylene:	0.91 ton
Chlorine:	2.0 tons
Lime (100% CaO):	1.6 tons
Electricity:	150 KWH
Steam:	12 tons
Water:	240 tons
By-products:	0.97 ton of calcium chloride
	0.10-0.15 ton of propylene dichloride

Process description

Propylene, chlorine and water are introduced into the bottom of a packed tower where reaction forms the chlorhydrin. The reaction mechanism is formation of hypochlorous acid (HOCl or HClO) which reacts rapidly with propylene. The chlorine/propylene ratio of gaseous feed to the tower is chosen so that the liquid effluent contains 4-5% chlorhydrin. The reaction is exothermic and maximum tower temperature is held at 50°C. A conversion of

25-40% is attained, the ultimate yield of chlorhydrin is 87-90% with most of the by-product as propylene dichloride. Unreacted propylene is scrubbed with caustic to remove HCl, which also forms, and then recycled.

Manufacture of Propylene Oxide by Direct Oxidation of propylene

Processes for the direct manufacture of propylene oxide by the single step oxidation of propylene have been the subject of intensive commercial development.

One possible route which is now being commercialized involves peroxidizing an organic compound, reacting the peroxide with propylene to give propylene oxide and then recovering a useful organic product from the peroxide.

The first commercial plant for direct manufacture of propylene oxide from propylene was built at Bayport, Texas.

A flowsheet of propylene oxide manufacture by chlorhydrin route is given below:

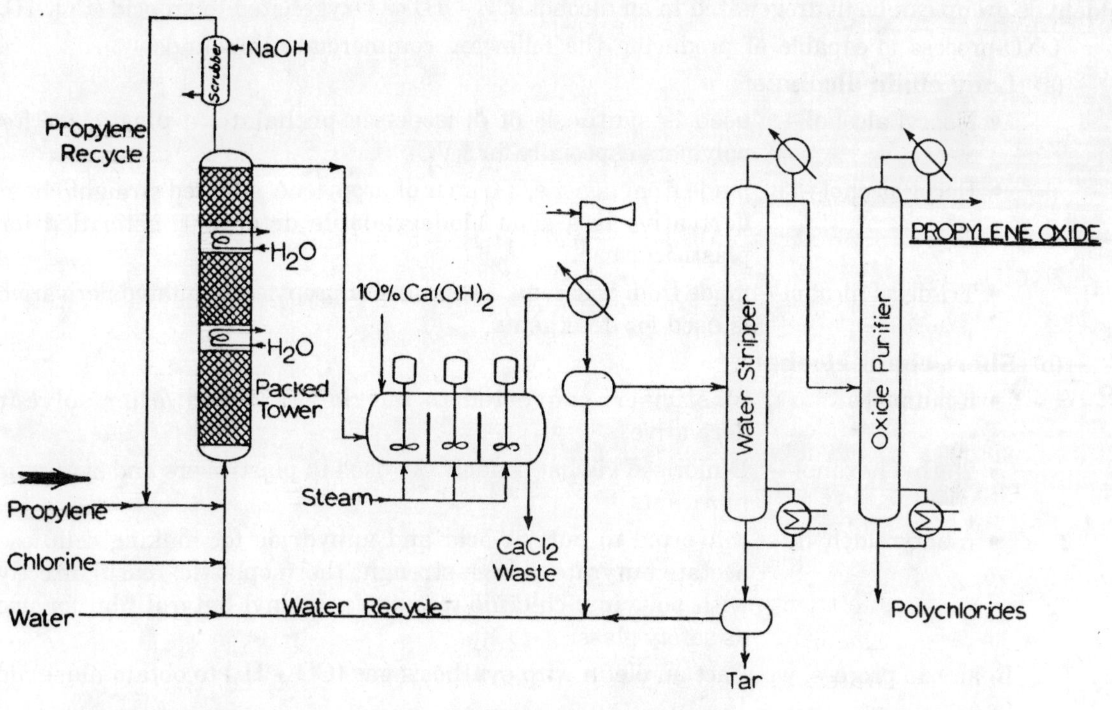

Figure 17.1: Manufacture of Propylene oxide by chlorhydrin route

PRODUCTION OF ALDEHYDES, ALCOHOLS & CORBOXYLIC ACID BY OXY-PROCESSING OF OLEFINS

Chapter 18

An oxo-process involves the substitution of H-CHO across a double bond, hence it is also called hydroformylation. The reactants in oxo-process are an olefin (eg. propylene) and synthesis gas (CO + H_2). The primary product is an aldehyde with one additional carbon atom. The aldehyde group can be hydrogenated to an alcohol (CH_2OH) or Oxygenated to an acid (COOH).

OXO-process is capable of producing the following commercial compounds:

(a) **Long chain alcohols:**

- Isoocyl alcohol— used is synthesis of di-isooctyle phthalate, a plasticiser for polymers especially for PVC.
- Decyl alcohol— made from nonene, a trimer of propylene, sulfated straight chain derivative is a good biodegradable detergent; esterified for plasticiser use.
- Tri-decyl alcohol— made from dodecene, a tetramer of propylene, sulfated derivative is used for detergents.

(b) **Short chain alcohols:**

- n-butanol— plasticiser; converted to butylacetate and other solvent derivative.
- 2-ethyl hexanol— Dimerised via batyraldehyde; used in plasticisers and synthetic lubricants.
- n-butyraldehyde— converted to butyric acid and anhydride for making cellulose acetate butyrate, a high strenght thermoplastic; reacts directly with polyvinyl chloride to make polyvinyl butyral film for use as safety glass.

In an oxo process, we react an olefin with synthesis gas (CO + H_2) to obtain aldehyde i.e.,

$$\text{Olefin} + CO + H_2 \longrightarrow \text{Aldehyde}$$

This reaction is called as hydroformylation

Dimerization (Option): Alternatively, if desired, the aldehyde dimerization reaction can be also carried out. In that case, two molecules of aldehydes react to obtain a higher order aldehyde.

The oxo processing can be targeted to generate both short chain and long chain aldehydes.

Hydrogenation (Option): When short and long chain aldehydes are subjected to hydrogenation then one can produce alcohol or an acid.

Oxygenation (Option): When short and long chain aldehydes are subjected to reaction with oxygen or air, then carboxylic acids can be formed.

Thus the oxo processing of olefins can produce:

- Short and long chain aldehydes.
- Short and long chain alcohols
- Short and long chain carboxylic acids

The oyo-process used for propylene can be suitably extended to other alkenes such as butylenes or ethylene.

Chemical Reactions & Operating conditions

- **Addition reaction ($Cn \rightarrow C_{n+1}$): $R \bullet CH = CH_2 + HCHO$**

$$\text{(or, CHOH)} \xrightarrow{\text{Cosalt}} R \bullet CH_2 \bullet CH_2 \bullet CHO$$

$$\text{(or, CHOH)} \longrightarrow R{-}\underset{\displaystyle CHO}{\underset{|}{CH}}{-}CH_3$$

  - Catalyst: Cobalt naphthanate
  - Operating conditions: Pressure of 100 - 400 atms and temperature of 100-200°C.
  - The reaction is exothermic.
  - Reaction occurs on a packed bed reactor where the packed bed acts as a surface to deposit the catalyst i.e., Cobalt.
  - In due course of reaction, the Cobalt enters the product stream and therefore needs to be recovered.

- **Dimerization reaction**

$$\left(C_{n+1} \longrightarrow C_{2n+2}\right): 2R \bullet CH_2 \bullet CH_2 \bullet CHO \xrightarrow[H_2O]{Z_n\,salt} R \bullet (CH_2)_3R(CH_2)_3CHO$$

  - Catalyst: Zinc salt
  - Operating conditions: Higher pressure and moderate temperatures.

- **Hydrogenation reaction**
  - Catalyst: Nickel
  - Operating conditions: 100 atms and 150°C

Chemical Reactions in oxo-process for butanol production

(a) Aldehyde step:

$$\underset{\text{Propylene}}{C_3H_6} + CO + H_2 \longrightarrow CH_3 \bullet CH_2 \bullet CH_2CHO \quad (\sim50\%)$$

$$\underset{\text{Propylene}}{C_3H_6} + CO + H_2 \longrightarrow (CH_3)_2CH \bullet CHO \quad (\sim50\%)$$

(b) Alcohol step:

$$C_3H_7CHO + H_2 \xrightarrow[\substack{150°C \\ 100\ \text{atms.}}]{\text{Ni catalyst}} C_3H_7CH_2OH$$

Raw material Requirements

Basis: 1 ton of butanol (99 + % purity and 85% yield)

Propylene :	0.70 ton	equimolar ratio
Hydrogen:	0.033 ton	

Process Description of oxo-processing of olefins & synthesis gas (CO + H_2):

Propylene is compressed to 250 atms. and cobalt naphthenate added to give 0.5-1% Co in solution. This stream is passed co-currently through a packed tower containing a porous carrier with 2% metallic cobalt deposited. The reaction is highly exothermic and the temperature of 170° C is controlled by recycle of a portion of the product stream after cooling.

The liquid fraction is mixed with steam at 180°C and a relatively low pressure of 20 atms. to decompose the cobalt carbonyl and naphthenate, depositing the cobalt on a porous carrier as the oxide. This cobalt is dissolved periodically in an acid wash and converted to the naphthenate for reuse. The unconverted synthesis gas from the Oxo converter is recompressed and recycled.

The crude butyraldehyde can be fractionated for product sales or continuously hydrogenated using a fixed bed nickel catalyst, 100 atms., H_2 pressure, and 150°C. The resulting butanols are fed to a distillation section comprising several fractionating columns in series. Light and heavy ends as by-products are obtained in addition to the purified alcohol.

A flowsheet of oxo-processing of olefins is given below:

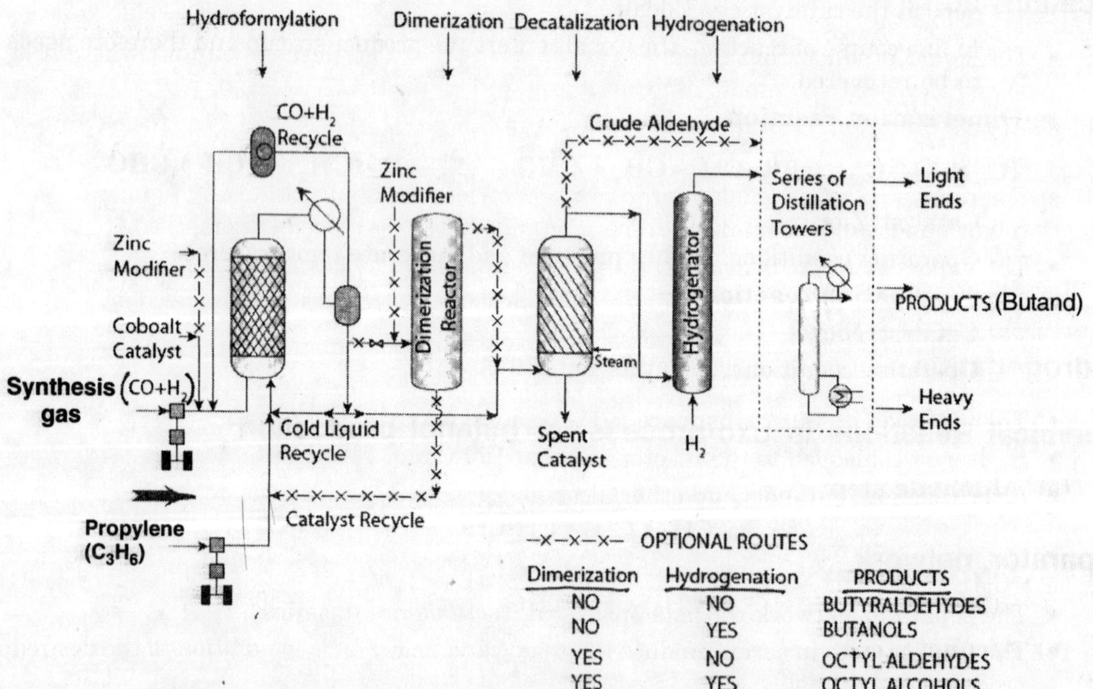

Figure 18.1: Flowsheet of oxo processing of olefins & synthesis gas

To produce alcohols, the process consists of four basic steps namely hydroformylation, decatalization, hydrogenation and a separation network.

Hydroformylation

- Propylene and synthesis gas are separately compressed and sent to the hydroformylation reactor along with the Cobalt catalyst (i.e., Cobalt napthanate).
- The concentration of Cobalt Naphthenate is about 1 wt % Cobalt in the feed mixture.
- During the reaction at the existing operating conditions some cobalt gets deposited on the porous carrier.
- Since the reaction is highly exothermic, cold liquid stream obtained after phase separation is recycled for cooling purpose.
- The product liquid from the reactor is sent to a partial condenser followed by phase separator. The phase separator separates the unreacted synthesis gas (by considering the propylene in the feed stock to be limiting) and the liquid product stream. The liquid product stream is partially recycled back to the reactor to serve as a diluent to control the temperatures in the packed bed reactor.
- The liquid product stream consists of dissolved cobalt and hence it needs to be extracted back as the salt to enter the hydroformylation reactor.
- Therefore, the product stream is partially sent to the decatalization process so as to remove the catalyst from the stream.

Decatalization

- The liquid product from the phase separator is sent to a packed bed unit where it is mixed along with steam at 180°C at a pressure of 20 atms.
- At this condition, Cobalt salt in the hydrocarbon liquid gets transferred to the porous carrier as cobalt oxide.
- For reuse, the cobalt oxide on the porous carrier is treated with acid to generate the cobalt naphthanate and this will be sent back to the hydroformylation reactor.
- The vapors generated from decatalization unit consist of the aldehyde products.
- These are sent to the hydrogenation unit next to convert aldehydes to alcohols.

Hydrogenation

- The aldehyde product vapors are fed to a nickel catalytic bed reactor.
- Hydrogen is also fed to the reactor at about 150°C and 100 atms.
- After reaction, butanols and other alcohols are sent to a separator network.

Separator network

- The separator network consists of several fractionators in series.
- The fractionators in series produce light ends and heavy ends in addition to the desired product i.e., butanol.

Chapter

19

MANUFACTURE OF BUTADIENE (C_4H_6)

Properties of Butadiene ($CH_2 = CH \bullet CH = CH_2$):

Mol. wt.:	54.09
M.P.:	–108.9°C
B.P.:	–4.41 °C
Sp. gr. @ 20°C:	0.621

Soluble in alcohols, ether; insoluble in water

Explosive limits:	Lower = 2.0% in air
	Upper = 11.5% in air
Toxic limit:	5,000 ppm

The major use for butadiene is in the manufacture of synthetic rubber (styrene butadiene rubber)

Methods of Production

- Dehydrogenation of butane
- Dehydragenation–dehydration of ethanol
- Dehydrogenation of butylene
- Steam cracking of hydrocarbons

Chemical intermediates and products derived from C_4 & C_5 hydrocarbons are shown below:

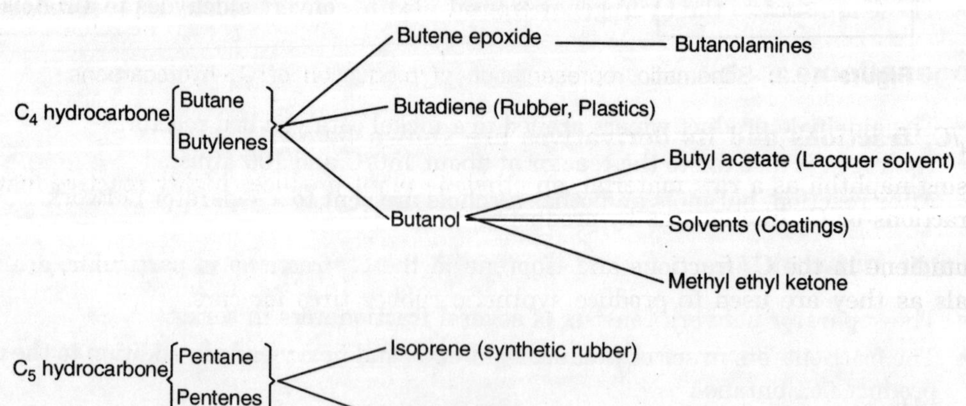

Figure 19.1: Production of chemical intermediates & products from C_4 - C_5 hydrocarbons.

C_4–Hydrocarbons with four C-atoms, primarily butane, butene and butadiene, are derived from three main sources:

- from natural gas and crude oil.
- steam cracking of higher hydrocarbons.
- from refinery gases.

Schematic representation of production of C_4–hydrocarbons is shown below.

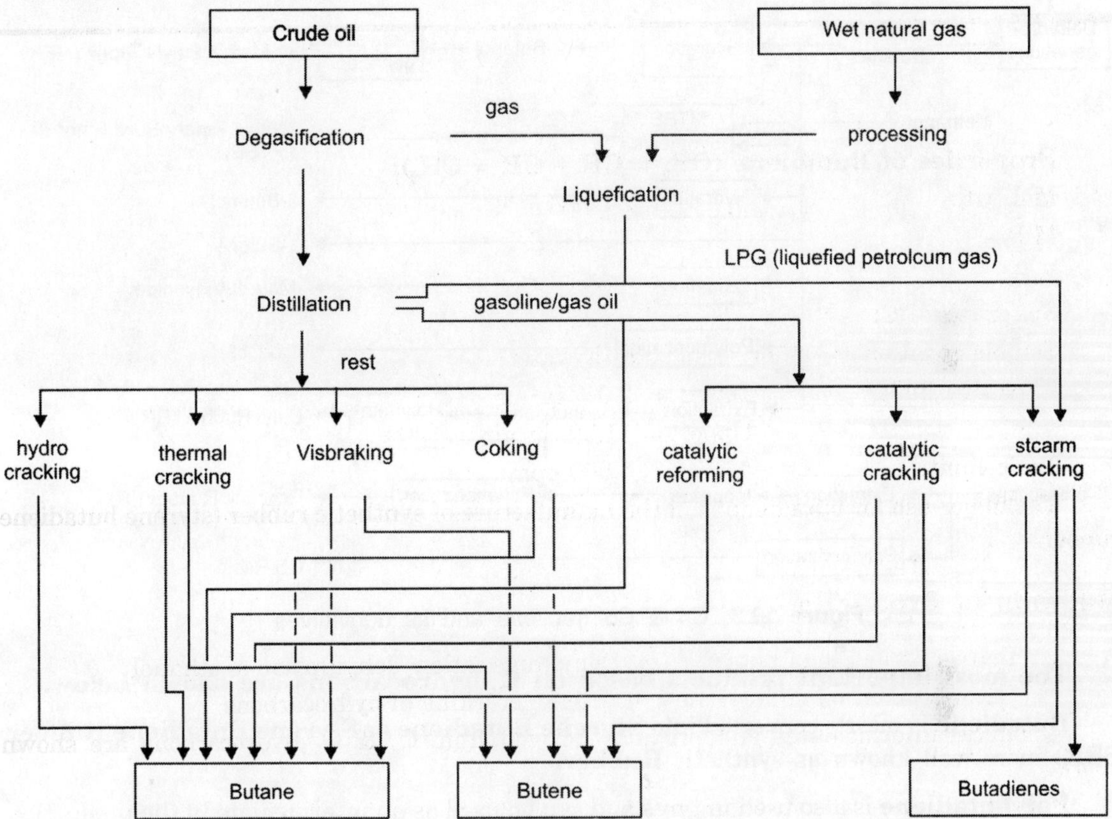

Figure 19.2: Schematic representation of production of C_4–hydrocarbons

C_4/C_5 fractions and its derivatives are shown below:

Using naphtha as a raw material, an ethylene plant produces highly reactive materials in C_4 fractions or C_5 fractions as by-products.

Butadiene in the C_4 fractions and isoprene in the C_5 fractions in particular, are useful chemicals as they are used to produce synthetic rubber tires for cars.

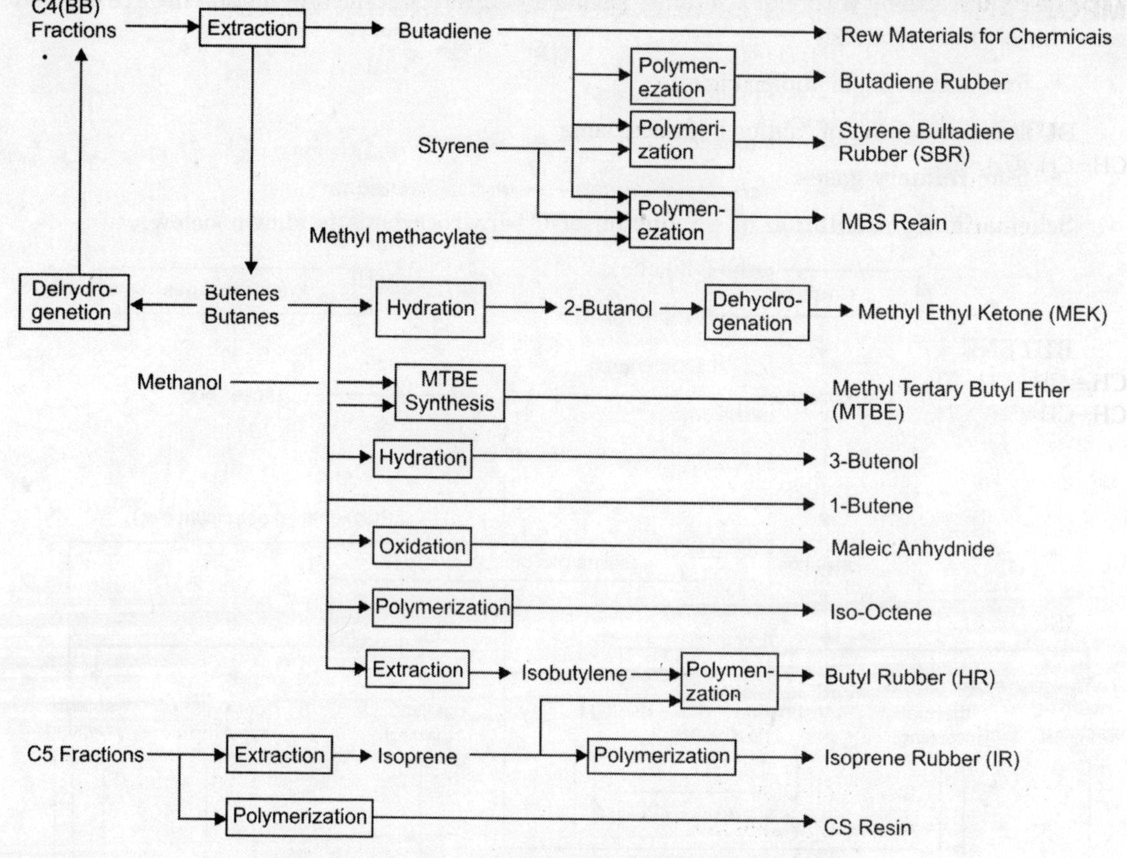

Figure 19.3: C4 & C5 fractions and its derivatives

The most important products based on C_4-hydrocarbons are shown below.

Butadiene is mostly converted into **Styrene Butadiene** or **Styrene Butadiene Rubber (SBR)** more well known as synthetic **Rubber.**

Polybutadiene is also used in tyres and can be used as an intermediate in the production of **acrylonitrile-butadiene styrene (ABS).**

IMPORTANT PRODUCTS BASED ON C-4 HYDROCARBONS:

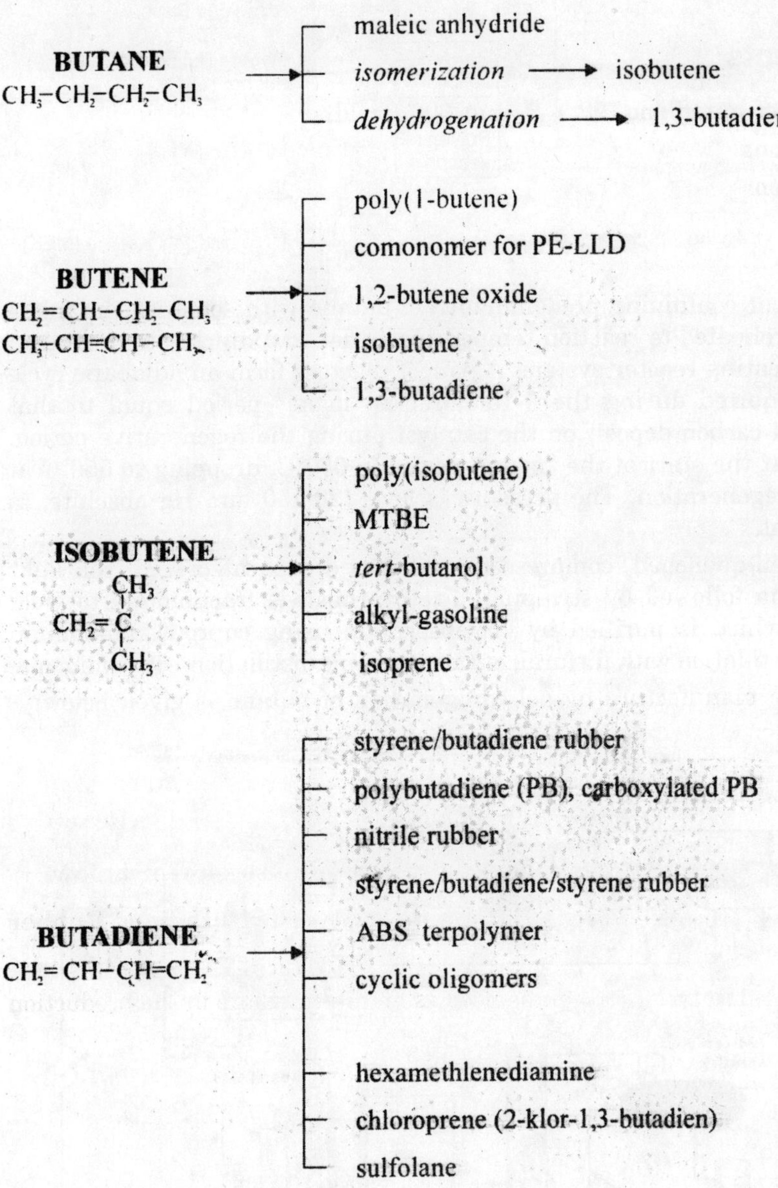

(a) Butadiene production by Dehydrogenation of Butane:

Chemical reactions :

(*a*) Main reaction:

$$C_4H_{10} \longrightarrow \underset{\text{Butadine}}{CH_2 = CH \bullet CH = CH_2} + 2H_2:$$

(*b*) Side reaction:

$$C_4H_{10} \longrightarrow \underset{n-butylenes}{CH_4 + H_2}$$

Row material requirements

Basis:	1 ton of butadiene (98 + % and 60% yield)
n-butane:	1.80 tons
By-products:	0.65 ton

Process description

A refinery gas of C_4/C_5 cut containing predominantly n-butane with some isopentane is mixed with recycle gas and preheated to reaction temperature prior to contact with a catalyst in a fixed bed, regenerative-heating reactor system. A pair of reactors form an adiabatic cycle with the heat of reaction required during the 5-15 minutes "make" period equal to that supplied by the combustion of carbon deposit on the catalyst during the regenerative period. The temperature of reaction at the start of the "make" period is 650°C, dropping to 550°C at the end before switching to regeneration. The pressure is low, 120-150 mm Hg absolute, to force the reaction to the right.

The product gases are oil-quenched, compressed, cooled and separated from the light ends by absorption in naphtha followed by stripping. The overhead is fractionated to yield crude butadiene at the top which is purified by (1) absorption using cuprous ammonium acetate (CAA), (2) extractive distillation with furfural, or (3) azeotropic distillation with ammonia.

A flowsheet of butadiene manufacture by dehydrogenation of butane is given below:

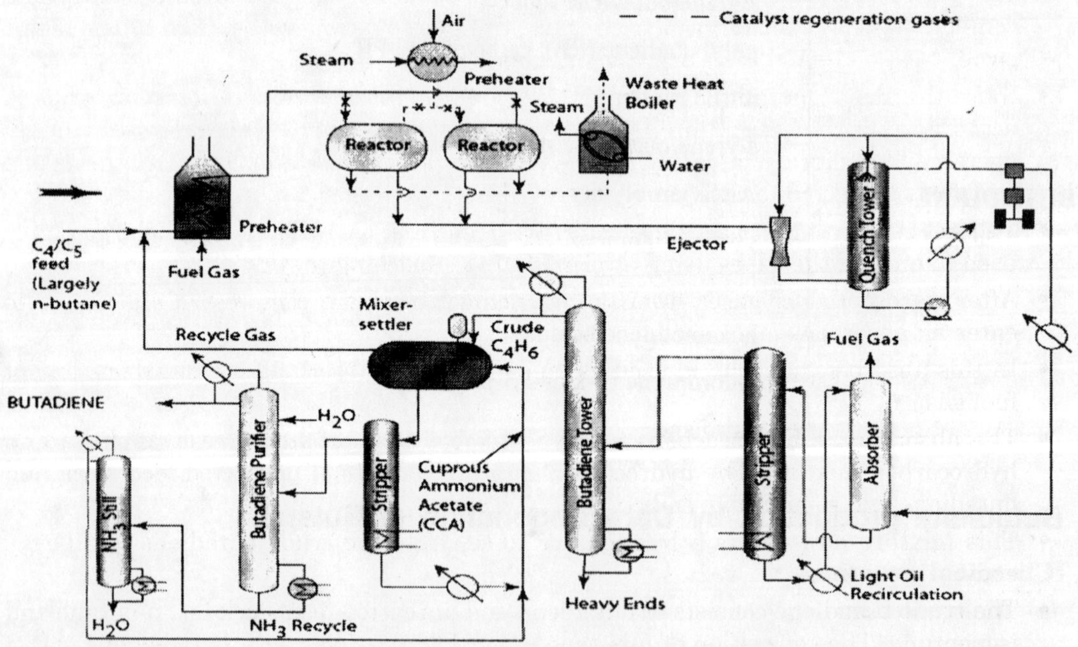

Figure 19.4: Flowsheet of Butadiene mannufacture by dehydrogenation of butane

The more common absorption process involves contact of the close-boiling butadiene-butene fraction with lean CAA solution which dissolves butadiene. A desorption step at higher temperature is followed by distillation, compression, and liquefaction of butadiene to give 98-99% product purity. Ammonia is recovered in the distillation tower by water addition; it is then separated as substantially anhydrous NH_3 by fractionation.

Salient features of process technology of butadiene manufacture is follows:

- The process technology for Butadiene manufacture consists of a reactor separator-reacycle system.
- The separation network is extremely complex and involves quenching, absorption, distillation and extractive distillation process.
- First, the feed stock is pre-heated in a furnace along with unreacted gases that have been recovered in the process using the separator network.
- After pre-heating in a furnace to desired temperature, the gases enter the catalytic packed bed reactors loaded with the catalyst (chromium oxide on alumina).
- After the specified residence time, the product is withdrawn and the feed to the unit is stopped.
- The coked catalyst is subjected to combustion using pre-heated air. Air pre-heating is done using steam in an extended area heat exchanger equipment. Therefore during regeneration, another set of valves operate to allow the pre-heated air in and enable the product withdrawal after the combustion.
- The pre-heated air not only removes the coke as CO_2, but increases the reactor temperature to 650 °C.
- The flue gases are sent to a waste heat recovery boiler to generate steam from water.
- The entire operation of a feed entry, product withdrawal, pre-heated air entry and combustion gases withdrawal from the packed bed reactor corresponds to one single cycle.
- Since the above operation is a batch operation, to make the operation continuous in accordance to the separation network, two reactors are used and these reactors are operated in cyclic fashion i.e., when the first reactor is subjected to reaction, the second reactor is subjected to catalyst regeneration and vice-versa.
- The hot reactor outlet gases are sent to a quenching operation where light gas oil is used to quench the gases using a recirculating quenching tower.
- After wards product gases from the quenching tower are compressed and cooled to enter an absorber.
- In this absorber, naphtha is used as an absorbent to absorb all hydrocarbons except fuel gas.
- The absorbent + hydrocarbons enter a stripper that produces fresh naphtha and hydrocarbon mixture. The hydrocarbon mixture consists of unreacted feed stock and butadiene and some heavy ends.
- This mixture now enters a fractionator to separate the crude butadiene and heavy ends.
- The crude butadiene consists of butadiene and unreacted feed stock i.e., n-butane and isopentane. The separation of *n*-butane, other hydrocarbons with butadiene is one of the difficult separations and they cannot be separated using ordinary distillation.

Therefore, a complicated route of separation is followed next that involves azeotropic distillation using ammonia.

- The crude butadiene is mixed with ammoniated cuprous ammonium acetate solution in a mixer settler. This solution is generated by absorbing ammonia into fresh cuprous ammonium acetate solution.
- The ammoniated cuprous ammonium acetate is sent to a mixer settler unit where the butadiene dissolves in the ammoniated solution. The gas from the mixer settler unit is recycled to mix with the feed stock and enter the pre-heater.
- The ammoniated cuprous ammonium acetate solution is thereby stripped to separate butadiene + ammonia from the ammonium acetate solution. The regenerated fresh solvent is allowed to absorb NH_3 and thereby enter the mixer-settler unit.
- The ammonia + butadiene mixture enters a fractionator fed with water. Here, water interacts with ammonia and generates the ammonium hydroxide product as the bottom product and butadiene is obtained as the top product.
- The ammonia solution is subjected to stripping to separate water and ammonia. The water is recycled back to the butadiene purifier and ammonia is allowed to get absorbed into the fresh cuprous ammonium acetate solution.
- In India butadiene is manufacture also from ethanol by catalytic cracking at 400 - 450ºC over metal oxide catalyst.

Production of Butadiene by Oxydehydrogenation of butane (C_4H_{10})

Since the equilibrium in the reaction

$$C_4H_{10} \rightleftharpoons C_4H_6 + 2H_2$$

is reversible, it is obvious to add enough oxygen to the reacting mixture to pull the equilibrium far to the right and increase conversions from 20% or so (requiring much expensive recycle) to nearly 100%.

Chemical reaction:

$$C_4H_{10} + O_2 \longrightarrow C_4H_6 + 2H_2O$$

Process description

Butanes or butenes are fed with a carefully controlled amount of oxygen, together with a small amount of a halogen (bromine or iodine) promoter into a reactor containing a heterogeneous catalyst. The higher yields simplify the subsequent product separation step greatly. Decrease in plant size for equivalent capacity is made possible by adding oxygen to the dehydrogenation process.

Peroduction of Butadiene from Ethanol

Chemical reactions:

(*a*) Dehydrogenation ;

$$CH_3CH_2OH \longrightarrow CH_3CHO + H_2$$

(*b*) Dehydration:

$$CH_3CH_2OH + CH_3CHO \longrightarrow CH_2 = CH{\bullet}CH = CH_2 + 2H_2O$$

A Comparative layout of butadiene manufacturing plant by oxydehydrogenation vs. straight dehydrogenation of butane is shown below in figure.

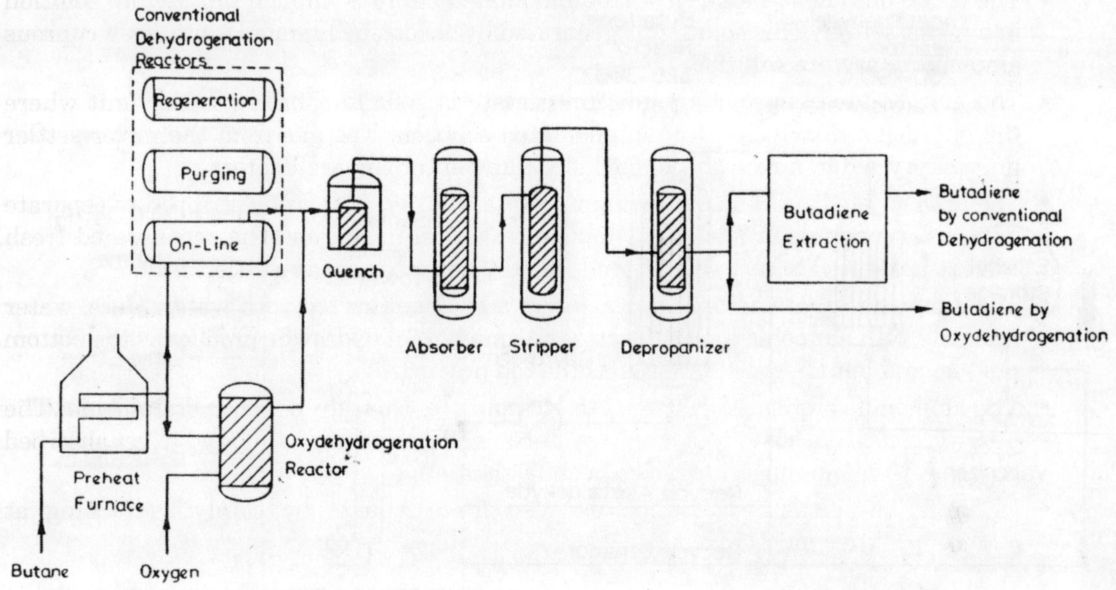

Figure 19.5: Comparative plant layout for making butadiene from butane by oxydehydrogenation versus straight dehydrogenation

Raw material requirements

Basis: 1 ton of butadiene (98+ % purity and 70% yield)
Ethanol: 2.88 tons
By-products: ethylene, butene, ethyl ether, ethyl acetate, butanol

Process description

Ethanol (92-95 % grade) is vaporized and passed over a dehydrogenation catalyst at 260-290°C to produce acetaldehyde: 90-94% yield on a 30-50% per pass basis.

Excess alcohol is mixed with acetaldehyde (3:1) and fed as a liquid to a fixed bed converter containing tantala-silica catalyst. Pressure is essentially atmospheric, the temperature is 325-350°C'. A conversion of 28-30 per cent based on acetaldehyde is obtained with an ultimate yield in this step of 75%.

The reaction gases contain numerous compounds in addition to unreacted ethanol-acetaldehyde. These are separated by a series of fractionating columns with the crude butadiene ultimately being purified.

Oxidation of ethanol to acetaldehyde using air is also a fully commercialized process. It has the disadvantage that H_2 cannot be recovered, but this is offset by not having to supply external heat for this exothermic reaction.

A one-step route is being sought to give butadiene directly from ethanol. So far, the correct dehydrating catalyst has not been found.

A flowsheet of butadiene production by dehydrogenation of ethanol (ethyl alcohol) is given below:

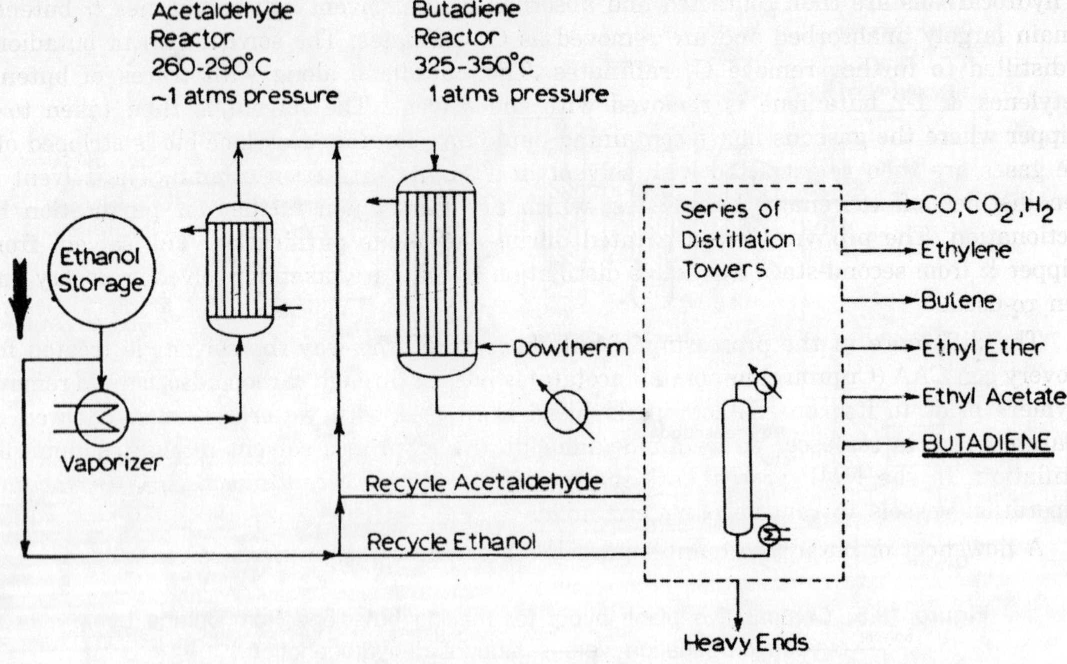

Figure 19.6: Production of butadiene from ethanol dehydrogenation.

The future of the butadiene business depends on synthetic rubber production. The SBR plant at Bareilly, started in 1963, is based on butadiene from alcohol, but no further alcohol-based plants are contemplated. This plant will be uneconomic as soon as ethylene, butadiene, and benzene from petroleum raw materials become available at a central petrochemical supply base located closer to end user customers.

Utilisation of butadiene is mainly for SBR, polybutadiene rubber (BR), adiponitrile, chloroprene, ABS polymers, nitrile rubber (NBR) etc.

Production scheme of Butadiene in an Indian Petrochemical Plant

Butadiene are produced by solvent extraction or extractive distillation process wherein the butadiene is extracted by a solvent viz. acetonitrile, cuprous ammonium acetate (CAA), n-methyl pyrolidone (NMP), furfural, dimethyl formamide (DMF) or dimethyl acetamide. No single solvent has dominated the process worldwide and this suggests that there are no significant differences in both operation and economics using most of these solvents. The solvents used in India are Acetonitrile (NOCIL, TTC) and N-Methyl pyrolidone (IPCL, Vadodara & HPL, Haldia).

Process Description

The C_4 stream is first taken to feed preparation where oxygen if any present in it is removed from feedstock by washing with sodium nitrite solution. Washing solution is then distilled to remove C_3 hydrocarbons, which are sent back to cracker. The vapour phase mixed C_4 hydrocarbons are then contacted and absorbed in the solvent, where butanes & butenes remain largely unabsorbed and are removed as C_4 raffinates. The solvent rich in butadiene is distilled to further remove C_4 raffinates. The butadiene along with traces of butene, acetylenes & 1-2 butadiene is removed with the solvent. The solvent is then taken to a stripper where the gaseous phase containing butadiene, butane, acetylene etc is stripped off. The gases are then re-extracted with solvent in a second extraction column. The solvent is then distilled off to remove butadienes, which are then taken further for purification by fractionation. The propynes are separated during butadiene purification. The solvent from stripper & from second-stage extractive distillation column are taken for solvent recovery and then re-used.

The difference in the processing of butadiene lie in the way the solvent is treated for recovery e.g. CAA (Cuprous ammonium acetate) is passed through carbon adsorbers to remove polymers prior to its reuse. Acetonitrile bleed is diluted with water, polymers allowed to separate as oil in coalescer & acetamide/ammonia removed in a solvent recovery column by distillation. In the NMP process, solvent is regenerated on a continuous basis in vacuum evaporation vessels to remove polymeric solids.

A flowsheet of butadiene manufacture in an Indian plant is given below:

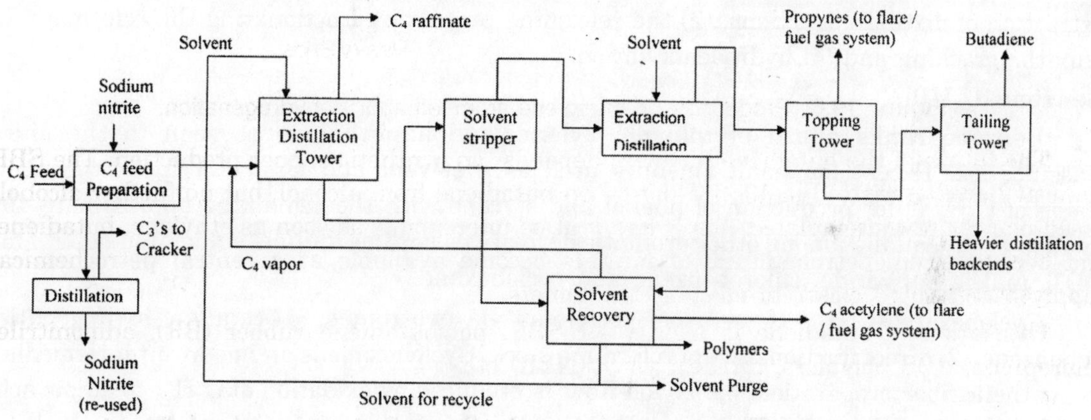

Figure 19.7: Production of butadiene from ethanol dehydrogenation.

IMPORTANT AROMATIC COMPOUNDS-BENZENE, TOLUNE & XYLENE (BTX)

Chapter 20

Important aromatics for chemical systhesis are Benzene (B) Toluere (T) & xylene (X) (Called BTX) and naphthalene (N).

Coking Coal carbonization is used for metallurgical coke production. The volatiles (coke oven gas) are stripped of tar by electrostatic precipitators, scrubbed with dilute H_2SO_4 to remove ammonia. The remaining crude benzole fraction is absorbed in a stable oil such as wash oil. This benzol has a proximate composition of 70% B, 15% T, 7% X, 4% N, and solvent naphtha 4% and is fractionated to separate these components in benzol rectification plant.

The production of aromatic raw materials from petroleum is derived from (1) solvent extraction of distillate fractions, (2) the reforming process by fractionating the reformate, (3) naphtha cracking and (4) hydrodealkylation.

Benzene (C_6H_6)

Benzene ranks ahead of toluene, xylene, and naphthalene as seen in the above classification. Except for small amounts used as a solvent and for motor fuels, the major usage of C_6H_6 is for production of phenol and styrene with the remaining approximate 40% divided about equally among other aromatic derivatives such as aniline, sulfonated detergents, DDT, maleic anhydride, chlorobenzene and cyclohexane.

Cyclohexane (C_6H_{12}) comes from petroleum via two routes: (1) catalytic hydrogenation of benzene, (2) direct fractionation of reforming stock. Cyclohexane is an important intermediate in synthetic fiber preparation. 6,6-Nylon fiber is produced by oxidation of C_6H_{12} to adipic acid, which is then condensed with hexamethylene diamine.

Uses of benzene is for the production of:

(*a*) Ethylbenzene (styrene)

(*b*) Cyclohexane

(*c*) Nitrobenzene

(*d*) Maleic anhydride

(*e*) Alkylbenzenes

(*f*) Miscellaneous (*e.g.,* chlorobenzene, benzenesulfonic acid)

End uses of various aromatic conversion products are given below is figure:

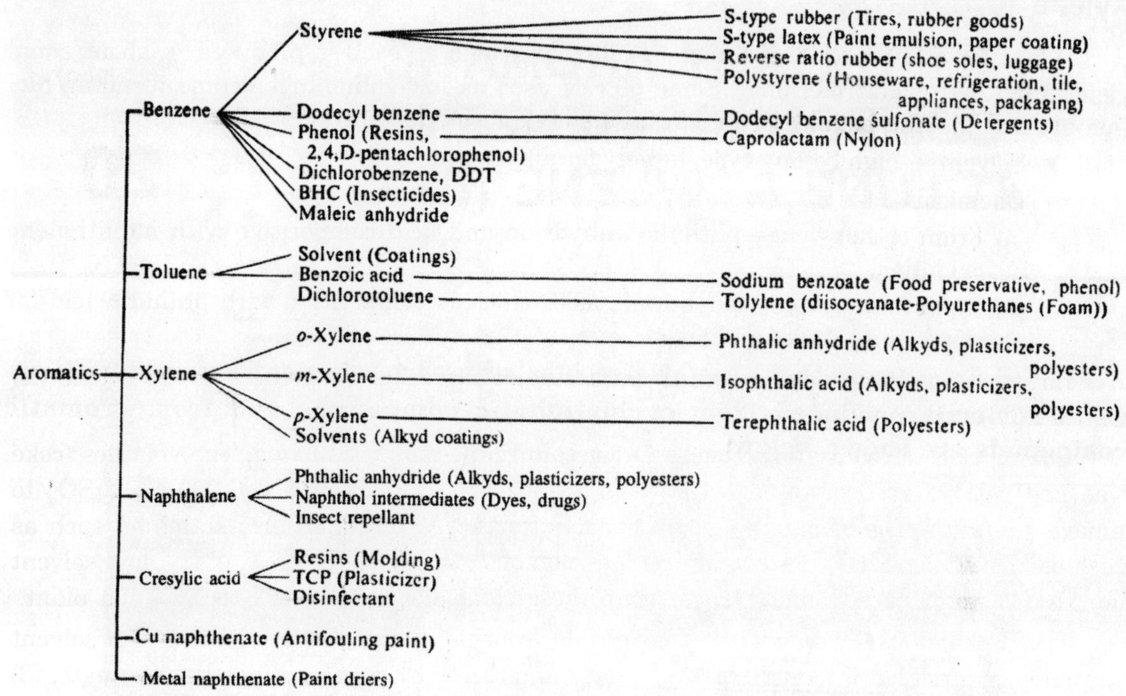

Figure 20.1: End uses of chemicals & intermediates prodcued from aromatics.

Toluene ($C_6H_5 \bullet CH_3$)

Toluene is produced in yields about equal to that of benzene and of xylene in petroleum reforming operations. The demand for toluene is not as much as for benzene.

The major end uses of toluene are:

- Solvents—paints, rubber and plastic cements.
- Chemicals—benzyl chloride and its peroxide for polymerization catalysts; benzoic acid; as plasticizer; phenol from benzoic acid; toluene disocyanate in urethane foams.
- Detergents— toluene sulfonates.

These uses cannot absorb all of the toluene production; so other uses plans for using up the excess toluene are:

- Blending in motor petrol to upgrade octane number; this produces low income yield as fuel value.
- Production of phenol
- Dealkylation with H_2 to yield benzene.

The principal source of toluene is catalytic reforming of refinery streams. This source accounts for 87% of the total toluene produced. An additional 9% is separated from pyrolysis of gasoline produced in steam crackers during the manufacture of ethylene and propylene. Other sources are an additional 2% recovered as a by-product from styrene manufacture and 2% entering the market from separation from coal tar.

Xylene [$C_6H_4 \cdot (CH_3)_2$]

A similar story to that of toluene exists for xylene as each is produced in about equal yields from petroleum reforming. Some 60% is used as low-value, high-octane blending fuel (mainly in U.S.) the balance of important high-value products are as follows:

- Solvents—high boiling type, largely for alkyd resins.
- Chemicals
  (a) From orthoxylene—phthalic anhydride and acid competitive with naphthalene oxidation.
  (b) From metaxylene—isophthalic acid which is competitive with phthalic acid for reinforced plastics and plasticizers.
  (c) From paraxylene—dimethyl terephthalate used in polyester fibers and films.

Chemicals produced from cycloaliphatic compounds and from aromatic compounds are shown below:

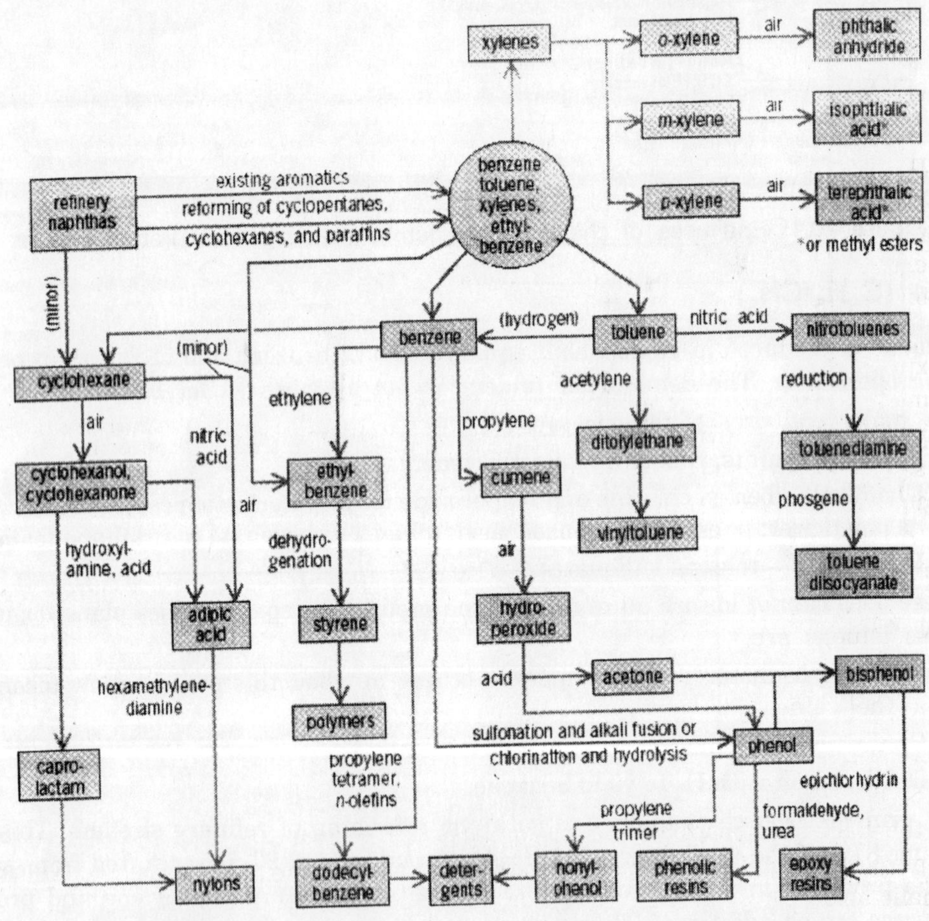

Naphthalene ($C_{10}H_8$)

Bulk of naphthalene supply derived form coke oven gas goes into the production of phthalic arhydride which is mainly used in the production of alkyd resin & plasticisers and partly for dyestuffs. Naphthalene can also be produced by hydrodealkylation of petroleum products (e.g. alkylnaphthalene)

Aromatics and Its Derivatives produced by various unit operations & processes are shown below:

Using naphtha as a raw material, an ethylene plant produces aromatic derivatives (BTX fractions) such as benzene (B), toluene (T), and xylenes (X) as cracked gasolines.

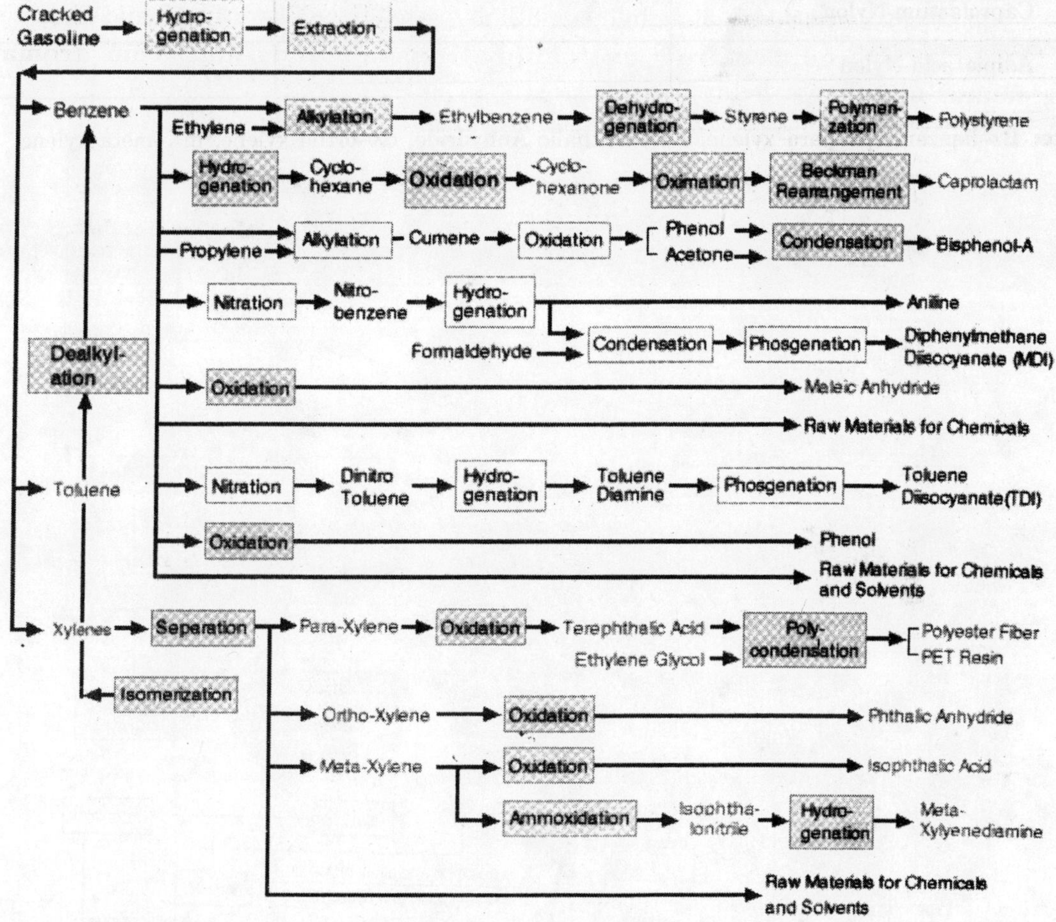

Of the BTX fractions, benzene and xylenes are particularly used in large quantities in general-purpose resins and fibers after being processed with polystyrene, caprolactam, and terephthalic acid.

Since cracked gasolines produced from ethylene plants only are not enough to meet the demand for these products, supply from reformulated gasolines in the oil refining industry is increasing.

Important Benzene, tolune & Xylene (BTX) Derivatives are listed belows:

Benzene	toluene	Xylene
EB-Styrene	Toluene diisocynate	pX-PTA-Polyester
Cumene-Phenol-phenolic resins	Motor Fuel-Octane Enhancement	oX-Phathalic anhydride
LAB	Solvent	mX-isophthalic acid
Cyclohexane	Conversion to Bz-pX	
Cyclohexnol		oX/mX Conversion to pX
Caprolactum-Nylon		
Adipic acid-Nylon		

Note: Bz–Benzene, px–para xylene, PTA–Phthalic Anhydride, oX–ortho xylene, mX–meta xylene

Chapter 21

PRODUCTION OF BENZENE AND TOLUENE

BENZENE AND TOLUENE ARE PRODUCED BY VARIETY OF PROCESSES VIZ:

- By extraction/fractionation from Pyrolysis Gasoline.
- Reforming of Naphtha (in Petroleum Refineries).
- Distillation of coke oven light oil (in coke oven plants).

Benzene can also be produced by toluene de-alkylation and transalkylation/disproportionation.

Manufacture of benzene and toluene from pyrolysis gasoline

The pyrolysis gasoline (pygas) is a stream produced by the high temperature, short residence time cracking of naphtha in petrochemical plants. According to cracking severity, the pygas contains typically 20-40% benzene, 16-20% toluene, 10-13% higher aromatics (e.g. ethyl benzene, xylenes, styrenes etc) with a balance of non-aromatic hydrocarbons (mainly olefins & diolefms).

The benzene extraction from pyrolysis gasoline is carried out in the following steps:

- Hydro-treating
- Secondary Hydrogenation
- Solvent Extraction
- Fractionation

Simplified flow diagram is presented in Fig. 21.1 and each of the process steps are described below:

(a) Hydro-treating

Hydro-treating is the process by which the quality of pyrolysis gasoline stream is improved by subjecting them to mild or severe conditions of hydrogen pressure in the presence of a catalyst. The process flow diagram is depicted in Fig. 21.2.

The pyrolysis Gasoline feed is heated in a furnace and combined with recycled hydrogen gas. The combined feed is passed through a reactor containing a catalyst bed where the hydrogenation reaction takes place. Upon leaving the reactor, the stream is cooled and moved to a separator vessel where recycle or net hydrogen is removed. The liquid then moves to a stabilizer or stripper, which removes, hydrogen, hydrogen sulfide, ammonia, water and light hydrocarbons, dissolved in the separator liquid.

(b) Secondary Hydrogenation

The pyrolysis gasoline normally undergoes second stage hydrogenation (generally carried out at high temperature of 300°C) in which olefins are saturated, organic sulfur forms hydrogen sulfide, combined nitrogen is converted to ammonia, and oxygenated compounds are reduced to hydrocarbons and water. After these parallel reactions have been completed, the gases and liquid are separated. The liquid is then stripped of gaseous impurities, such as hydrogen sulfide and remaining light hydrocarbons.

(c) Separation of Benzene & Toluene from Hydrogenated Pyrolysis Gasoline (pygas)

The hydro-treated pyrolysis gasoline is rich in aromatics such as benzene (typically about 40%) toluene (typically 20%) and has small amount of xylene as well as non-aromatics of similar boiling points. Therefore, separation of these aromatics from other process streams is achieved by extractive distillation and liquid-liquid extraction. Liquid-liquid extraction is the most commonly used extraction method. If only one aromatic (e.g., benzene, toluene or xylene) is to be recovered in pure form, extractive distillation is preferred as the capital and operating expenses are less than those associated with liquid-liquid extraction.

Liquid-Liquid Extraction

The solvents commonly used in liquid-liquid extraction are sulfolane and tetraethylene glycol. It is adopted by three units in India (Reliance-Hazira, IPCL-Vadodara and NOCIL-TTC). Processes using other solvents such as tetraethylene glycol, di-ethylene glycol, dimethylsulfoxide (DMSO) are not adopted in India. The hydrogenated pyrolysis gasoline stream is fed in the extractor. Where it is contacted with the solvent in the extractor, the non-aromatics (the raffinate phases) leaves through the top while a mixture of aromatics, solvent & light non-aromatics leaves through the bottom. The raffinate is taken to a raffinate wash column where it is washed with water. Raffinate obtained is sent to storage, the bottom is sent to a water stripper. The aromatics/solvent & light non-aromatic stream is transferred from the extractor directly to a stripper. The stripper removes solvent, leaving behind the desired aromatic, which is then sent to a solvent recovery column for recovering the solvent if any. The aromatics phase is passed through a clay tower (to remove olefins) and then taken for fractionation to produce pure benzene & toluene.

Extractive distillation

Extractive distillation methods are preferred when only one aromatic (e.g. benzene, toluene, or xylene) is to be recovered in pure form. Extractive distillation methods separate components by introducing a solvent to a mixture. The solvent suppresses the relative vapour pressure of some components compared to other components in the mixture allowing the desired component to be extracted. Hydro-treated PGH (pyro gasline hydrocarbons) is processed to remove lighters (less than C_5) & heaviers (greater than o-xylene) hydrocarbons. The crude BTX stream and the extractive solvent are fed to an extractive distillation column where non-aromatics (the raffinate) are separated. The aromatic/solvent stream is transferred to a stripper, which removes pure Benzene at the top leaving behind solvent, which is reused. Some of the solvents commonly, used in extractive distillation units are di-methyl-formamide, n-formyl-morpholine, n-methyl-pyrrolidone and sulfolane. HPL, Haldia uses n-methyl pyrrolidone to extract benzene from pyrolysis gasoline. The non-aromatics (raffinate) maybe further distilled in a small column to recover any solvent. Thus the extractive distillation is simpler in design comprising primarily of the two columns (one for extractive distillation/one for stripping). The pressure in the coloums is close to atmospheric or vacuum and the working temperature is 175°C. Flowsheet of extractive distillation process is given in figure 21.3.

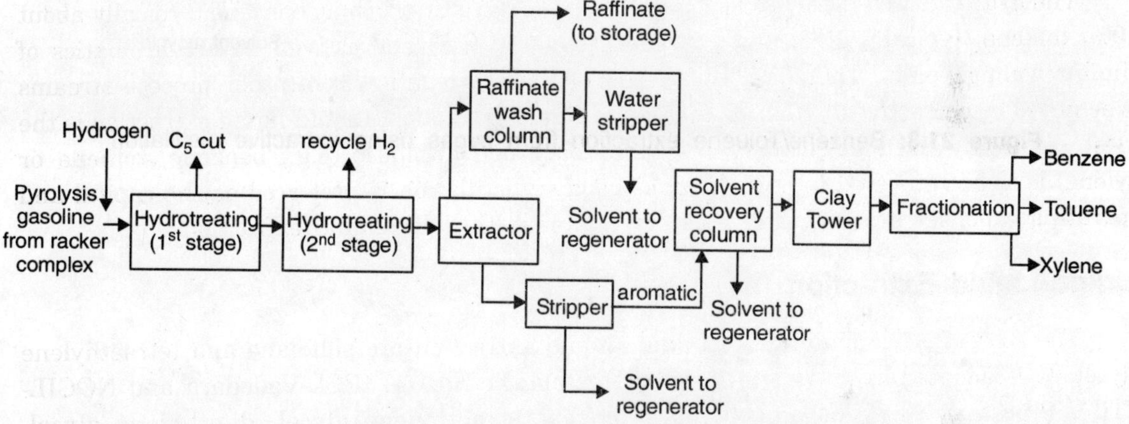

Figure 21.1: Benzene/Toluene Extraction From Pyrolysis Gasoline (using Liquid–liquid Extraction)

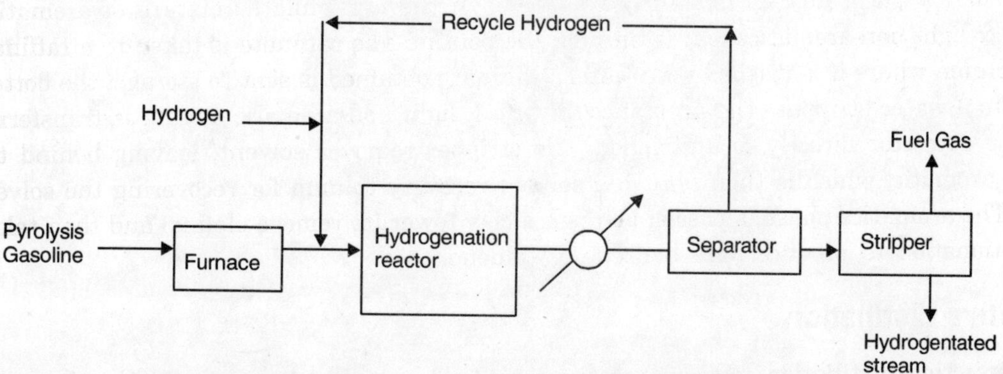

Figure 21.2: Process flowsheet for hydro-treating to improve the Quality of pyrolysis gasoline (pygas)

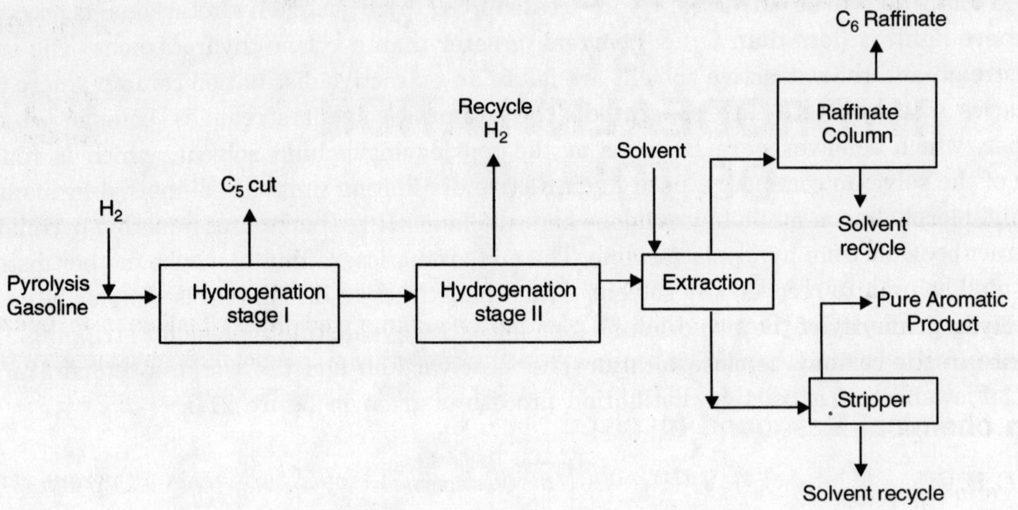

Figure 21.3: Benzene/Toluene extraction from pygas using extractive distillation

PRODUCTION OF BENZENE BY HYDRODEALKYLATION OF TOLUNE

Chapter 22

Hydrodealkylation is a method of converting toluene and dialkylbenzenes to benzene. Toluene is reacted with H_2 to produce benzene & methane as loluene does not have much market value.

Main chemical Reactions involved

$C_6H_5CH_3 + H_2 \rightarrow C_6H_6 + CH_4$ (600 - 650°C & 35 - 40 kg/cm^2, chromia on porous carrier catalyst)

$C_6H_4(CH_3)_2 + 2H_2 \rightarrow C_6H_6 + 2CH_4$

Side reactions

Toluene $\rightarrow$ diphenyl + H_2

Toluene + benzene $\rightarrow$ Methyl diphenyl

Toluene + H_2 $\rightarrow$ Alkanes (cracking reaction)

Process Description

Hydrogen-rich make-up gas and recycle gas are compressed & mixed with the liquid alkyl aromatic feedstock plus recycle. The combined feed is preheated by means of heat exchange with reactor effluent and finally in a fired preheater for charge to the reactor which contains either a fixed bed catalyst or is tubular, non-catalytic. The effluent is cooled via reactor feed preheater, then with water, and separated in a high pressure stage to gas & liquid fractions. A major portion of the gas is recycled, the balance going to a methane separation unit for H_2 enrichment and make-up. It may also be used for fuel. The liquid fraction is stripped of residual gas at low pressure and fractionated to produce high purity benzene (less than 0.5 ppm thiophene) and recycle alkyl aromatics.

Yields are 95-98% for toluene conversion.

Hydrodealkylation has proven a low-cost, feasible method to convert unwanted toluene to benzene, thereby avoiding downgrading of toluene to fuel values. The process is readily handled within the refinery operations so that high grade benzene is available to consumer plants in the immediate vicinity.

A flowsheet of benzene production by hydrodealkylation of tolene is given below:

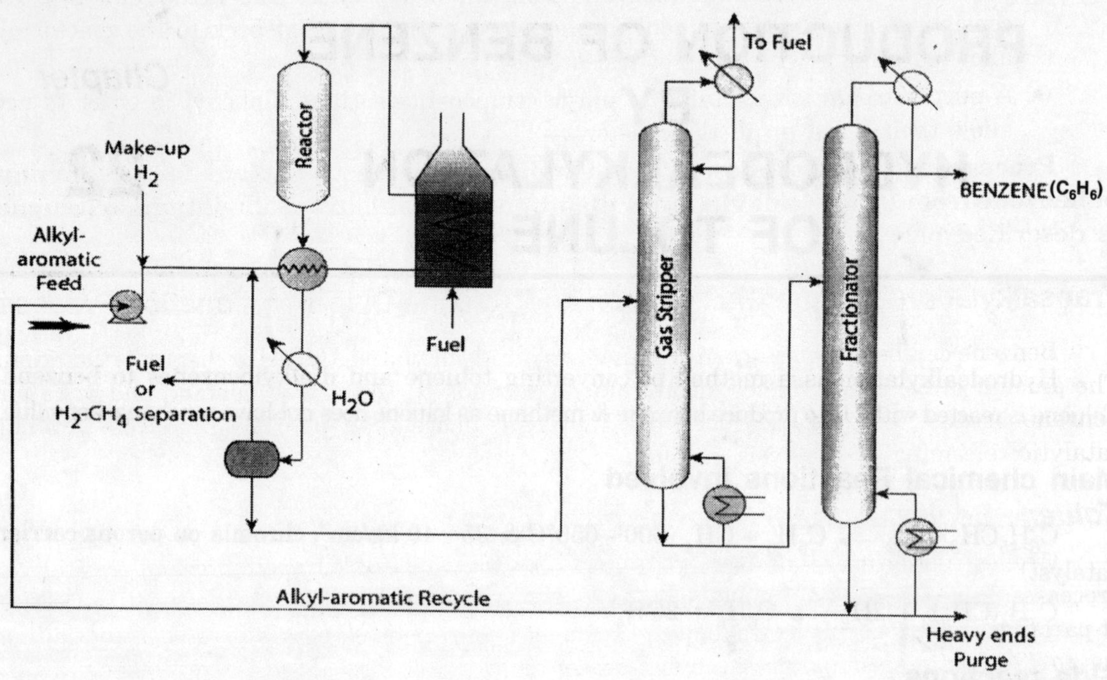

Not: For better heat economy in thermal process, reactor effluent can be used for stripper and fractionator reboiler heat supply.

Figure 22.1: Flowsheet of benzene production by hydrodealkylation of tolune

Salient features of the process is listed below:

- The process corresponds to a simple reactor-separator-recycle system facilitated with heat integration.
- Alkyl-aromatics (Toluenes and other higher order alkyl aromatics) along with makeup H_2 and recycle stream consisting of unreacted toluene is allowed to enter a feed pre-heater.
- The feed pre-heater increases the temperature of the reactants at the desired pressure.
- After the reaction, the reactor products are cooled using heat integration concept by exchanging heat with the cold feed stream. Further cooling of the stream is carried out using a water based cooler.
- The stream enters a phase separation unit where the fuel gas components such as H_2 and CH_4 are removed as vapour stream. The liquid stream consists of H_2, CH_4, Light ends, Benzene, Toluene, Diphenyl and other higher order aromatics.
- The vapour stream is partially purged and to a large extent recycled as the stream has good amount of H_2. The gas phase purge stream is for controlling the concentration of methane in the reactor.
- The liquid stream enters a gas stripper which removes the light ends as a gaseous product stream from the top tray. The bottom product is then sent to a fractionator.

- The fractionator separates benzene from all heavy ends. The heavy end product consisting of unreacted toluene, diphenyl etc. is largely sent back to the reactor by allowing mixing with the fresh feed.
- A purge stream is facilitated to purge components such as diphenyl in order to not allow their build up in the reactor.

Process technology followed in an Indian petrochemical plant for production of benzene from hydrodealkylation & disproportionation (transalkylation) of toluene is described below:

Transalkylation/Toluene Dealkylation and Toluene Disproportionation Process

Benzene can be produced from toluene by hydrodealkylation (HDA) or disproportionation. The amount of benzene produced from toluene depends on the overall demand and price for benzene because benzene produced by HDA costs more than benzene produced through catalytic reforming or pyrolysis gasoline.

Toluene Dealkylation for benzene production

Hydrodealkylation of toluene (HDA process) is accomplished through thermal or catalytic processes. Pure toluene (92 to 99%) or toluene (85 to 90%) mixed with other heavier aromatics or paraffins from the benzene fractionation column is heated together with hydrogen-containing gas to 730°C at a specified pressure and is passed over a dealkylation catalyst in the reactor. Toluene reacts with the hydrogen to yield benzene and methane. The benzene may be 'separated' from methane in a high-pressure separator by flashing off the methane-containing gas.

The product is then stabilized and benzene is recovered by distillation in the fractionation column. Recovered benzene is sent to storage. Un-reacted toluene and some heavy aromatic by-products are recycled. About 70 to 85 percent conversion of toluene to benzene is accomplished per pass through the system, and the ultimate yield is 95 percent of the theoretical yield. Because there is a weight loss of about 23 percent, the difference in toluene and benzene price must be high enough to justify use of the HDA process. Reaction is represented below:

$$\underset{\text{Toluene}}{C_6H_5CH_3} + \underset{\text{Hydrogen}}{H_2} \longrightarrow \underset{\text{Benzene}}{C_6H_6} + \underset{\text{Methane}}{CH_4}$$

This process is employed by NOCIL, IPCL - Vadodara and HPL - Haldia for conversion of Toluene to maximize benzene yield.

Toluene Disproportionation/Transalkylation to produce benzene

Toluene disproportionation (or transalkylation) catalytically converts two molecules of toluene to one molecule each of benzene and xylene. The basic process is similar to toluene hydro-de-alkylation, but can occur under less severe conditions. Transalkylation operates at lower temperatures, consumes little hydrogen and no loss of carbon to methane occurs as with HDA. Hydrogenated toluene is sent to a separator for removal of off-gases. The product is then stabilised and sent through clay towers. Benzene, toluene & xylene are recovered by distillation, and un-reacted toluene is recycled. Then HDA is a more economical and feasible process. Reaction is represented below:

$$\underset{\text{Toluene}}{2C_6H_5CH_3} \xrightarrow{\text{Transalkylation}} \underset{\text{Benzene}}{C_6H_6} + \underset{\text{Xylene}}{C_6H_4(CH_3)_2}$$

If benzene is the only product required, the TATORY process of UOP is adopted at Reliance, Hazira and Reliance, Patalganga for converting the toluene to benzene and xylene.

A flowsheet of toluene hydrodealkylation unit (HDA) for benzene production is shown below:

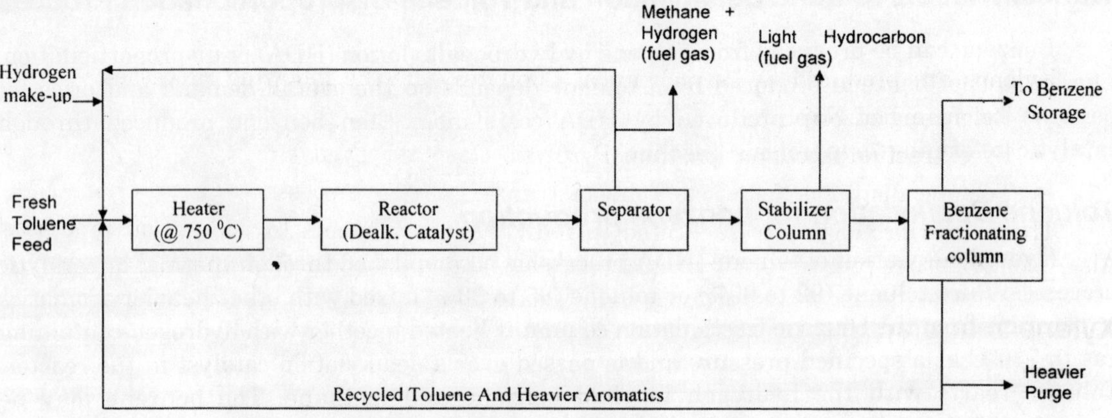

Figure 22.2: Flowsheet of A Toluene Hydro Dealkylaton Unit (HDA) to produce benzene

A flowsheet of toluene disproportionation/transalkylation process for benzene production is given below:

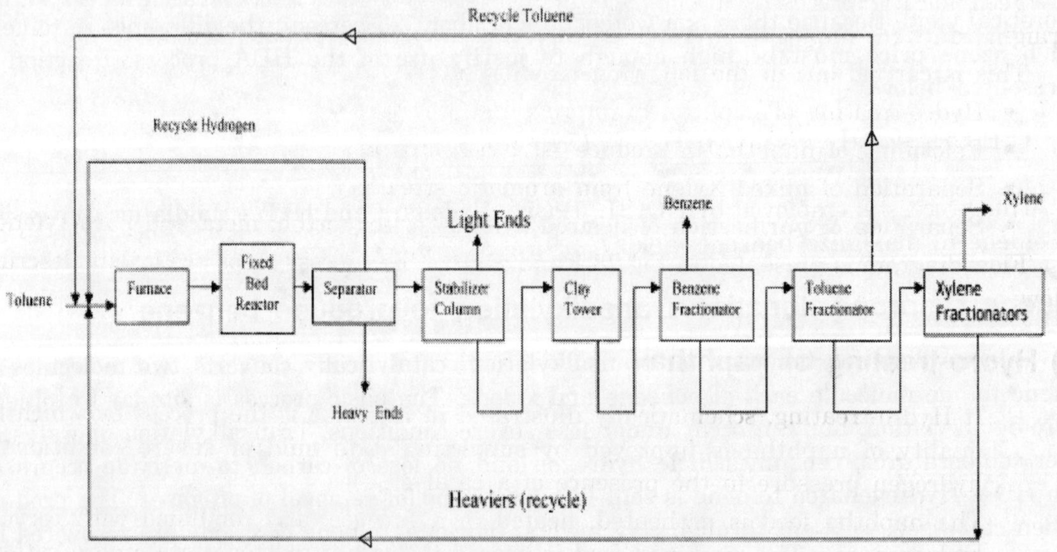

Figure 23.3: Flowsheet of A Toluene disproportionation/transalkylation Process (Tatoray process) for benzene production.

Chapter 23

MANUFACTURE OF XYLENE

Xylenes are produced in Petrochemical Complexes by the following processes:

- Reforming of Naphtha
- By extraction/fractionation from Pyrolysis Gasoline (pygas)
- Toluene disproportionation /Toluene transalkylation process:

The xylene produced are separated in individual components by separation processes. Also, isomerisation processes are used to increase the yield of particular xylene isomer.

Xylene manufacture by naphtha reforming:

Two types of processes are observed.

- optimised to produce both ortho-xylene & para-Xylene (IPCL, Vadodara & RIL, Jamnagar)
- optimised to produce para-Xylene only (RIL-Patalganga)

The process adopted for production of both ortho & para-Xylene is described below:

Feed for the processes are petroleum hydrocarbons with boiling range of 105-170°C (straight run light naphtha) or hydro cracked naphtha having similar boiling range.

This is carried out in the following reaction steps:

- Hydro-treating of naphtha to improve its quality
- Reforming of naphtha to produce BTX rich stream
- Separation of mixed Xylene from aromatic streams
- Separation & purification of desired xylene Isomer (ortho, meta and para-xylene)

Flow diagram is presented in Fig.23.1 and each of the process steps are briefly described below:

(a) Hydro-treating of naphtha

It Hydro-treating, schematically illustrated in Fig. 23.2 is the process by which the quality of naphtha is improved by subjecting it to mild or severe conditions of hydrogen pressure in the presence of a catalyst.

The naphtha feed is preheated, heated in a furnace, and combined with recycled hydrogen gas. The combined feed is passed through a reactor containing a catalyst bed where the hydrogenation reaction takes place. Upon leaving the reactor, the stream is cooled and moved to a separator vessel where recycle or net hydrogen is removed. The liquid then moves to a stabilizer or stripper, which removes, hydrogen, hydrogen sulphide, ammonia, water & light hydrocarbons, dissolved in the separator liquid.

(b) Reforming of naphtha to produce BTX rich stream

Reforming is a platinum catalysed high temperature vapour phase process which converts a relatively non-aromatic C_6 - C_{12} hydrocarbon mixture (naphtha) (typically containing less than 1% aromatics) to an aromatic product called reformate.

The main process units are depicted in Fig. 23.3 & described below:

- Reactors, which contain fixed bed catalysts.
- Heaters to bring the naphtha and recycle gas to reaction temperature and to supply heats of reaction.
- A product cooling system and a gas-liquid separator.
- A hydrogen-gas recycle system.
- A stabilizer to separate light hydrocarbons dissolved in the receiver liquid.

The naphtha is combined with recycled hydrogen, preheated, heated to the reaction temperature in a fired heater and then transferred to a series of catalyst-containing reactors. Because the reaction is endothermic, a series of three or four reactors with inter-stage reheat furnaces are necessary to achieve the required conversion. The reactors normally contain increasing amounts of catalyst in each stage.

The effluent from the last reactor is cooled and transferred to a receiving unit (e.g., the flash drum) where the hydrogen is separated from the liquid reformate. The hydrogen gases are compressed and recycled to the reactors while the reformate is moved to a stabilizer fractionator. The fractionator removes lighter hydrocarbons (hepton) to produce a stabilized reformate. The stabilized reformate is used as a feedstock for further recovery of benzene, toluene & xylene.

(c) Separation of mixed xylene from reformate

The stabilized reformate is then taken for fractionation (ortho-meta splitter). The bottom product is a mixture of o-Xylene & C_9 aromatics. The top product is a mixture of Ethyl benzene, o-,m-,p-Xylene is sent to PAREX for separation of p-Xylene isomer.

(d) Separation and purification of desired xylene isomer

Typical mix of xylene isomers from a catalytic reformate stream consists of the following: m-xylene (40%), o-xylene (24%), p-xylene (19%) and ethyl benzene (17%). The demand for mixed xylene is low in comparison to the demand for pure isomers, especially p-xylene. Hence, separation of isomers of xylene has to be resorted to. Isolation of individual isomers through conventional distillation is difficult for xylene isomers because of the closeness of their boiling points.

Hence, separation of xylene isomers is generally based on differences in adsorptive properties. In PAREX unit, p-Xylene is selectively adsorbed on the adsorbent. After the bed is saturated, p-Xylene is desorbed by using another organic such as toluene or p-diethyl benzene. The PAREX process adopted at IPCL, Vadodara and RIL Patalganga is developed by UOP and separates the p-xylene isomer using molecular sieves.

(e) Isomerisation

The paraxylene lean stream from adsorber (PAREX unit) is heated to reaction temperature & treated with hydrogen to convert ethyl benzene & m-xylene into p-xylene. Top product, which is a mixture of benzene, toluene & other light hydrocarbons is separated in the stabilizer sent to Gujarat refinery as gasoline blending stock & bottom product is sent back to ortho-meta-splitter.

XYLENE PRODUCTION FROM PYROLYSIS GASOLINE

Generally xylene content in pyrolysis gasoline is very low & hence only Petrochemical complexes having very large capacity &/or wanting xylene for downstream uses make xylene from pyrolysis Gasoline. In India, only RIL-Hazira which has a 7,50,000 TPA naphtha cracker makes xylene from pyrolysis gasoline.

A flowsheet of xylene production at Indian petrochemical Co. Ltd. (IPCL), Vadodara is given below:

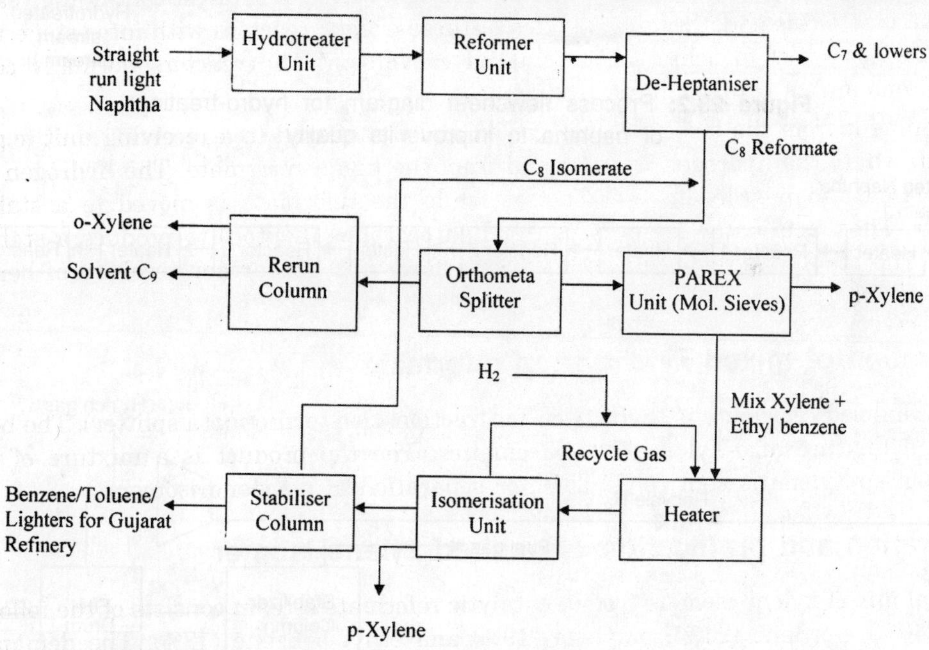

Figure 23.1: Flowsheet of Xylene Process (IPCL, Vadodara)

The pyrolysis gasoline (pygas) is a stream produced by the high temperature, short residence time cracking of naphtha in petrochemical plants. According to cracking severity, the pygas contains typically 20-40% benzene, 16-20% toluene, 10-13% higher aromatics (e.g. ethyl benzene, xylenes, styrenes etc.) with a balance of non-aromatic hydrocarbons (mainly oleflns & diolefins)

The xylene extraction from pyrolysis gasoline (pygas) is carried out in the following reaction steps:

- Hydro-treating
- Secondary hydrogenation
- Solvent extraction
- Fractionation

The details are already explained earlier.

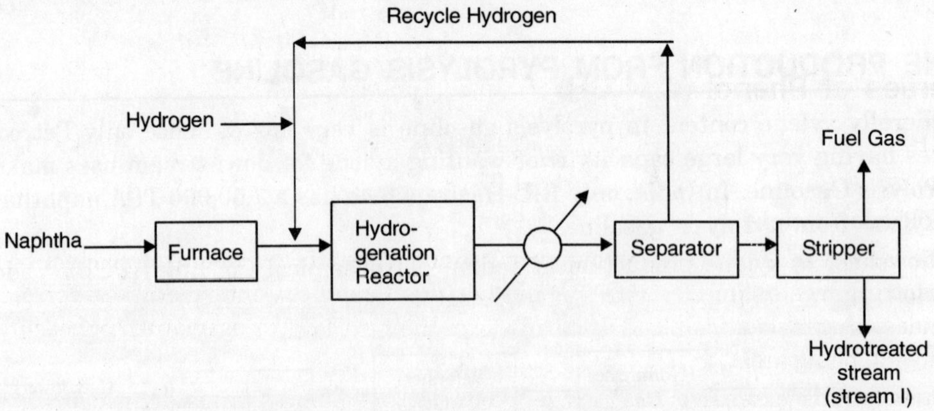

Figure 23.2: Process flowsheet diagram for hydro-treating of naphtha to improve its quality

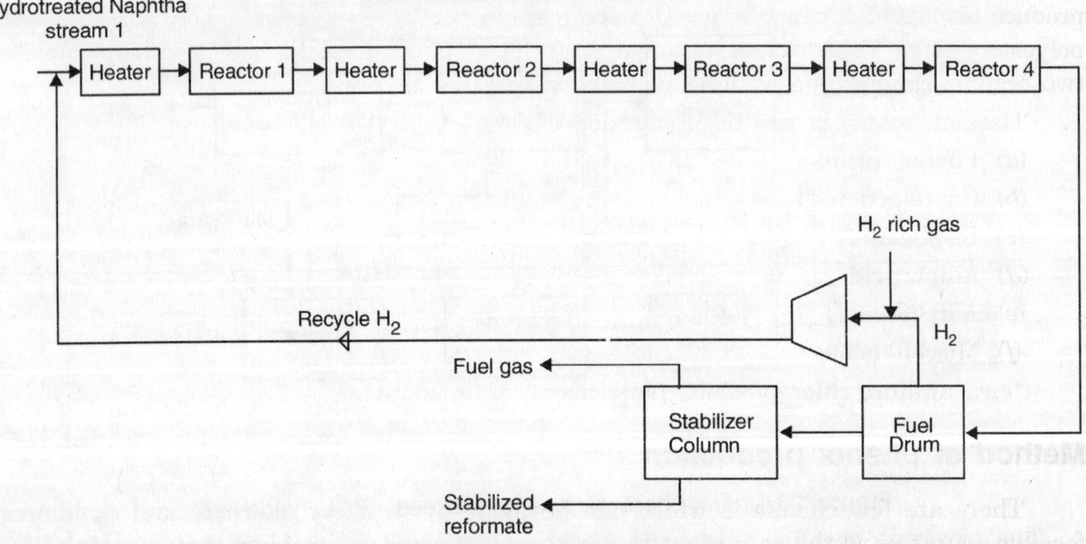

Figure 23.3: Process flowsheet diagram for naphtha reformer unit

MANUFACTURE OF PHENOL FROM CUMENE (ISOPROPYL BENZENE)

Chapter 24

Properties of Phenol (C_6H_5OH)

B.P.:	181.4°C
Sp. gr. @ 25°C	1.07
Molecular weight:	94

The largest user of phenol is the plastics industry with the chief end-product being cheap thermosetting resins made with formaldehyde. Chemical intermediates for herbicides, insecticides, pharmaceuticals and dyestuffs form the next largest usage with petroleum refining as a solvent being next.

Phenol finds extensive use in the production of phenol formaldehyde resin (bakelite) which is a thermoset resin. It is used in industrial & decorative laminates, molding powders for electrical appliances, plywood & particle board, abrasives, foundry molds, textile auxiliaries and varnishes. Both acetone and phenol are condensed in the presence of a catalyst to produce bisphenol-A which is used in the manufacture of epoxy resins, polycarbonate resins, polyester resins, rubber chemicals and fungicides. Phenol formaldehyde and bisphenol-A are two organic chemical-based intermediates with good demand.

Uses of phenol is for the production of :

(*a*) Phenol resins

(*b*) Caprolactam

(*c*) Bisphenol-A

(*d*) Adipic acid

(*e*) Alkylphenols

(*f*) Miscellaneous*

**e.g.,* aniline, chlorophenols, plasticizers, antioxidants.

Method of phenol production

There are few chemicals which can be made by so many alternate and commercially feasible routes as is the case of synthetic phenol. When one considers that less than 15% of all phenol production comes as natural phenol from coal carbonization plants, the importance of synthetic phenol is apparent.

Seven processes for synthetic phenol are available of which six are still competitive.

(a) Competitive processes

- Cumene peroxidation-hydrolysis
- Toluene two-stage oxidation
- Raschig: vapor phase hydrochlorination and hydrolysis
- Chlorobenzene-caustic hydrolysis
- Benzene sulfonate-caustic fusion
- Benzene-direct oxidation

(b) Processes not commercially feasible

- Chlorobenzene-soda ash hydrolysis

1. Cumene Process for Phenol Production

Chemical reactions involved:

(a) **Peroxidation:** (emulsification agent is NaOH)

$$C_6H_5CH(CH_3)_2 + \text{air} \xrightarrow[\text{emulsion, Normal T \& P}]{\text{aqueous}} C_6H_5C(CH_3)_2{-}O{-}OH$$

Cumene or isopropyl benzene → cumene hydroperoxide

(b) **Hydrolysis:**

$$C_6H_5C(CH_3)_2{-}O{-}OH + \text{aq } H_2SO_4\ (10-25\%) \xrightarrow[\text{1 atm.}]{60°C} C_6H_5OH + (CH_3)_2CO$$

phenol, acetone

Raw materials requirements

Basis: 1 ton of phenol (92% yield) + 0.6 ton acetone

Cumene:	1.40 tons
Air:	1.53 tons

Small quantities of sulfuric acid and emulsifying agents.

Co-product: Acetone

Process description

Cumene (isopropyl benzene) made by alkylation of benzene with propylene is mixed with recycle cumene (1 : 4 ratio) and sent to the hydrogenator. Unsaturates and methyl styrene are converted to saturated materials to avoid undesirable decomposition of the peroxide during the oxidation step. Hydrogen over nickel catalyst at 100°C in a batch reactor is the usual process for purification.

Oxidation is carried out in the presence of air in an aqueous emulsion stabilized by an alkali such as sodium hydroxide in the 8.5-10.5 pH range. Vent gases are passed through a condenser to recover hydrocarbons.

A flowsheet of phenol production from curene (isopropyl benzene) is given below:

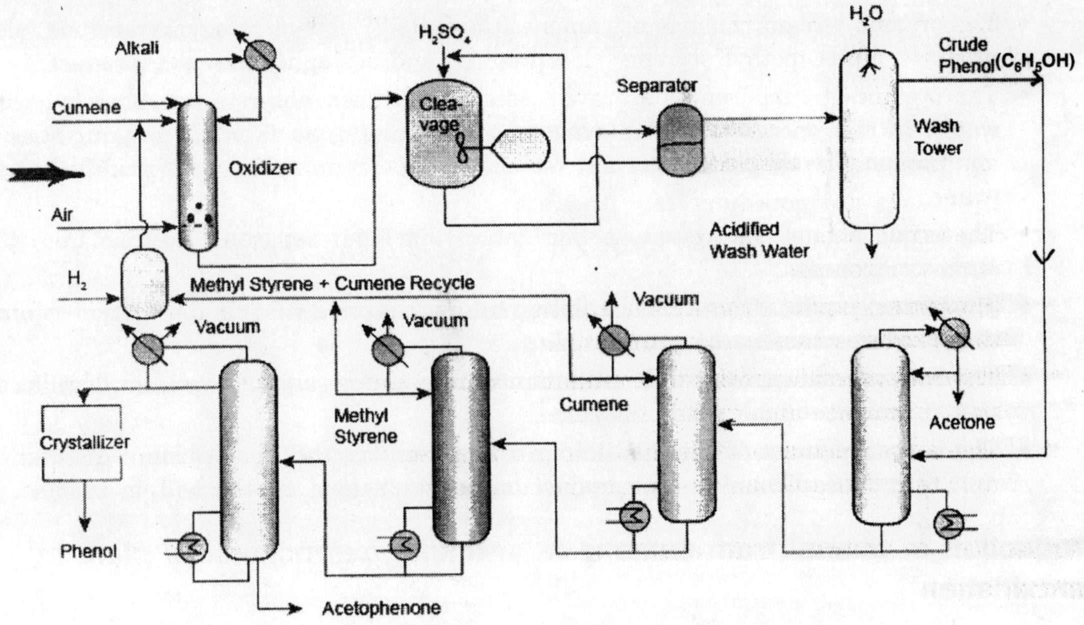

Figure 24.1: Flowsheet of Phenol production from cumene

Salient Features of process technology is listed below:

- Fresh cumene is mixed with hydrogenated unreacted cumene, comenem + alpha methyl styrene (recycle streams).
- The pre-purification step involves converting unsaturates such as alpha methyl styrene and others to saturates. This is done by using nickel catalyst at 100 °C and feeding hydrogen to the pre-purification reactor. The product from this unit then mixes with the fresh cumene.
- The fresh cumene and processed cumene and alpha methyl styrene are fed to the oxidation reactor. The oxidation reactor refers to a gas liquid reaction between air (Oxygen) and the cumene.
- An emulsion of cumene is prepared in the oxidation reactor by adding alkali to it.
- pH is maintained in the range of 8.5 - 10.5 to suit good emulsification conditions.
- After reaction, vent gases are condensed & recycled back and the product is sent to a cleavage unit.
- The cleavage unit consists of a stirrer and is fed with fresh & recycled H_2SO_4 aqueous solution to enable the hydrolysis of cumene hydroperoxide.
- The product streams from the cleavage unit enter a settler (phase separator) which

upon gravity settling yields two streams namely the acid rich aqueous stream and the phenol rich organic stream.

- The aqueous stream consists of the sulphuric acid and is sent back to the cleavage unit as a recycle stream.
- The organic stream consists of cumene (unreacted), phenol (product), acetone (side product), alpha methyl styrene (side product) and acetophenone (side product).
- The organic stream from the gravity settler unit then enters a scrubber fed with water. In this operation, water extracts the remaining acids in the organic stream and produces crude phenol stream. Water leaving the unit consists of acidified wash water.
- The crude phenol then enters a distillation unit that separates acetone from the other components.
- The bottom product from this distillation column enters a vacuum distillation column that produces cumene as a top product.
- The bottom product from the distillation column enters another vacuum distillation unit to produce alpha methyl styrene.
- The bottom product of this distillation column enters the final vacuum distillation unit to produce phenol as top product and acetophenone as the bottom product.

Production of phenol from cumene in an Indian petrochemical plant by peroxidation

Here in peroxidation process, phenol is obtained by the decomposition of cumene hydroperoxide derived from cumene (which is produced from benzene & propylene)

In the two-stage process, pure Cumene is fed into an oxidation vessel where it is mixed with a dilute solution of sodium carbonate. Air is introduced and the mixture, at a temperature of 110-130°C, is left in contact until 25-30% of the Cumene is oxidized to the hydroperoxide. If the conversion is allowed to proceed beyond this point, there is an increased risk of by-product formation. Lower operating temperatures favour the yield of hydroperoxide but conversion is lower.

In the second stage, the crude mixture is concentrated to about 80% Cumene hydroperoxide before being fed in to a cleavage reactor. The reaction is carried out at a temperature of 70-80°C and a pressure of 0.3 bar in the presence of a dilute acid, such as a 10% solution of Sulphuric acid. Both the oxidation and cleavage reactions are very exothermic. Temperature control can be effected either by the evaporation of water which is present or by using cooled resultant mixture to dilute the hydroperoxide. The major reaction products are Phenol and Acetone, with small quantities of Acetophenone, alpha Methyl styrene and Cumene.

The reaction products are separated by distillation. Acetone is flashed from the top of the first column and purified by distillation. The bottoms from the column are further distilled to remove any unreacted Cumene and alpha Methyl styrene. By- product alpha Methyl styrene can be recovered and sold or converted to Cumene by catalytic hydrogenation before being recycled to the first-stage reactor. Further distillation separates by-product Acetophenone from Phenol, which is recovered overhead. The flow diagram is depicted in Fig.24.2

Reactions:

$$\underset{\text{cumene}}{C_6H_5CH(CH_3)_2} + O_2 \longrightarrow \underset{\text{cumene hydroperoxide}}{C_6H_5C(CH_3)_2OOH}$$

$$C_6H_5C(CH_3)_2OOH \longrightarrow \underset{\text{Phenol}}{C_6H_5OH} + \underset{\text{Acetone}}{CH_3COCH_3}$$

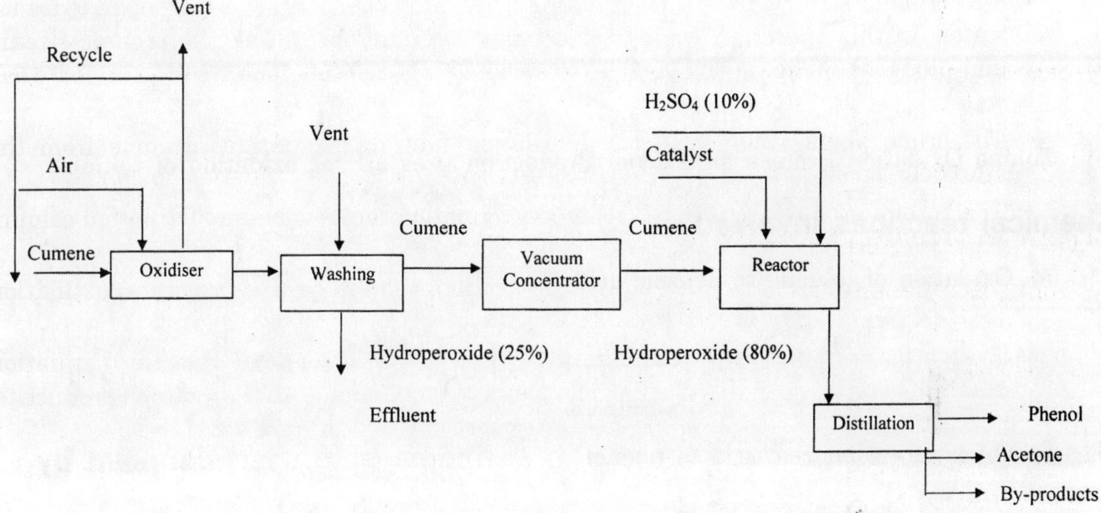

Figure 24.2: Flowsheet of diagram for phenol/Acetone production by peroxidation cumene

MANUFACTURE OF PHENOL FROM TOLUENE OXIDATION

Chapter **25**

Tolune Oxidation Process for Phenol Production uses air for oxidation of tolune.

Chemical reactions involved

(a) Oxidation of toluene to benzoic acid

$$\underset{\text{Toluene}}{C_6H_5CH_3} + 1\text{-}1/2O_2 \xrightarrow[\text{cobalt naphthenate}]{150^\circ C} \underset{\text{Benzoic acid}}{C_6H_5COOH} + H_2O$$

(b) Oxidation of benzoic acid to phenol

$$\underset{\text{Benzoic acid}}{C_6H_5COOH} + 1/2O_2 \xrightarrow[\text{cupric benzoate}]{220^\circ C} \underset{\text{phenol}}{C_6H_5OH} + CO_2$$

Raw materials requirements

Basis: 1 ton of phenol (80% yield)

Toluene: 1.34 tons

Air: 8.8 tons

Small quantities of cobalt naphthenate and cupric benzoate catalyst

Process description

A two-stage air oxidation process is used. In the first stage, fresh plus recycle toluene are mixed with a small quantity of cobalt naphthenate catalyst and charged to the reactor which is a liquid-filled tower through which air is sparged. Cooling tubes are provided to remove the exothermic heat of reaction.

The reactor is run at 150°C and 3 atms. Excess air is used, but toluene conversion is limited to 40% to avoid excessive side reactions. These give by-products such as benzaldehyde, benzyl alcohol, benzyl benzoate, CO and CO_2. With conversion of toluene at 40%, the ultimate yield of benzoic acid is about 90%.

Off-gases from the reactor are vented through a water-cooled condenser to remove water and to allow return of toluene. Liquid from the reactor continuously passes to a distillation column which strips the toluene and other volatile by-products from the acid fraction in the bottoms. Purified benzoic acid is separated by extracting the bottoms with hot water, then crystallizing and filtering the crude benzoic acid. The latter can be recrystallized to meet specifications as a market outlet for benzoic acid.

To make phenol, the crude acid is melted, mixed with cupric benzoate catalyst, then charged to an air-sparged tower containing cooling tubes and mechanical agitation. Reactor conditions are 220°C and 1.3-1.7 atms. Excess air is again necessary to get a 70-80% conversion of benzoic acid with a yield of 90% phenol. The overall process yield for the two steps is about 80%.

Phenol product is obtained by continuously distilling the reactor liquor into a fractionating column where unreacted benzoic acid is returned to the reactor. Non-condensibles such as N_2, O_2 and CO_2 are vented through a condenser along with the condensible fraction of phenol-water. Phenol is withdrawn as the bottom layer in a separator. This crude phenol is again fractionated with purified phenol coming off as bottoms and the overhead phenol-water azeotrope sent to another column for splitting.

The heavy ends in the benzoic acid oxidation tower are water-extracted to recover phenol and benzoic acid which are then recycled, after concentration, to the second stage oxidation tower.

A flowsheet of phenol manufacture from toluene oxidation is shown below:

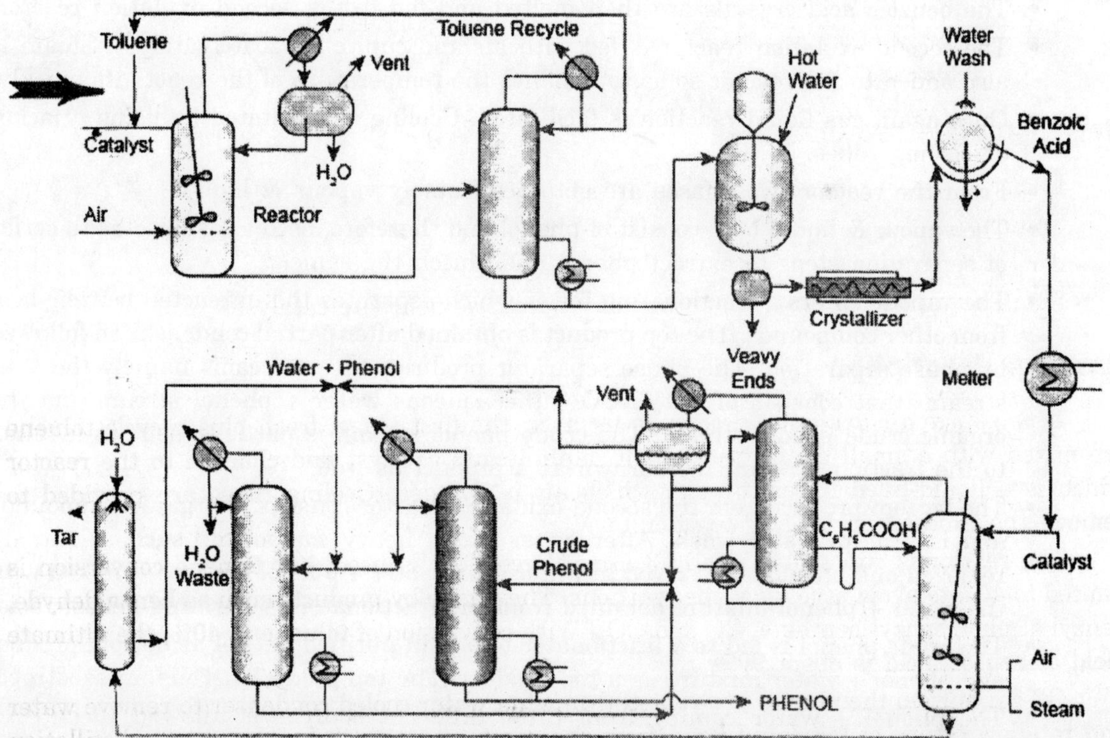

Figure 25.1: Flowsheet of Phenol manufacture from Toluene Oxidation

Salient features of process is listed below:

- Fresh and recycled toluene first enter the oxidation reactor.
- The oxidation reactor is a gas-liquid reactor in which air is sparged through a column of liquid (toluene).
- To reduce side reactions, toluene is used as a limiting reactant.
- The catalyst is fed to the reactor along with the toluene stream.
- After reaction, several products are formed including undesired impurities such as benzaldehyde, benzyl alcohol, benzyl benzoate, CO and CO_2.
- The reactor is facilitated with vent after cooling and partial condensation of the air mixed hydrocarbon vapour stream.

  The liquid product from the reactor then enters a distillation column which separates toluene from the other organics.
- The organics then enters a hot water wash unit where hot water extracts benzoic acid. All other impurities don't dissolve in water and leave the wash unit as heavy ends.
- The benzoic acid rich hot water stream is sent to a crystallizer where benzoic acid crystals are formed.
- The benzoic acid crystals are once again washed with water to obtain benzoic acid crystals.
- The benzoic acid crystals are then melted and fed to the second oxidation reactor.
- The second oxidation reactor is fed with air and cupric benzoate catalyst. Steam is also added to the reactor so as to enhance the temperature of the reactants quickly.
- Once again gas liquid reaction is facilitated. Cooling is facilitated by using a jacket or cooling tubes.
- From the reactor two phases are obtained namely vapour & liquid.
- The vapour & liquid both consist of phenol and therefore, both are subjected to series of separation steps to extract phenol and enrich the same.
- The vapour enters a fractionating tower which separates the unreacted benzoic acid from other compounds. The top product is obtained after partial condensation followed by phase separation. The phase separator produces three streams namely the vent stream (that consists of CO & CO_2), the aqueous water + phenol stream and the organic crude phenol stream. The crude phenol stream is partially sent as a reflux to the fractionators and withdrawn as a product as well.
- The bottom product from the second oxidation reactor consists of organic compounds and is sent to water wash. After water wash, heavy compounds such as tar are removed and phenol is dissolved in water. This water + phenol mixture is mixed with the water + phenol mixture obtained from the fractionators separating benzoic acid.
- The crude phenol is fed to a fractionator to obtain purified phenol as bottom product and phenol + water mixture as a top product. The top product here is an azeotrope.
- The phenol + water coming from three different sources is allowed to enter a distillation column that generates water waste as a top product and a crude phenol stream as a bottom product.

MANUFACTURE OF PHENOL FROM HYDROCHLORINATION OF BENZENE

Chapter **26**

Phenol can be manufactured from Benzene using several ways as given below:

- Benzene hydrochlorination to form Benzyl chloride followed by hydrolysis of benzyl chloride to form phenol.
- Benzene chlorination to form benzyl chloride which is transformed to sodium benzoate and eventually to phenol using NaOH and HCI.
- Benzene sulfonate process: In this process, benzene is convered to benzene sulfonate using sulphuric acid and eventually through neutralization, fusion and acidification, the benzene sulfonate is gradually transformed to phenol.
- However manufacture of phenol from following two processes are importent:
  - Benzene hydrochlorination route.
  - Benzene from chlorobenzene route.

Manufacture of Phenol using Hydro Chlorination Route of Benzene

Chemial Reactions involved:

(a) **Hydrochlorination:**

– Benzene + HCI + Oxygen ⟶ Benzyl chloride + Water

$$C_6H_6 + HCl + \frac{1}{2}O_2 \longrightarrow C_3H_5Cl + H_2O$$

– Catalyst: $FeCl_3 + CuCl_2$

– Operating conditions: 240°C and atmospheric pressure

(b) **hydrolysis of benzyl chloride:**

– Benzyl chloride + water ⟶ Phenol + HCI

$$C_6H_5Cl + H_2O \longrightarrow C_3H_5OH + HCl \text{ (recycled)}$$

– Catalyst: SiO_2

– Here, HCI is regenerated and is recycled.

– Operating conditions: 350°C and atmospheric pressure

Raw materials requirements:

Basis: 1 ton of phenol (75% yield)

Benzene: 1.1 tons; HCl : 0.19 ton; Air: 2.4 tons

Process Description

Raschig process of hydrochlorination of benzene to produce phenol has two vapor phase Catalytic stages. Purified benzene is fed to a heated, packed reactor containing ferric chloride & cupric chloride catalyst. Chlorination with HCl – O_2 at 240°C occurs to produce 10 - 20 & benzene. Fractionation separates unreacted benzene from chlorobezene and polychlorobenzene.

Crude polychlorobenzene is scribbed with phenol, water washed and sent to the second catolytic stage, where it is hydrolysed in a tubular furnace with either SiO_2 or $Ca_3\ (PO_4)_2$ as the catalyst. phenol from hydrolyser is washed with water, then exlrocted by benzene and finally purified by distillation. HCl vapor from hydrolys is recycled to the hydrochlorination stage.

A Flowsheet of manufacture of phenol using hydrochlorination of benzene is given below:

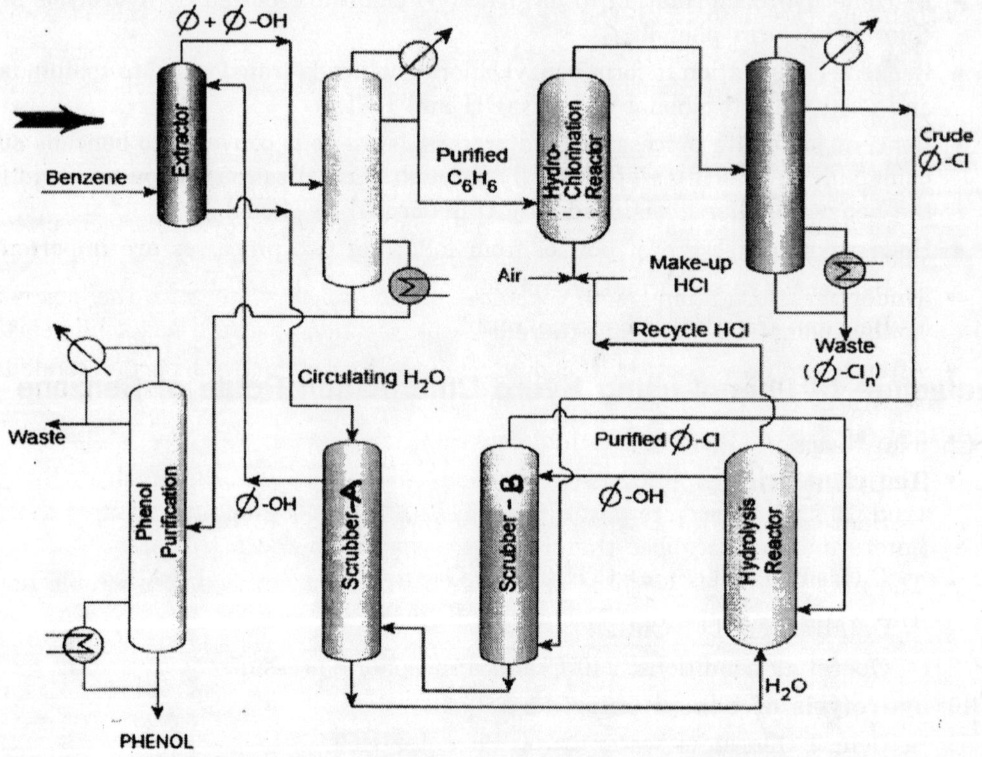

Figure 26.1: Flowsheet of manufacture of phenol using hydrochlorination of benzene

Salient Features of process technology is listed below

- In this process, Benzene is used to extract phenol from phenol +water mixture. This unit is termed as an extraction unit (liquid liquid extraction principle). Therefore, this unit takes up fresh benzene and phenol + water mixture and produces two streams namely water stream (bottom product) and benzene + phenol stream (top product). The water stream is fed to a scrubber unit.

- Then onwards, the organic mixture is fed to a distillation column that produces purer benzene as the top product. The bottom product is phenol with other impurities.
- The bottom phenol rich product is sent to the phenol fractionator to obtain waste product as top product and pure phenol as bottom product.
- The purer benzene then enters the hydrochlorination reactor in which a mixture of HCL & O_2 is fed at 240°C. Under these conditions, Benzene will be also in vapour state.
- Therefore, the reactor is a gas solid reactor.
- The conversions are pretty low and not more than 20% of the benzene is converted to benzyl chloride.
- Eventually, the products are sent to two fractionators that separate unreacted benzene, crude benzyl chloride and poly benzyl chlorides as various products. The unreacted benzene is sent back to the hydrochlorination reactor as a recycle stream.
- The crude benzyl chloride then enters an absorber unit a where phenol is used to purify the benzyl chloride from other organic compounds (such as benzene and polybenzyl chlorides).
- The purified benzyl chloride stream then enters the hydrolysis reactor in which water is passed along with benzyl chloride over the silica catalyst. The reactor itself is a furnace with catalyst loaded in the tubes and hot fuel gases are circulated in the shell to obtain the desired higher temperature.
- Under these conditions, both reactants are in vapour state (with the benzyl chloride boiling point of 179°C) and therefore, the reaction is also a gas solid reaction.
- After hydrolysis reaction, the product vapors are sent to a partial condenser that separates the HCl from the organic phase.
- The HCl is recycled to the hydrochlorination reactor.
- The phenol rich product stream is sent as a solvent for the scrubber that purifies crude benzyl chloride to purer benzyl chloride. The bottom product from the scrubber enters another scrubber that receives water from the extractor.
- The unit B enables washing of the phenol to remove any water soluble impurities. The water from the unit B enters the hydrolysis reactor.

MANUFACTURE OF PHENOL FROM CHLOROBENZENE

Chapter **27**

CHEMICAL REACTIONS INVOLVED

There are three reactions to convert benzene to phenol using chlorination route:

(a) **Chlorination:**

- Benzene + Cl_2 ⟶ monochloro benzene $\left[C_6H_6 + Cl_2 \longrightarrow C_6H_5Cl\right]$
- Operating temperature: 85°C
- Catalyst: Fe or $FeCl_3$ catalyst

(b) **Causticization**

- Benzyl chloride + NaOH ⟶ sodium benzoate

  $\left[C_6H_5Cl + NaOH(aq) \longrightarrow C_6H_5ONa\right]$
- NaOH is in aqueous media
- Operating conditions: 425°C and 350 atms
- Exothermic reaction

(c) **Hydrolysis:**

- Sodium benzoate + HCl (aq) ⟶ Phenol + NaCl (aq)

  $\left[C_6H_5ONa + HCl\,(aq) \longrightarrow C_6H_5OH + NaCl\right]$
- Operating conditions: Nothing specific

Raw materials requirements

Basis: 1 ton of phenol (95% yield)

Benzene:	0.90 ton
chlorine:	0.82 ton
NaOH (100% basis):	0.67 ton
HCl:	0.51 ton

Process description

Dry benzene and a catalyst of iron turnings or anhydrous ferric chloride are charged continuously into a chlorinator. The partially chlorinated mixture boils up into a fractionating column. Benzene is fractionated from the top and returns as recycle while monochlorobenzene is withdrawn near the bottom plate of the column.

Chlorobenzene and dilute caustic soda (10% solution) are mixed in a pump in a mole ratio of 1 : 1.25. Diphenyl oxide (about 10 wt.%) is added to repress the formation of more diphenyl oxide and the mixture is pumped through a preheater, then to a multitube reactor where causticization occurs at 425°C and 350 atms. Residence time is around 15 minutes. Heat is removed from the reactor efflux by exchange in the feed preheater. The cooled hydrolyzate is acidified in a neutralizer to liberate phenol and sodium chloride which must be separated by distillation. Salt is recycled to an electrolysis unit to produce the needed chlorine and caustic.

Phenol is stripped of diphenyl oxide and other minor by-products in a vacuum. A portion of the diphenyl oxide is returned to causticization tower with the remainder redistilled under vacuum for commercial marketing of this plus other derivatives.

A flowsheet of manufacture of phenol from chlorobenzene is shown below:

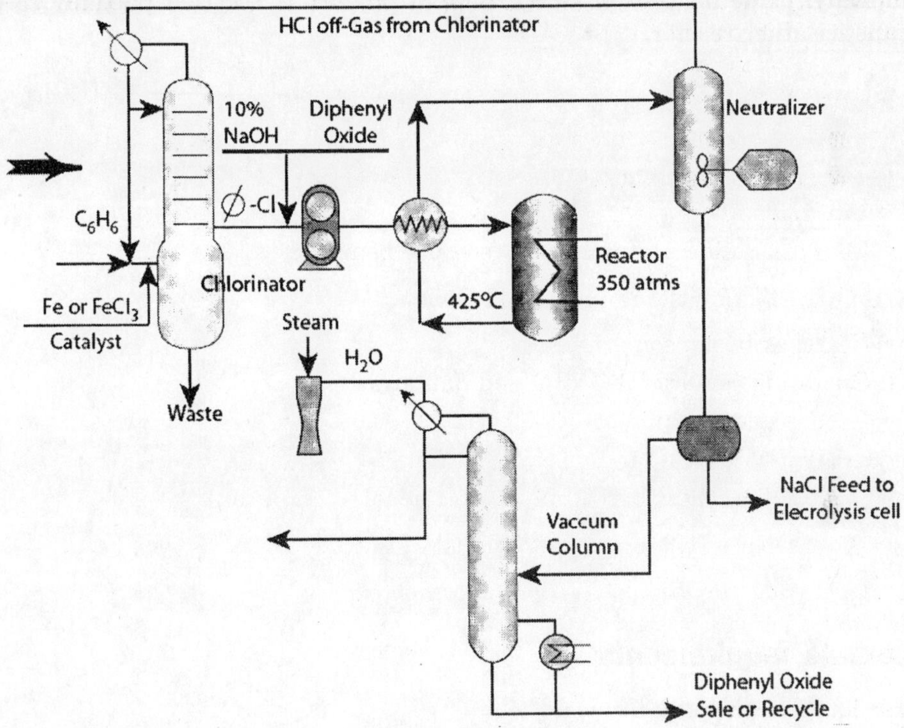

Figure 27.1: Flowsheet of manufacture of phenol from chlorobenzene

Salient Features of process is listed below:

- Benzene is first dried. Dry benzene and $FeCl_3$ catalyst enters the chlorination reactor.
- After the solid-gas-liquid reaction, the products are further heated up to enter a fractionator.
- The fractionator separates benzene as a top product and monochloro benzene as the bottom product. The top product also consists of HCl off gas that is obtained as the vapour stream from the partial condenser.

- The benzene is further cooled and sent back to the reactor and also as a reflux to the fractionator.
- The bottom product monochloro benzene is mixed with 10 % NaOH solution and diphenyl oxide to enter a high pressure pump followed with heat integrated heat exchanger that pre-heats the feed to higher temperature.
- The feed then enters the causticization reactor which has cooling water tubes to control the temperature.
- The product stream is cooled using heat integrated exchanger and then enters a neutralizer that is fed with the HC1 obtained from the fractionator partial condenser.
- After neutralization, the product phenol is separated from the aqueous phase using gravity settling principle.
- Organic layer rich in phenol is sent to a vacuum column to separate phenol from diphenyl oxide (bottom product). Bottom product is recycled partially to enter the causticisation reactor.

Chapter 28

PRODUCTION OF STYRENE FROM BENZENE

Styrene is produced from benzene via the ethylbenzene route followed by its dehydrogenation.

Reactions involved

(a) Alkylation of Benzene:

- Benzene + ethylene ⟶ Ethyl benzene
- Catalyst: $AlCl_3$ granules
- C_2H_5Cl provides hydrogen and chlorine free radicals
- Operating conditions: 95°C and 1 atm pressure
- Reaction is exothermic

(b) Dehydrogenation of ethylbenzene:

- Ethylbenzene ⟶ Styrene + Hydrogen
- Reaction is endothermic
- Catalyst: SnO or FeO
- Operating conditions: 800°C

Properties of styrene ($C_6H_5 \cdot CH \cdot CH_2$)

Moleculer weight:		104
M.P.:		—30.6°C
B.P.:		145.2°C
Sp. gr. (at 20°C)		0.903
Explosive limit:	Lower =	1.1 volume% in air
	Upper =	6.1 volume% in air
Toxicity limit:		400 ppm

The impetus to growth of styrene production was the synthetic rubber requirements of World War II. The copolymerized styrene-butadiene rubber (SBR) is developed for such process. Demands for styrene in the elastomer field continue to rise. Polystyrene molding resin in solidified form and foamed for insulation is of greater importance.

Styrene Use is shown below

Styrene is used for the production of:

(a) Polystyrene
(b) Styrene-butadiene rubber (SBR)
(c) Acrylonitrilc-butadiene-styrene (ABS)
(d) Styrene-acrylonitrile (SAN)
(e) Polyester resins (unsaturated)

Methods of Production of_styrene:

- Dehydrogenation of ethylbenzene.
- Hydrogenation-dehydration of acetophenone
- Dehydrogenation of Ethylbenzene is the preferred method of styrene production which is described below.

Raw materials requirements

Basis: 1 ton of styrene (86% yield)

Benzene:	0.87 ton
Ethylene:	0.32 ton
$AlCl_{3:}$	10 -11 kg

Process description of styrene producton from benzene

Benzene is alkylated with ethylene by Friedel-Craft method using aluminum chloride or other acid-type catalyst. The resulting ethylbenzene is catalytically dehydrogenated in steam or excess benzene atmosphere to yield styrene.

Recent refinery operation has accomplished the separation of ethylbenzene (with a 2.1°C lower boiling point than *p*-xylene) from the xylene fractions in petroleum reformates. The recovery of this high purity ethylbenzene overhead fraction avoids alkylation procedures. The bulk of the plants, however, use the latter since ethylene and benzene are more plentiful.

The alkylation operation must be done under very dry conditions with high purity feedstock to avoid activity loss of the catalyst. All benzene feed must be dried by azeotropic distillation. Ethyl chloride, which serves ultimately as a source of hydrogen and chlorine free radicals for catalyst promotion, is added to ethylene which is fed continuously along with benzene to the alkylation tower operating at 95°C and 1 atm. pressure. Granular $AlCl_3$ catalyst is fed continuously at the top of the reactor. The reactant mole ratio is 0.6 C_2H_2/1.0 C_6H_6 and no ethylene is recycled.

The alkylation tower is water-cooled to control the exothermic reaction. The alkylated products are pumped to a cooler at 40°C where the aluminum chloride complex is separated and a split stream is fed to the alkylator. The bleed-off fraction is pumped to a high temperature (200°C) dealkylator to break down the polyethylbenzenes to benzene and ethylbenzene which are returned to the system. The tar residue plus aluminum chloride is water-extracted to recover 80-85% of $AlCl_3$.

Crude acidic ethylbenzene from the coolers is neutralized with 50% NaOH solution, stripped to remove polyethylbenzenes and the overhead sent to the benzene column which separates wet benzene from ethylbenzene. A final distillation followed by 20% caustic wash, then drying by percolation in a flake caustic bed, produces 99% ethylbenzene for dehydrogenation in yields averaging 95%.

The polyalkylbenzenes must be removed as bottoms in the atmospheric stripper to avoid formation of polyfunctional, highly reactive products during the dehydrogenation step. The stripper bottoms fraction is distilled at 50 mm Hg pressure with the overhead combined with bottoms from the ethylbenzene column and returned to the alkylation unit for dealkylation of these higher molecular weight compounds. The highest molecular weight polybenzenes as bottoms from the vacuum still can only be split in the high temperature dealkylator.

Dehydrogenation of ethylbenzene produces styrene. Steam is injected with the hydrocarbon in a mole ratio of 15 H_2O/1 $C_6K_5C_2H_5$. The mixed feed passes through the preheater to achieve an input temperature of 500°C to the reactor which operates at 800°C. The dehydrogenation catalyst is promoted zinc, chromium, iron, or MgO on activated carbon, alumina, or bauxite. Conversion is 35-40% with an average yield of 91 %.

Reaction product is cooled in the feed preheater, then by steam quenching. A water-cooled condenser and separator operation yields crude styrene-ethylbenzene with impurities of 1 % toluene, 0.7% benzene, and 0.3% tar. Sulfur stabilizer is added at this point and the hydrocarbon mixture is passed through a series of vacuum distillation columns to allow the separation of impurities at low enough temperature to avoid styrene polymerization. Benzene and toluene are removed at 160 mm Hg and a reboiler temperature of 90°C. Temperatures higher than this cause autopolymerization of benzene. The second column at 35 mm Hg and 90°C reboiler temperature separates styrene from ethylbenzene. The styrene, stabilized at this point with *p*-tertiary catechol, is vacuum distilled again to remove traces of tar and sulfur. The final overhead (99.8% styrene) is refrigerated at 10°C or less and either stored in insulated tanks or used immediately for polymerization.

Salient Features of process technology are listed below

- The process consists of two separate reactor-separator-recycle networks in which one corresponds to the ethylbenzene flow sheet and the other corresponds tos tyrene flowsheet.
- In the flowsheet for ethylbenzene production we find that Benzene (wet) is sent first to an azeotropic distillation unit that separates water and produces dry Benzene. Dry Benzene is required so as to avoid unnecessary reactions in the alkylation reactor as well as damage to the catalyst as alumina can get formed.
- Dry Benzene + Ethylene + Ethyl chloride + $AlCl_3$ enter the alkylator catalyst.
- The reactor could be a jacketed tower where water is used as a cooling fluid in the jacket to control the reactor temperature.
- The reactor produces two products namely uncondensible gases and the liquid product in which $AlCl_3$ complex is available. This complex needs to be regenerated and sent back to the alkylator.
- The alkylator product is sent to a cooler which upon cooling to 40°C separates the

A Flowsheet of styrene production from benzene is given below in figure.

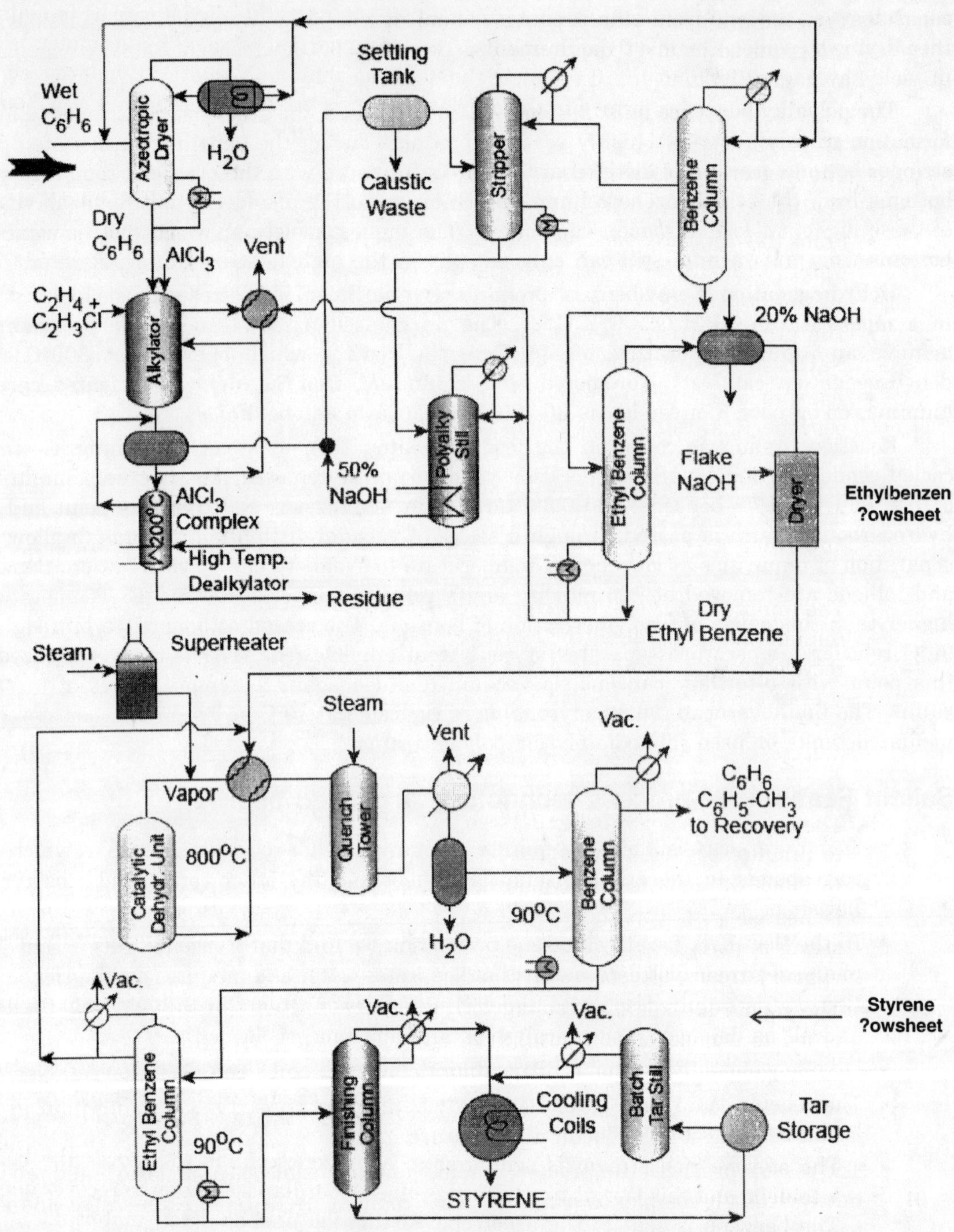

Figure 28.1: Flowsheet of manufacture of styrene from benzene

aluminium chloride complex stream from the product stream. The other stream from the cooler is the ethylbenzene rich product stream.

- The aluminium chloride stream is partially recycled to the alkylator so as to maintain the required catalyst requirements. The other portion of the $AlCl_3$ complex is sent to a dealkylator unit in which the feed is heated to 200°C. By doing so, the polyethylbenzenes formed in the alkylator are converted to benzene and ethylbenzene (cracking reaction).
- The benzene and ethylbenzene are returned to the cooler.
- The delkylator produces a residue product consisting of tar + $AlCl_3$ mixture.
- From this mixture, $AlCl_3$ is recovered using water extraction as $AlCl_3$ is soluble in water. From there $AlCl_3$ is recovered from the water and returned back to the alkylation reactor.
- The product stream from the cooler consisting of ethylbenzene is mixed with 50% NaOH to remove acidic impurities. Eventually, after settling waste is eliminated.
- The purified ethylbenzene then enters a stripper that separates ethylbenzene + benzene from the polyalkylbenzenes. The polyalkylbenzenes are sent to a polyalkyl still that separates the benzene + ethylbenzenes from the polyalkylbenzenes (bottom product). The polyalkyl still is operated under vacuum. The polyalkylbenzenes are fed to the dealkylator and the benzene + ethylbenzene rich stream is sent to a heat integrated exchanger that extracts heat from the vent gases and then eventually enters the alkylation reactor.
- The top product from the stripper is ethylbenzene + benzene and it enters a benzene column that separates wet benzene from crude ethylbenzene. The wet benzene is recycled to the azeotropic dryer where it is mixed with fresh wet benzene to enter the azeotropic dryer.
- The crude ethylbenzene is further purified in a fractionator where the bottom product (with benzene) is mixed with the top product of the polyalkyl still. Thereby, the stream enters the heat integrated exchanger.
- The ethylbenzene is further subjected to caustic wash and finally it is sent to a dryer to produce dry ethylbenzene.
- In dehydrogenation of ethylbenzene flowsheet, we find that, the ethylbenzene (dry) is heated with superheated steam to enter the catalytic dehydrogenator. Excess steam is used in this process. The feed pre-heating is carried out using the product vapour stream. The reaction is gas phase catalytic reaction.
- The vapour stream after cooling with the feed stream in a heat integrated exchanger is fed to a quench tower using steam quenching.
- After quenching, partial condensation of the quenched vapors produces three streams one being the vapour vent, the other being water and the third being the organic phase rich with styrene.
- The styrene rich stream is sent first to a benzene column to recover the benzene + toluene and this is sent to a benzene-toluene distillation column to recover benzene. The benzene is sent to the azeotropic distillation unit as a raw-material.
- The bottom product from the benzene column enters an ethylbenzene column which

separates ethylbenzene from the styrene stream. The ethylbenzene stream is mixed with the dry ethylbenzene to enter the catalytic dehydrogenator.

- The bottom product from the ethylbenzene column is the styrene enriched stream and this is sent to the finishing column where styrene is further purified from unwanted impurities such as tar. The tar is further batch distilled to recover styrene from the tar. The styrene finishing column also produces styrene product. Both styrene products from batch still and styrene finishing column are mixed and cooled to store as styrene product.
- All three columns namely benzene, ethylbenzene and finishing columns are operated under vacuum.

Styrene is mainly used for making plastic toys and model kits. Moreover, housing for machines as well as refrigerator doors and air conditioner cases are made of polystyrene (a polymer of styrene).

PRODUCTION OF PHTHALIC ANHYDRIDE FROM NAPHTHALENE OR O-XYLENE

Chapter

29

Properties of Phthalic Anhydride,

Mol. wt.:	148.1
M.P.:	130.6°C
B.P.:	284.5 °C
Sp. gr. @ 4°C:	1.53

Slightly soluble in hot water and ether; sublimes below the melting point.

Ignition temperature: 588°C

Major outlets for phthalic anhydride are in alkyd resins and plasticizers. The alkyd industry has been given a boost by the development of solution and emulsion types of alkyd coatings. Phthalate derivatives continue as a standard plasticizer in many resin formulations, particularly vinyl resins.

METHOD OF PRODUCTION

Oxidation of Naphthalene and/or O-xylene

Chemical reactions:

(a) **Naphthalene oxidation**

$$\text{Naphthalene } (C_{10}H_8) + 4\tfrac{1}{2}O_2 \text{ (air)} \xrightarrow[400°C]{V_2O_5} \text{Phthalic ahydride (PA)} + 2CO_2 + 2H_2O\,(g)$$

(b) **o-xylene oxidation:**

$$\text{o-xylene} + 3O_2 \text{ (air)} \xrightarrow[480\text{-}600°C]{V_2O_5} \text{PA} + 3H_2O$$

(c) Major side reaction—complete combustion (15-20%)

- $C_{10}H_8 + 12O_2$ (air) $\longrightarrow$ $10CO_2 + 4H_2O$
- $C_8H_{10} + 10^1/_2O_2$ (air) $\longrightarrow$ $8CO_2 + 5H_2O$

(d) Minor side reaction (3-5% of yield)

- $C_{10}H_8 + 9O_2$ (air) $\longrightarrow$ $C_4H_5O_3$ (maleic anhydride) + $6CO_2 + 3H_2O$
- $C_8H_{10} + 7^1/_2O_2$ (air) $\longrightarrow$ $C_4H_2O_3$ (maleic anhydride) + $4CO_2 + 4H_2O$

All side reactions are much more endothermic than the principal reaction and a high degree of temperature control is needed to optimize yields.

Raw materials requirements

Basis:	1 ton of phthalic anhydride (75% yield)
Naphthalene:	1.15 tons
Air:	22.6 tons
o-xylene:	0.96 ton
Air:	20.1 tons

Process description

Naphthalene, *o*-xylene, or mixtures thereof are melted and vaporized by bubbling primary preheated air through the molten material. Secondary air to make up the 22 kg air/kg feed requirement is added downstream and the reacting mixture is passed through a preheater to a fixed bed converter. This is a multitubular catalytic reactor filled with supported V_2O_5. The exothermic heat of reaction is removed from the outside of the tubes by a molten salt circulating system. Conditions for oxidation within the converter are 380° - 450°C.

Reacted effluent gases are passed through a waste-heat steam boiler and thence to a preheater where the final temperature is dropped to a few degrees above the dew point of phthalic anhydride. Condensation of the solid product can be effected by pairs of switching fin-tube condensers, a cyclone, a fluidized bed condenser, or a water scrubber. The crude solid product is melted, distilled under vacuum, and condensed with a flaking or pelletizing operation completing the production scheme. The major impurity is maleic anhydride in 0.2-0.4% quantities.

Maleic anhydride and an easily converted derivative, fumaric acid, can be recovered in 3-5% yield from the off-gas scrubbing liquor of the crude phthalic anhydride condensers.

A process modification which is competitive is the substitution of a fluid bed catalytic reactor for the fixed bed multitubular reactor. Conditions at the head end are somewhat different. The molten naphthalene and/or *o*-xylene is sprayed directly into the air-fluidized catalytic reactor which operates at 600°C. The reacted gases rise upwards through the dense phase into a wider calming-settling zone and thence through the internal cyclone to dust filters. Condensation is the same as above, but no vacuum distillation is required.

A comparison of reactor conditions for the two cases shows that the fluid bed process operates at 140-220°C higher, a longer contact time of 10 sec versus 0.1 to 0.4 sec, a lower air-feed ratio of 15 kg air/kg feed. The yields are equivalent (75-80%) and almost no maleic acid is produced, thus giving a very high purity phthalic anhydride product.

A flowsheet of manufacture of phthalic anhydride (PA) by oxidation of naphthalene or o-xylene is given below:

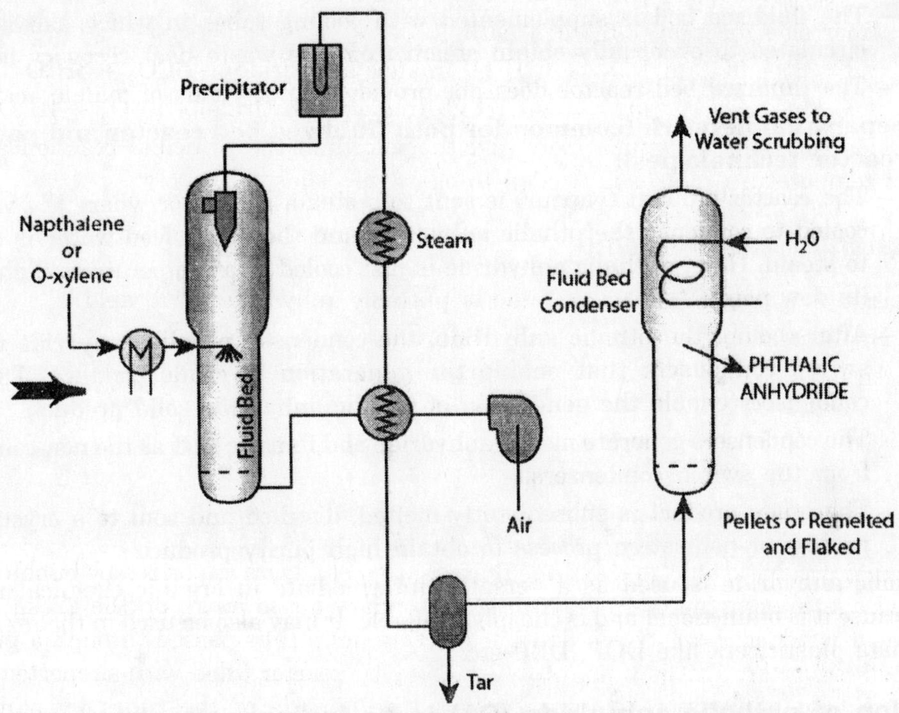

Figure 29.1: Flowsheet of manufacture of phthalic anhydride by oxidation of naphthalene or o-xylene

Salient features of process technology is listed below

- The usage of both multitiubular reactor or fluidized bed reactor for the production of pthalic anhydride is practised.
- While multitubular reactor is demonstrated for o-xylene, the fluidized bed reactor is demonstrated for naphthalene oxidation.
- **Ortho xylene to Phthalic anhydride (Fixed bed reactor):**
  - Air is filtered and compressed to enter a air pre-heater.
  - O-xylene is also separately vaporized to generate the o-xylene vapour.
  - The o-xylene and compressed air enter a multitubular reactor. The reactor design is essentially a shell and tube set up where the catalyst is held in the tubes.
  - Salt solution is circulated to cool the reactor in the shell side. The salt solution that gets heated up is sent to a waste heat recovery boiler to generate process steam.
- **Napthalene to phthalic anhydride (Fludized bed reactor):**
  - Napthalene and air are allowed to enter the fluidized bed. Napthalene in molten form is sprayed on to the fluidized bed of catalyst in air.

– The operating temperature is about 600 °C in this case.
– Eventually, the vapors (+ catalyst dust) enter a cyclone unit to separate the catalyst particles and the vapors from the cyclone separator.
– The fluidized bed is supplemented with cooling tubes in which salt solution is circulated to eventually obtain steam from the waste heat recovery boiler.
– The fluidized bed reactor does not provide higher yields of maleic acid.

- **Separation network (common for both fluidized bed reactor and packed bed reactor technologies):**
  – The reactor effluent (vapour) is sent to a steam generator where the vapors are cooled to condense the pthalic anhydride and the boiler feed water is converted to steam. Here, phthalic anhydride is just cooled to a temperature slightly above its dew point.
  – After cooling the pthalic anhydride, the condensed pthalic anhydride is sent to switch condensers that enable the generation of crude product. The switch condensers enable the generation of pthalic anhydride solid product.
  – The condensers generate maleic anhydride and fumaric acid as the non-condensibles from the switch condensers.
  – The crude product is subsequently melted, distilled and sent to a crystallization process or pellization process to obtain high purity product.

Phthalic anhydride is used as a versatile intermediate in organic chemical reactions, mainly because it is bifunctional and is cheaply available. It may also be used in the manufacture of phathalate plasticizers like DOP, DEP etc.

Production of phthalic anhydride (PA) in an Indian petrochemical plant

Phthalic anhydride is manufactured by partial oxidation of o-xylene with air in presence of catalyst. Air from process air blower is preheated to 180°C and feed stock o-xylene to 130°C and both are intimately mixed at certain proportion. The air, o-xylene vapour mixture is highly explosive in nature. This mixture is fed to top of fixed bed catalytic reactor where steady catalyst temperature of about 380°C is maintained. Under this condition partial oxidation of o-xylene to phthalic anhydride takes places with evolution of large quantity of heat. The heat of reaction is utilised for captive power generation and also for process heating.

The hot product gas coming out of reactor at around 380°C containing PA & other by-products and excess air is cooled in gas cooler to around 160°C. The gases from gas cooler are fed to oil cooled finned tube type switch condenser. These operate as alternate cooling-heating cycles, whereby during the cold cycle, PA desublimates and deposits on fins. During the heating cycle, hot oil is passed to melt and remove the PA. The gases from switch condenser are subjected to extensive scrubbing with water spray with re-circulation and finally let out through chimney. The PA is subjected to heat treatment in pre-treatment vessel under atmospheric conditions in continuous distillation unit. The preheated PA is led to fore running column, which operates, under vacuum and low cuts are removed from top.

PA is continuously fed to final distillation column, where pure PA is taken as top product. Pure PA is stored is sludge form is a rotating drum, which is cooled by circulating water. PA solid are taken off and bagged.

A flowsheet of production of phthalic anhydride by oxidation of o-xylene is shown below:

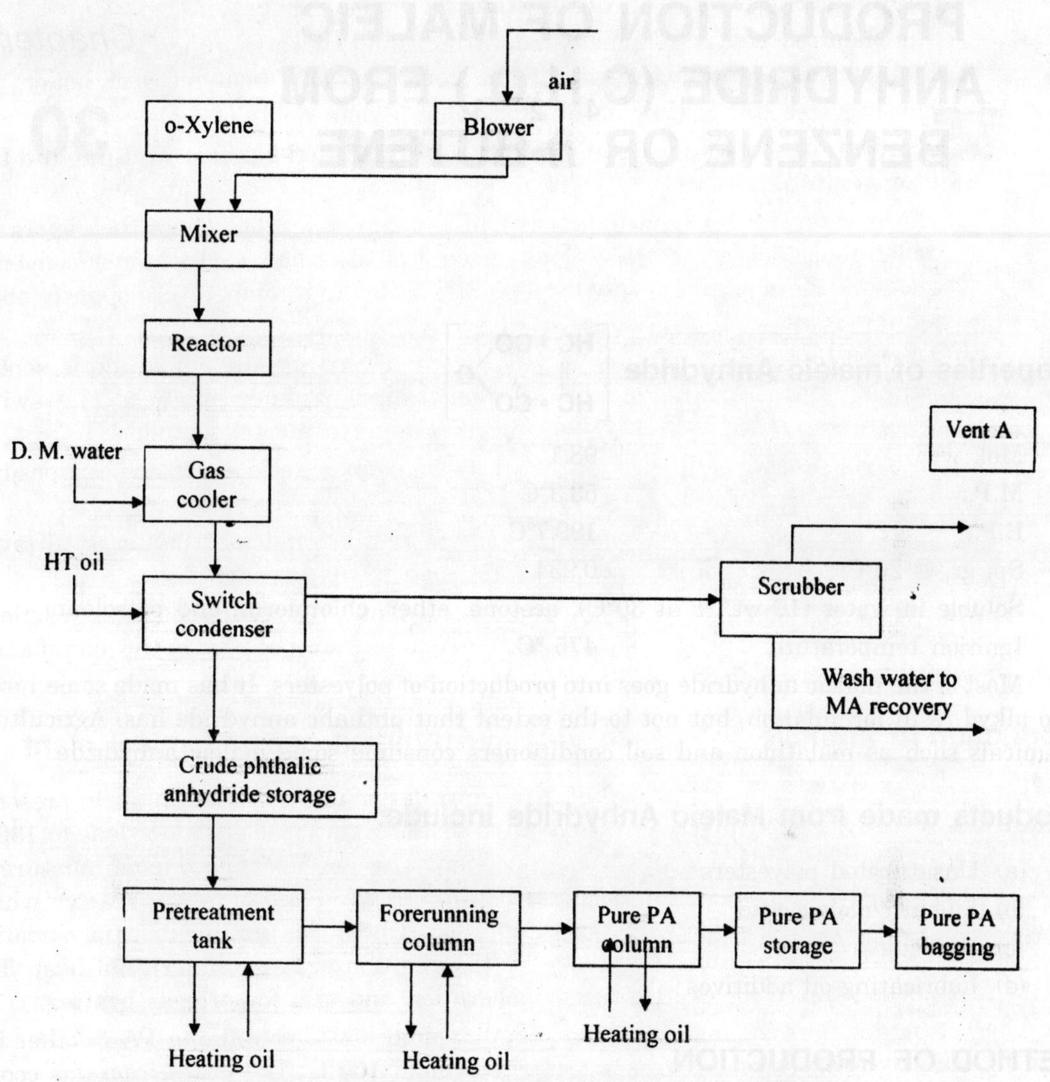

Figure 29.2: Flowsheet of phthalic anhydride (PA) manufacture by oxidation of o-xylene

PRODUCTION OF MALEIC ANHYDRIDE ($C_4H_2O_3$) FROM BENZENE OR *n*-BUTENE

Chapter 30

Properties of maleic Anhydride $\left[\begin{matrix} HC \cdot CO \\ \| \\ HC \cdot CO \end{matrix} \diagdown\!\!\!\!\diagup O\right]$

Mol. wt.:	98.1
M.P.:	53.3°C
B.P.:	199.7°C
Sp. gr. @ 20°C:	0.934

Soluble in water (16 wt. % at 30°C), acetone, ether, chloroform, and petroleum

Ignition temperature: 475 °C

Most of the maleic anhydride goes into production of polyesters. It has made some inroad into alkyd resin formulation, but not to the extent that phthalic anhydride has. Agricultural chemicals such as malathion and soil conditioners consume some maleic anhydride.

Products made from Maleic Anhydride include:

(a) Unsaturated polyesters
(b) Fumaric/maleic acid
(c) Pesticides
(d) Lubricating oil additives

METHOD OF PRODUCTION

Oxidation of Benzene or n-butene

Chemical reactions

(a) **Benzene pxidation:**

$$C_6H_6 + 4\tfrac{1}{2}O_2 \xrightarrow[400\text{–}500^\circ C]{V_2O_5} C_4H_2O_3 + 2H_2O + 2CO_2 :$$

(b) **n-Butene oxidation:**

$$C_4H_8 + 3O_2 \longrightarrow C_4H_2O_3 + 3H_2O :$$

Quantitative, requirements:

Basis:	1 ton of maleic anhydride (60% yield)	
Benzene:		1.33 tons
Air:		20-22 tons
Basis:	1 ton of maleic anhydride (50% yield)	
Butylene:		1.07 tons
Air:		38-40 tons

Process description:

This oxidation process is very similar to the fixed bed catalytic operation in making phthalic anhydride. Benzene is vaporized with a large excess of air at 1-1.5 atms, gage in a steam coil to avoid explosive compositions. After preheating, it is fed to a fixed bed converter where the contact time is about 0.1 sec, the temperature is 400-500°C and the pressure 0.8-1 atm. gage.

The exit gases from the reactor are passed successively through a waste heat boiler, feed preheater, tar knock-out drum, and absorbed in water to yield a 40 percent maleic acid solution. This is reconverted to anhydride by a hydrocarbon azeotropic distillation system. The crude product is purified by vacuum distillation.

A flowsheet for the production of maleic anhydride from benzene or n-butene is given below:

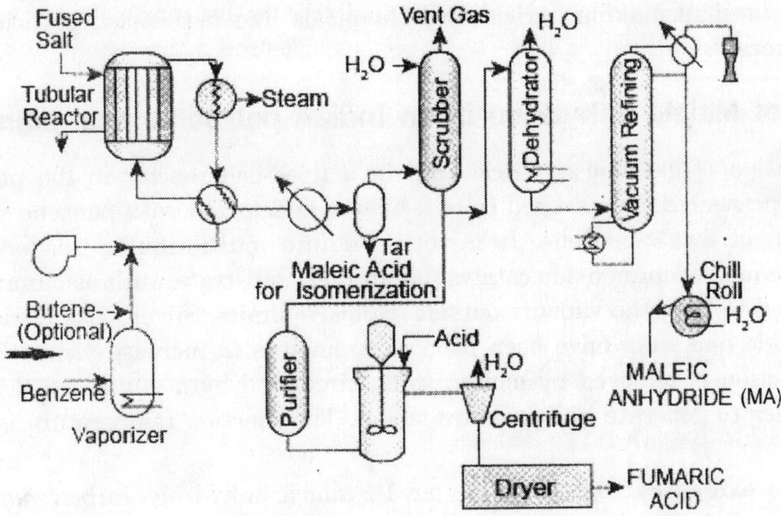

Figure 30.1: Flowsheet for manufacture of meleic anhydride from benzene or n-butene

Salient feature of the process technology are listed below:

- The process technology is similar to phthalic anhydride production.
- Benzene or butene is first vaporized in excess air.

- Then the mixture enters the catalytic tubular reactor (with shell arrangement). The catalytic reactor is circulated with salt solution so as to absorb the heat generated using the reaction.
- The reactor products enter a waste steam recovery boiler where the boiler water is converted to steam.
- The product vapors eventually enter an integrated heat exchanger to exchange heat with the pre-heated feed stream.
- Eventually, the vapors are condensed followed with gravity settling to remove tar as a product in the gravity settling. The vapors then enter a water scrubber to obtain 40 % maleic acid solution. The leaving gases consist mainly of CO_2. The maleic acid from the absorber is partially sent to a dehydrator that removes water using azeotropic distillation principle. The purified maleic anhydride product is further sent to vacuum distillation to obtain the maleic anhydride product.
- The dilute maleic acid solution is partially sent to an isomerisation unit where HCl is used as the isomerisation catalyst.
- After reaction, the product is sent to a centrifuge that separates water from the fumaric acid.
- Eventually, fumaric acid is sent to a drier to obtain dry fumaric acid. For fumaric acid (an isomer of maleic acid), the reaction is maleic acid → fumaric acid.

Maleic anhydride can be used as a highly reactive and versatile raw material. It can be used in the manufacture of alkyd resins, which in turn are used for making paints & coatings. It can also be used in making agricultural chemicals like herbicides, pesticides and plant growth regulators.

Production of Maleic anhydride in an Indian petrochemical plant

The oxidation of benzene is carried out in a fixed-bed reactor in the presence of air. Filtered air is preheated, compressed from a blower and mixed with benzene vapour to give a concentration of 3wt% benzene. It is then fed into multi-tubular reactor containing a vanadium oxide-molybdenum oxide catalyst on an inert substrate, such as alumina. An excess of air is required to keep the vapours outside explosive limits. Silver, cobalt, nickel, titanium and sodium oxide and salts have been used as promoters to increase yields. Heat from the exothermic reaction is removed by molten salts, circulated by a pump, which pass through a heat exchanger to generate highpressure steam. The reaction temperature is kept at 350-450°C.

The reactor exit gases (containing arc .nd 1% maleic anhydride, carbon monoxide, carbon dioxide and traces of phenols and carboxylic acids) are cooled to 55-65°C. About 60% of the maleic anhydride is condensed as a liquid and removed from the vapour stream. It is separated from the gases and sent to the crude maleic anhydride tank. The remaining gases are scrubbed counter currently with water in an absorption tower. The resultant maleic acid is dehydrated by azeotrophic distillation with o-Xylene, and water is removed overhead.

The dehydrated product is combined with the condensed anhydride before being recovered by distillation. Purification of the crude maleic anhydride is carried out by vacuum distillation or sublimation. Maleic anhydride is either sold in the molten state or formed into pellets and flakes and bagged.

Reactions:

$$\underset{\text{Benzene}}{C_4H_8} + 4.5O_2 \longrightarrow \underset{\text{Meleic Anhydride}}{C_4H_2O_3} + 2H_2O + 2CO_2$$

A flowsheet for the production of maleic anhydride from benzene or n-butene is given below:

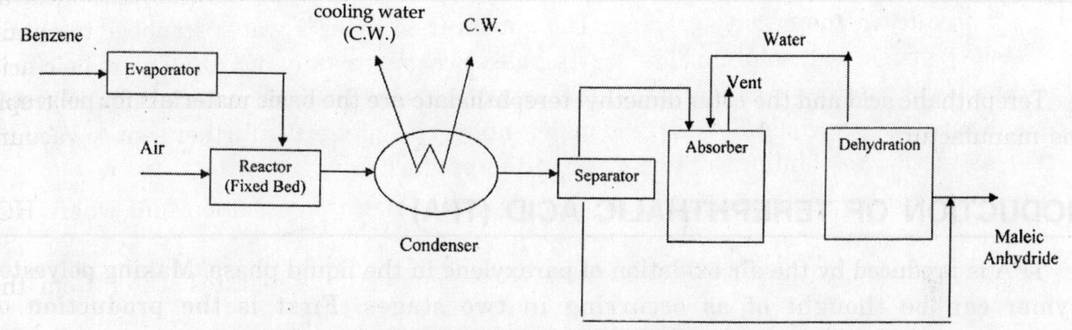

Figure 30.2: Flowsheet of production of maleic anhydride by oxidation of benzene

PRODUCTION OF TEREPHTHALIC ACID (TPA) & DIMETHYL TEREPHTHALATE (DMT)

Chapter 31

Terephthalic acid and the ester dimethyl terephthalate are the basic materials for polyester films manufacture

PRODUCTION OF TEREPHTHALIC ACID (TPA)

TPA is produced by the air oxidation of paraxylene in the liquid phase. Making polyester polymer can be thought of as occurring in two stages. First is the production of bis-(β-hydroxyethyl) terephthalate (bis-HET) by causing ethylene glycol to esterify with TPA or DMT:

$$ROOC-C_6H_4-COOR + 2\ \underset{\text{EG}}{HOCH_2-CH_2OH} \rightleftharpoons \underset{\text{bis-HET}}{HOCH_2CH_2OOC-C_6H_4-COOCH_2CH_2OH} + 2ROH \qquad R = H.\ CH_3$$

A number of oligomers with varying values of n are also formed:

$$n\ \underset{\text{bis-HET}}{HOCH_2CH_2OOC-C_6H_4-COOCH_2CH_2OH} \rightleftharpoons \underset{\text{PET}}{\left[-OOC-C_6H_4-COOCH_2CH_2-\right]_n} + n\ \underset{\text{EG}}{HOCH_2-CH_2OH}$$

Dimethyl terephthalate (DMT) can be used as a substitute for purified terephthalic acid (TPA) as polyester raw material. All over the world, there has been a clear trend in the preference of TPA to DMT for plyester production.

Production of Poly(Ethylene) Terephthalate

In the second stage, polycondensation of bis-HET and the mixture of oligomers yield poly(ethylene) terephthalate of high molecular weight and ethylene glycol.

Poly(ethylene) terephthalate may be used to produce filament yarns and staple fiber for textile applications.

Nitrobenzene

The production of nitrobenzene is done by the direct reaction of benzene with nitric acid (usually in the presence of sulfuric acid as a dehydrating agent). It is a commercially important method of making nitrobenzene as a raw material for aniline & aniline dyes; also the same technique can be applied to the nitration of toluene to give trinitrotoluene (TNT) as an explosive.

Nitrobenzene is used as a chemical intermediate in the manufacture of aniline, benzidine & many dyestuffs; it is also used widely as an industrial solvent in polishes, etc.

The reaction involved in nitrobenzene manufacture is:

$C_6H_6 + HNO_3(H_2SO_4) \longrightarrow C_6H_5NO_2 + H_2O(H_2SO_4);$

Production of Di-methyl terephthalate (DMT) in an Indian petrochemical plant:

Di-methyl terephthalate is produced by the reaction of paraxylene with oxygen & then with methanol. The oxidation reaction takes place in the presence of catalyst. The Un-reacted paraxylene from oxidation products are stripped completely with steam. The oxidate mixture is then reacted with methanol to produce esters (DMT & Methyl Para toluate). The product esters are then distilled to separate low boiling residue and dibasic acids from DMT. The DMT is further purified by crystallisation to remove impurities like isomers & aldehydes. The purified DMT in the form of pellets is sent to storage.

Chemical reaction

$$CH_3\text{-}C_6H_4\text{-}CH_3 \;(\text{P-xylene}) + O_2 \longrightarrow HOOC\text{-}C_6H_4\text{-}COOH \;(\text{PTA}) + 2CH_2OH \longrightarrow CH_3OOC\text{-}C_6H_4\text{-}COOCH_3 \;(\text{DMT})$$

A flowsheet for manufacture of dimethyl terophthalate (DMT) from oxidation of p-xylene is given below:

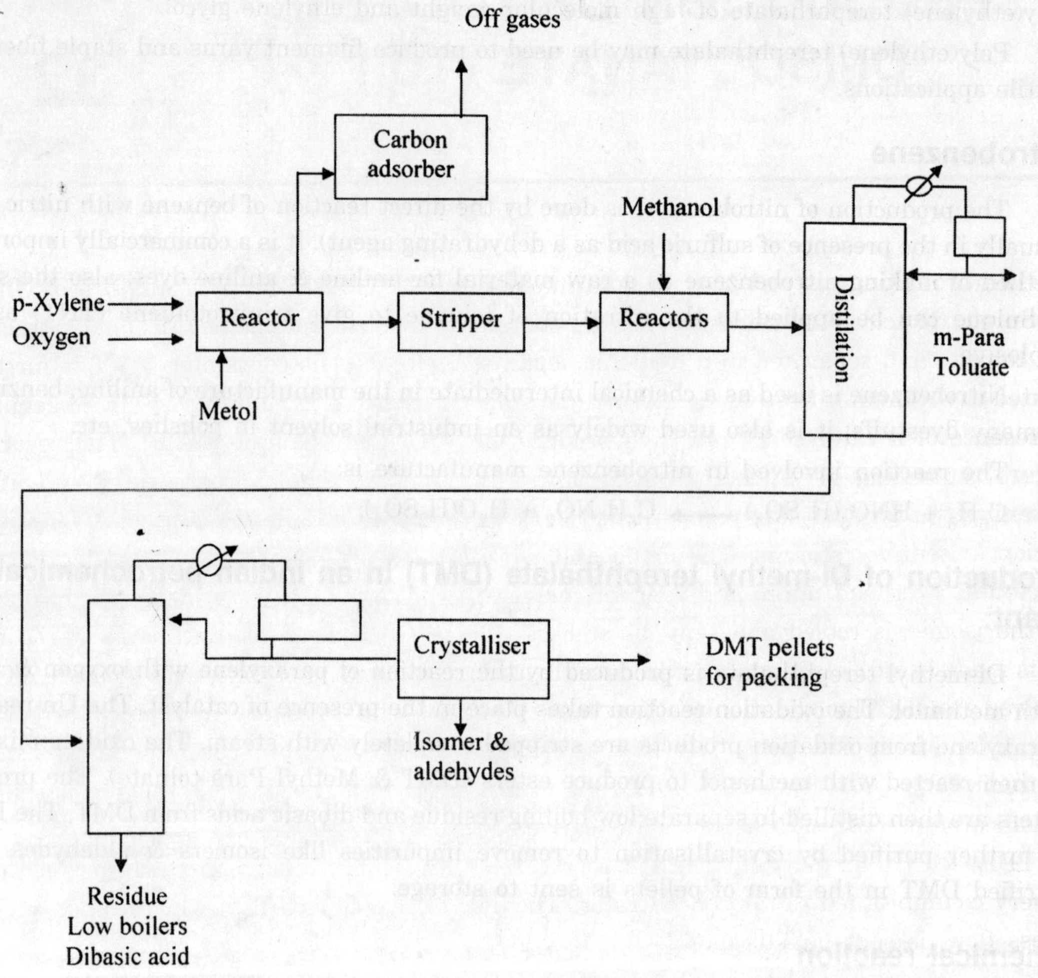

Figure 31.1: Flowsheet of Manufacturing Process of DMT (Dimethyl Terephthalate)

PRODUCTION OF TOLUENE DI-ISOCYANATE (TDI)

Chapter 32

Di-nitro-toluene is produced by continuous nitration of Toluene by a mixture of Nitric and Sulphuric acids. The nitration mixture containing mono nitro-toluene (MNT) is extracted with toluene and separated in a dynamic separator. The toluene containing MNT is further nitrated to obtain di-nitro-toluene. After three stages washing, it is continuously hydrogenated in presence of a catalyst at 115°C to produce meta Toluene diamine (MTD). The reaction mixture containing MTD-water-catalyst slurry is removed from hydrogenerator and after thickening the catalyst slurry is returned back to the reactor while the filtered MTD water solution is sent for storage. The ortho and para-isomers are separated by distillation and collected as overhead purge and sent for incineration. The MTD with a maximum of 0.25% of ortho isomer is cooled and sent for storage. Phosgene required for converting MTD into TDI is made in situ reacting CO, generated by controlled burning of coke, with chlorine. MTD dissolved in ODCB (ortho dichloro benzene) is reacted with Phosgene in a pipeline reactor using 100% excess Phosgene. The off-gas containing HCl and Phosgene are separated from the crude TDI - ODCB solution and sent to absorbers for separating Phosgene and HCl. Degasificaton of TDI - ODCB solution is carried out to remove remaining Phosgene and HCl. The TDI - ODCB is vacuum distilled to recover ODCB. The bottom from primary ODCB recovery column is fed to tar concentrator. The TDI rich and pure TDI, which comes out as process flow, is sent for storage.

CHEMICAL REACTION

$$\underset{\text{Toluene}}{C_6H_5CH_3} + 2HNO_3 \xrightarrow{H_2SO_4} \underset{\text{Di-nitro-toluene}}{CH_3C_6H_3(NO_2)_2} + 2H_2O$$

$$\underset{\text{Toluene}}{CH_3C_6H_3(NO_2)_2} + 6H_2 \xrightarrow{\text{Catalyst}} \underset{\text{m-Toluene diamine (MTD)}}{CH_6C_6H_3(NH_2)_2} + 2H_2O$$

$$CH_3C_6H_3(NH_2)_2 + 2COCl_2 \xrightarrow{\text{ODCB}} \underset{\text{TDI}}{CH_3C_6H_3(NCO)_2} + 4HCl$$

A flowsheet for production of tolune di-isocynate (TDI) is given below:

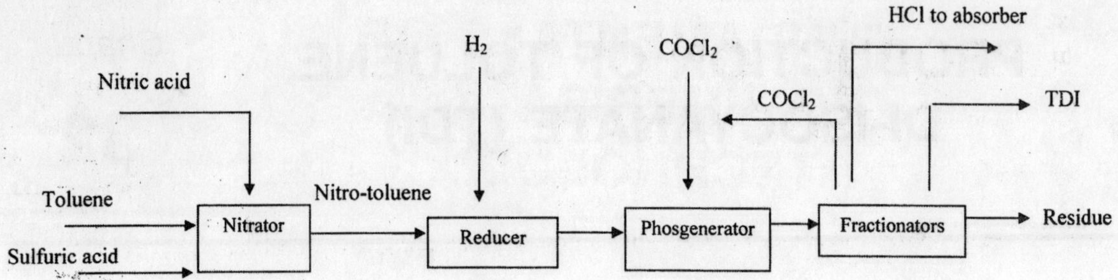

Figure 32.1: Flowsheet of manufacture of TDI(Toluene di-isocynate)

Chapter 33

FUNDAMENTALS OF POLYMERISATION TECHNOLOGY

Polymers are large chain molecules having a high molecular weight in the range of 10^3 to 10^7. These are made up of a single unit or a molecule, which is repeated several times within the chained structure.

Monomer: A monomer is the single unit or the molecule which is repeated in the polymer chain. It is the basic unit which makes up the polymer.

Thermosetting Polymer: There are some polymers which, when heated, decompose, and hence, cannot be reshaped. Such polymers have a complex 3-D network (cross-linked or branched) and are called Thermosetting Polymer's. They are generally insoluble in solvents and have good heat resistance quality. Thermosetting polymers include phenol-formaldehyde, urea-aldehyde, silicones etc.

ThermoplasticPolymer: These polymers are composed of monomers which are linear or have moderate branching. They can be melted repeatedly and casted into various shapes structures. They are soluble in solvents, but do not have appreciable thermal resistance properties. Vinyls, cellulose derivatives, polythene and polypropylene fall into the category of thermoplastic polymers.

Some monomers and Corresponding polymers & Repeat Units are tabulated below:

S.N.	Monomer	Polymer	Repeat Ok Unit
1.	Ethylene $CH_2 = CH_2$	Polyethylene $-(CH_2 - CH_2)_n-$	$- CH_2 - CH_2 -$
2.	Propylene $CH_2 = CH(CH_3)$	Polypropylene $-(CH_2 - CH(CH_3))_n-$	$- CH_2 - CH(CH_3) -$
3.	Styrene $CH_2 = CH(C_6H_5)$	Polystyrene $-(CH_2 - CH(C_6H_5))_n-$	$- CH_2 - CH(C_6H_5) -$
4.	Vinyl chloride $CH_2 = CH(Cl)$	Poly (vinyl chloride) $-(CH_2 - CH(Cl))_n-$	$- CH_2 - CH(Cl) -$

5.	Vinylidene chloride Cl $CH_2 = C$ Cl	Poly (vinylidene chloride) Cl $-(CH_2 - C\ \ )_n -$ Cl	Cl $- CH_2 - C -$ Cl
6.	Vinyl acetate $CH_2 = CH$ $\vert$ $OCOCH_3$	Poly (vinyl acetate) $-(CH_2 - CH)_n -$ $\vert$ $OCOCH_3$	$- CH_2 - CH -$ $\vert$ $OCOCH_3$
7.	Tetrafluoroethylene $CF_2 = CF_2$	1 Poly (tetrafluoroethylene) $-(CF_2 - CF_2)_n -$	$- CF_2 - CF_2 -$
8.	Methyl acrylate $CH_2 = CH$ $\vert$ $COOCH_3$	Poly (methyl acrylate) $-(CH_2 - CH)_n -$ $\vert$ $COOCH_3$	$- CH_2 - CH -$ $\vert$ $COOCH_3$
9.	Methyl methacrylate CH_3 $CH_2 = C$ $\vert$ $COOCH_3$	Poly (methyl methacrylate) CH_3 $-(CH_2 - C\ \ )_n -$ $\vert$ $COOCH_3$	CH_3 $- CH_2 - C -$ $\vert$ $COOCH_3$
10.	Acrylonitrile $CH_2 = CH$ $\vert$ CN	Polyacrylonitrile $-(CH_2 - CH)_n -$ $\vert$ CN	$- CH_2 - CH -$ $\vert$ CN
11.	Formaldehyde $CH_2 = O$	Polyformaldehyde or Poly (oxymethylene) $-(CH_2 - O)_n -$	$- CH_2 - O -$
12.	Caprolactam $CH_2)_5$ / \\ $O = C$ —— NH	Polycaprolactam (Nylon 6 polyamide) $[- C - (CH_2)_5 - N -]_n$ $\Vert \qquad\qquad\quad \vert$ $O \qquad\qquad\quad H$	$- C - (CH_2)_5 - N -$ $\Vert \qquad\qquad\quad \vert$ $O \qquad\qquad\quad H$
13.	Butadiene $CH_2 = CH - CH = CH_2$	Polybutadiene $(- CH_2 - CH = CH - CH_2 -)_n$	$- CH_2 - CH = CH\ CH_2 -$
14.	Chloroprene $CH_2 = C - CH_2) = CH_2$ $\Vert$ Cl	Polychloroprene $(- CH_2 - C = CH - CH_2 -)_n$ $\Vert$ Cl	$- CH_2 - C = CH - CH_2 -$ $\Vert$ Cl

Exaple: Polymerization of monomer styrene to polystyrene may be represented by:

$$nCH_2 = CH(C_6H_5) \xrightarrow[\text{and/or catalyst (Polymerization)}]{\text{Heat/light}} -(CH_2 - CH(C_6H_5))n-$$

Styrene (monomer) — Polystyrene (polymer)

Nomenclature of Vinyl Polymers are tabulated below:

S.N.	Monomer Source	Monomer Name	Polymer Repeating Unit	Source Name	IUPAC* Name
1.	$CH_3CH{=}CH_2$	Propylene	$-[CH_2CH(CH_3)]-$	Polypropylene	Poly(propylene)
2.	$CH_2{=}CHCl$	Vinyl chloride	$-[CH_2CH(Cl)]-$	Poly(vinyl chloride)	Poly(1-chloro ethylene)
3.	$CH_2{=}C(CH_3)_2$	Isobutylene	$-[CH_2C(CH_3)_2]-$	Polyisobutylene	Poly(1-1, dimethylethylene)
4.	$CH_2{=}CHCN$	Acrylonitrile	$-[CH_2CH(CN)]-$	Polyacrylonitrile	Poly(1-cyano ethylene)
5.	$CH_2{=}C(CH_3)CO_2CH_3$	Methyl methacrylate	$-[CH_2C(CH_3)(CO_2CH_3)]-$	Poly(methyl methacrylate)	Poly[1-(methoxycarbonyl)-1-methylethylene]
6.	$CH_2{=}CHOC({=}O)CH_3$	Vinylacetate	$-[CH_3CH(OC({=}O)CH_3)]-$	Poly(vinyl acetate)	Poly(1-acetoxyethylene)

.7	(pyridine ring, N) $CH=CH_2$	Vinyl acetate	$-[CH_2CH-]-$ (pyridine ring, N)	Poly(2-vinyl pyridine)	Poly[1-(2 pyridinyl)-ethylene]
8.	CH_3 $CH_2=C—CH=CH_2$	Isoprene	$-[CH_2 \quad CH_2]-$ $C=C$ $CH_3 \quad H$	*cis*-1,4-poly-isoprene	cis-Poly(1-methyl-1-butene-1,4-diyl)
9.	Cl $CH_2=C—CH=CH_2$	Chloroprene	$-[(CH_2CH-]-$ $ClC=CH_2$	3,4-Polychloro-prene	Poly[1-(1-chloro vinyl) ethylene]

* IUPAC. International Union of Pure and Applied Chemistry.

Nomenclature of Nonvinyl Polymers are tabulated below:

S.N.	Monomer Source	Monomer Name	Polymer Repeating Unit	Source or Common Name	IUPAC Name
	Polyethers:				
1.	O $CH_2=CH_3$	Ethylene oxide	$-[CH_2CH_2O]-$	Poly(ethylene oxide)	Poly(oxyethylene)
2.	$HOCH_2CH_2OH$	Ethylene glycol	$-[CH_2CH_2O]-$	Poly(ethylene glycol)	Poly(oxyethylene)
3.	CH_2O	Formade-hyde	$-[CH_2O]-$	Polyformal-dehyde	Poly(oxyethy-lene)
4.	CH_3CHO	Acetalde-hyde	$-[CHO]-$ CH_3	Polyacetal-dehyde	Poly(oxyethy-lidene)
	Poly esters:				
5.	(β-lactone ring: C=O, O)	β-Propiol-actone	$-[OCH_3CH_2C(=O)]-$	Poly(b-propiol-actoe) or poly (3-propionate)	Poly[oxy(1-oxopro pane-1,3-diyl)

6.	$HO(CH_2)_9CO_2H$	10-Hydroxy-decanoic acid	$\left[O(CH_2)_9\overset{O}{\overset{\Vert}{C}} \right]$	Poly(10-decanoate)	Poly[oxy(1-oxo-decane-1,10-diyl)]
7.	$HOCH_2CH_2OH$ +	Ethylene glycol +	$\left[OCH_2CH_2O\overset{O}{\overset{\Vert}{C}}-C_6H_4-\overset{O}{\overset{\Vert}{C}} \right]$	Poly(ethylene terephthalate)	Poly(oxythylene oxyterephthaloyl)]
8.	$HO_2C-C_6H_4-CO_2H$	Terephthalic acid			
9.	(caprolactam ring: C=O, NH)	Caprolactam	$\left[NH_2(CH_2)_5\overset{O}{\overset{\Vert}{C}} \right]$	Polycaprolactam or nylon 6	Poly[imino (1-oxo-hexane-1, 6-diyl)]
10.	$H_2N(CH_2)_{10}CO_2H$	11-Aminoundecanoic acid	$\left[NH_2(CH_2)_{10}\overset{O}{\overset{\Vert}{C}} \right]$	Poly(11-undecanoamide) or nylon 11	Poly[imino(1-oxo-undecane-1,11-diyl)]
11.	$H_2N(CH_2)_6NH_2$ +	Hexamethylenediamine +	$\left[NH(CH_2)_6NH\overset{O}{\overset{\Vert}{C}}(CH_2)_8\overset{O}{\overset{\Vert}{C}} \right]$	Poly(hexamethylene sebacamide) or nylon 610	Poly[iminohexane-1, 6 diyliminosebacoyl)
12.	$HO_2C(CH_2)_8NO_2H$	Sebacic acid			
13.	$Cl\overset{O}{\overset{\Vert}{C}}-C_6H_4-\overset{O}{\overset{\Vert}{C}}Cl$ +	Terephthaloyl chloride +	$\left[NH-C_6H_4-NH\overset{O}{\overset{\Vert}{C}}-C_6H_4-\overset{O}{\overset{\Vert}{C}} \right]$	Poly(m-phenyleneterephthalamide)	Poly(imino-1,3-phenyleneiminoterephthalyyl)
14.	$H_2N-C_6H_4-NH_2$ (meta)	*m*-phenylenediamine			
15.	$H_2N(CH_2)_4NH_2$ +	Tetramethylenediamine +	$\left[NH(CH_2)_4NHO_2S-C_6H_4-SO_2 \right]$	Poly(tetramethylene-*m*-benzenesulfonamide)	Poly(sulfonyl-1,3-phenylnesulfnylimino-butane-1, 4-diylimino)
16.	$ClO_2S-C_6H_4-SO_2Cl$ (meta)	*m*-Benzene-disulfonyl chloride			

Polymer Classification

Polymers are generally classified on the basis of:

- Physical and chemical structures.
- Preparation methods,
- Physical properties.
- Applications.

CLASSIFICATION ACCORDING TO PHYSICAL & CHEMICAL STRUCTURES

(A) On the basis of functionality or degree of polymerization (DP)

The functionality of a monomer or its degree of polymerization determines the final polymer that will be formed due to the combination of the monomers. The number of reactive bonds or groups that are available for coupling will determine whether the monomer will be mono-, bi-, tri-, or polyfunctional.

(a) **Monofunctional or unifunctional:** When a single reactive group is present in the monomer molecule, then it is termed as monofunctional or unifunctional. However a monofunctional group cannot lead to the propagation of a polymer chain. For example , in carboxylic acid, CH_3COOH, the –COOH group is the monofunctional group and obviously a polymeric chain can not be built from this acid alone.

(b) **Bifunctional:** When two reactive groups are present in the monomer molecule, then it is termed as bifunctional. Polymerization reaction with bifunctinal groups occur when a double bond splits to couple with another double bonded monomer.If a double bonded molecule is present, then the polymer would be–

$$nR = R \longrightarrow \text{———} (R - R)_n \text{———}$$

If only ethylene ($CH_2 = CH_2$) is present, the polymer (polythene) will be $nCH_2 = CH_2 - (CH_2 - CH_2)n-$

(c) **Trifunctional:** When a monomer contains three groups that are reactive, then it is termed as trifunctional. A trifunctional polymer may be a mixture of monofunctional and bifunctional monomer structures.

Let us consider the reaction of phenol and formaldehyde to form trimethyl phenol.

Trifunctional

(3) (1) (2)

$$+ H-\overset{\overset{O}{\|}}{C}-H + H-\overset{\overset{O}{\|}}{C}-H + H-\overset{\overset{O}{\|}}{C}-H$$

O - H
O O
→ C — O — (3) (1) — O — C
H H
(2)
O
H — C — H

In the figure, the three formaldehyde molecules contain three reactive groups in the form of –OH groups. These reactive groups get attached to the positions 1, 2 and 3 in the phenol molecule.

(B) Classification On the basis of physical structure

(a) **Linear polymers:** These polymers have a linear structure. Only bifunctional monomers can form linear polymers.

• • • • — R — R — R — R — R — R — R — R — • • •

They are generally thermoplastic polymers, and except for very high molecular weight materials, they are soluble in solvents. These polymers creep flow at > 150°C without chemical change.

(b) **Cross -linked polymers:** These polymers have a 3-D network structure. Only bi-functional and tri-functional monomers can form cross-linked polymers. Elastomeric properties are generally endowed in polymers have cross-linkage across the linear chains.

• • • —R—R—S—R—S—R—S— • • •

• • • —S—R—S—R—S—R—R— • • •

Thermosetting polymers having high thermal resistance and mechanical strength may be formed if the cross-linking among the monomer groups is extensive.

(c) **Brancheth-chain polymer:** When there is no cross-linking in a polymer molecule yet, branches arise from the parent polymer chain, then the polymer formed is called a branched - chain polymer.

A small amount of trifunctional impurities in a bifunctional monomer may cause branching. Branching may create several physical properties in a polymer such as decrease of solubility in solvents, rise in the softening point and also the reduction in thermoplastic properties.

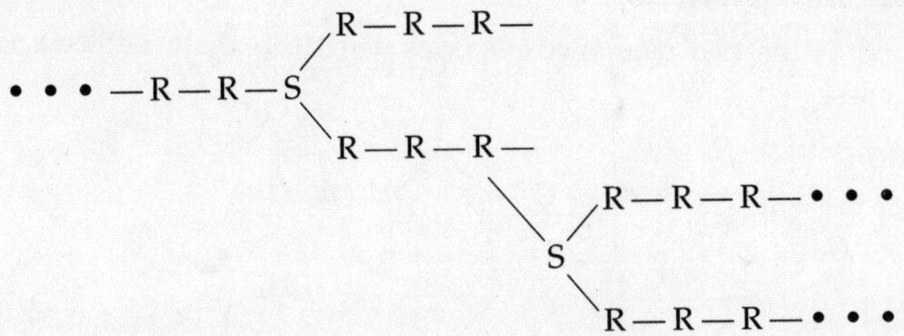

```
              R—R—R—
             /
• • • —R—R—S
             \
              R—R—R—
                    \      R—R—R—• • •
                     \    /
                       S
                          \
                           R—R—R—• • •
```

(C) Classification On the basis of polymeric chemical reactions

Depending upon the basis on which monomers undergo chemical reactions to form polymers, polymers can be further subdivided.

(a) **Polycondensation Reaction:** It is a stepwise process in which the monomers react by eliminating small molecules like H_2O, NH_3, CH_2O and NaCl in a repetitive manner. Indefinite growth of the polymer occurs at both ends. The condensation reaction terminates when one of the reacting molecules is completely consumed in the reaction. In general, bifunctional monomers with different functional groups undergo such reactions. One common example of a condensation reaction is the esterification reaction. Benzoic acid reacts with ethanol undergoing condensation reaction to give ethyl benzoate, expelling one molecule of water (H_2O). The polymer formed after the reaction (in this case, ethyl benzoate) is called the condensation polymer.

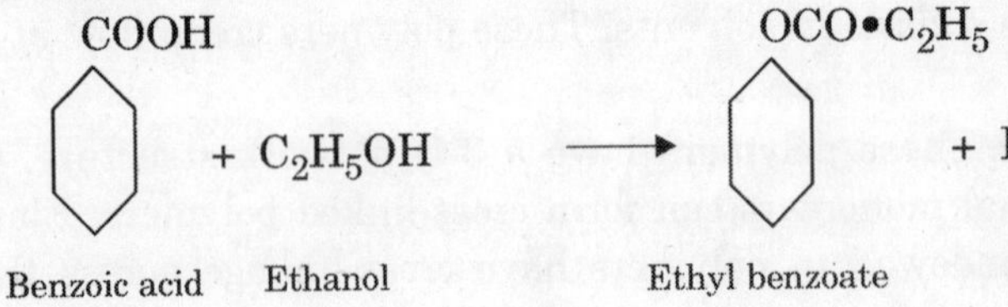

$$C_6H_5COOH + C_2H_5OH \longrightarrow C_6H_5OCO \cdot C_2H_5 + H_2O$$

Benzoic acid   Ethanol   Ethyl benzoate

(b) **Addition Reaction:** These reactions occur by the formation of rapid formation of rapid chains and no elimination of small molecules occur. Such reactions generally occur among double or triple-bonded molecules, or oxo-ring compounds.

Addition reactions can be further of three types:

(i) Step Reactions

(ii) Chain Reactions

(iii) Ionic Reations

(i) **Step growth addition reactions:** In step reactions, a π bond is broken and two new α bonds are formed. The substrate of an electrophilic addition reaction must have a double bond.

(ii) **Chain growth addition reactions:** In chain addition reactions, the polymer is formed in a single step and within the fraction of a second. The chain kinetics occur in 4 steps - Initiation, Propagation, Transfer and Termination.

Variations of Chain Addition Reaction :

If we consider 'M' and 'N' be two monomeric species, then the chain addition reaction can be of the following types:

Homopolymerization

If only M and N are present,

$$M \xrightarrow{M} MM \xrightarrow{M} MMM \xrightarrow{M} M_n$$

$$N \xrightarrow{N} NN \xrightarrow{N} NNN \xrightarrow{N} N_n$$

Copolymerization

If M and N are present together,

Random type

MM MN NNN MM N M MM N N....

Alternating type

M N M N M N M N M N M.....

Block Polymerization

In this type, multimers resulting from straight polymerization are coupled together in blocks.

— — • M M M • N N N N • M M • N N N N N • — —

Graft Polymerization

These have a structure which differs from block polymers ; and are formed from M and N multimers.

```
                                   /N N N N — — — graft
                                  /
M M M M M M M M M M M M M — — — main chain
    |
    N N N N — — — graft
```

In Graft polymerization, β - or X - radiation or chemically reactive functional groups ensue the formation of reactive radicals on the main chain or on the grafted section.

Reaction Kinetics:

Let,

A - radical source

B - monomer molecule

* - activated

B_m, B_n - growing monomer molecule chain of lengths 'm' and 'n'.

X - any solvent, impurity, etc. which can act as transferable energy species.

Initiation

It is the rate controlling step in the chain Addition Reaction. This stage is relatively slow and requires energy for breaking and forming the bonds.

The energy is supplied as thermal energy or in the form of radiations (α-, β-, or γ - rays). However , by using a catalyst, the activation energy required can be substantially lowered.

$$nA \xrightarrow{\text{Energy}} A^*, 2A^*, ...$$

Propagation

These reactions occur between the already formed free radicals and an inactive monomer. With low activation energy, there is more probability of reaction or collision between the species. As a result, rapid reactions occur to attain the given degree of polymerization.

$$A^* + B \longrightarrow B^* + A$$

$$B^* + B \longrightarrow BB^*$$

$$BB^* + B \longrightarrow BBB^* \longrightarrow B^*_n \, [B^*_n$$

in Graft polymerization, β - or X - radiation are chemically reactive.

Chain Transfer Reaction

In this reaction, one radical of the multiplying chain terminates and without the loss of any number of free radicals, a new chain begins. These reactions control the degree of polymerization and the rate of branching of the polymers.

Branching usually occurs at higher temperature, with the reaction of a multimer with a dead chain, followed by lengthening of the branch due to successive monomer addition.

$$B^*_n + B \longrightarrow B_n + B^* \text{ (with monomer)}$$

$$B^*_n + B_m \longrightarrow B_n + B_m^* \text{ (with multimer)}$$

$$B^*_n + X \longrightarrow B_n + X^* \text{ (with other molecules)}$$

Chain Termination

A chain terminates with the net loss of one free radical. It may occur by the collision of growing chains with coupling or disproportionation. However , if a growing chain collides with an initiation radical, undesirable impurities or the wall of the reaction vessel, then it might also terminate.

$$2 B_m^* \longrightarrow B_{2n}$$

$$B^*_n + B_m^* \longrightarrow B_m + B_n.$$

$$B^*_n + Y^* \longrightarrow B_n + Y$$

(iii) Ionic Reaction

These are similar to chain addition reaction with free radicals , however ; instead of neutral, unpaired electrons, they have a positive or negative charge. A co-catalyst may be used, which is usually a trace of H_2O, which is necessary to produce ionization, as the reaction cannot occur in absolute dry conditions.

An example of ionic addition reaction is the preparation of polyisobutylene with boron trifluoride catalyst.

Initiation:

$$BF_3 + H_2O \text{ (Trace)} \longrightarrow F_3BOH^- + H^+$$

$$H^+ + CH_2 = C(CH_3) \longrightarrow (CH_3)_3C^+$$

Propagation:

$$(CH_3)_3 \bullet C + CH_2 = C(CH_3)_2 \longrightarrow (CH_3)_3 \bullet C - CH_2 - \overset{\displaystyle CH_3}{\underset{\displaystyle CH_3}{\overset{|}{\underset{|}{C^+}}}}$$

Termination:

$$R - \overset{\displaystyle CH_3}{\underset{\displaystyle CH_3}{\overset{|}{\underset{|}{C^+}}}} \begin{cases} \nearrow R = C(CH_3)_2 + H^+ \\ \searrow R - C \begin{matrix} \diagup CH_3 \\ \diagdown\!\!\diagdown CH_2 \end{matrix} + H^+ \end{cases}$$

MAIN POLYMERIZATION REACTION TYPE

Chain Growth (addition) Reaction

- Reaction of unsatureated or cyclic monomers which link to form a continuously increasing chain. No release of any co-products.
- Initiation is a slow step. Propagation & termination steps are rapid.
- Monomer concentration decreases continuously during reaction.
- Icreased reaction time has no effect on average moleculer weight (M.W.) or chain length.
- Intermediates are transient & non-isolable.
- Examples are: polythylene, polypropylene, poly(vinyl chloride), polystyrene.

Steep Growth (condensation) Reaction

- Reacton of molecules containing specific functional groups that undergo condensation to a larger molecule, with release of H_2O, ROH HCl] etc.
- Concentrations of monomer(s) decrease rapidly; concentration of polymer increasese gradually.
- Increased reaction time results in increased average M.W. and chain length.
- Isolable reaction intermediates.
- Exapmples are: polyesters, polyamides, polyurethanes, polycarbonates.

Characteristics of step growth & chain growth polymerizations are tabulated below

Step growth polymerisation	Chain grwoth polymerisation
• Polycondensation process: any two species (monomer or growing chain) can reacd as long as functional groups are present.	• Chain reaction: Initiation propagation and/or transfer, terminationt; addition occurs during propagation to the ends of growing chains, upon transfer the polymer product is unreactive.
• Branched and network polymers can be formed.	• Straight-chain and branched-chain polymers are formed.
• Catalyst is very desirable but not strictly necessary.	• Catalyst, co catalyst and/or initiator are required.
• Monomer disappears early in reaction.	• Monomer concentration decreases steadily with time.
• Product oligomers are steadily converted to higher MW Product; intermediates can be isolated at any time. MW increases steadily with time.	• High MW polymer formed immediately: since product molecules do not undergo further reaction MW changes very little with time.
• MW is independent of temperature.	• MW decreases with increasing temperature.
• Monomer concentration decrease is second or higher order.	• Monomer concentration decrease is first order.
• Long reaction times are essential to obtain high MW.	• Long reaction time improves yield but does not affect MW.
• Medium viscosity increases rapidly rear reaction end.	• Medium viscosity increases monotonically.
• Typical catalysts are organic acids, amines, or metal salts: active sites are cations acid aleohol groups.	• Active sites are unpaired electrons, anions, cations and coordination bonds with transition metal complexes.

Commercially Important Vinyl Polymers Prepared by Free Radical Polymerization are given below:

S.N.	Polymer or copolymer	Major uses of the polymers
1.	Polyethylene, low-density (LDPE)	Packaging film, wire and cable insulation, toys, flexible bottles, housewares, coatings.
2.	Poly (vinyl chloride)	Construction, rigid pipe, flooring., wire and cable insulation, film and sheet.
3.	Polystyrene[b]	Packaging, foam insulation, appliances, house wares, toys.
4.	Polychloroprene (neoprene rubber)	Tires, wire coatings, belting, hoses, shoe heels, coated fabrics.
5.	Poly(methyl methacrylate)	Automotive parts, decorative panelling. skylights, glazing.

6.	Polyacrylonitrile	Food packaging.
7.	Poly(vinyl acetate)	Water-based paints, adhesives, poly(vinyl alcohol) precursor
8.	Poly(v inylidenc chloride)[C]	Food packaging
9.	Poly(acrylic acid) and poly(methacrylic acid)	Adhesives, thickening agents, lonomers
10.	Poly (ethyl acrylate) and poly (methyl acrylate late)	Inks, varnishes
11.	Polyacrylamide	Thickening agent, flocculant
12.	Polytetrafluoroethylene	Machine parts, gaskets, nonstick cooking utensils
13.	Polytrichlorofluoroethylene	Gaskets, tubing, wire insulation
14.	Poly(vinylidene flouride)	Protective coatings, gaskets, pipe
15.	Poly(vinyl fluoride)	Protective coatings
16.	Allyl resins[d]	Lenses, electronics parts
17.	Styrene-butadiene rubber (SBR)	Tires, belling, flooring, molded goods, shoe soles, electrical insulation
18.	Acrylonitrile-butadiene copolymer (ntrile rubber)	Fuel tanks, gasoline hoses, adhesives, impregnated paper, leather and textiles
19.	Acrylonitrile-butadiene-styrene copolymer (ABS)	Appliances, business machines, automotive parts, luggage, telephones, electronics components
20.	Styrene-acrylonitrile copolymer (SAK)	Automotive parts, battery cases, kitchenware, appliances, medical supplier, furniture
21.	Styrene-maleic anhydride copolymer (SMA)	Automobile interior parts, trays, tumblers
22.	Ethylene—vinyl acetate copolymer (EVA)	Shrink-wrap food packaging, adhesives, coatings, drum liners, carpet backing, gaskets
23.	Acrylonitrile—vinyl chloride copolymer	Textile fibers

a. Numerous copolymers are. manrufactured besides the high-volume ones listed.

b. Poly(p-methyl styrene) & is also available.

c. Including copolymers, principally vinyl chloride

d. Crosslinked diallyl and triallyl esters.

Commercially Available Polymers Synthesized with Complex Coordination Catalysts are listed below:

S. N.	*Polymer*	*Principlal Stereochemistry*	*Typical Uses*
	Plastics		
1.	Polyethylene, high density (HDPE)	—	Bottles, drums, pipe, conduit sheet film, wire and cable insulation
2.	Polyethylene, ullrahigh molecular weight (UHMWPE)	—	Surgical prostheses, machine parts, heavy-duty livers
3.	Polypropylene	Isotactic	Automobile and appliance parts, rope cordage, webbing, carpeting, Film
4.	Polyt(1-butene)	Isotactic	Film, pipe
5.	Poly(4-melbyl-1-pentcac)[a]	Isotactic	Packaging, medical supplies, lighting
6.	Polystyrene pentene	Syndiotactic	Specialty plastics
7.	1,4-Polybutadiene	trans	Metal can coatings, potting compounds for transformers
8.	1-4-Polyisoprene	trans	golf ball covers, orthopedic devices
9.	Ethylene-1-alkene[b] coloymer (linear low-density polyethylene LLDPE)	—	Blending with LDPE, packaging film, bottles
10.	Ethylene-propylene block copolymers (polyallomers)	Isotactic	Food packaging, automotive trim, toys, bottles, film, heat-sterilizable containers
11.	Polydicyclopentadiene[c]	—	Reaction injection molding (RIM), structural plastics
	Elastomers:		
12.	1,4-Polybutadiene	cis	Tires, conveyer belts, wire and cable insulation, footware
13.	4-Polyisoprene	cis	Tires, footware adhesives, coated fabrics
14.	Poly(1-octenylene)	trans	Blending with other elastomers
15.	Poly(l,3-cyclopentenylevin-ylene) (norbornene polyncr)[c]	trans	Molding compounds, engine mounts, car bumper guards
16.	Polypropylene (amorphous)	—	Asphalt blends, sealants, adhesives, cable coatings
17.	Ethylene—propylene copolymar (EPM, EPR)	—	Impact modifier for polypropylene, car bumper guards
18.	Ethylene-propylene diene copolymer (EPDM)	—	Wire and cable insulation, weather stripping, tire side walls, hose, seals

(a) Usally copolymerized with small amount of l-pentene.

(b) 1-Butene, 1-hexene, and 1-octene.

(c) Synthesized by ring-opening metathesis polymenzation of the corresponding cycloalkene.

Commercially Important Polymers Prepared by Step-Reaction Polymerisation are listed below:

S.N.	Polymer Type	Repering Funcional Unit
1.	Polyether	—Ar—O—
2.	Polyether [poly(phenylene oxide)]	(2,6-disubstituted phenylene ring, R groups at both ortho positions)—O—
3.	Polyether (epoxy)[a]	$—CH_2CH(OH)CH_2OAr—$
4.	Polysulfide	—ArS—
5.	Poly(alkylene polysulfide)	$—RS_3—$
6.	Polysulfone	$—ArSO_2—$
7.	Polyester	—RC(=O)O—
8.	Polycarbonate	—ROC(=O)O—
9.	Polyamide	—RC(=O)NH—
10.	Polyurea	—RNHC(=O)NH—
11.	Polyurethane	—ROC(=O)NH—
12.	Phonol-formaldehyde[b]	(phenol ring with OH, CH_2 bridges)

13.	Urea-formaldehyde[c]	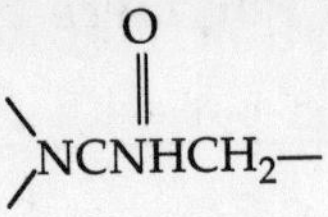 $>NC(=O)NHCH_2-$
14.	Velamine-formaldehyde[d]	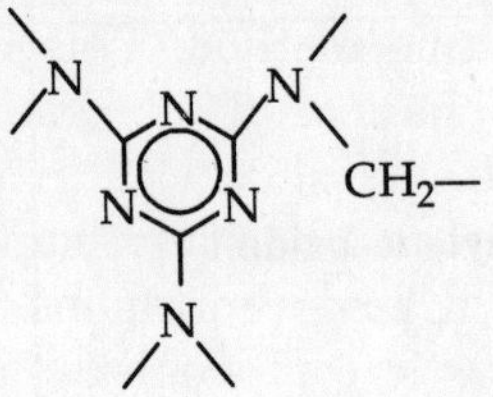 triazine ring with three $>N-$ groups, CH_2-
15.	Polyimide	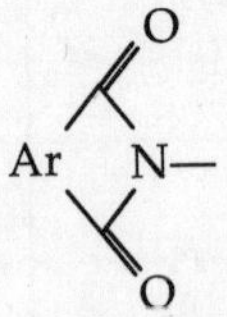 Ar(C=O)(C=O)N—
16.	Polybenzinidazode	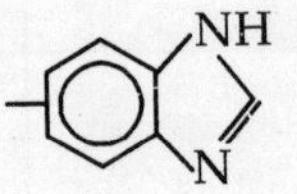 benzimidazole ring (NH, N)
17.	Polyphenylene	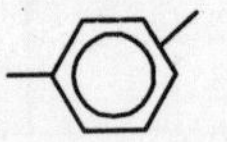 phenylene ring
18.	Poly(p-xylylene)	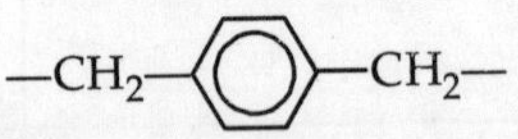 $-CH_2-C_6H_4-CH_2-$

(a) Formed from epoxide terminated oligomer.
(b) Complex network soucture
(c) Network linear.

Polymer Classification According To Preparation Methods:

Preparation methods for polymers may be roughly divided into two categories:

(A) Condensation Polymerization Methods.

(B) Addition Polymerization Methods.

(A) Condensation Polymerization Methods:

These methods are usually employed for low molecular weight functional group reactions, where the stoichiometric proportions of the reactions are fixed for the desired final products. During processing, solvent addition may or may not be important. These reactants are usually mixed in a batch reactor with controlled temperature and addition of catalysts. Vacuum processing or AzeotropicDistillation may be employed to remove condensation products such as H_2O to obtain a higher degree of polymerization.

(B) Addition Polymerization Methods:

These methods are carried out with controlled thermal and catalyst conditions. They may be further classified as—

(i) **Bulk Polymerization:** This mode of polymerization may be employed to obtain the purest form of polymer, and the greatest yield of polymer per unit volume may be obtained using this method. This method involves only the monomer molecule, an initiator and a chain transfer agent (if necessary). In a large scale batch form, the process is to be run slowly or in continuous mode with large heat transfer area. Casting of objects of various shapes may be accomplished using the Batch Bulk Polymerization. Using continuous bulk polymerization processes, polystyrene and other thermoplastic compounds may be moulded.

(ii) **Solution Polymerization:** An inert solvent is added to the reacting components in the reaction vessel in this process. The solvent enhances the heat capacity, thereby reducing the viscosity and facilitating heat transfer. Some of the solvent may be refluxed to remove heat from the reaction vessel. But, much of the reactor space is taken up by the solvent. As such, compared to Bulk Polymerization, Solution Polymerization reduces both the reaction rate and the molecular weight of the compounds. Additional batch & continuous processes are used to separate the final polymer product and to recover and store the solvent used.

(iii) **Suspension Polymerization:** In order to control the enormous amount of heat release in Bulk Polymerization, Suspension Polymerization method was developed. The reaction mass is dispersed as minute droplets of size 0.01 - 1 mm in diameter in a continuous aqueous phase. Each of these droplets act as tiny bulk reactors. Heat transfer occurs from the droplets to the water having large heat capacity and low viscosity. Cooling jackets are used to facilitate heat removal. Agitators are used along with suspending agents in the aqueous phase inorder to maintain a specific droplet size and dispersion.

The Suspension Polymerization process cannot be run in a continuous mode, since, such a system has stagnant corners where polymer accumulation may occur. On a commercial scale, Suspension Polymerization is carried out in jacketed, stainless steel or glass - lined stirred tanks, which may have a capacity of 75 m^3.

Suspension Polymerization produces small, uniform polymer spheres. These are used directly, or may be extruded and chopped to form larger, moulded pellets.

(iv) **Emulsion Polymerization:** This is the most widely used method of polymerization. This process overcomes the difficulty of heat control and low degree of polymerization. The organic reaction mass is emulsified with soap in a continuous aqueous phase. The dispersed particles are smaller in size than in Suspension Polymerization (< 0.1 mm) . In addition, due to stabilizing action of soap, the emulsion obtained is stable and agitation may not be necessary. In some methods, a water soluble initiator may be used. The main product of Emulsion Polymerization is latex, which forms the basis of the popular latex paints. By coagulating the latex with ionic salts and acids, solid rubber may also be obtained.

(v) **Homogeneous Polymerization:** In case of homogeneous bulk polymerization, the feed is a gas , liquid or solid monomer. No initiators or additives are used. For homogeneous Solution polymerization, the monomer is completely dissolved in a solvent.

(vi) **Heterogeneous Polymerization:** In heterogeneous Emulsion polymerization, the monomer molecules are emulsified in aqueous media in the form of micelles. For heterogeneous Suspension polymerization, the monomer is suspended in an aqueous or other type of media as large droplets.

Free Radical Polymerization Technique—Their advantages & disadvantages are tabulated below:

S.N.	Method	Advantages	Disadvantages
1.	Bulk	Simple; no contaminants added	Reaaction exotherm difficult to control; high viscosity
2.	Suspension	Heat readily dispersed; low viscosity; polymer obtained in granular form and may be used directly	Washing and/or drying required; agglomeration may occur contamination by stabilizer
3.	Solution	Heat readily dispersed; low viscosity; may be used directly as solution	Added cmt of solvent: solvent difficult to remove; possible chain transfer with solvenl; possible; environmental pollution
4.	Emulsion	Heat readily dispersed; low viscosity high molecular weight obtainable; may be used directly as emulsion: works on Tacxy polymers	Contamination by cmulsifier and other ingredients; chain transfer agents often needed to control DP: washing and drying necessary for bulk polymer

Commercially Important Polymers Prepared by lonic Polymerization are listed below:

S.N.	polymer or copolymer	Major Uses
	Cationic[a]:	
1.	Polyisobutylene and polybutenes[b] (low and high molecular weight)	Adhesive, sealants, insulating oils, lubricating oil and grease additives, moisture barriers
2.	Isobutylene-isoprene copo1ymer[c] ("butyl rubber")	Inner tubes, engine mounts and springs, chemical tank linings, procective clothing, hoses, gaskets, electrical insulation
3.	Isobutylene~Cyclopentadiene copolymcr	Ozone-resistant rubber
4.	Hydrocarbon[d] and polyterpene resins Coumarone-indone resins	Inks, varnishes, paints, adhesives, sealants, Flooring, coatings, adhesives
5.	Poly(vinyl ether)s	Polymer modifiers, tackificrs, adhesives
	Anuibuc[f]	
6.	cis-1,4-Polybutadiene	Tires
7.	cis-1,4-Polyisoprene	Tires, footware, adhesives, coated fabrics

8.	Styrene-butadiene rubber (SBR)[g]	Tire trads, belting, hose, shoe soles, flooring, coated fabrics
9.	Strene-butadiene block and star copolymers	Flooring, shoe Soles, artificial leather, wire and cable insulation
10.	ABA block copolymers (A – styrene, B = butadiene or isoprcnc)	Thermoplastic elastomers
11.	Polycyanoacrylate[h]	Adhesives

(a) $AlCl_3$, and BF_3 most frequently used co-initiators.

(b) polybutenes are copolymers based on C_4 alkenes and lesser amounts of propylene and C_5 & higher alkenes from refinery streams.

(c) Terpolymers of isobutylene isoprene and divinylbenzene are also used in sealanr and adhesive formulations.

(d) Aliphatic and aromatic refinery products.

(e) Coumarone (benzofuran) and indene (benzocyclopentadiene) are products to coal tar.

(f) n-Butyllithium most common initiator.

(g) Contairs higher cis content thar SBR prepared by free radical polymerisation.

(h) Monomer polymerized by adventitius water.

Polymer Classification According to physical properties

Polymers can also be classified according to physical properties as:

(i) Thermoplastic

(ii) Thermosetting

(iii) Elastomers

(iv) Fibers

(i) **Thermoplastic:** The polymers in this category are composed of monomers which are linear or have moderate branching. They can be melted repeatedly and casted into various shapes and structures. They are soluble in solvents, but do not have appreciable thermal resistance properties. Vinyls, cellulose derivatives, polythene and polypropylene fall into the category of thermoplastic polymers.

(ii) **Thermosetting:** There are some polymers which, when heated, decompose, and hence, cannot be reshaped. Such polymers have a complex 3-D network (cross-linked or branched) and are called Thermosetting Polymers. They are generally insoluble in solvents and have good heat resistance quality. Thermosetting polymers include phenol-formaldehyde, urea-aldehyde, silicones, allyls, polycondersation products, protein-aldehydes etc.

(iii) **Elastomers:** These are resistant solids which have considerable flexibility. They are composed of polymers with glass transition temperature below room temperature. One major difference between elastomers and plastics is that the elastomer is in a liquid state, while plastics are in the glassy state. Examples of elastomers are Butadiene, Butadiene co-polymers and their derivatives, polycondensation products, silicones, thiokols, isobutylenes etc.

(iv) **Fibers:** These are solids which can form thread like structures and have high tensile strength.

Examples of fibers are Polyamides, Polyesters, Polyurethanes, Protein derivatives, polycarbamides, polyvinylidine chloride and cellulose & polyvinyl derivatives.

Polymer Classification According to Applications:

On the basis of applications, polymers can be further classified as:

(i) **Adhesives:** Some polymers can be used for bonding. They are usually of the resin type and require some water resistance . Some common adhesives are:

Cellulose adhesives ; which consist of cellulose derivatives dissolved in a solvent. eg,Duco cement.

Vinyls ; these are rubber base type water-emulsified latex adhesives. Apart from these, some cheap natural products such as starch, dextrins, proteins and natural rubber may also be used for adhesive formulations.

(ii) **Coatings and films:** A large bulk of the polymers produced are used for manufacturing coatings and films. Free films of polyethylene and cellulosic types, protective & decorative coatings are the products of the polymer industry.

Coatings can be manufactured by solvent evaporation followed by polymerization. Emulsion and casting or extrusion of free films by mechanical methods can also produce coatings and films.

(iii) **Fibers:** These are formed by extrusion or spinning of linear monomer molecules into thin sections of diameter in the range of 10 - 50 microns.

Fibers have excellent tensile strength, creep and resilience. Nylon, dacron, rayon are examples of synthetic fibre.

Fibers are extensively used in the textile industry . Cotton, wool etc. are the examples of some natural fibers.

A summary of Classification of Polymers is tabalated below:

Basis of Classification	Polymer Types	Examples
1. Origin	(a) Natural (as available in nature)	Natural rubber, natural silk, cellulose, proteins, starch etc.
	(b) Semi - synthetic (Man - made)	Hydrogenated, halogenated and cyclo (natural) rubber; cellulosics (cellulose esters/ethers), etc.
	(c) Synthetic (Man - made)	Polyethylene, polypropylene, polystyrene, polybutadiene, nylon poly-amides, polycarbonates, phenolics, amino resins, epoxy resins etc.
2. Thermal response	(a) Thermoplastics (they soften or melt on heating and harden on cooling over	Polyethylene, polypropylene, polystyrene, nylon polyamides, linear polyester [poly (ethylene terephthalate)], etc.

	many cycles of heating & cooling and retain solubility & fusibility).	
	(b) Thermo setting (they usually soften or melt initially on heating, but fast undergo chemical changes to finally turn insoluble and infusible).	Phenolic resin, amino resins, epoxy resins, diene rubbers (vulcanized), unsaturated polyesters.
3. Mode of formation	(a) Chain - growth or addition	Polyethylene and other polyolefins, Polystyrene and related vinyl polymers etc.
	(b) Step - growth or condensation	Polyesters and polyamides, polycarbonates, phenol (urea, melamine)-formaldehyde resins, epoxy resins etc.
4. Line Structure	(a) Linear (having no branches)	High density polyethylene (HDPE), polyvinyls, bifunctional (polyesters and polyamides) etc
	(b) Branched (having branches)	Low density polvethylene (LDPE), higher poly (α -olefins), phenolic resoles and resitols, poly (3-hydroxy alkanoates) etc.
	(c) Cross linked or network (having a complex network structure)	Phenolic C-stage (resite) resin, C-stage amino (urea / melamine-formal-dehyde) resins, cured epoxy resin and unsaturated polyester resin etc.
5. Application and Physical properties	(a) Rubbers (showing long range elasticity)	Natural rubber, (1, 4 cis poly isoprene) synthetic rubbers (polybutadiene, SBR, nitrile rubber, polychloroprene rubber, polylacrylate rubber, polyurethane rubbers, silicone rubbers etc.)
	(b) Plastics (shapable under pressure, aided by heat)	Polyethylenes, polypropylene (isotactic), polystyrene, poly (vinyl chloride), nylon polyamides, linear aromatic polyesters and polyamides, polycarbonates, acetal resins etc
	(c) Fibres (available in fibrillar or filamentous form)	Cotton (cellulose), natural silk, artificial silk (rayons), poly (ethylene terephthalate) fibre, nylon polyamide fibres etc.

6. Tacticity	(a) Isotactic (stereoregular)	Poly (α-olefins) and related polymers.
	(b) Syndiotactic (do)	
	(c) Atactic (stereo irregular)	
7. Crystallinity	(a) Crystalline (crystallinity, $\geq 50\%$)	Polyethylene (HDPE and LDPE), polypropylene (isotactic), stretched nylon polyamides, polyoxymethylene etc, cellulose (cotton) fibre.
	(b) Semi - crystalline (crystallinity, 30 - 50%)	Polybutene, cellulosics [cellulose esters (rayons) particularly if stretched], Gutta percha (1,4 trans poly isoprene) etc.
	(c) Amorphous or non-crystalline (crystallinity $< 25\%$)	Natural rubber and most synthetic rubbers,N-alkylated($>$15%alkylation) nylon polyamides, poly (methacrylates and acrylates) poly (vinyl acetate), polystyrene etc.

Physical Properties of Polymers

Polymers satisfy need for materials of unique physical & chemical properties and/or materials which are competitive with other sources such as wood & metals.

Research scientists are interested in certain of the physical properties of polymers to help explain structural modes. In particular, the colligative behavior of polymers in solvents is important in molecular weight determinations.

Thus, the physical properties of polymers is very importan.

List of Important Physical Properties of polymers:

An evaluation of a polymer for a specific requirement will encompass many of the property measurements listed below:

Optical Properties:

- Odor
- Hardness
- Density
- Weatherability

Mechanical Properties:

- Tensile strength
- Compressive strength
- Flexural strength
- Shear strength
- Impact resistance
- Rigidity and brittleness
- Ductility
- Creep
- Fatigue
- Dimensional stability
- Sound transmission

Thermal Properties

- Thermal diffusivity
- Heat resistance
- Coefficient of expansion
- Flammability

Electrical Properties

Resistivity

Dielectric

Power factor

Arcing resistance

Solvent Properties

Solubility

Plasticizer compatibility

Viscosity

Chemical Resistance to:

Chemical solutions

Oils and fats

Solvents

OUTLINE OF CERTAIN PHYSICAL PROPERTIES OF POLYMERS

(i) Molecular weight

(a) **Determination of molecular weight distribution which dictates many of the physical properties:** Fractional precipitation from solvents gives most accurate distribution curves.

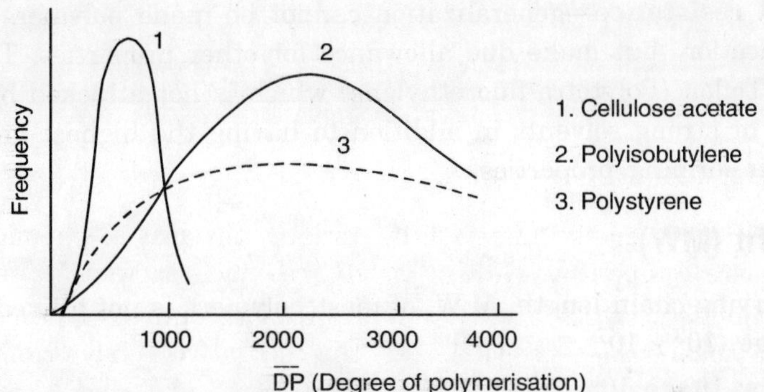

In general, polymers with narrow band distribution of molecular weights have more predictable physical properties.

—average molecular weights can be determined from empirical formula:

$$N = K(M_v)^u,$$

where M_v is somewhere between the number average and weight average molecular weight; this must be considered in solvent polymerization processes.

(b) **Plasticizer compatibility**—high boiling esters are frequently compounded with polymers to decrease brittleness and increase plastic flow.

(ii) Optical Properties—luster, clarity, reflectivity, and permanent coloring are real advantages of polymers.

(iii) Hardness—most plastics have low scratch resistance, can be improved by lamination with glass fibers.

(iv) Density—gravity range from 0.85 to 2.5 with majority around 1.0; have weight advantage over most materials of construction.

(V) Weatherability—varies, but in general most plastics will darken when exposed to sunlight; moisture resistance excellent.

(vi) Mechanical properties:

Plastics exhibit a wide variety of mechanical properties and generalities cannot be used safely to predict performance. Polymers should be tested under expected conditions to guarantee performance. It is safe to say that Plastics are not as strong as metals and much more sensitive to temperature changes. Improper mechanical design using plastics is largely responsible for the poor service record of polymers.

(vii) Thermal properties:

(a) **Thermal diffusivity**—low, giving good thermal insulation characteristics, particularly when gas-foamed.

(b) **Heat resistance**—poor, ranging from 35 to 260°C in continuous service.

(c) **Flammability**—resistance to ignition & burning poorer than for metals and ceramics.

(viii) Electrical properties—insulators with high dielectric strength, low dielectric constant and low power factor. The last two minimize power losses; polyethylene, Teflon and polystyrene are superior to glass & ceramics in UHF insulation applications.

(ix) Chemical resistance—generalization cannot be made polymers can be found for each application, but make due allowance for other properties. The most resistant plastic is Teflon (Polytetra-fluorethylene) which is not attacked by highly corrosive chemicals or strong solvents in addition to having the highest heat resistance, but the poorest forming properties.

Molecular weight (MW)

Because of varying chain length, M.W. of most polymers is not a fixed value, but varies over a certain range (10^3 - 10^7).

Vinly polymers: 10^5 to 10^6

Very polar functional groups: 15,000 - 20,000

Crystallinity

Because of varying chain length and high M.W., most polymers are amorphous and only semi-crystaline polymers with higher crystalinity are generally tougher, stiffer, more opaque, more resistant to solvents, and of higher density than polymers of low crystallinity. Also have more sharply defined melting points. Higher the degree of crystallinity, more pronounced the properties.

The superior mechanical properties of highly crystalline polymers are a reflection of greater cohesive strength arising from more effective intermolecular forces among the closely packed molecules. Opaqueness arises from scattering of light by the crystallites.

Polymer crystallinity is increased by:

- Highly stereoregular structure
- Little or no chain branching
- Highly polar groups

Glass Transition Temperature (Tg)

At sufficiently low temperatures, even amorphous polymers are hard and brittle (glass-like) As temperature is increased kinetic energy of molecules increases, but motion is still restcicted to short-range vibrations and rotations, so long as the polymer retains its glass-like structure.

At a certain temperature, called the Glass Transition Temperature, T_g, the polymer loses its glass-like properties and becomes softer & more elastomeric, but does not melt. Above T_g,molecular motions are more long-range, involving segmental motion of 20–50 atom chains.

If heating continues further, polymer will eventually lose its elastomeric properties and melt to a flowable liquid. This melting point (M.P) is usually not sharply defined.

Since T_g is a function of rotational freedom, any structural feature that limits rotation will result in a higher T_g:

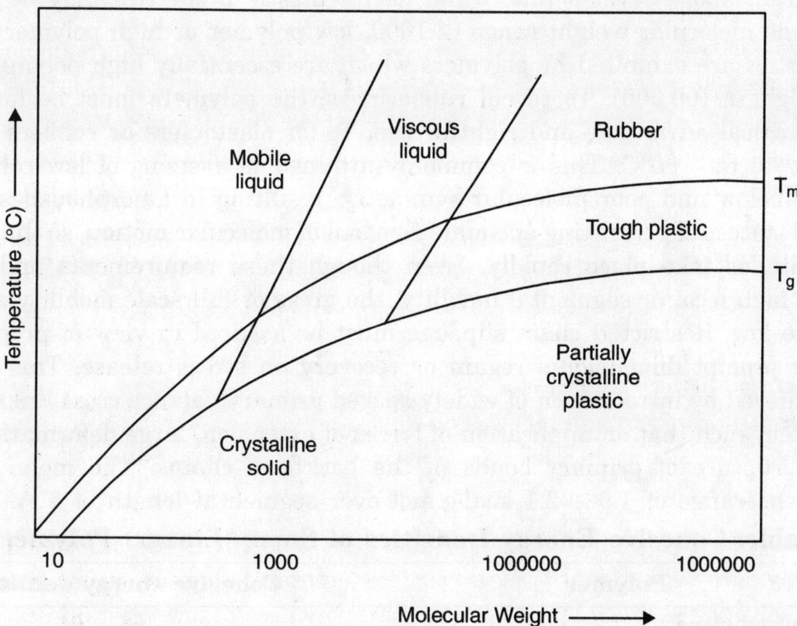

Figure 33.1: The dependence on molecular weight of melting & glass transition temperatures and physico-chemical properties of polymers is shown above.

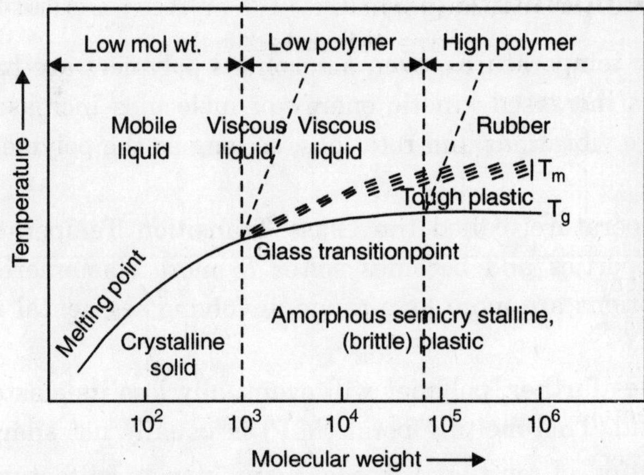

Figure 33.2: Temperature-molecular weight plot of polymers.

The diagram above reveals that while useful plastic properties may be exhibited by polymers of wide molecular weight range (≥ 1000, low polymer or high polymer), good useful rubbery properties are exhibited by polymers which are essentially high polymeric in nature (molecular weight ≥ 100,000). To reveal rubberiness, the polymers must be far above their T_g, and for practical advantage and rightful use, T_g for elastomers or rubbers should be in the range of – 40 to – 80°C. This is commonly attained in systems of low cohesive energy density (Table below and poor molecular symmetry, resulting in amorphousness, at least in the unstrained state and permitting adequate freedom of molecular motion, so that deformation of high magnitudes take place rapidly. Even though these requirements and features for rubbers imply high local or segmental mobility, the gross or full scale mobility as in the flow region must be low. Restricted chain slippage must be assured in view of property demand in the form of prompt dimensional regain or recovery on stress release. This restriction is technically achieved by introduction of widely spaced primary valance cross links in the chain molecular system such that on application of forces of extension, large deformations may take place without rupture of primary bonds of the backbone chains. The molar cohesion for rubbers is in the range of 1.9 – 2.1 kcal/g.mol over segmental length of 5 Å.

Table: Cohesive Energy Densities of Some (Linear) Polymers

S.N.	Polymer	Cohesive energy density, cal/cm.
1.	Polyethylene	56 – 64
2.	Polystyrene	75 – 88
3.	Polyisobutylene	56 – 64
4.	Polyisoprene (natural rubber)	60 – 67
5.	Poly (vinyl acetate)	82 – 94
6.	Poly (vinyl chloride)	85 – 95
7.	Poly (methyl methacrylate)	78 – 85
8.	Poly (ethylene terephthalate)	90 – 115
9.	Poly (hexamethylene adipamide), nylon 66	180 – 220
10.	Polyacrylonitrile	150 – 230

Table: Glass Transition Temperature (T_g) of Representative Vinyl Polymers are tabalated below.

S.N.	Polymer $\left[CH_2CH(R) \right]$	R	T_g(°C)
1.	Polyethylene	H	–20
2.	Polypropylene	CH_3	5
3.	Poly(1-butene)	CH_3CH_2	–24
4.	Poly(1-pentene)	$n\text{-}C_3H_7$	–40
5.	Poly(1-hexene)	$n\text{-}C_4H_9$	–50
6.	Poly(1-heptene)	$n\text{-}C_5H_{11}$	–31
7.	Poly(1-decene)	$n\text{-}C_8H_{17}$	–41
8.	Poly(1-dodecene)	$n\text{-}C_{10}H_{21}$	–6
9.	Poly(3-methyl-1-butene)	$CH(CH_3)_2$	50
10.	Poly(4-methyl-1-pentene)	$CH_2(CH_3)_2$	29
11.	Poly(3,3-methyl-1-butene)	$C(CH_3)_3$	64
12.	Poly(5-methyl-1-hexene)	$CH_2CH_2CH(CH_3)_2$	–14
13.	Poly(4,4-dimethyl-1-pentene)	$CH_2C(CH_3)_3$	59
14.	Poly(vinyl *n*-butyl ether)	$O\text{-}C_4H_9\text{-}n$	–55
15.	Poly(vinyl *t*-butyl ether)	$O\text{-}C_4H_9\text{-}t$	88
16.	Poly(vinyl chloride)	Cl	81
17.	Poly(vinyl alcohol)	OH	85
18.	Polystyrene	C_6H_5	100
19.	Poly(2-vinylnaphthalene)	2-naphthyl (structure)	151
20.	Poly(4-vinylpyridine)	4-pyridyl (structure), N	142

Table: Glass Transition Temperature (T_g) of Representative Nonvinyl Polymers are tabulated below.

S.N.	Polymer	T_g (°C)
1.	$-\overset{\overset{O}{\|}}{C}(CH_2)_4\overset{\overset{O}{\|}}{C}O(CH_2)_2O-$	–63
2.	$-\overset{\overset{O}{\|}}{C}(CH_2)_4\overset{\overset{O}{\|}}{C}O(CH_2)_4O-$	–118
3.	$-\overset{\overset{O}{\|}}{C}-C_6H_4-\overset{\overset{O}{\|}}{C}O(CH_2)_4O-$	69
4.	$-\overset{\overset{O}{\|}}{C}-C_6H_4-\overset{\overset{O}{\|}}{C}O(CH_2)_2O-$	80
5.	$-O-C_6H_4-\underset{CH_3}{\overset{CH_3}{\overset{\|}{\underset{\|}{C}}}}-C_6H_4-O\overset{\overset{O}{\|}}{C}-$	149
6.	$-\overset{\overset{O}{\|}}{C}(CH_2)_4\overset{\overset{O}{\|}}{C}NH(CH_2)_6NH-$	57
7.	$-\overset{\overset{O}{\|}}{C}(CH_2)_4\overset{\overset{O}{\|}}{C}NH(CH_2)_{10}NH-$	50
8.	$-NH(CH_2)_5\overset{\overset{O}{\|}}{C}-$	77
9.	$-NH(CH_2)_{10}\overset{\overset{O}{\|}}{C}-$	46
10.	$-O-C\langle^{N}_{NH}\rangle C_6H_3-C_6H_4-C_6H_3\langle^{NH}_{N}\rangle C-$	429

Table: Glass Transition Temperature (T_g) of some DienePolymers are tabulated below.

	Polymer	Tg (°C)	
		cis	trans
1.	1, 4-Polybutadiene	–102	–58
2.	1, 4-Polyisoprene	–67	–70
3.	1, 4-Polychloroprene	–20	–40

Properties of common thermoplastics and thermosets are tabulated below.

S.N.	Polymer	Mer structure	Structural state	Tensile strength (MPa)	Impace strength ($J\ m^{-1}$)[a]	Density ($g\ cm^{-3}$)	T_s (°C)[b]	T_{max} (°C)[c]	Applications
1.	*Thermoplastics:* ABS	copolymer of acrylonitrile,	glassy copolymer with rubbery domains	40	200–600	1.03	–	71-93	refrigerator linings, lawn and garden equipment
2.	Polyamides: (e.g. nylon 6,6)	NH–$[CH_2]_6$–NH O O | | –C–$[CH_2]_4$–C–	highly crystalline, spherulitic structure, H-bonding	80	100	1.15	57	82–150	bearings, gears, cams, bushings, handles, wire coatings
3.	Polycarbonate	CH_3 O | | O–ϕ–C–ϕ–O–C– | CH_3	stiff molecular structure	62	650–850	1.20	150	120	machine parts, car parts propellers, instrument panels, CDs
4.	LD polyethylene[d]	[–CH_2–CH_2–]	branched (60% C)[r]	14	50–700	0.92	–110	82–100	flexible sheet, bottle, toys
5.	HD polyethylene[c]		linear (95% C)[r]	28	25–100	0.95	-90	80–120	batteries, tanks, ice trays
6.	Polypropylene	[–CH_2 –C(CH_3)H–]	isotactic (amorphous)	34	25–100	0.91	–10	107–150	bottles, packing, TV cabinets
7.	Polystyrene	[–CH_2–C(ϕ)H–]	atactic (amorphous)	50	12–20	1.09	100	104	styrofoam, wall tile battery cases, toys, lighting
8.	Polyvinyl chloride (PVC)	[–CH_2–C(Cl)H–]	largely amorphous	40	50–125	1.4–1.5	105	110	pipe, garden hose electrical insulation
9.	Polyethylene icrephthalate (PET)	O–CH_2–CH_2 O O | | –O–C–ϕ–C	crystalline	54	12–35	1.15	70–80	80–120	fiber, film, magnetic carpet, clothing, tires

10.	*Thermosets:* Phenolic	Benzene ring with three CH_2 groups	amorphous, wood and glass reinforced	52	12–900	1.3–1.9	150–288	electrical equipment, motor housings, insulators, brake and ransmission parts
11.	Polyester	$OH–CH_2$ and $OH–CH_2$ joined to $CH–O–C(=O)–(CH_2)_x–C(=O)–OH$	amorphous, glass reinforced (fiberglass)	28–138	400–800	1.7–2.3	150–177	car panels and body parts, boat hulls, indoor construction panels, pipes, tanks, ducts

[a] Based on notched izodtest.
[b] Glass to liquidtransition temperature.
[c] Maximum use temperature (no load).
[d] LD = low density.
[e] HD = high density.
[f] C = crystallinity.

Effect of Variation of Polymer Structure on Properties and Application Prospects are tabulated below.

S.N.	Polymer	Repeat Unit	T_g, °C	T_{m},°C	Application prospects/ remarks
1.	Polyethylene	$-CH_2-CH_2-$	– 115, – 60	137	General purpose, low modulus polymer with good degree of toughness; excellent electrical insulator.
2.	Polystyrene	$-CH_2-CH-$ (with phenyl ring)	80 to 100	240	Rigid, brittle transparent polymer; excellent insulator.
3.	Poly vinyl naphthalene PVN	$-CH_2-CH-$ (with naphthyl ring)	150	360	Highly, rigid, brittle polymer; excellent insulator.
4.	Poly (para-phenylene) PPP	–(p-phenylene)–	---	530	Tough, thermostable, intracable polymer with semi-conducting or even good conducting (when doped) properties with electrical conductivity in the range 10^{-18} to 10^{2} $(\Omega\ cm)^{-1}$
5.	Poly(tetramethylene Oxide), PTMO	$-(CH_2)_4-O-$	80	60	Weak, flexible polymer.
6.	Poly (phenylene Ether), PPE	–(p-phenylene)–O–	80	300	A good engineering thermoplastic having high strength and toughness.
7.	Poly phenylene Oxide, PPO	–(2,6-dimethyl phenylene, CH_3, CH_3)–O–	209	---	Very low water absorption, more resistant (chemically, mechanically and thermally) than PPE.
8.	Poly tetramethylene	$-(CH_2)_4-S-$	---	70	Weak, flexible polymer.
9.	Poly (phenylene	–(p-phenylene)–S–	90	295	Strong, tough, thermally resistant polymer. Normally a good insulator But when properly doped, say with 2-5% AsFs, it assumes a semiconducting character.

10.	Poly tetramethylene	$-(CH_2)_4-SO_2-$	---	270	Presence of strongly polar sulphone ($-SO_2-$) links makes the polymer mechanically strong and thermally more stable, as can be realized by comparing PTMSO with PTMS or PTMO.
11.	Poly phenylene sulphone, PPSO	$-C_6H_4-SO_2-$	350	525	Mechanically very strong and thermally stable upto 500°C; high performance plastic.
12.	Poly ether sul-Phone, PES	$-C_6H_4-SO_2-C_6H_4-O-$	230	---	Thermostable, high performance transparent thermoplastic with improved processability in comparison with PPSO.
13.	Poly ether ketone PEK	$-C_6H_4-C(=O)-C_6H_4-O-$	155	365	High performance thermoplastic with excellent environmental stress cracking resistance; continuous use temperature nearly 275°C
14.	Poly ether ether ketone, PEEK	$-C_6H_4-C(=O)-C_6H_4-O-C_6H_4-O-$	140	343	High performance, chemical resistant thermoplastic, showing a high continuous use temperature (250°C)

Rigid, bulky aromatic ring substituents make a vinyl polymer more rigid and brittle. When aromatic rings constitute the chain backbone, the polymer becomes rigid, stiff, strong and tough, -S- and –O- inter unit linkages make the polymer weak, relatively soft and flexible, while highly polar $-SO_2-$ linkages along the chain backbone make the polymer more rigid, strong and thermostable.

Incorporation of rigid, bulky groups, such as aromatic rings as side groups in the polymer chain renders it more stiff and higher melting. Comparison of relevant features of polyethylene, polystyrene and poly (vinyl naphthalene) brings home the point.

Polymer additives used for modifying it svarios properties are tabulated below.

S.N	Additive Type	Functions of additives
	Mechanical property modifiers:	
1.	Plasticizers	Increase flexibility
2.	Impact modifiers	Improve impact strength
3.	Reinforcing fillers	Increase strength properties
4.	Nucleatingagents	Modify crystalline morphology
	Surface property modifiers:	
5.	Slip and antiblocking agents	Prevent film and sheet sticking
6.	Lubricants	Prevent sticking to machinery
7.	Antistatic agents	Prevent static charge on surfaces
8.	Coupling agents	Improve bonding between polymer and filter
9.	Wetting agents	Stabilize dispersions of filter
10.	Anti forgging agents	Disperse moisture droplets on films
	Chemical property modifiers:	
11.	Flame relardants	Reduce flammability
12.	Ultraviolet stabilizers	Improve light stability
13.	Antioxidants	Prevent oxidative degradation
14.	Biocides	Prevent mildew
	Aesthetic property modifiers:	
15.	Dyes and pigments	Impart color
16.	Odorants	Add fragrance
17.	Deodorants	Prevent development of odor
18.	Nuclcating agents	Improve light transmission
	Processing modifiers:	
19.	Plasticizers	Reduce melt viscosity
20.	Slip agents and lubricants	Prevent sticking to processing machinery
21.	Low-profile additives	Prevent shrinkage and warpage
22.	Thickening agents	Increase viscosity of polymer solutions or dispersions
23.	Heat stabilizers	Prevent degradation during processing
24.	Defoaming agents	Reduce foaming
25.	Blowing agents	Manufacture stable foams
26.	Emulsifiers	Stablize polymer emulsions
27.	Cross linking (curing) agents	Crosslink polymer
28.	Promoters	Speed up crosslinking (curing)

Polymer additives used for modifying it svarios properties are tabulated below.

S.N	**Aromatic**
1.	Di-2-ethylhexyl phthalate
2.	Di-*n*-octyl phthalate
3.	Di-*i*-octyl phthalate
4.	Di-*i*-decyl phthalate
5.	Di-*n*-undecyl phthalate
6.	Di-*n*-tridecyl phthalate
6.	Tri-2-ethyhexyl trimellitate
	Aliphatic:
8.	Di-2-ethylhexyl adipate
9.	Di-2-ethylhexyl sebcate
10.	Di-2-ethylhexyl azelate
	Epoxy:
11.	Epoxidized linseed oil
12.	Epoxidized soya oil
	Polymeric:
13.	Poly(alkylene adipates, sebacates, or azelates)
	Fire retardant:
14.	Chlorinated paraffins
15.	Phosphate esters

BINARY COMMON SYNTHETIC POLYMERS

Thermoplastics: The left panels below show the polymer name & synonyms, structural formula, glass transition temperature (T_g) and melting point/decomposition temperature (T_m).

1.	**Polycarbonate** (Lexan(R) O—⟨benzene ring⟩—C—⟨benzene ring⟩—O—C(=O)— T_g = 145°C, T_m = 250°C.	This polymer was discovered independently in Germany and the U.S. in 1953. Lexan is exceptionally hard and strong; we see it most commonly in the form of **compact disks.** It was once widely used in water bottles, but concerns about leaching of unreacted monomer (bisphenol-A) has largely suppressed this market.
2.	**Polyethylene terphethalate** (PET, **Mylar**) $(O=C$—⟨benzene ring⟩—$C(=O)$—O—CH_2—$CH_2^-)_n$ T_g = 76°C, T_m = 250°C.	Thin and very strong films of this material are made drawing out the molten polymer in both directions, orienting the molecules into a highly crystalline state becomes "locked-in" on cooling. Its many applications include food packaging (in foil-laminated drink container and microwaveable frozen-food containers), overhead projector film,

		weather balloons, and as aluminum-coated reflectivematerial in spacecraft and other applications.
3.	**Nylon** (*a polyamide*) 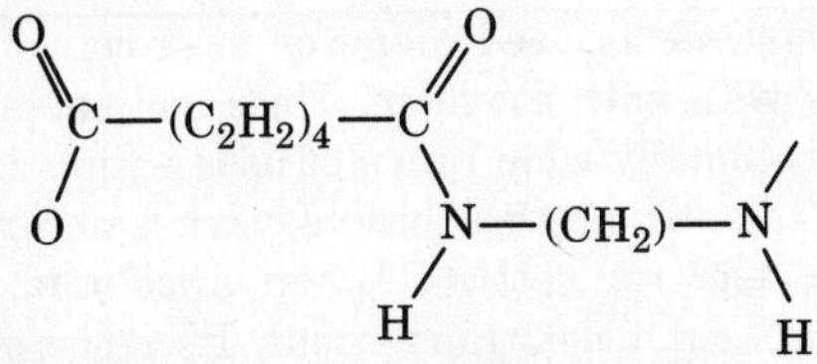 O　　　　　　　O C—$(C_2H_2)_4$—C O　　　N—(CH_2)—N H　　　　　　H T_g = 145°C, T_m = 255°C.	Nylon was invented by *DuPont* chemist *Wallace Carothers.* The comman form **Nyclon 6.6** has six carbon atoms in both parts of its chain; there are several other kinds. The two copolymer sub-units are held together by *peptide*bonds, the same kinds that join amino acids into proteins. Nylon 6-6 has good abrasion resistance and is self-lubricating, which makes it a good engineering material is also widely used as a fiber in carpeting, clothing, and tire cord.
4.	**Polyacrylonitrile** (Orlon, Acrilan, "acrylic" fiber) (–CH(CN)–CH_2–) T_g = 85°C, T_m = 318°C.	Used in the form of fibers in rugs, blankets, and clothes especially cashmere-like sweaters. The fabric is very soft but tends to "pill" i.e., produce fuzz-like blobs. Owing to its low glass transition temperature, it requires careful treatment in cleaning and ironing.
5.	**Polyethylene** (–CH_2–CH_2–) T_g = –78°C, T_m = 100°C.	Control of polymerization by means of catalysts and additives has led to a lage varietyof materials based polyethylene that exhibit differences in densities, degree of chain branching and crystallinity, and cross-linking. Some major types are low-density (LDPE), linear low density (LLDPE), high-density (HDPE). LDPE was the first commercial form (1933) and is used mostly for ordinary *"plastic bags"*, but also for food containers and in six-pack soda can rings. Its low density is due to long-chain branching that inhibits close packing. LLDPE has less branching; its greater toughness allows its use in those annoyingly-thin plastic bags often found in food markets. A "very low density" from (VLDPE) with extensive short chain branching is now used for plastic stretch wrap (replacing the original component of Saran Wrap) and flexible tubing. HDPE has mostly straight chains and is therefore stronger. It is widely used in milk jugs and similar containers, garbage containers, and as an "engineering plastic" for machine parts.
6.	**Polymethylmethacrylate** (Plexiglass, Lucite, Perspex)	This clear, colorless polymer is widely used in place of glass, where its greater impact resistance, lighter weight, machineability are advantages. It is normally copolymerized with other substances

	$\left(-\overset{CH_3}{\underset{COOCH_3}{C}}-CH_2-\right)$ T_g = 114°C, T_m = 130–140°C.	to improve its properties. Aircraft windows, plastic signs, and lighting penels are very common applications. Its compatibility with human tissues has led to various applications, such as replacement lenses for cataract patients.
7.	**Polypropylene** $\left(-\overset{CH_2}{\underset{O}{C}}-CH_2-\right)_n$ T_g = –10°C, T_m = 173°C.	Polypropylene is used alone or as a copyolymer, usually with with ethylene. These polymers have an exceptionally wide range of uses – rope, binder covers, plastic bottles, staple yarns, non-woven fabrics, electric kettles. When uncolored, it is translucent but not transparent. Its resistance to fatigue makes it useful for food containers and their lids, and flip-top lids on bottled products such as ketchup.
8.	**Polystrene** $-\overset{H}{\underset{C_6H_5}{C}}-\underset{H}{C}-$ T_g = 95°C, T_m = 240°C.	Polystyrene is transparent but rather brittle, and yellows under uv light. Widely used for inexpensive packaging materials and "take-out trays", foam "packaging peanuts", CD cases, foam-walled drink cups, and other thin-walled and moldable parts.
9.	**Polyvinyl acetate (PVA)** $\left(-CH_2-\underset{\underset{O=C\;\;CH_3}{O}}{CH}-\right)_n$ T_g = 30°C	PVA is too soft and low-melting to be used by itself. it is commonly employed as a water-based emulsion in points, wood glue and other adhesives.
10.	**Polyvinyl chloride** ("PVC") $\left(-\overset{Cl}{\underset{H}{C}}-CH_2-\right)_n$ T_g = 85°C, T_m = 240°C.	This is one of the world's most widely used polymers. By itself it is quite rigid and used in construction materials such as pipes house siding, flooring. Addition of plasticiser make it soft and flexible for use in upholstery, electrical insulation, shower curtains and waterproof fabrics. There is some effort being made to phase out polymer owing to environmental concerns.
11.	**Synthetic rubbes** $\left(-CH_2-CH=\underset{Cl}{C}-CH_2-\right)_n$ Neoprene (polychloroprene)	Neoprene, invented in 1930, was the first mass-produced synthetic rubber. It is used for such things as roofing membranes and wet Suits. Polybutadiene substitutes a hydrogen for the chlorine; is the major component (usually admixed with other rubbers) of tires.

	T_g = –70°C $\left(-CH_2-CH=\underset{\displaystyle H}{\underset{\vert}{C}}-CH_2- \right)_n$ **Polybutadiene** T_g< –90°C	SBS(styrene-butadiene-styrene) rubber is a block copolymer whose special durability makes it valued for tire treads.
12.	**Polytetrafluroethylene (Teflon,** PTPF) $\left(-\underset{\displaystyle F}{\underset{\vert}{C}}-\underset{\displaystyle F}{\underset{\vert}{C}}- \right)_n$ Decomposes above 350°C.	This highly-crystalline fluorocarbon is exceptionally inert to chemicals and solvents. Water and oils do not wet it, which accounts for its use in cooking ware and other anti-stick applications, including personal care products. It is also employed in *fabric* for rainwear. These properties – non-adhesion to other materials, non-wetability, and very low coefficient of friction ("slipperyness") – have their origin in the highly electronegative nature of fluorine whose atoms partly sheield the carbon chain. Fluorine's outer electrons are so strongly attracted to its nucleus that they are less available to participate in dispersion force interactions.
13.	**Polyaramid (Kevlar)** H N N O NH Sublimation emperature: 450°C.	Kevlar is known for its ability to be spun into fibres that have five times the tensile strength of steel. It was used in the 1970s to replace steel tire cords. Bulletproof vests are one of it more colourful uses, but other applications include boat hulls, drum heads, sports equipment, and as a replacement for asbestos in brake pads. It is often combined with carbon or glass fiber composite materials. The high tensile strength is due in part to the extensive hydrogen bonding between adjacent chains.

POLYMER PROCESSING TECHNOLOGIES

Chapter **34**

All the processing stages between manufacture of the crude polymer and the final molding step are called the compounding processes. These include storage, transportation, metering, mixing, plastication and granulation. Sometimes additives may be added to the polymer mix to allow production on an industrial scale. Addition of plasticizers renders flexibility & ductility to rigid and brittle polymers. In a similar way, fillers can enable to enlarge the volume and also has influence upon the polymer properties.

The various compounding processes are:

(a) Mixing

Mixing allows obtaining a homogeneous mixture of polymer and additives. The additive is distributed uniformly throughout the crude polymer, without any unnecessary stressing. Mixing can occur under the action of gravitational forces in a tumble mixer ; or in a ribbon mixer, where materials are mixed in a horizontal barrel in a spiral manner. Such mixers are collectively known as cold mixers.

Another class of mixers are the hot mixers, wherein the polymers are heated by contact with heated metal components. Such mixers are also known as high – speed or turbo mixers. This method is widely used in the plastic industry, especially in the compounding of poly vinyl chloride (PVC).

(b) Rolling and Kneading

Before processing, the plastics are compacted, melted, homogenized and modified with additives. These include rolling, in which the plastic is passed between two counter rotating, heated rolls wherein it is pressed into a thin layer. It is an open process in which additives can be added, and other scrap materials may be reintroduced. PVC is one plastic which extensively employs this process. Moreover, the material temperature can be controlled precisely and cleaning of the machine is easy.

(c) Pelletizing

It is the method of producing granules of equal size and shape. This enables the optimum feed behaviour on the processing equipment. Pelletizing can be done in hot or cold conditions.

Using a pelletizing die, strands, ribbons or sheets are formed ; and after solidification, they are chopped into the desired shapes. This is one procedure strictly followed in cold pelletizing. However, in hot pelletizing, the plasticated material is sent through a die, and a blade that rotates on the surface of the die plate chops the exiting strands. The melt is then cooled ; and subsequently the water is removed.

(d) Shredding and Grinding

In thermoplastics processing, a variety of semifinished and finished products are generated. These products, in order to be made suitable for processing, must be cut to produce a pellet form. A range of particle sizes is obtained in the process.

Regranulating is achieved through cutting mills. The material is added through a feed system wherein a rotor revolves with high momentum. The system is further equipped with multiple cutting heads that operate against stationary cutting edges. The rotating cutting heads are divided and displaced with respect to each other in some other machines. As soon as the desired particle size is obtained, the material is ejected through a sieve, the pores of which determine the size of the particles. The housing and rotor may be equipped with water cooling apparatus to remove the excessive heat generated.

(e) Storage and Transportation

The storage and transportation of base polymers and intermediate products are achieved through automatic equipments. These equipments have an edge over the manually operated ones as they ensure improved working conditions, lower work place, lower contamination of materials and subsequent reduction losses. Moreover, accident hazards are also reduced. Large external silos are used for the storage of polymers, either in powder form, or in the pellet form. By incorporating vibrating frames, bunker cushions, slotted shelves or similar techniques as ventilation, the formation of granule bridges is avoided. The materials are transported via road or rail. Reserve materials and additives are transported in ships in bulk containers, drums or sacks. Pneumatic equipments are generally preferred for conveying powders and pellets. The conveyors used may be high pressure conveyors, medium pressure conveyors or low pressure conveyors. Fill level sensors help in controlling the transport interval.

Inorder to meter liquids such as plasicizers or other additives for compounding, oscillating displacement pumps are used. Metering of soilds can be done using gravimetric metering equipments or volumetric metering equipments.

TECHNIQUES FOR PRESSURELESS PROCESSING OF POLYMERS

The Pressureless Processing Techniques are used for starting materials such as monomers, plastic solutions and dispersions that can flow and that can be formed without application of external forces. The pressureless processing techniques include - ;

- Casting
- Dipping
- Coating
- Foaming

(a) Casting (Figure 34.1)

By polymerizing monomers in molds, thick walled articles, embeddings, sheets and blocks can be produced. These are preferentially made up of polystyrene, polyamides and polymethylmethacrylate (PMMA). Inorder to avoid thermal stresses, adequate temperature regulation is required. The mold material choice is dependent upon the use and surface quality of the moldings.

Film casting is mainly used in the photographic industry for the manufacture of cellulose acetate films. In this method of casting, polymer solution, melts or dispersions are cast directly into a metal drum or a rotating belt; and from a slot die into precipitation bath.

Slush molding enables the production of hollow articles which are made from PVC pastes. These are made using two part metal molds, which are heated and filled with the paste. The paste gels on the outermost edges forming a skin ; and subsequently the excess paste is poured off. The remaining layer is gelled in an oven and after cooling, a hollow mold is obtained.

Slush molding can be further modified into Rotational casting for the production of hollow articles, made from PVC. The only difference is that in rotational casting, the amount of PVC paste is metered to the mold, which makes the removal of excess paste and post gelation techniques redundant. The mold is rotated about two perpendicular axes, while heating, and this enables uniform wetting of the cavity surface. Roatational casting has the advantage of low investment cost over slush molding. In the manufacture of thick-walled, symmetrical structures such as pipes and sockets, centrifugal casting may be used. A partially filled mold is rotated rapidly about an central axis. Centrifugal force press the material outward in the mold, and in the process, the material gets compacted.

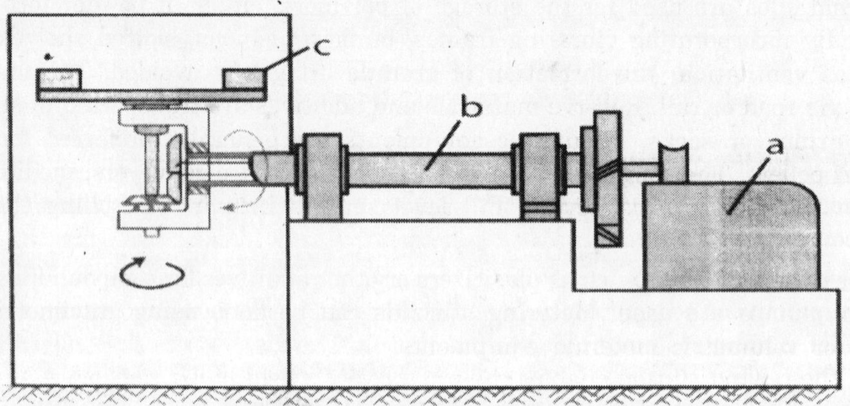

Figure 34.1: Casting Machine
(a) Motor; (b)Drive Shaft; (c) Molds

(b) Dipping (Figure 34.2)

Sometimes, molds or articles are dipped into solutions, dispersions, pastes, melts or powders. Later on, the adhering material on the surface is solidified by heat treatment.

For manufacturing articles such as gloves, boots etc. in which one side remains open, appositive mold is immersed and subsequently the casting is gelled. A series of identical molds mounted on dipping frames are used for the mass production of such articles. During the dipping process, degassed PVC paste placed in a vessel is raised and lowered. Sometimes, the dipping frame is moved, while the vessel is stationary. The immersion and removal stages are carried out slowly to allow the material to flow homogeneously from the mold and to prevent the introduction of air into the paste. The gelation is usually carried out in circulating air oven, and during the process the paste temporarily becomes fluid. The molds containing the gelled coating are immersed in water for cooling. It is cooled to around 50°C and at this temperature, the moldings can be easily removed from the molds. Cables and wires are extensively coated using this process.

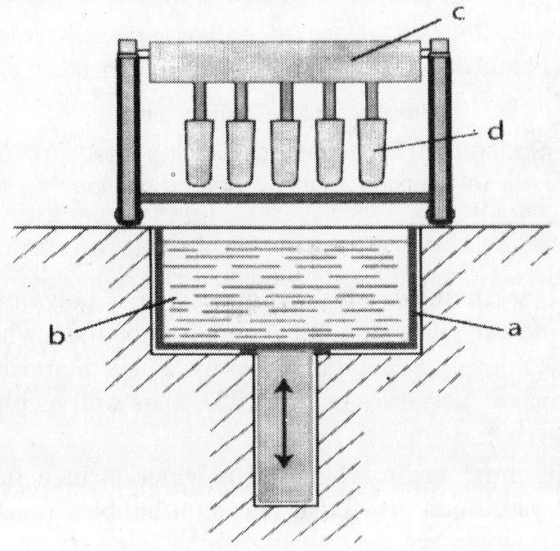

Figure 34.2: Dipping Unit.
(a) Vessel; (b) PVC paste; (c) Dipping frame; (d) Molds

(c) Coating (Figure 34.3)

This process is used in the production of self-supporting films by coating sheet materials such as paper and textiles. The most important materials that are used for the production of coatings are PVC pastes.

The equipment for coating consists of an unwinding station, an expander, coating head, gelling channel, cooling system, and a rewind station. The disk brakes slows down the unwind station, and the rewind station enables the product to be held under tension. A smooth finish to the sheet is provided by the expanders. With the help of the coating head, the PVC paste is spread upon the moving sheet manually or mechanically. Thereafter, a doctor blade is employed to spread the sheet to required thickness. A revolving roller supports the sheet, due to which an exactly defined contact area is maintained. The rolls distribute the escaping paste and form a film.

The rotation speeds of the rolls can be adjusted.

Hot air or infrared dryers supply the heat required for gelation of the paste.

However, circulating air heating in combination with infrared heaters is used in some cases.

The cooling effect must be high enough so that no adhesion between the individual layers occurs. The cooling is attained by two water - cooled, corrosion – resistant hollow drums.

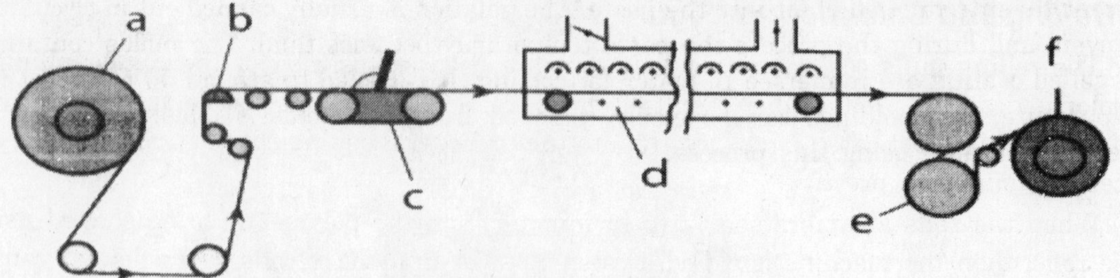

Figure 34.3: Coating Unit
(a) Unwinding Station; (b) Expander; (c) Coating head; (d) Gelling Tunnel; (e) Cooling Roll; (f) Rewind Station

(d) Foaming

By adding physical or chemical blowing agents to a polymer, foams are generated. Physical blowing agents include gases and low-boiling compounds, whereas ; chemical blowing agents generally are gas releasing compounds. Only a few materials are employed in the manufacture of foams , such as polystyrene, PVC, LDPE as well as phenolic and polyurethane resins.

The foaming process must begin when the polymer is in a flowable form and allows bubbles to develop. The conditions are fixed, once the bubbles reach the optimum size.

POLYMER PROCESSING UNDER PRESSURE

Sometimes the molding process is carried under high pressure in which the solid thermoplastic materials are melted, molded and cooled subsequently. Some of the commonly used techniques for processing under pressure are:

- Compression molding
- Rolling and Calendering
- Extrusion
- Blow Molding
- Injection Molding

(a) Compression Molding

For the manufacture of thick-walled components and those that cannot be plasticized in extruders, compression molding is used. It plays an important role in the processing of elastomers and thermosets.

A press consisting of fixed and moving platens is used. The plunger and cavity components are mounted on these plates, and the parting plane is kept horizontal. The cavity is filled with the cold plastic and is melted by heating the mold under low pressure. When the melting is complete, the mold is closed and pressure applied. The molding is removed by releasing the press once the ejection temperature is reached.

Compression molding is uneconomical due to the fact that the process has a long cycle time ; as the material in the mold is heated and cooled by conduction via the mold surface.

(b) Rolling and Calendering (Figure 34.4)

The rolling mills consist of two rolls that are used exclusively for the compounding of thermoplastics. The product that results from rolling is a sheet that can be granulated after cooling or fed in plasticated form directly into the processing equipment. PVC and rubber are processed using this process.

When the rolls are more than two in number, and the polymer is to be shaped using these rolls; then the machine is called a calendar. The number of rolls in a calendar can be upto seven ; however, most commonly, four rolls are used. The rolls are made of polished cast iron, and are driven individually & heated. The heating medium is usually oil or water. The temperature and speed increases from one roll to the next. The roll mill must be able to withstand the extreme forces to which it is exposed, and hence, be made extremely rigid.

The calenders are one of the most expensive types of processing equipments and are employed for the manufacture of a few special products like floor coverings and films made of plasticised PVC.

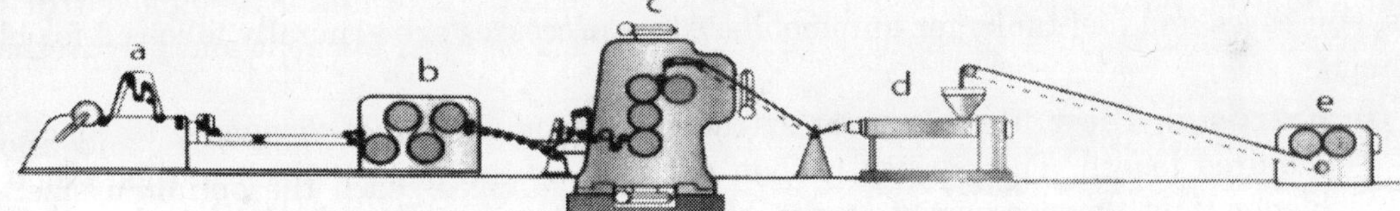

Figure 34.4: Calendering line (a) Winder and Edge Cutter; (b) Cooling Rolls; (c) Four-roll calendar; (d) Extruder; (e) Mixing Roll Mill.

(c) Extrusion (Figure 34.5)

This method enables the attainment of a wide range of film thickness, and large scale processing. Moreover, there is no necessity for expensive solvents in the recovery systems.

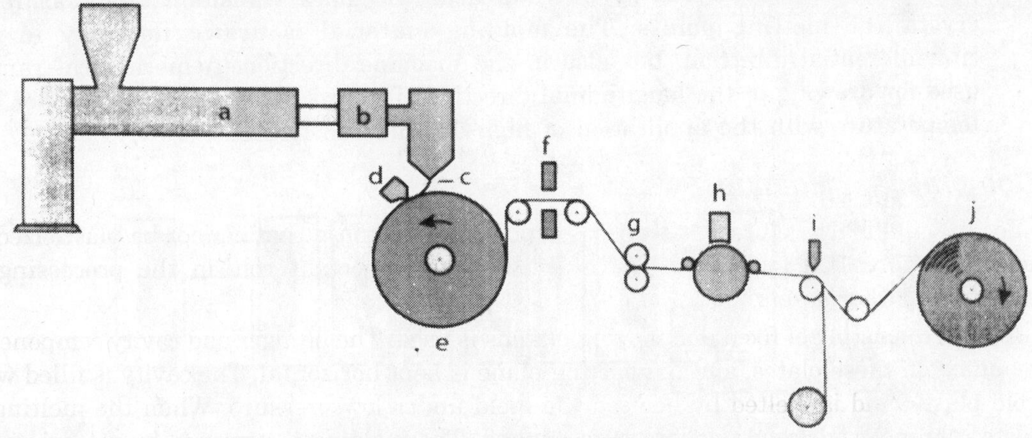

Figure 34.5: Extrusion Molding

(a) Extruder; (b) Filter; (c) Slot Die; (d) Air Knife; (e) Cooling Roll; (f) Device for measuring film thickness; (g) Tension rolls; (h) Corona Treatment; (i) Edge Treatment; (j) Winding

A device known as extruder is used to melt the polymer, which is given the specific shape by a die. Later on, a cooling system is used to solidify it.

At the onset, the raw material is dried using heated air or vacuum, before it is melted in the extruder. The extruders may be single – screw extruders, or twin screw extruders, the single screw extruder being the most common. The melt is filtered depending upon the desired purity, and then the filtered raw materials are sent through the slot die. The width and thickness of the film are determined by the width of the die and the die gap respectively. The die gap can be further adjusted during production.

The melt film emerging from the die is cast on to a highly polished roll. A vacuum device prevents the entrapment of air between the melt and the roller surface. The final film that is obtained is wound onto a roll after trimming off the edges. In a separate process, the film is cut into the designate width.

(d) Blow Molding (Figure 34.6)

This is the most common method for manufacturing hollow articles from thermoplastics. The most simple products and articles that are made by blow molding are bottles, ventilation ducts, suit cases and fuel tanks for automobiles. Two processes are generally followed for blow molding:

(i) **Extrusion Blow Molding:** An extruded tube or parison is elongated to reach the required length. The parison is then enclosed by the mold , which is then cut by a blade. It is then conveyed to the blowing station. In the blowing station, a blow mandrel is inserted into the mold and the actual process is subsequently carried out. It is then cooled and the finished product is removed by opening the mold. The open mold is then transferred to the position below the extrusion die in order to receive a new parison. The process then continues.

(ii) **Stretch Blow Molding:** This is a special form of blow molding in which the mechanical properties are considerably improved. A high degree of orientation is introduced in the plastics by drawing near the glass transition temperature or crystalline melting points. The molding material is drawn not only in the circumferential direction, but also in the machine direction. A mechanical ram is used for drawing in the longitudinal direction. The process is carried out under low temperature with the application of high deformation forces.

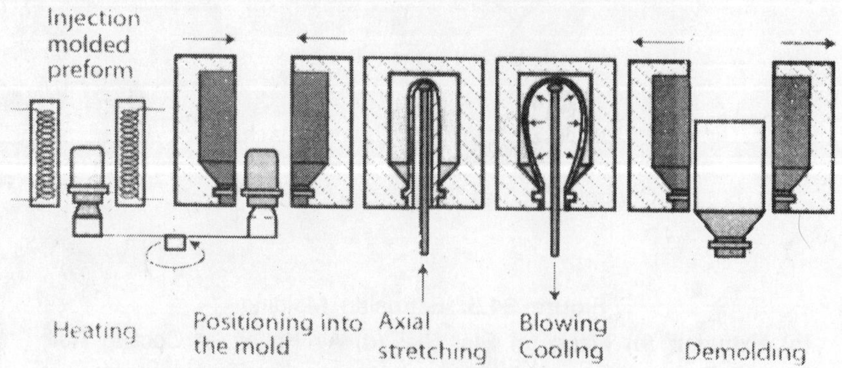

Figure 34.6: Blow Molding Process

PVC. polypropylene and poly ethyl terephthalate (PET) are some of the materials for use with this process.

(e) Injection Molding (Figure 34.7)

The Injection Molding technique is significant because of its ability to manufacture complex molding geometries in a single stage operation. The process is suitable for mass production operations because of the high degree of automation.

The Injection molding technique is a discontinuous process. The plastic pellets or granules are melted and injected under pressure into the mold cavity, wherein it is thermally cross-linked or solidified by cooling.

Through a runner system the hot melt prepared in the injection cylinder is injected into the cavity of the mold. The pressure is increased towards the end of the filling stage because of the increasing length of the flow path. After the injection phase, the holding pressure stage follows. It reduces the volume losses due to cooling of the melt, and thus, avoids marks and voids in the molding. After this, the molding is cooled sufficiently so that it is dimensionally stable for ejection. When the holding pressure phase is finished, the plastic is prepared for the next cycle by the plastication unit by rotating the screw. The melt is transported to the area in the front of the screw by the helical backward movement of the screw in the plastication cylinder against a back pressure. After this the molding is cooled and is mechanically removed from the mold via a handling equipment.

The entire process is fully automated and is coordinated by a control unit.

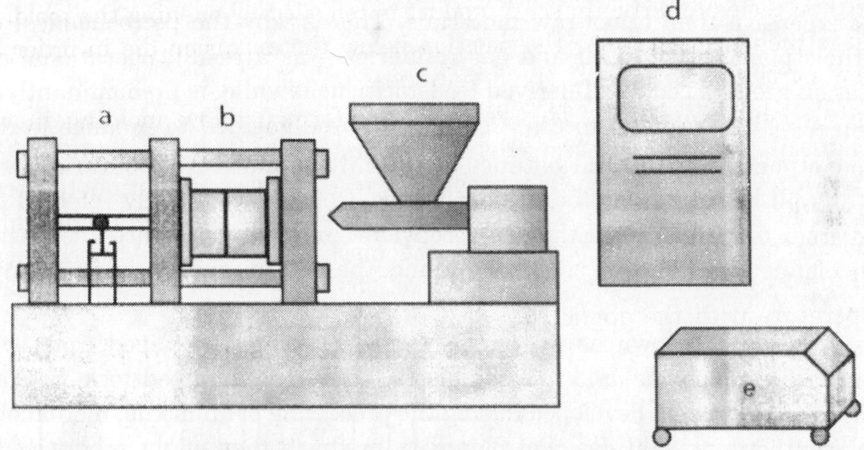

Figure 34.7: Injection Molding Machine
(a) Clamping unit; (b) Mold; (c) Plastication Unit;
(d) Control Unit; (e) Temperature Control Unit

Appendix

A

PETROCHEMICALS FEEDSTOCKS, PRODUCTS & APPLICATIONS

Petrochemical feedstocks are the various hydrocarbons derived from the refining of oil, gas and coal. These are then further refined to produce petrochemical products. They are called the building blocks of the Petrochemical industry. These building blocks are converted into a wide range of chemical products with a wide array of uses. At the feedstock stage, they are usually known as intermediates – then the intermediates are processed into plastics, liquids and resins which ultimately are turned into useful products. Some feedstocks, however, are used directly to produce petrochemicals, such as methane and BTX (Benzene, Toluene & Xylene). But ethane, propane, butanes, naphtha and gas oil are optional feedstocks for steam crackers that produce intermediate petrochemical feedstocks. Other examples of intermediate feedstocks include ethylene, propylene, butenes and butadiene. Gas & oil are the common starting points for feedstocks because they are readily available, can be processed efficiently and are less expensive than other raw materials. This is why the petrochemical companies often build their plants close to oil and gas refineries – as a result, operational costs, such as transportation are reduced. Coal-derived feedstock, meanwhile, is predominantly methanol, obtained from a coal to liquids process. Coal can also be gasified to produce feedstocks.

Methane, ethane, propane and butanes are mainly obtained from natural gas. Naphtha and gas oil, as well as benzene, toluene and xylenes, (a group commonly referred to as BTX) are obtained from oil refineries. Ethylene, propylene and butadiene are the basic building blocks of all olefins (also known as alkenes) and these form the basis for many common products.

Synthesis gas, also known as syngas, is the term for gas obtained from synthesising hydrogen and carbon monoxide, and this can also be converted into feedstock. Syngas can also be an intermediate by-product developed during the processing of ammonia, methanol, synthetic petroleum or synthetic natural gas. Petroleum by-products that might otherwise end up as waste can be conserved as feedstock. During the gasfication refining process, any material which contains carbon can also be converted. Since feedstocks and the end products vary, there are many different production methods. For example, an ethylene-producing plant is most likely to use catalytic cracking, a technique that uses high pressure and high temperatures to crack natural gas. But in a methanol producing plant, a reforming process, using high temperature steam, medium pressure and a catalyst, will produce the product.

Petrochemicals are the derivatives of mostly crude oil and natural gas. Olefins (ethylene, propylene & butadiene) and Aromatics (benzene, toluene & xylenes) are the major building blocks from which most Chemicals and Petrochemicals are produced. They are used in dyes, synthetic fibres, rubbers, plastics, pharmaceutical bulk drugs, industrial appliances, packaging industry, detergents (surfactants).

Petrochemicals production process consists of primarily two stages. In the first stage naphtha, produced by refining, crude oil or natural gas is used as a feedstock and is cracked. Cracking (breaking of long chain of hydrocarbon molecule) produces olefins and aromatics. In stage two, these building blocks are polymerized (made to undergo chemical processes) to produce downstream petrochemical products (polymers, polyesters, fibre intermediaries and other industrial chemicals).

Olefins are the key building blocks of the petrochemical industry. Of the main, while India has sufficient capacities for ethylene and propylene, styrene continues to be in deficit. Ethylene capacity grew in the five year plan due to partial de bottlenecking of capacity by GAIL, HPCL and the start-up of IOC complex at Panipat. Propylene production boosted due to extraction of Propylene by RIL at its Refinery Complex. For Butadiene, RIL and Haldia are the only two domestic producers.

India has petroleum-refining capacity of over 200 Million Tonnes per Annum (MTPA).

Petrochemicals cover basic chemicals like Ethylene, Propylene, Benzene and Xylene. The other major components are the intermediates like MEG, PAN and LAB etc. Synthetic fibres like Nylon, PSF and PFY, Polymers like LDPE/HDPE, PVC, Polyester & PET etc. and Synthetic rubbers like SBR & PBR.

The petrochemical industry in India has been one of the fastest growing industries in the country. Indian petrochemical industry grew at a rate of ~11% in 2014-15. This is being led by strong growth in polymers, fibre intermediates, synthetic fibre and elastomers. Per capita plastic consumption in India is still hovering at 7.0 kgs as compared to 46 kgs in China and 65 Kgs in Europe. This signifies huge potential for future growth going by current global average per capital consumption.

Structure of the Petrochemical Industry is shown below:

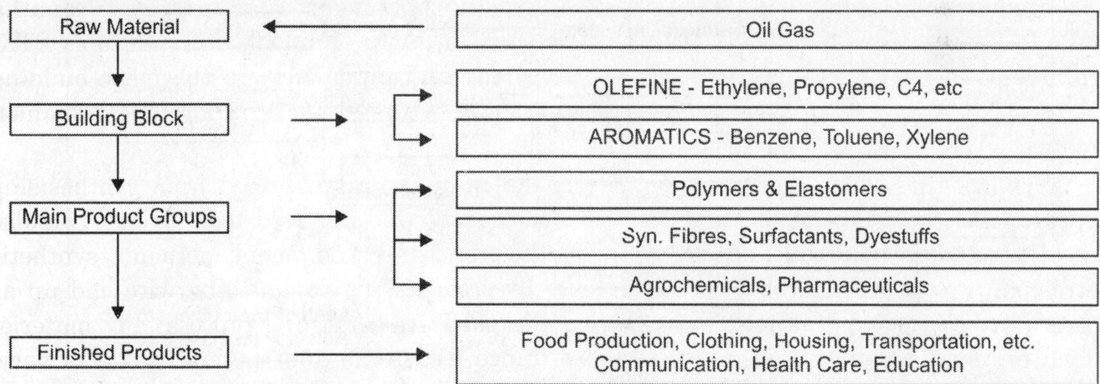

Ethylene capacity reached 3.1 MMT in 2011 and is expected to reach 6.5 MMT by 2016. Propylene capacity reached 3.9 MMT in 2011 and is expected to be 5.3 MMT by 2016. The demand for Automobiles, Packaging, White good, Medicare, Agriculture and Building & onstruction is likely to remain strong. Combining all the petrochemical sectors, demand in India is expected to remain robust in coming years. This industry provides the foundation for manufacturing industries like construction, packaging, pharmaceuticals, agriculture, textiles etc. The sector has huge unrealised potential.

Chemicals derived from petroleum or natural gas - petrochemicals - is an essential part of the chemical industry today. Petrochemicals are a fairly young industry; it only started to grow in the 1940s, more than 80 years after the drilling of the first commercial oil well in 1859.

Crude oil to Petrochemicals tree is shown below:

Ans. 1.Fast delivery at almost no cost.

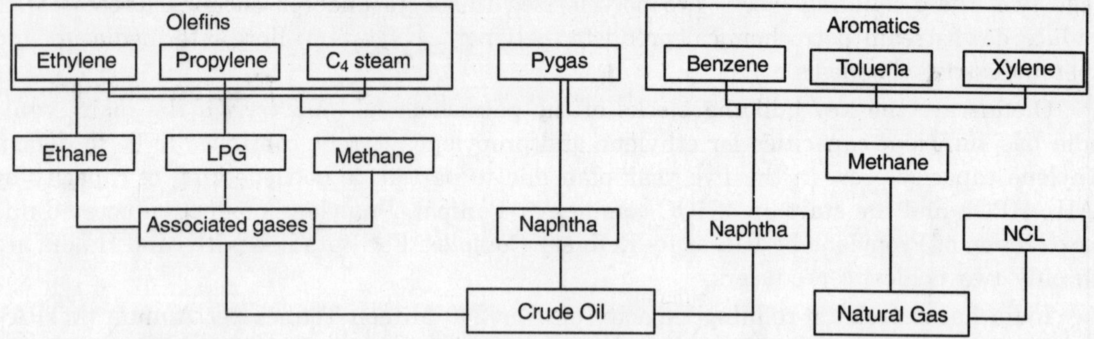

Petrochemicals do not reach directly to the final consumer. They are first sold to customer industries, undergo several transformations, and then go into products that bear no relation to the initial raw material.

The starting point for petrochemicals is Naphtha or Gas. Olefin rich Naphtha is cracked to produce Ethylene, Propylene and Butadiene, popularly known as C2, C3 & C4 streams. Aromatic rich Naphtha is preferred to produce Benzene, Toluene and Xylenes, popularly known as C6, C7 & C8 streams. Associated gases are rich source of Methanol i.e. C_1 hydrocarbons. These materials i.e. C1 to C8 are the basic building blocks for a host of day to day consumer and industrial products. The figure below indicate products which can be produced and their applications.

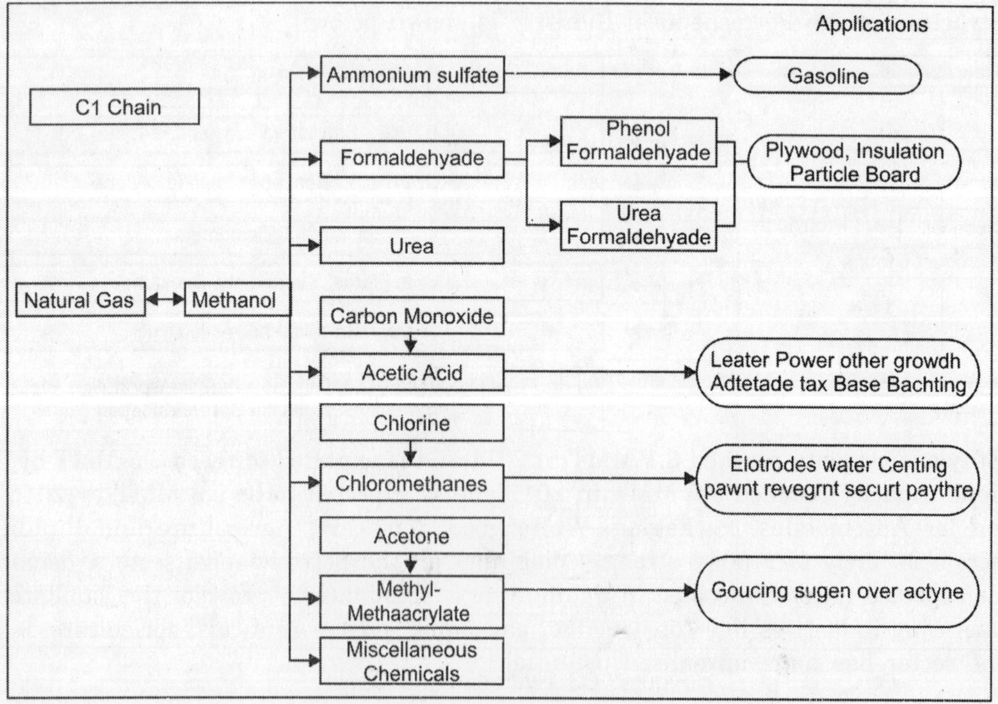

Figure: C1 Hydrocarbon Chain

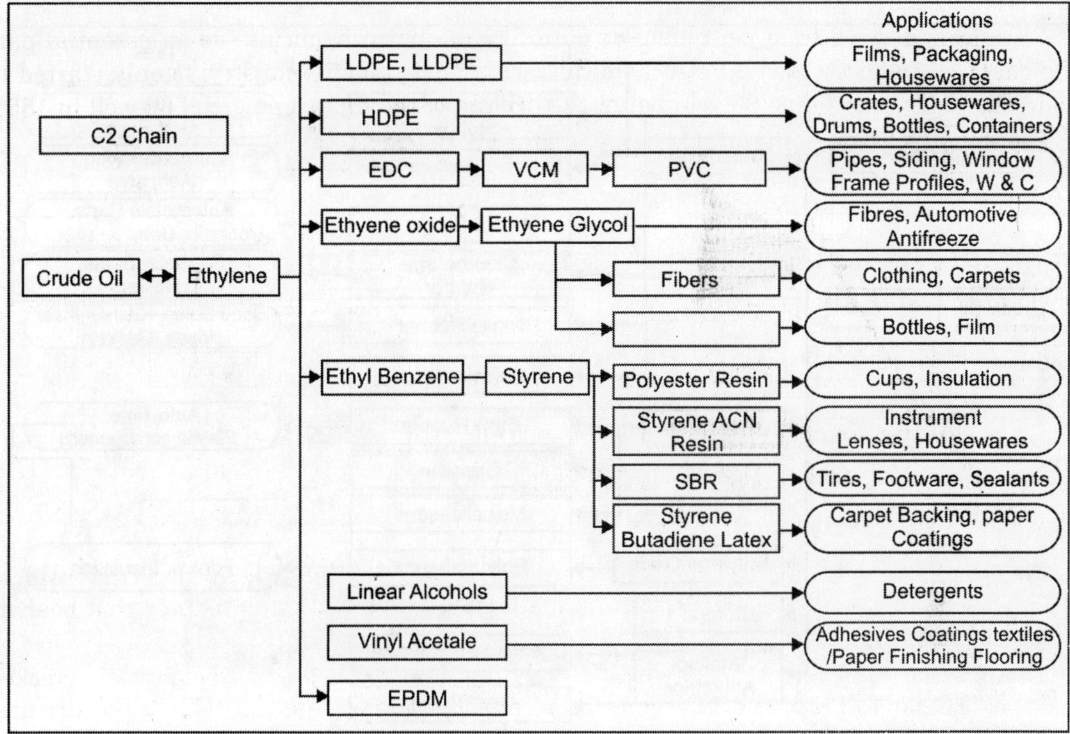

Figure: C2 Hydrocarbons Chain

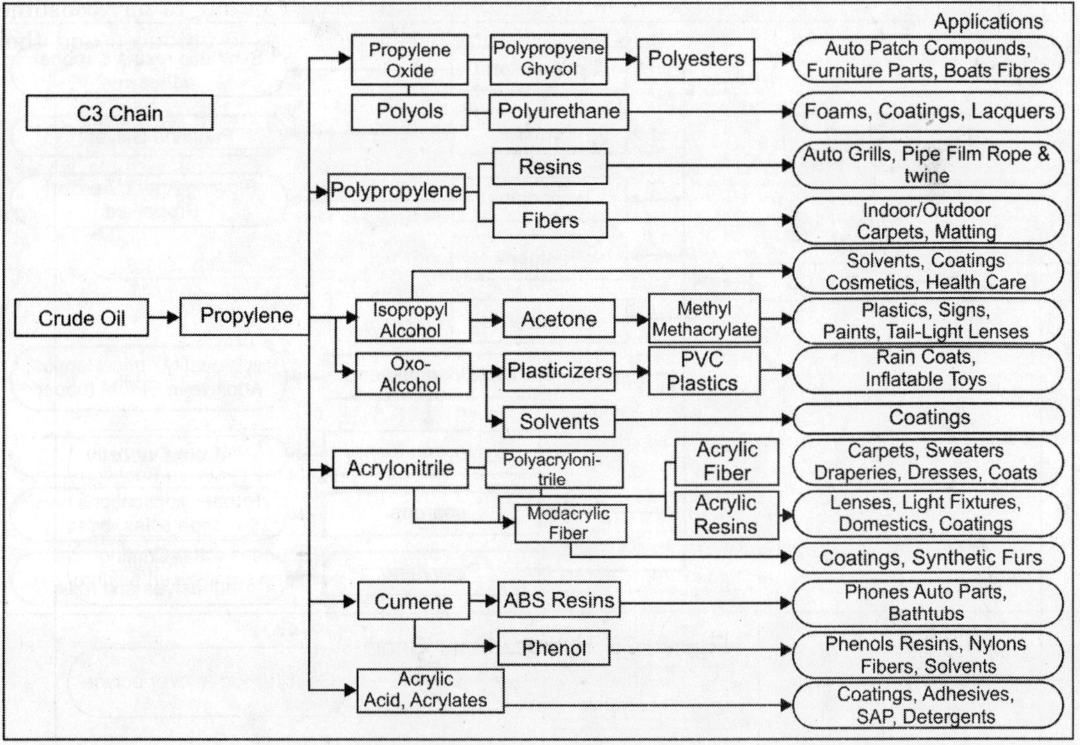

Figure: C3 Hydrocarbons Chain

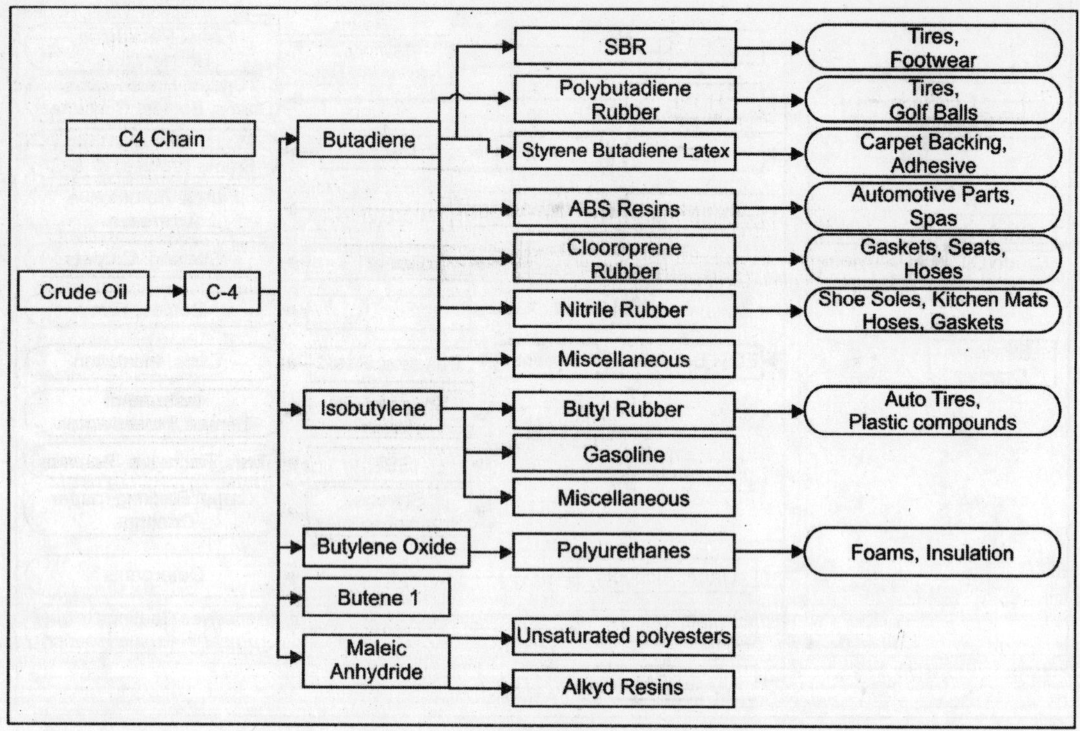

Figure: C4 Hydrocarbons Chain

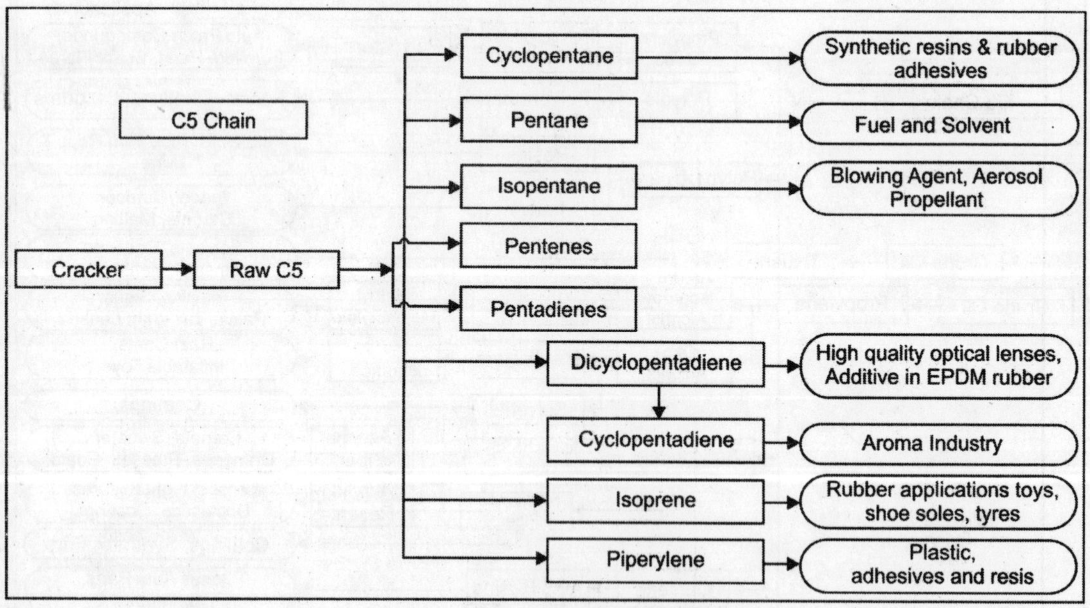

Figure: C5 Hydrocarbons Chain

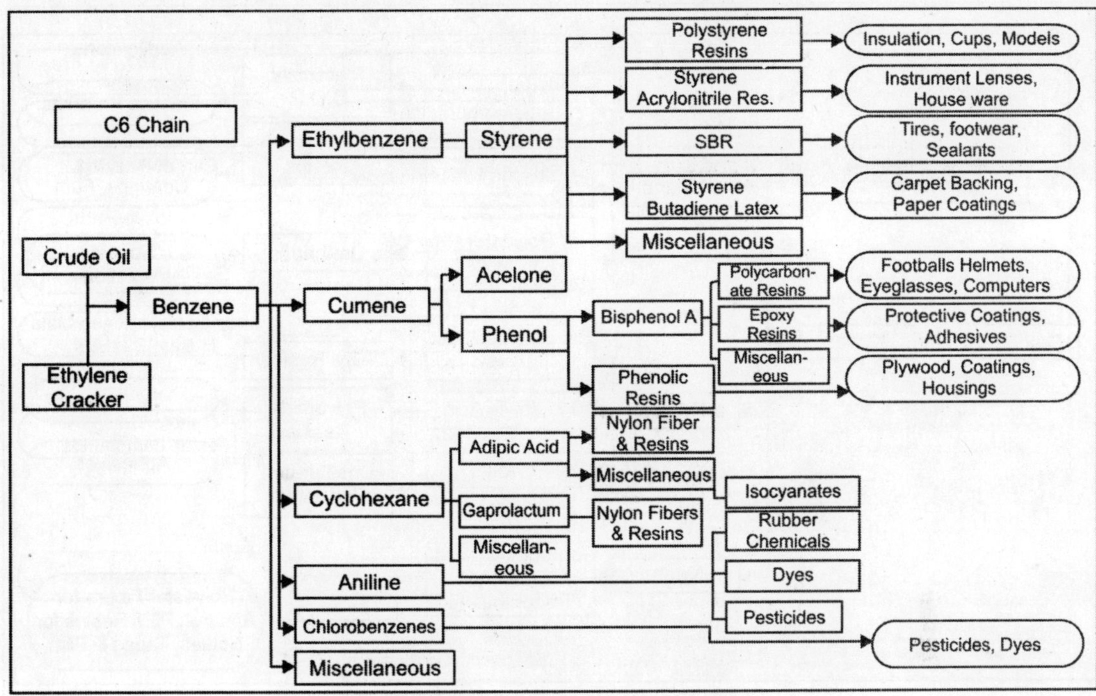

Figure: C6 Hydrocarbons Chain

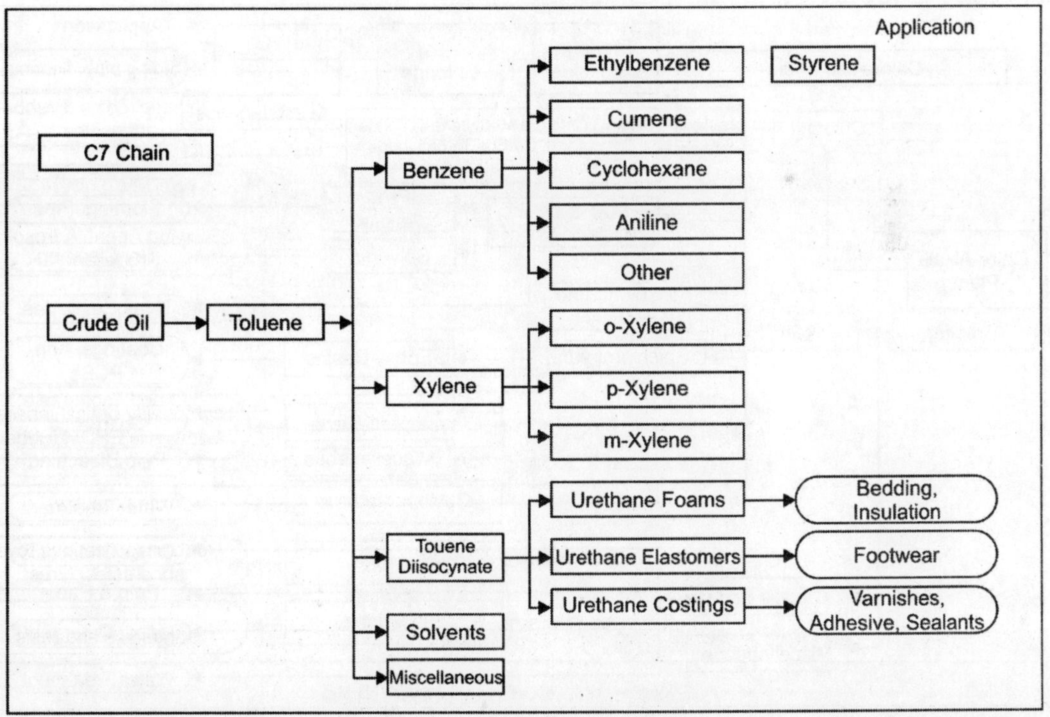

Figure: C7 Hydrocarbons Chain

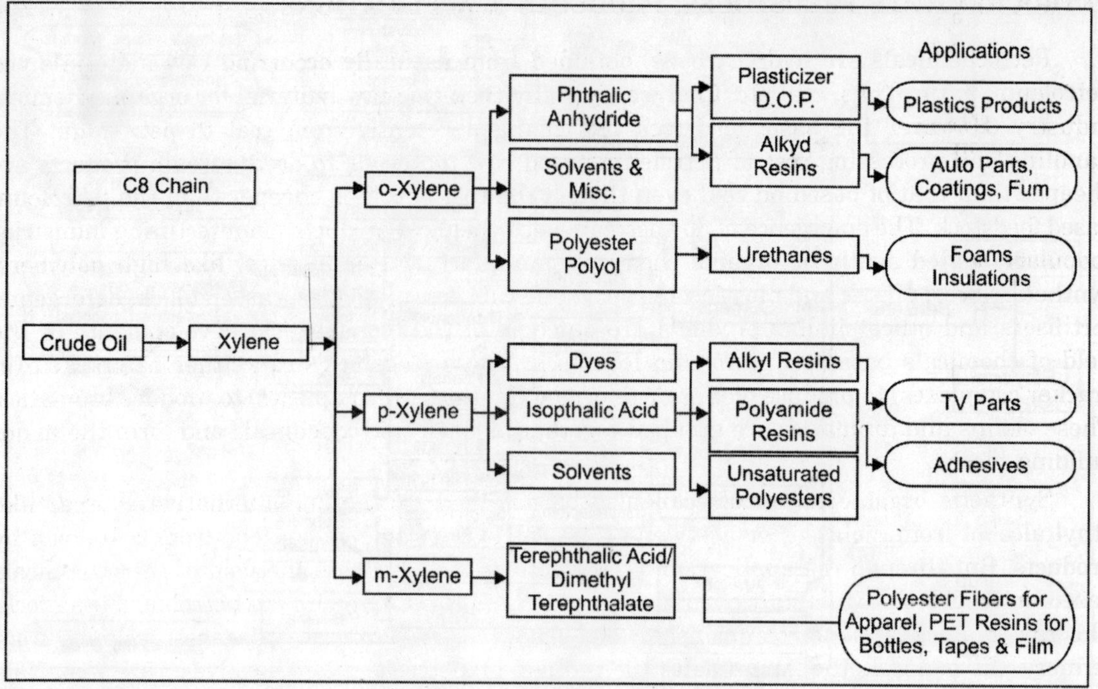

Figure: C8 Hydrocarbons Chain

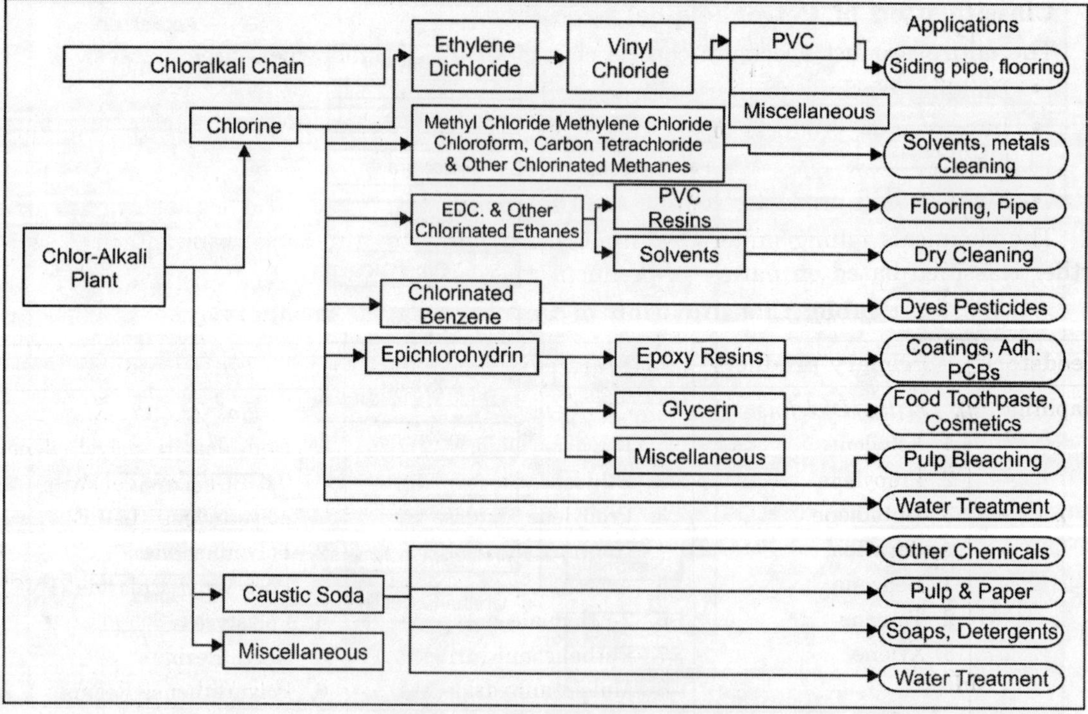

Figure: Chlor Alkali chain

Petrochemicals are hydrocarbons, obtained from naturally occurring raw materials viz. petroleum, natural gas, coal etc. Coal was initially the basic raw material for organic chemical industry. However the basic feedstock has changed recently from coal to petroleum. The handling and processing cost of petroleum based raw materials to down-stream products are cheaper than that of based on coal even though the cost of coal is cheaper than the petroleum based feedstock. The emergence of downstream petrochemical products manufacturing industries (popularly called synthetic organic chemicals manufacturing industries) like high polymers, synthetic fibres, plastic and plasticizers, synthetic rubbers, pesticides, carbon black, detergents, fertilisers and other similar products are outcome of the technological developments in the field of chemicals based on petroleum feedstock. These feedstock's can either be cracked (in cracker complexes) to produce olefins or reformed (in aromatic complexes) to produce aromatics. These olefins and aromatics are grouped together as basic petrochemicals and form the major building blocks.

Synthetic organic chemicals can also be obtained from other alternative sources like ethyl alcohol from molasses or acetylene from calcium carbide or benzene from coke oven by products. But through the application of new process technology in the field of petrochemicals based on feedstock available from refineries, there is a positive shift to petroleum feedstock. The raw materials of petroleum origin are crude oil, natural gas, off gases residues from refinery. In general, the manufacturing process of petrochemicals involves raw materials undergoing one or more chemical reactions followed by different unit operations to separate the product from side products and co-products.

Classification of Petrochemical products:

The entire product spectrum can be classified into the following three classes:

- Building block or primary petrochemical products.
- Intermediate products or secondary petrochemical products, produced from building block.
- Final or end products, coming from intermediate products.

The chemicals falling under the three classes are listed in Table below. Each class is further classified based on nature of products.

Table: Classification of Petrochemicals products

Feedstock	Primary products	Intermediate products	Final Products
Naphtha	*Olefins / Diolefins* 1. Ethylene 2. Propylene 3. Butadiene **Aromatics:** 1. Benzene 2. Toluene 3. Xylene	*Organics* 1. Ethylene Oxide 2. Ethylene Glycols 3. Propylene Oxides 4. Isopropyl alcohol 5. Acetone 6. 2-Ethyl alcohol 7. Phthalic anhydride(PA) 8. Maleic anhydride(MA) 9. Phenol 10. Styrene 11. Polyethylene	*Plastics and resins* 1. High density polyethylene (HDPE)/ Low density polyethylene (LDPE) 2. Polypropylene 3. Poly Vinyl Chloride (PVC) 4. Polystyrene 5. Alkyl Resins 6. Polyurathene; Resins 7. Resins PF 8. Alkyl Benzene Sulphonate (ABS) Resins

		12. Chlorinated hydrocarbon 13. Isocynates 14. Cumene 15. Butanol ***Synthetic fibre*** 1. Caprolactum 2. Dimethyl Terephalate(DMT)/ Terephthalic acid(TPA) 3. Acrylonitrile	*Synthetic Fibre:* 1. Nylon filament Fibre 2. Nylon tyre cord and other fibre 3. Polyster Filament yarn 4. Polyster staple fibre 5. Acrylic Fibre 6. Polypropylene fibre ***Synthetic Rubber:*** 1. Styrene Butadiene rubber (SBR) 2. Poly Butadiene Rubber (PBR)

Major Petrochemicals category wise is listed below:

A. Basic major petrochemicals	
1. Synthetic Fibres	: Acrylic Fibre, Polyester Staple Fibrefill, Nylon Filament Yarn, Nylon Industrial Yarn/Tyre Cord, Polyester Filament Yarn, Polyester Staple Fibre, Polypropylene Filament Yarn, Polypropylene Staple Fibre.
2. Polymers	: Low Density Polyethylene, High Density Polythylene, Polyestyrene, Polypropylene (Inc. Co-Polymer), Expandable Polyestyrene, Poly Vinyl Chloride, Linear Low Density Polythylene, PVC Compound.
3. Synthetic Rubber	: Styrene Butadiene Rubber, Poly Butadiene Rubber, Nitrile Butadiene Rubber, Ethyl Propylene Dimers, Ethyl Vinyl Acetate, Butyl Rubber.
4. Synthetic Detergent Intermediates	: Linear Alkyl Benzene Sulfonate, Ethylene Oxide.
5. Performance Plastics	: Abs Resin, Nylon-6, Polymethyl Methacrylate, Styrene Acrylonitrile, Nylon 6,6.
B. Intermediates	
1. Fibre Intermediates	: Acrylonitrile, Caprolactum, Mono Ehylene Glycol, Purified Terephthalic Acid, Dimethyl Terephthalate.
2. Building Blocks	
(i) Olefms	: Ethylene, Propylene, Butadiene.
(ii) Aromatics	: Benzene, Toluene, Orthoxylene, Paraxylene Mixed Xylene.

C. Other Petro-based Chemicals	:	Diethylene Glycol, Diacetone Alcohol, Ethylene Dichloride, Butanol, Oxo Alcohol, 2-Ethyl Hexanol, Vinyl Chloride Monomer, Epichlohydrine, Iso Butylene, Metaxylene, Methyl Isobutyl Ketone, PBT, PET, PIB, Polycarbonate, Propylene Oxide, Propylene Glycol, Polyvinyl Acetate Resin, Unsaturated Polyster Resin, Methyl Methacrylate, Iso-Butanol, Ethyl Benzene, C4-Raffmate, Cellulose Acetate Butyrate, Cellulose Acetate Sheet, Cellulose Nitrate Sheet, Melamine Moulding Powder, Polyacetal Resin, Phthalic Anhydride, Styrene, Vinyl Actate Monomer, Isopropanol

Coverage of chemicals under the petrochemical complexes is shown in figure.

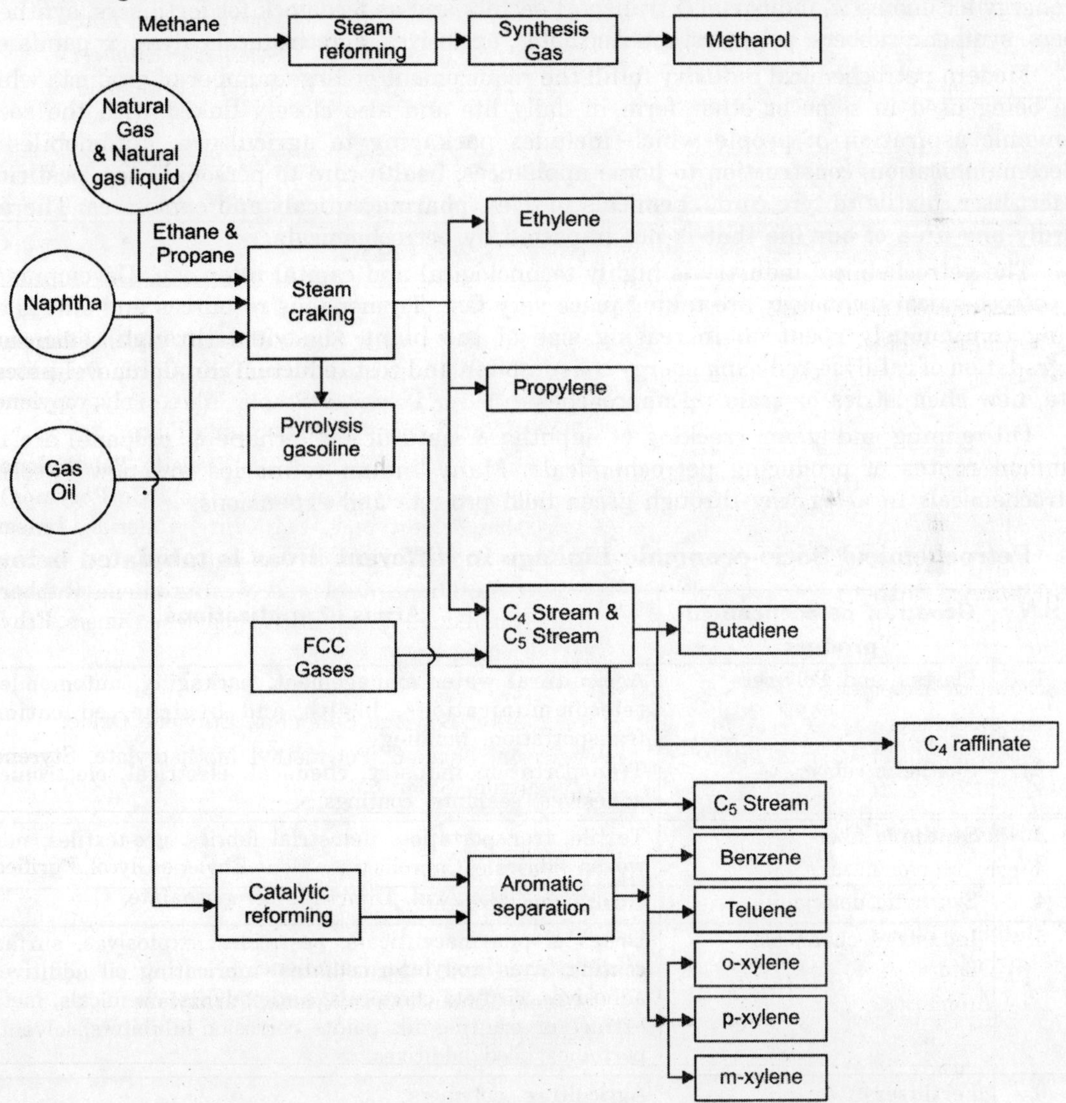

Figure: Coverage of Chemicals under Petrochemical Complexes Industry

PRODUCTION PATTERNS

Olefins (ethylene, propylene & butadiene) are made by cracking of natural gas, C_2/C_3 fraction of natural gas and naphtha. Currently, 59% of the country's ethylene capacity is based on gas and 40% on naphtha.

Natural gas results in higher yields of ethylene compared to other olefins. On the other hand, naphtha gives a wider range of output products (including propylene, butadiene derivatives). Use of alcohol from molasses as feedstock for ethylene manufacture is rapidly declining.

Aromatics (Benzene, Toluene & Xylene) are produced by number of processes, predominantly by naphtha steam reforming, as recovery products from coke oven gas and as refinery by-products.

Petroleum and petrochemical industries are providing the major basic needs as a source of energy for domestic, industrial & transport sectors and as feedstock for fertilisers, synthetic fibers, synthetic rubbers, polymers, intermediates, explosives, agrochemicals, dyes, & paints etc.

Modern petrochemical industry fulfill the requirement of large number of products which are being used in some or other form in daily life and also closely linked with the socio-economic aspiration of people which includes packaging to agriculture, automobiles to telecommunication, construction to home appliances, health care to personal care, pesticides to fertiliser, textile to tyre cord, chemicals to dyes, pharmaceuticals and explosives. There is hardly any area of our life that is not impacted by petrochemicals.

The petrochemical industry is highly technological and capital intensive. Developments in petrochemical technology are taking place very fast. Tremendous resources and efforts are being continuously spent on increasing size of the plant, the yield through continuous upgradation of catalyst, reducing energy consumption and cost reduction through novel process rate, new chemistries or scale up approaches.

Oil refining and steam cracking of naphtha & natural gas (ethane & propane) are the common routes of producing petrochemicals. Many Indian refineries are now entering petrochemicals in a big way through green field projects and expansions.

Petrochemical Socio-economic Linkage in different areas is tabulated below.

S.N	Group of petrochemical product	Areas of applications
1.	Plastics and Polymers	Agricultural water management, packaging, automobiles, telecommunications, health and hygiene, education, transportation, building.
2.	Synthetic rubber	Transportation industry, chemical, electrical, electronics, adhesives, sealants, coatings.
3.	Synthetic fiber	Textile, transportation, industrial fabrics, geo-textiles, non-woven fabrics.
4.	Synthetic detergents	Health and Hygiene.
5.	Industrial chemicals	Drugs & pharmaceuticals, pesticides, explosives, surface coating, dyes and intermediates, lubricating oil additives, adhesives, oil field chemicals, antioxidants, chemicals, metal extraction, printing ink, paints, corrosion inhibitors, solvents, perfumes, food additives.
6.	Fertilisers	Agriculture, polymers.

The petrochemical complexes involve one or a combination of the following operations:

- **The manufacture of basic raw materials** like syngas, methane, ethylene, propylene, acetylene, butadiene, benzene, toluene, xylenes, etc. The basic building processes include partial oxidation, steam reforming, catalytic and thermal cracking, alkylation, dealkylation, hydrogenation, disproportionation, isomerisation, etc. The commonly used unit operations are distillation, extractive distillation, azeotropic distillation, crystallisation, membrane separation, adsorption, absorption, solvent extraction, etc.
- **Manufacture of intermediate chemicals** derived from the above basic chemicals by various unit processes like oxidation, hydrogenation, chlorination, nitration, alkylation, dehydrogenation along with various unit operations like distillation, absorption, extraction, adsorption, etc.
- **Manufacture of target chemicals and polymers** that may be used in the manufacture of target products and chemicals to meet the consumer needs. It includes plastics, synthetic fibers, synthetic rubber, detergents, explosives, dyes, intermediates, and pesticides.

First Generation Intermediates	Hydrogen, Ammonia, Methanol Olefins and Dienic Hydrocarbons, Ethylene, Propylene, Butadiene, Isoprene, etc. Aromatic Hydrocarbons, Benzene, Toluene, Xylenes, Styrene, etc.
Second Generation Intermediates	Introduction of various hetro atoms into final molecule including oxygen, nitrogen, chlorine and sulfur by various unit process Intermediates.
Target Products	Plastics, Synthetic fibre, Fertilizers, Solvents, elastomer, Drugs, Dye stuff, Detergent, Explosive, Pesticides.

Various important Petrochemical feedstocks, products & their applications are discussed below:

Methane & Derivatives

Reaction solvent, textile dry cleaning, fire extinguishing, metal degreasing, grain fumigant. packaging, auto parts, appliance parts, plumbing & hardware, pens & pencils, lighters, shavers, moulding compound, particle board adhesive, foam insulation, automotive parts, furniture, dinnerware, plastics, gasoline component, single cell protein (use as additives and animal feeds), chemical intermediate (e.g. perfumery chemicals), windscreen wash, ethyl alcohol denaturing, latex paints, lacquer resins, enamel resins, oil additives, leather finishes, paper coating, textile coating, floor polishes, high octane gasoline component & oxygenate.

Butylenes & Derivatives

Sealants, caulking compound, adhesives, tyre products, lubricant viscosity improver, adsorbent fibre, pipe fabrication, solvent (e.g. vinyl coatings, acrylic coating), paper and ink manufacture, lubricant oil dewaxing.

Xylene & Derivatives

Plasticisers, dyes, chemical intermediate, clothing, carpet upholstery, tyre cord, conveyor belts, home furnishings, beverage bottles, coatings, adhesives, films & tapes, moulded products, hoses, wheels & rollers, insulation, recreational products, tubing, furniture, pipes, tanks, automotive parts, simulated marble, bowling balls, shower cubicles, cements, buttons, automobile parts.

Ethylene & Derivatives

Packaging film, plastic bags, milk bottles, oil cans, fuel tanks, caps & lids, insulation, pipe & tubing, diaper covers, homewares, toys, comonomer (linear low density polyethylene), synthetic lubricants, chemical intermediate, detergent, paper & textile, wax substitute, oil field chemical, mould release, plasticiser, leaf treating, cement additives, cosmetics, corrosion inhibiter, photo chemicals, non-ionic surfactant, fumigant, sterilising agent, pharmaceuticals, latex paints, antifreeze, polyester fibre & resin, brake fluid, hydraulic fluid, clothing fibre, household, non-woven, draperies, plastics, lacquer, aspirin, fragrances, chemical intermediate, starch modifier, solvents - coatings, plastics, toiletries, medical applications, chemical feed-stocks, gasohol, mouthwash, wood glue.

Butanes & Derivatives

Domestic (e.g. space-heating, cooking, lighting, hot water supply and refrigeration), leisure (e.g. caravans and boats, barbecues and green-houses, camping and outdoor activities, blow lamps and other hand tools), industrial (e.g. welding, jewelery making/soldering, aerosol propellants), automotive fuel (for fork lift trucks, cars, taxis and other vehicles which operate '.within a restricted radius of their base depots).

Benzene & Derivatives

Transparent sheets (e.g. aircraft windows, police car glass, storm doors, bath & shower doors, lighting fixtures), moulding & extruding (e.g. tail light lenses, dials & knobs, bathroom plumbing materials, bottles, medical & dental parts), pharmaceuticals, adhesives, paint thinner/ cleaner, solvent (e.g. fibre spinning, protective coatings, ink, paper coatings, magnetic tape, lubricant oil dewaxing, rubber processing), polycarbonates (e.g. transparent plastics), epoxy resins (e.g. floors & floor coatings, plastic tiles, & dispensing machines), flame retardant, resins, plywood adhesives, chemical intermediate (plastic bags, carpeting, cosmetics, perfumes, oil well drilling, food industry, paper industry, ceramics), disinfectant, fibre, automotive parts, film wire, film, cable coating, moulded products, recreational products, oil additives, agricultural chemicals, preservatives for fats and oil, latex paints, homewares, furniture, appliance parts, electronic components.

Butadiene & Derivatives

Tyres and tyre products, hoses and belts, rubber goods, footware, adhesives and cements, sealant additives, moulded products, latex foams, latex paints, wire and cable coating, coated fabrics, carpet backing.

Propylene & Derivatives

Rubbing alcohol, pharmaceuticals, coating solvents, personal care products, aerosols, chemical intermediates (e.g. plastics & rubber products), carpeting, brushes, rope, tape, toys, non-woven fibre, plastic bottles, boxes, appliance parts, protective coating in latex paints, adhesives, ink solvents, extracted solvents for oils, perfumes, drugs, plastics, leather solvents, coated paper & textiles, clothing, carpet; draperies & curtains, awnings & blankets, home furnishings, paint rollers, speaker grills, battery separators, pipe & fittings, automotive parts, appliance parts, business machines, telephones, packaging, luggage, toys, medical disposables, industrial parts, boxes & containers, polyurethane foam, fibreglass composites, food fumigant, chemical intermediate, detergents, solvent, antifreeze, humectant (creams, lotions, moisturisers), preservative, lubricant, softening agent, cosmetics, hydraulic fluid, cutting oil, suntan lotion, paper, cements, ceramics, disposable diapers, floor polish.

Toluene & Derivatives

Solvents, high octane gasoline component, explosives, flexible foams (beddings, cushions, car seats), rigid foams (insulators, refrigerators, trucks & rail cars), marine coatings, varnishes.

Petroleum product tree with areas of coverage for environmental emission standards is shown in figure.

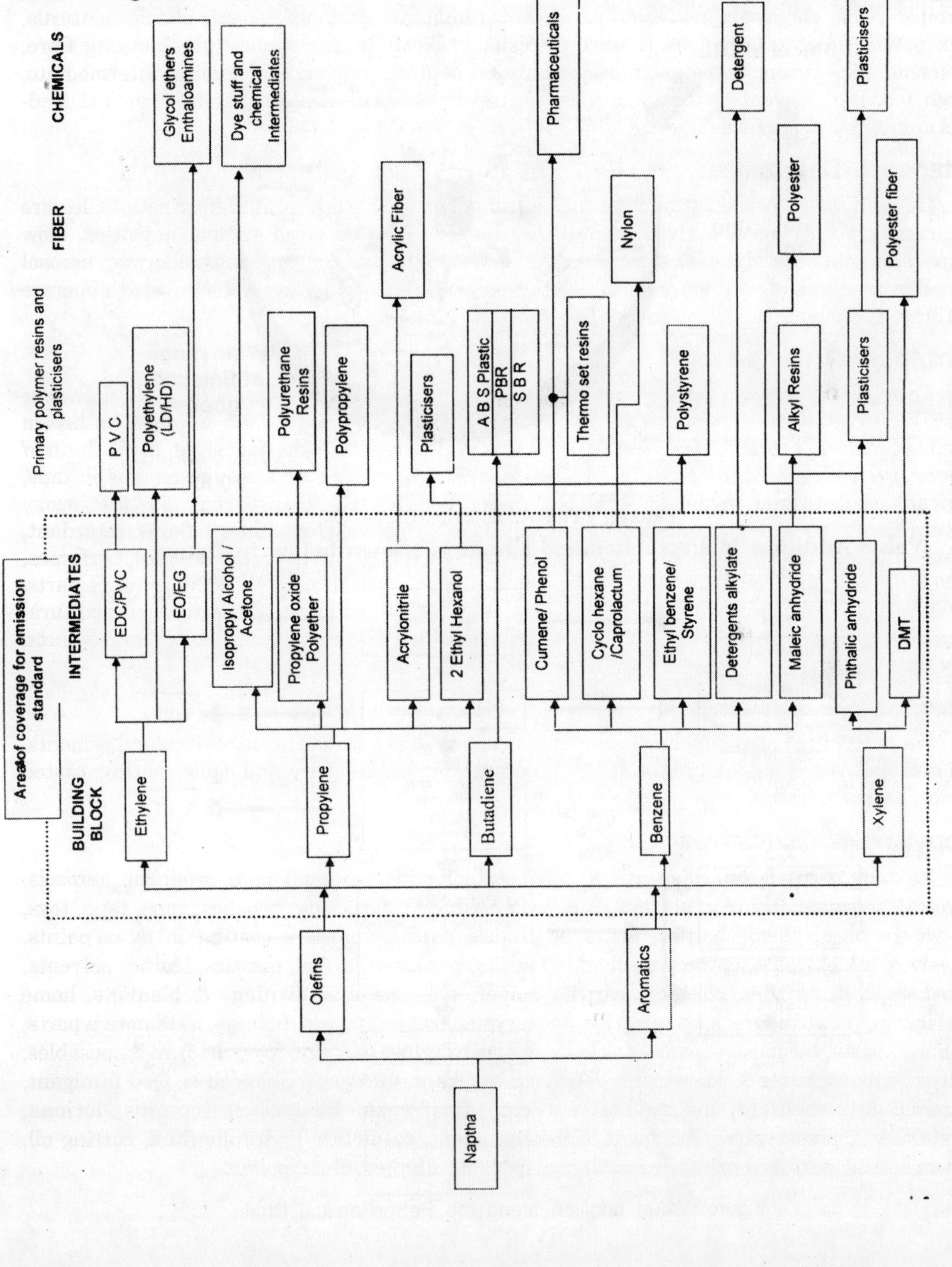

Figure: Petrochemical Products Tree

VALUE ADDITIONS

Petrochemical offer one of the highest value additions in manufacturing sector. Crude, which is predominantly used for fuel application is 200% value addition and converting to consumer end products is 2000% value addition. Figure below indicate what all can be produced with one barrel of oil.

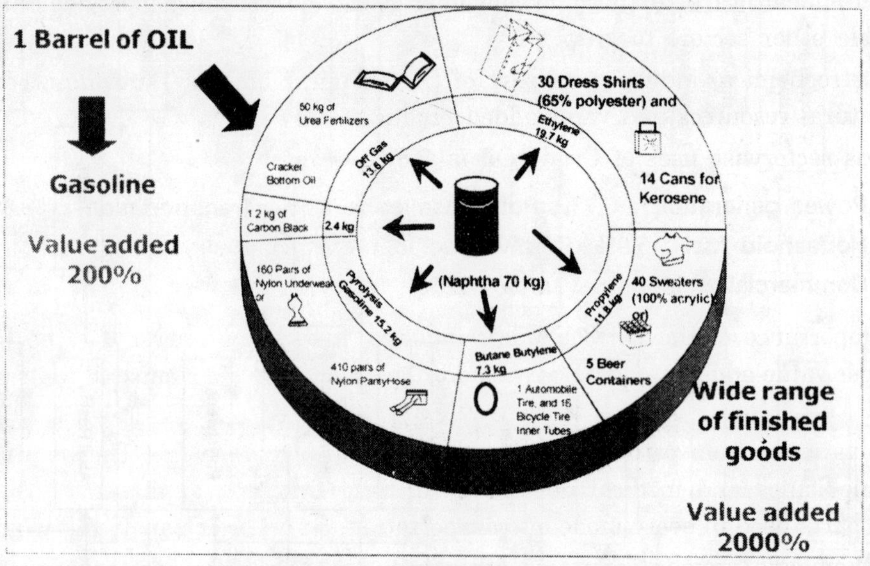

Figure: Value Addition to Crude Oil.

Value Addition in Petrochemical Chain is shown below in figure.

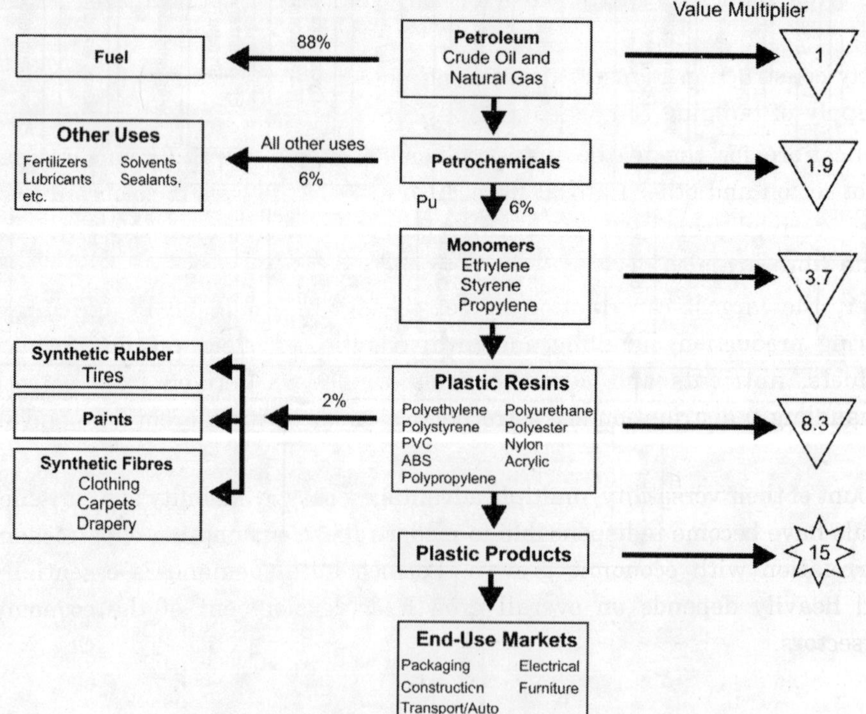

Figure: Value addition along the Petrochemical Chain

Petrochemicals industry plays a vital role in economic growth & development as one of the pillars of material industry. The three main materials include metal, inorganic material and synthetic polymers. Petrochemicals are the primary source of synthetic materials.

Due to its very nature Petrochemicals is an "enabler" industry playing a vital role in the functioning of virtually all key sectors in the economy including packaging, agriculture, infrastructure, healthcare, textiles and consumer goods. Petrochemicals provide critical inputs which enable other sectors to grow.

The petrochemical industry, as part of the chemical industry, provides the vital link between natural resources and value-added products.

Various sectorwise uses of Crude Oil is shown below:

Power generation : 5.7%; Petrochemicals: 4.5%; Transportation : 46.8%
Household fuel : 5.9%; Captive use in refining industry : 8-13%
Commercial & industrial uses : 28%

The importance of the petrochemical industry as a value creator for the economy is because these value-added products create additional wealth when compared to its alternative use as fuel.

Products made from petrochemicals are essential to modern life. These include plastic products, medicines and medical devices, cosmetics, furniture, appliances, TVs & radios, computers, parts used in every mode of transportation, solar power panels and wind turbines.

Petrochemicals provide inputs to other industry in the form of polymers, synthetic rubber and synthetic fibre. Automotive industry depends on synthetic rubber, tyre-cord carbon black for the tyres, numerous plastic parts for body, under-hood applications and paint for the body.

Similarly construction industry depends on petrochemical industry and its downstream sector for supply of building components, pipes, profiles, technical textile and paint.

Synthetic fibre for the textile industry has supplemented and complemented limited availability of cotton and other natural fibre. Also most of the dyes & colourants used by the industry are derived from petrochemicals. Polyester, Polyamide (Nylon), Acrylic and Polypropylene fibre are also derived from petrochemicals.

However, the largest contribution of petrochemicals would be in agriculture & food industry during production, handling and in distribution. Modern farming technology uses plastic products, nutrients and agro-chemicals which are derived from petrochemicals. Packaging, handling & distribution use petrochemical products in different form and at different stages.

On account of their versatility, multiple advantages, easy availability and cost-effectiveness, petrochemicals have become indispensable to modern life. Consumption of petrochemicals has a strong correlation with economic growth. Petrochemical demand is essentially derived demand and heavily depends on overall growth & development of the economy and its constituent sectors.

The common man is touched by petrochemical products across all walks of life. Mass products like tooth brush; tiffin box, low cost footwear, healthcare items, public transport, packaging of consumables like milk, etc are affordable for the common man for daily use, thanks to petrochemicals.

The food chain, from "farm-to-fork", to sustain a large growing population depends on petrochemicals to support the production, distribution and storage/shelf-life of food.

Farm productivity and yield enhancements are driven by modern agricultural techniques involving:

- *Herbicides reduce effects of non-productive plants.*
- *Pesticides minimize wastage to pests.*
- *Irrigation by plastic pipes (e.g. micro-irrigation) saves water and enhances production.*
- *Agro-machinery / Tractors (polymer parts, lubricants) improve productivity.*
- *Synthetic fertilizers (based on hydrogen from natural gas).*

Transportation / Distribution / Storage for reducing spoilage, wastage and maintaining quality of grains, fruits, vegetables, sugar, etc.

- *Plastic Woven sacks for grains, sugar.*
- *Cold storage refrigerants enhance the shelf life of produce making it available round-the-year.*
- *Plastic & Leno bags are convenient for storage, handling and retail maintenance" (wash n wear).*

Affordable, durable, low-maintenance and attractive clothing for the masses is made possible by petrochemicals based synthetic fibers providing alternatives to natural fibers like cotton and wool.

- *Polyester, Nylon and Acrylic fibers are affordable, durable and "low-cost" items.*
- *Synthetic dyes provide attractive and fast "non-fading" colors & designs.*
- *Synthetic detergents make life easier for the housewife.*
- *Other chemical auxiliaries help the feel, luster and durability of fabrics.*

Affordable, durable & attractive housing and an improved quality of life for all strata of society is made possible by the use of petrochemical products:

- *Construction chemicals enhance the strength and life of RCC buildings.*
- *Plastic sheets provide shelter for the poor from the elements.*
- *Paints / Coatings not only improve the attractiveness of a dwelling, but also protect the structure from corrosion and deterioration.*
- *Appliances and Consumer Electronics made with plastics reduce the drudgery of the housewife and entertain the family, enhancing their quality of life.*
- *Plastic pipes for water distribution and sewage disposal are lightweight and non-corrodible (durable).*

Use of Petrochemicals in Buildings is shown below in figure.

False ceiling, wall paneling, PVC wire/flooring/roofing sheet, electrical conduit pipes, wall covering, insulated pipe, chairs, sewerage pipes, septic tanks, casing pipe,water storage tank, canopy etc.

A PVC Home is shown below.

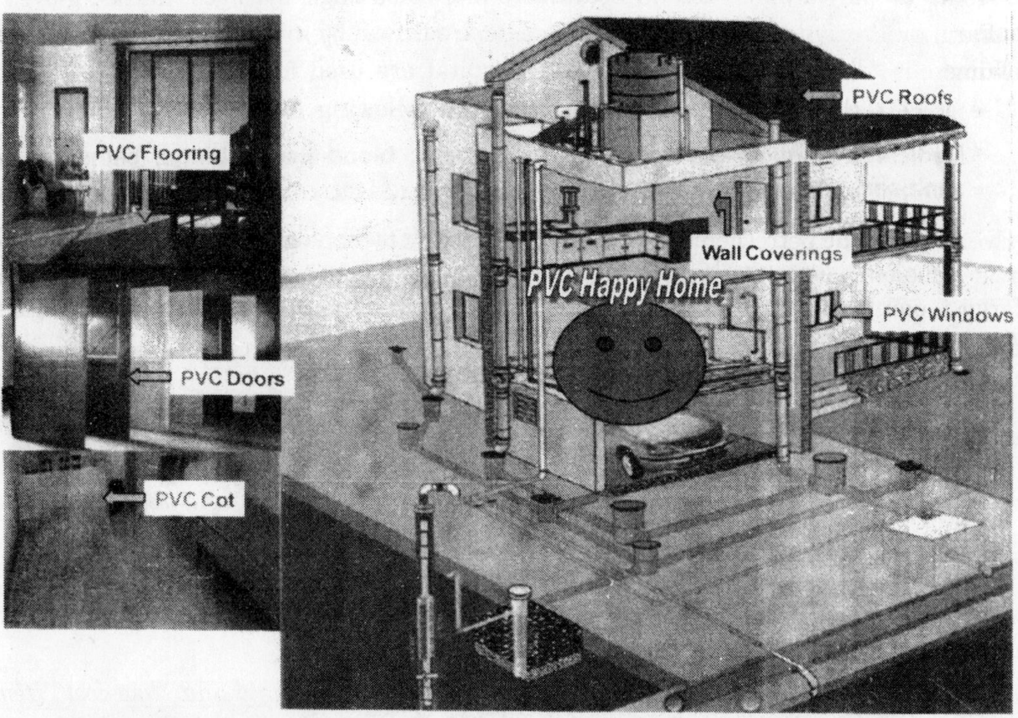

Use of Petrochemicals in various Consumer Electronics is shown below:

Cell phone, computer, CD/DVD player, refrigerator,micro chips, dish washer, water heater washing machine, television, air conditioner, telephone etc.

Petroleum not only provides the fuel for transportation, but more importantly, many petrochemicals which make the ride comfortable:

- *Plastic parts in automotives are light weight to enhance fuel efficiency.*
- *Lubricants help automotive engines run smoothly.*
- *Refrigerants make for a comfortable ride.*
- *Bitumen, a petroleum derivative, helps construct good quality all-weather roads.*
- *Numerous aircraft components are made from high performance polymers.*

Use of Petrochemicals in Automobiles is shown below.

Seat belt/cover/base/fabric, wheel cap, tail light, floor mat, mirror housing, bumper grill, speedometer front cover, air intake manifold, radiator grill, shroud fan, bumper, dash board, battery casing etc.

A whole lot of hygiene and cleaning products have petrochemicals as their main ingredients apart from packaging.

Shampoos for hair-care are formulated with petrochemical cleaning agents (detergents) Detergents make for a clean laundry and household.

High absorbing materials and polymer non-woven films are vital ingredients for Feminine Hygiene, Diapers, and Incontinence pads to help keep users dry.

Plastics do play a major role in healthcare like blood bags, syringes, masks, gloves and implants.

Numerous "disinfected" products in the hospital are used for health care.

Use of Petrochemicals in Healthcare is for the following items :

Medical & surgical gloves/gowns, face mask, blood bag, tubings, catheters, hand sanitisers etc.

Even though the petrochemical industry is capital & technology intensive, the downstream sector is a major avenue for large scale employment. The downstream plastic processing is the largest employer with over million people deriving their livelihood from this sector directly or indirectly.

Appendix

B

USES OF PETROCHEMICALS IN VARIOUS SECTORS

USES OF PETROCHEMICALS IN HEALTHCARE

Petrochemicals are used in healthcare products, from the most common place to the highly specialised. An everyday example is ASA – or Acetyl salicylic acid – an important part of many pain medications. While penicillin is manufactured via fungi & microbes, phenol & cumene are used as preparatory substances. These substances are also used in the production of aspirin, with acetyl salicylic acid being aspirin's main metabolite.

Other common medical products that use petrochemicals include antihistimines, anti bacterials, rectal suppositories, cough syrups, lubricants, creams, ointments, salves, analgesics and gels. Petrochemical resins have also been used in drug purification. These resins simplify mass production of medicine, thus making them more affordable to produce and then distribute. The resins have been used in the production of a wide range of medications including those for treating AIDS, arthritis and cancer. Plastics play an important role in healthcare too. Resins and plastics from petrochemicals are used to make artificial limbs and joints. They are also used in hospitals and other medical facilities for storing blood & vaccines, for use in disposable syringes and other items of medical equipment that are used once to prevent the threat of contagion. Specially created polymers are used extensively in healthcare, most notably during cardiac surgery or for auditory and visual stimulators.

Eye glasses have benefitted from the use of plastics in frames & lenses and contact lenses are also made of plastic. Even safety has improved thanks to the introduction of child-proof caps and tamper-proof seals for medication containers, all made using plastics. Surgical gloves are often made from pliable plastics, plastic petri dishes are essential to laboratories and at a larger level, for the housing of large diagnostic medical machinery.

USE OF PETROCHEMICALS IN COSMETICS

The production of cosmetic products makes up a very small proportion of global petrochemical use. However, they are products which many of us use on a daily basis . While there has been a push towards "all-natural" ingredients in cosmetics in recent years as well as much debate over whether petrochemicals are safe to include in cosmetics, petrochemicals are a common component of many popular beauty products.

Common petrochemical-based ingredients of cosmetics include phenoxyethanol, waxes and mineral oil. Phenoxyethanol is a very widely used ingredient in cosmetics but it is only used at less than 1% of the product composition, which is the legal maximum in most countries. It is an oily, slightly viscous fluid that smells faintly of roses. It is used in many skincare products, eye makeup fragrances, blushers, detergents, baby products, sunscreens and cuticle softeners. It is synthesized for use in commercial cosmetic products.

Waxes are an important group of ingredients for the cosmetics industry and while some, such as lanolin, are animal waxes and others, such as jojoba, are plant waxes, mineral waxes and synthetic waxes are derived from petrochemicals. Mineral waxes used in cosmetics include ozokerite, paraffin, microcrystalline wax, ceresin and petrolatum.

Ozokerite is a naturally occurring mineral wax but it is also produced as a mixture of microcrystalline wax, paraffin wax and other natural waxes that match the properties of ozokerite. It is used in creams because of its thickening and emulsifying properties.

Ceresin is usually a type of paraffin wax or it is made from a blend of petroleum waxes. It has similar properties to ozokerites so is generally used for similar applications.

Paraffin wax is a refined mixture of solid crystalline hydrocarbons. However, it is more brittle in texture than ozokerite and not very compatible with mineral oil so its cosmetic uses are generally limited to use as a moisturiser or emollient, particularly in regard to manicures and pedicures.

Microcrystalline wax is less brittle and more malleable than paraffin wax, is more compatible with oils and contains 1–4% mineral oil. As such, it is widely used in the cosmetics and personal care industry. It is especially useful in lipsticks and lip balms to prevent the product from "sweating" and is also commonly used in baby products, makeup, nailcare, skincare, haircare, fragrance and sunscreen.

Mineral oil has a broad range of applications both in and out of the cosmetics and personal care industry. It is used in products that condition or protect the skin and hair and as a solvent. Medically, it is used in laxatives, and it is also used as a release agent for non-stick bakeware. It is a mixture of liquid hydrocarbons produced from the distillation of petroleum and then refined so it is suitable for commercial use.

Petrolatum is best known in its commercial form as Vaseline, the popular petroleum jelly product. Petroleum jelly is a semi-solid wax that can be used as an emollient and lubricant and is also used as a bodifying agent to improve the viscosity of cosmetic products.

Synthetic waxes are made by blending low molecular weight polymers of ethylene. Synthetic beeswax is made from a blend of fatty esters, fatty acids, fatty alcohols and high molecular weight hydrocarbons.

Myreth, oleth, laureth and ceteareth ingredients are produced by reacting the starting material with ethylene oxide and are root names for groups of ingredients, usually followed by a number, such as "ceteareth-15. Some of these root molecules may also be derived from petrochemical sources. Combinations of these ingredients are used mainly in soap and cleansing products where creating a foamy lather is desirable and the aim is to clean and soften the skin.

Polysorbates also contain ethylene oxide chains and are widely used in the cosmetics industry. Main uses include skincare products such as fresheners and cleansers, makeup bases, foundations, shampoos, fragrances and products for creating permanent waves in hair.

Other petrochemical ingredients have been banned or restricted. Benzene, for example, has been identified as a carcinogen and banned as a cosmetic ingredient. Toluene, is only used in very tiny quantities in the cosmetics industry as an ingredient in some nail products, such as certain nail varnishes.

USES OF PETROCHEMICALS IN COMPUTERS & ELECTRONICS

So many of the electronic products are made of petrochemicals. These include mobile phones, television sets, CD players, radios, DVD and Blu-Ray players, games consoles, cords, cables and devices such as iPods.

Virtually no electronic need is too big or too small for plastics to play a role, whether it is the rugged plastic required for water resistant radios used by fire-fighters and light sockets on drilling rigs or more delicate applications, such as microwave transmission and heat-resistant automotive fuses.

There are many benefits to using petrochemical derived products to manufacture computers and electronics. The latest electronic devices utilise new generation plastics because of their special features as described below:

Resource-efficiency: Liquid crystal display (LCD) flat screens, commonly used in televisions and computer monitors, are made of liquid crystalline plastics which use at least 65% less electricity than screens with cathode ray tubes. Polymers are also useful for storing energy for longer. Modern refrigerators are insulated with thermal-efficient plastic foam with interiors made of durable, easy-to-clean plastics. Other large appliances, such as dish washers, are sporting more and more plastic components.

Light yet strong: Plastics are lightweight as well as being durable, attractive and cost effective. As such, they are used in nearly all small appliances, such as coffee makers, irons, electric kettles, mixers, hair drye rs, food processors, microwave ovens and shavers.

Films of polycarbonate are used to create the touchscreens used on products such as tablets and smartphones - this eliminates the need for a heavy keyboard. Such plastics are also used for phone handsets and MP3 players.

Also, keyboards and computer housings are made from styrenic plastics which have the advantage of being light and strong. While there is a strong drive today towards storing data online, CDs and CD-ROMs are still widely used. Most CDs consist of a clear polycarbonate, derived from benzene, cumene or bisphenol. This polycarbonate is easy to mould and thermoform. It is impressed with tiny bumps which are arranged as a long spiral track of data. This is then covered with an aluminium layer and then a thin layer of sprayon acrylic is added for protection. Finally, a label can be printed on the finished product. The plastic boxes in which CDs are usually stored are made from polystyrene, another petrochemical-based product.

Smaller size: Plastics are useful for making smaller components, especially those which cannot be seen. Nearly half of all plastics used in the electronics sector are for cable sheathing and small internal components. An use of petrochemicals in computers is microchips, made using plastics which are cheaper to produce than sand-based silicone. The miniaturisation of circuit boards and components such as computer chips, relies on high-performance plastics to provide tough, dimensionally stable parts that can withstand the stress of assembly and use. With plastics, electronics designers decrease size while increasing the functionality of circuitry in consumer, business and industrial applications.

The extensive use of plastics in the computer industry is also due to the miniaturisation of many products - they need to be light, strong and durable. Mobile phones, digital cameras and laptops are just three products which have become steadily smaller and this has been aided by advances in plastics technology.

Safety: Plastics with thermal and insulating properties insulate almost all building wiring today and are used in switches, connectors and receptacles. Fires can start from electronic sources so flame-retardant materials are important to manufacturers of electronic goods. A polymer is formulated especially for each application for maxinum resistance to fire and heat. Plastics are also used to improve safety when used for insulation on items such as cords and cables as well as the devices and appliances themselves. On a basic level, smoke detectors made with plastics, have become standard fixtures in modern homes, cutting down on fire deaths and property losses.

Resistance: Resistance to electromagnetic radiation, mechanical shocks & stress, combined with durability and flexibility, make plastics ideal for applications such as safe, efficient power supplies.

Cost benefits: Plastics have improved the durability, weatherability and chemical resistance of many electrical products. As a result, cost savings are made with increased product life. Plastics make electronic products practical and economical. Use of plastic has made products such as television sets far more accessible to more people globally with the cheaper production costs and corresponding drop in prices.

More efficient manufacturing: With computers playing a bigger role in the automated production lines of many modern factories, manufacturers rely heavily on a wide range of plastics for control panels, housings, wiring boards, sensors and robotic components.

Medical electronic applications: Electronics play an important role in the field of medicine – devices such as scanners and X-rays which allow early detection of medical conditions without invasive exploratory surgery rely on radiation-transparent plastic materials used in X-ray tables. These permit lower strength radiation to be used effectively.

Innovation and new advances: Plastics in the electronic sector are constantly evolving. An example of this is lithium batteries which can now be made from recycled plastic bags. Plastic batteries that are made from conductive polymers offer high power and low weight. The optical properties of polymers also help with facilitating the flow of data over long distances.

Mobile phones which have become smaller, lighter and more practical thanks to improvements in micro-technology and plastics.

Home computers play multiple roles for families, including budgeting, corresponding, playing games and working from home. All these machines depend on plastic housings, circuit boards, components and packaging.

A new era for plastic electronics is based on inherently conductive polymers (ICPs) and flexible substrates. Products which will use ICPs, include anti-static coating, rechargeable batteries, radio frequency identification (RFID) and organic light-emitting diodes (OLEDs) for computers and mobile phones, flexible solar panels that can be laminated to walls and ceilings or used to as a source of energy for portable equipment.

The multifaceted uses of petrochemicals have proven to be a great boon to the electronics sector. The electrical goods and electronic industries will continue to rely heavily on petrochemical products and industries for many important components and to ensure the products we take for granted constantly improve.

End of Life Measures for Plastics in Electronics

In response to concerns about plastic in obsolete electronics, infrastructure for recycling plastics used in electronics has been developed in many countries. The market for recovered plastics is growing with many product and industry applications under development. These include: compact disc covers, camera casings, automotive parts, construction, shipping and concrete.

Types of Plastic used in Electronics are Listed below:

Different types of plastics perform different roles in the manufacture of electrical products. Common plastics types and their main uses are listed below:

Acrylonitrile butadiene styrene (ABD): Computer housings, monitors, keyboards, telephone handsets.

Alkyd resins: Circuit breakers, switch gear.

Amino resins: Lighting fixtures. Useful because of low conductivity.

Epoxy resins: Electrical components.

Ethylene vinyl acetate (EVA): Handle grips, fridge-freezer door seals, vacuum cleaner hoses.

Liquid crystal polymer (LCP): Moulded interconnect devices. Replaces ceramics in high performance electrical components used in high-frequency applications.

Phenol formaldehyde: Fuseboxes, switches, handles.

Polyacetal: Also known as polyoxymethylene (POM), used in electrical appliance housings.

Polyamide: Adaptors, food processor bearings.

Polycarbonate: Telephones.

Polyesters: Switches, electrical insulation.

Polyethylene: Electrical insulation.

Polymethyl pentane (PMP): Circuit boards.

Polyphenylene oxide (PPO): TV housings, coffee machines.

Polyphenylene sulphide (PPS): Transformers, hair dryer grilles, element bases, moulded interconnect devices. Replaces ceramics in high-performance electrical components used in high-frequency applications.

Polypropylene: Kettles.

Polystyrene: Refrigerator trays and linings.

Polysulphone: Microwave grilles.

Polyvinyl chloride (PVC): Cable trunking, cable and wire insilation.

Styrene acrylonitrile (SAN): Outer covers of electrical devices, battery housings, air conditioning components.

Urea formaldehyde: Fuse boxes, switches.

TRANSPORT AND AUTOMOTIVE USES OF PETROCHEMICALS

The petroleum industry has an important role to play in the world of transport by providing the fuels that power vehicles. But oil also has a vital role to play in terms of the petrochemical products which are essential to the modern transport industry. Indeed, the cars, buses, motorcycles, trucks and aeroplanes of today are all heavily reliant on the use of petrochemical-derived products.

Petrochemical in Car Production

Automotive bodies and components benefit from the light weight, design flexibility and durability of many petrochemical products.

Petrochemical products are generally easy to process and more cost-efficient for the manufacturer because of availability and reduction in vehicle assembly time. The light weight is a particular bonus because this translates into energy savings for the end user. With this increased use of strong, lightweight plastics in cars, the environmental outcomes are impressive - the lighter weight of plastic components means fuel economy can be improved by 25-35% and for every kilogram of weight lost, the car will emit 20kg less of carbon dioxide over its operating life.

An example of a lightweight petrochemical derived product with multiple uses is Chevron's Ryton PPS. Made from polyphenylene sulphide (PPS), it is a lighter weight, high performance engineering thermoplastic. It is used in parts exposed to high temperatures, automotive fluids and mechanical stress, such as brake systems, fuel systems, transmission systems, coolant systems & induction systems. It is also used in car lighting, for power train components and electrical components. Ryton PPS compounds are also used in automotive fluids, such as brake fluid and power steering fluid.

High density polyethylene gasoline tanks have helped prevent countless fires. This is because of their excellent shape-resistance. In the event of a collision, the tank holds its shape, petrol does not leak and a fire cannot start. Other uses for poly ethylene in the manufacture of cars include reservoirs for wiper fluid, brake fluid and coolant, as well as flexible parts, such as suspension dust covers.

Other plastics that are commonly used in car production include dodecyl mercaptan, used for headlight lens covers, and dodecyl disulphide, used for brake light covers. Petrochemical-derived synthetic rubber tyres have improved car safety with their excellent road-holding abilities, especially in wet conditions. Synthetic rubber is derived from petroleum, as opposed to natural rubber, which is obtained from certain tropical trees and this is commonly used on aircraft and bicycle tyres as well as automotive vehicles and motorcycles. Styrene butadiene rubber (SBR) is a common synthetic rubber used for tyres, belts and hoses because of its excellent abrasion resistance and durability. Another synthetic rubber widely used in the automotive industry is thermoplastic olefin rubber. It is used for parts such as interior & exterior trims, door seals, arm and head rests, consoles, airbag covers and instrument panels.

Cyclohexyl mercaptan is used in the automotive world to vulcanise rubber for tyres. Phenolic resins are used in binders for friction materials in brakes & clutches, and car booster seats made from petrochemical-based products, such as styrene-based plastics, have been developed to meet crash test standards.

Styrene is used to make acrylonitrile butadiene-styrene (ABS), a heat-resistant thermoplastic widely used in automotive parts, including headlight covers and moulded plastic interior parts. Seat belts and air bags also rely on petrochemical-derived products for their manufacture. Styrene-butadiene latex is used to make carpet backing for vehicles. Polyalphaolefins (PAOs) are base oils which are not only used to manufacture fluids, oils & lubricants when mixed with other additives, but can also be used in seat covers and other automotive fabrics that require low volatility. Other fabric-related uses for petrochemicals include cyclohexane, a feedstock for nylon. This is found in seat belts and carpets, but nylon resins are also used in engine covers, air intake manifolds, airbags, reservoirs and in air conditioning components, such as fans.

Paraxylene is a feedstock for polyester, found in seat fabrics, headliners and carpet, as well as radial tyre cord. Products used in car maintenance also rely heavily on petrochemicals, such as antifreeze and radiator coolant, which is produced using ethylene gtycol. Hydraulic brake fluids and petroleum demulsifiers are also made with ethylene glycol. Other petrochemicals used to make car maintenance products include polysulphides for engine lubricants and sulfolene for transmission fluids and other base oil formulations.

Petrochemicals uses in Road Construction

Roads themselves are also created using petrochemicals. Asphalt, a by-product of crude oil refining, has long been used on roads, motorsport race tracks and footpaths. A combination of standard asphalt with elastomers provides road surfaces that are elastic, cohesive and adhere well to the ground. Kromatis, is used for urban roads and it can be cold-mixed or hot-mixed and bulk coloured - its use helps mark traffic areas clearly, such as bus lanes, improving safety.

Uses of Petrochemicals in Motor sport

Motorsport is heavily reliant on a range of petrochemical-derived products, such as carbon fibre, used to make lightweight body parts. To improve the power-to-weight ratio, carbon fibre and other plastics are used extensively. This is an example of how motorsport technology transfers to passenger cars, which are featuring more plastic components than ever before and, especially at the luxury end of the market, carbon fibre is being used more extensively.

In chassis design one team uses carbon fibre for the top & bottom wishbones of the suspension and moulded carbon fibre is used in conjunction with an aluminium composite for strength and lightweight efficiency. The car also features a synthetic rubber fuel cell reinforced with Kevlar, another petrochemical-derived product, and a seat made from an anatomically formed carbon composite.

Aeronautical Uses of Petrochemicals

Like the automotive and motorsport industries, the aeronautical industry relies heavily on petrochemicals. Composites derived from petrochemicals can save more than 30% of the weight of an aircraft structure. Keeping components light is essential on aircraft and the weight of every item counts, right down to the plastic used in the miniature spirit bottles. Since the 1970s, the use of plastics in the manufacture of aircrafts has increased from 4% to almost 30% and is expected to reach 50% in 2017. Ethylene and propylene glycols are used in products, similar to car antifreeze, to help aeroplanes take off in winter weather conditions.

USES OF PETROCHEMICALS IN CONSTRUCTION INDUSTRY

Petrochemicals, in particular plastics, have revolutionized the construction industry. After packaging, the construction industry is now the world's second highest user of plastics. This is because of their durability, versatility, excellent strength to weight ratio, cost-effectiveness, corrosion resistance and low maintenance. The main uses for plastics in construction include seals, windows, doors, pipes, cables, floor coverings and insulation. Indeed, plastic can be used to create entire buildings.

Plastics

Pipes and conduits are the largest uses for polymers for the construction industry. Often used in place of copper or lead, which can both be problematic, PVC and polyethylene are generally used for such applications as large pipes for sewage, drainage & potable water, as well as cabling. Windows and doors were commonly clad and profiled with wood before the rise of the plastics sector but PVC-U has now become a common material for these purposes, as well as coving and skirting. PVC can also be used as a membrane for roofing and linings. Exterior cladding and roofing using phenolic resin rather than timber is also popular. Another advantage these materials have over wood is reduced fire risk.

Insulation, which has great energy conservation advantages, has also advanced thanks to petrochemical developments. It is commonly manufactured from rigid polystyrene foam which can be incorporated into panels or fitted into the construction of walls and roofs. Polystyrene-based insulation is light, strong and easy to install.

Another example of a useful plastic are high density polyethylene materials which are commonly used as house wraps. Such housewraps provide breathable protection from water for buildings. It is applied between a building's outer cladding and frame.

Elastomer Epoxy and Paints

Elastomer is used for seals and gaskets. In construction, these are mostly used for weather strips, aperture seals, gaskets and expansion joints. The polymers used in these items are chloroprene and ethylene propylene diene monomer. These are chosen because they are weather-resistant, deformation-resistant and retain elasticity. Large quantities of polymers are also used in construction as a base for adhesives and sealants.

Epoxy resins are flexible resins made using phenols (aromatic alcohols) 'Epoxy' refers to a chemical group consisting of an oxygen atom bonded to two carbon atoms that are already bonded in some way. They are predominantly used in coatings, adhesives, electrical laminants and composites. This is because of useful properties such as adhesion, strength, chemical resistance and resistance to environmental degradation. It is because of this that as well as being used in construction, they play a major role in the manufacture of aircraft components and boat hulls.

Epoxy resin coatings are also used in non-slip floor coatings, which are commonly used in warehouses, food preparation areas and factories. These coatings are hard-wearing and resistant to abrasion, chemical spillage and impact. Similarly, polyurethane coatings are used on floors in places such as mortuaries, breweries, vehicle maintenance workshops and abbatoirs, where floors need to be able to withstand high temperature cleaning and chemical resistance.

Paints are used externally & internally in construction and many synthetic-based paints rely on petrochemical products during their production. The film-forming component of paint is known as the binder – this helps paint adhere to surfaces and influences the paint's longevity, gloss, strength and flexibility. Synthetic binders include resins like acrylics, vinyl-acrylics, vinyl acetate/ethylene (VAE), polyurethanes, polyesters, melamine resins and epoxy resins. Emulsion paint (known as latex paint) is a term for all paints that use synthetic polymers such as acrylic, vinyl acrylic and styrene acrylic as binders. The solvent (or diluent) component of paint serves to dissolve the polymer and adjust the paint's viscosity. Solvent-borne paints, commonly marketed as "oil-based paints" can contain different combinations of organic solvents including aromatics, alcohols and ketones. Specific examples include petroleum distillate, glycol ethers, synthetic resins and esters.

Other petrochemical-based additives for paint can be used to modify surface tension, improve flow, impart anti-freeze properties, change the texture, improve adhesion, reduce levels of glossiness and improve the finished appearance.

Concretes

Concrete – a man-made material that combines cement with aggregate and water – has been an important building material for centuries. The ancient Egyptians used a concrete like material more than 5,000 years ago, using gypsum mortars, mortars of lime and mud mixed with straw to bind bricks when building the pyramids.

Since ancient times, concrete has come a long way and modern concrete often depends on petrochemical-based additives. Plasticisers, also known as water-reducing admixtures, reduce the quantity of water needed in a concrete mix. Newer water-reducing admixture products are based on polycarboxylic ethers, naphthalene formaldehyde condensates or hydroxylated polymers.

Accelerating admixtures are used to speed up the drying time of concretes. While calcium chloride-based accelerators can corrode steel used in concrete reinforcing, triethanolamine can be used instead. Conversely, retarding mixtures are used in hot weather to delay drying times. They can also act as water reducers.

Petrochemicals can play a part in the production of sealers for concrete. Sealers improve the water tightness of concrete, make it easier to clean and protect concrete from damage caused by toxic spills. Film-forming sealers are commonly produced using acrylics, polyurethanes and epoxies. They are easy to use, dry quickly and are cost-effective.

By-products of the refining process can also be used in concrete production, such as fly ash, a byproduct from the coal sector. Fly ash can reduce up to 30% of the cement needed in a mix, it improves workability and reduces the heat generated by concrete. Ground granulated slag from blast furnaces can also be used like fly ash in concrete production.

Petrochemicals are used to create a wide range of new materials which will continue to play an important role in sustainable construction projects. Polymeric composites and glass-reinforced plastic materials are two examples. Plastics are advantageous during construction for a number of reasons, such as versatility, resistance to corrosion, light weight, strength and potential for Creating more load-bearing structures. As such, their use in construction has a long future.

USES OF PETROCHEMICALS IN FERTILISERS

Petrochemical based fertilisers still dominate the international market and demand remains strong.

To understand the role of petrochemicals in fertilisers, it is important to understand the three main nutrients that benefit plants. They are nitrogen (for strong growth), phosphorus (for root growth) and potassium (for improved plant metabolism & disease resistance). The production of nitrogen for fertiliser is based on a process developed by Haber for ammonia production, which is then processed into a number of different nitrogen fertilisers including urea, nitric acid and ammonium nitrate, Also, urea & ammonia can be blended with water to make liquid nitrogen fertiliser. However, ammonium nitrate fertiliser is heavily regulated because it can be made to explode.

Gasified coal and gasified petroleum coke can also be used to produce nitrogen fertilisers, but these make up a minority of global synthetic fertilizer production. In modern synthetic fertiliser production based on coal, there are two common processes used. One is the Winkler process which uses a fluidized bed in which powdered coal is agitated with reactant gases. The other is the Koppers-Totzek process which operates at a higher temperature - powdered coal is reacted while it is entrained in gases passing through a reactor. Ash is removed as molten slag at the bottom of the reactor.

Synthetic Versus Organic Fertilisers

Synthetic fertilisers are chemically identical to organic fertilisers, but fertiliser companies have acknowledged the damage careless use of fertilizer can do, such as causing algae to grow and depleting waterways of oxygen. However, animal manure also contributes to such pollution. Another attraction of petrochemical fertilisers is that they are generally more cost-effective to produce and cheaper to buy than many organic alternatives. Dependence on chemical fertilisers can be reduced by combining its use with organic fertilisers. Unlike organic fertilisers, chemical fertilizers can be easily blended to provide more precise levels of nutrients. Along with more eco-friendly manufacturing processes, many nitrogen fertiliser producers claim they are using ingredients, such as monoethanolamine and potassium carbonate, because of their purity and to ensure that compositions of synthetic fertilizer are accurate so they can be used efficiently.

Petrochemical companies involved in the production of fertiliser, are being encouraged to diversify, to investigate alternative feedstocks and to ensure best practice techniques are used along the entire supply chain to minimise the environmental impact of nitrogen fertiliser production.

PETROCHEMICALS USAGE IN FOOD

The petrochemicals used in foods are usually derived from either petroleum (crude oil) or natural gas. These resulting intermediates and derivatives from the petrochemical processes, such as thermal cracking, catalytic cracking and steam reforming, are used as food additives. Petrochemicals can take forms as synthetic food additives that are not differentiated from their natural counterparts.

Food Additives

Food additives are "any substance not normally consumed as a food in itself and not normally used as a characteristic ingredient of food, whether or not it has nutritive value, the intentional addition of which to food for a technological purpose results in it or its by-products becoming directly or indirectly a component of such foods" Generally, most food additives could be categorized into one or more of these sub-classes: **food flavourings, food colourings and food preservatives.**

Food Flavouring

Food flavourings are substances that are added to foods to develop tastes and smells. A widely used petrochemical for artificial food flavouring is butyric acid. Although butyric acid has an unpleasant odour and acrid taste, it can be converted into butyric esters that have pleasant aromas or tastes that are used as fragrant and flavouring agents in food and beverages.

Butyric acids are produced industrially via chemical synthesis. This involves the oxidation of butyraldehyde that is produced using propylene derived from crude oil. Despite alternative methods, such as extraction from butter and natural fermentation, chemical synthesis is still the preferred method as it is low in production cost and there is a higher availability of starting materials. Butyric acid is found in a wide range of food products including sweet & savoury biscuits, coffee whitener, chocolate, confectionery and processed dairy products.

Food Preservation

Food spoilages are characterised as the deterioration of varying degrees in organoleptic properties (how a food is perceived via the senses, such as taste, touch and smell), nutritional value, safety and aesthetic appeal of foods to an extent that the foods are deemed to be unfit for human consumption. These spoilages can be triggered by many factors, however, the growth of microorg anisms remains the main contributor. Through the understanding of how micro organisms are able to proliferate and thrive in our foods, measures can be taken to optimise shelf life. Food preservatives are substances that are introduced into foods to retard or eliminate the growth of micro organisms. One of the most commonly used petrochemicals in the food industry as a preservative is benzoic acid. Benzoic acid occurs naturally in sources such as cranberries, prunes and plums, or can be synthesised via petroleum-derived intermediates such as phthalic anhydride or toluene.

Benzoic acid and its salts & esters (benzoates) are commonly used in fruit juices and soft drinks in an acidic pH range. Different forms of benzoic acid and benzoates used include benzoic acid, sodium benzoate , potassium benzoate and calcium benzoate. The antimicrobial property of benzoic acid is attributed to its ability to be absorbed by the microorganisms. Inside the cell, benzoic acid lowers the pH of the internal environment and inhibits the metabolism of the microorganisms, resulting in retardation of growth and ultimately death of the microorganism. As benzoic acid produces a sour taste, they are used as food flavourings to some extent. They are used to flavour the pickles used in burgers.

Mineral oils are also used as food preservatives. They are manufactured from crude mineral oils though various refining steps and encompass a subgroup of the family of mineral oils and waxes. There are two different types of waxes derived from petroleum that are used in the food industry – paraffin waxes that are distinguished by large, well-formed crystals and microcrystalline waxes that have a higher melting point with smaller and irregular crystals.

The microcrystalline wax is non-toxic and approved for use internationally for surface treatment of non-chocolate confectionery, chewing gum, melons, papayas, mangoes, and avocado. The wax coating improves the appearance and shelf life of the foods by inhibiting dehydration and protecting from oxidation. The protective barrier of wax also attenuates the risks of microbial contamination from the environment. In addition to coatings, microcrystalline wax is also an essential ingredient in the production of chewing gum. Another petrochemical that is used in the manufacture of chewing gum is resin.

One of the many side effects of microbial growth in foods is the secretion of the digestive enzyme lipase that breaks down fats. Unsaturated fats in particular, are susceptible to oxidative damage. The presence of water and oxygen may also cause oxidation of these unsaturated fats. Effects of the oxidation of fats include decreased nutritional quality, increased toxicity, development of rancid flavours & odours, and altered texture & colour. Petrochemicals like butylated hydr oxyanisole (BHA) & butylated hydroxytoulene (BHT) are added to spare the oxidation of fats. As these substances are fat soluble and have antioxidant properties, they are able to extend the shelf-life of these products.

Foods commonly associated with the use of BHA & BHT are those containing fats such as vegetable oils, animal fats, flavourings, spices, nuts, processed meats and snack products. Despite the similar antioxidant properties of vitamin E, BHA and BHT are preferred for their superior thermal stability. BHA has greater heat tolerance compared to BHT. Since the 1970s, BHA has been steadily replacing BHT due to toxicity studies. In addition to their functions as antioxidants in food, BHA and BHT are also used in food packaging materials.

Antioxidants applied indirectly into foods, for example, via food packaging, are known as secondary antioxidants. In recent years the use of packaging materials that exert antimicrobial and anti-oxidative properties has garnered considerable interest. In the USA, cereal industries have been utilising wax paper liners applied with BHA and BHT to contain breakfast cereals & snack food products. BHA-and BHT-treated liners exert anti-oxidative effects on the foods via evaporative migration into the foods.

Plastics have long been used as a protective barrier for foods and are derived from petroleum. Examples of the different types of plastics that are frequently used in the food industry include polyethylene terephthalate (PET) in food-contact films and high density polyethylene in food containers.

Additives for Flavour and Preservation

Some petrochemical-derived additives perform more than one role when included in food products. Glycerol or glycerine is an example of an approved, petrochemical-derived food additive that has the combined effect of acting as a food flavouring and food preservative. Glycerine is an odourless, colourless, oily, viscous liquid with a sweet taste. Among its desirable effects in foods are: being used as a humectant (retains moisture), emulsifier for fats, smoothing agent, heat transfer media for frozen foods, solvent for food flavourings and colourings, and as a sweetener.

The multi-step production of synthetic glycerol via the petrochemical propylene is known as the epichlorohydrin process. Natural routes of glycerol production such as oil and fat processing dominates the industry as synthetic means became economically unviable.

Food Colouring

The earliest discovery of petrochemicals as synthetic dyes dates back to 1856, when the purple dye, mauve, from coal tar was discovered. The discovery of mauvin initiated a race to discover other synthetic dyes that are present in the myriad of chemicals in coal tar. Dyes from coal tar were quickly adopted to colour food, drugs and cosmetics. Food dyes in the present, however, are derived from petroleum. As food consumers, we are attracted to products that are aesthetically pleasing. We form judgments using colours to perceive the quality and attractiveness of a product that is being marketed to us. Consumers associate colouration with superior flavour. Hence, most food manufacturers incorporate food colourings or food dyes in order to enhance the visual quality of their products.

Synthetic dyes are cheaper to produce, more stable and produce brighter colours than most natural colourings. The prevalence of petroleum-derived food dyes is such that they are used in products including breakfast cereals, candy, snacks, beverages, vitamins, and even on the skin of citrus fruits.

Alternatives to the use of Petrochemicals in Food Industry

Notwithstanding the many advantages of using petrochemicals in the food industry, there have been rising concerns over the safety and long term consumption effects on human health. A clear cut consensus on the usage of the different petrochemicals cannot be made, but efforts and measures have been taken by relevant government bodies to regulate levels of usage. With the increasing demand for organic foods, some manufacturers have opted for natural substitutes. As with most petrochemicals used in the food industry, similar substances could be found in naturally occurring sources. However, due to efficiency, availability and cost issues, this can then result in increased costs being passed on to consumers. Natural sources lack stability and there are still certain properties of synthetic sources that are unrivalled, such as brighter colours from dyes & heat stable antioxidants. Furthermore, in the area of plastics, researchers have been looking into substituting petroleum derived plastics with plant-based plastics. However, there is still a lack of technological advances to ensure that naturally sourced plastics are on par with petroleum-derived plastics in terms of effectiveness, cost and production efficiency.

TEXTILE INDUSTRY USES OF PETROCHEMICALS

Without petrochemicals, we would not have the advances that have allowed for affordable mass production of clothes, durable fabrics for a wide range of uses, the variety of fabric dyes we have available and fabrics that are easier to care for than natural fibres, such as cotton and silk. Synthetic fibres possess qualities that are not easily found in natural fibres. The synthetic textiles industry has also offered economic benefits to many countries While all synthetics fibres are man-made using petrochemical-derived polymers, other man-made fibres can be composed of a range of materials, including silk, cotton and wool as well as plant polymers, such as cellulose. Synthetic fibres have a wide variety of uses that go beyond familiar items of clothing, such as nylon stockings. This includes industrial uses, such as tyre cord and flame-proof linings, as well as home furnishing textiles, automotive textiles and sports & leisure uses, such as wetsuits and waterproof tents.

When magnified, synthetic fabrics look like plastic that has been spun together. Generally, synthetic fabrics are made from chemically produced fibres. The chemicals used to make these fibres include sodium hydroxide and carbon di-sulphide, which are derived from coal, oil or natural gas. Polyester, acrylic and nylon are all examples of fabrics that come from these sources. Viscose, however, can be sourced from either pine trees or petrochemicals.

Initially, the fibre-forming polymers are solid. The solids are converted into a liquid state. This is done by melting, if the polymers are thermoplastic synthetics (i.e: they soften and melt when heated), or dissolving them in a solvent if they are non thermoplastic cellulosics.

If the polymer cannot be melted or dissolved, they are chemically treated to form soluble or thermoplastic derivatives. The chemicals in liquid form are then forced through a device called a spinneret in a process called extrusion. At this stage, the liquid is thick and viscous with a consistency similar to cold honey. A spinneret is similar to a bathroom shower head and it may have from one to several hundred holes. As the liquid emerges from the spinnerets and is exposed to air, it cools and forms tiny threads. These threads are dyed before being woven or knitted together to make fabric. This can be done through four methods of spinning: wet, dry, melt and gel-spinning.

Wet spinning is used on fibre-forming substances that have been dissolved in a solvent. The spinnerets are submerged in a chemical bath and as the filaments emerge, they precipitate and solidify. Acrylic, rayon and Spandex are produced via this process.

In dry spinning, solidification is achieved by evaporating the solvent in a stream of air or inert gas. The filaments do not come into contact with a precipitating liquid so there is no need for drying. This process can be used to make acetate, acrylic, Spandex and vinyon. In melt spinning, the fibre-firming substance is melted for extrusion through the spinneret and then directly solidified by cooling. This process can be used to make nylon and polyester products. Melt-spun fibres can be extruded from the spinneret in different cross-sectional shapes, such as round, trilobal, pentagonal and octagonal. Trilobal-shaped fibres reflect light to give a sparkling effect to the fabric. Fibres that are spun in a pentagonal shape or are hollow are used in carpets to show less dirt. Hollow fibres also trap air and create insulation. Octagonal-shaped fibres offer glitter-free effects.

Gel-spinning is a process used to make highstrength fibres. In this process, the polymer is not in a true liquid state during extrusion. The polymer chains are bound together at various points in a liquid crystal form instead - this produces strong inter-chain forces in the resulting filaments that can significantly increase the tensile strength of the fibres. Stronger fabrics can also be created by drawing the filaments while they are solidifying or afterwards - this pulls the molecular chains together and orients them along the fibre axis to create stronger yarns. Sometimes, additional chemicals can be added to the fabric for specific purposes. Chemicals can make the fabric softer, free of wrinkles, more stretchy, flame-resistant, stain-resistant, waterproof and mothrepellant.

Textiles created from petrochemicals include fabrics commonly used in clothing, such as rayon, acetate, nylon, acrylic, polyester, elastane, which is commonly marketed as Lycra or Spandex. Other more specialist materials created from petrochemicals include Kevlar and neoprene. Synthetic fabrics often have multiple uses, some of which cannot be easily achieved with natural fibres alone. Adding further chemicals to synthetic fibres can further improve their versatility.

Fabrics with excellent stretch are commonly used in swimwear, lingerie and hosiery as well as garments that benefit from stretchy fabric, including T-shirts and tracksuits. Synthetics can be mixed with natural fibres to create improved fabrics. For example, polyester, when mixed with cotton, is a common example of a synthetic-natural blend, whereby a garment such as a T-shirt benefits from the cooling, breathable nature of cotton and the stretch, durability, easy care & quick drying properties of polyester.

Polyester is also used in raincoats, fleece jackets, children's nightwear, medical textiles & working clothes. Other blends of natural & synthetic fibres include cotton/Lycra blend, which is especially useful for manufacturing denim which is more comfortable and fits better than 100% cotton jeans, and acrylic/wool blend, which is used to manufacture trousers that are less expensive and easier to care for than 100% wool.

Acrylic is widely used for jumpers, fleece jackets and blankets because it has similar properties to wool, but dries much faster and is more easily washed in machines.

Nylon, has a wide range of uses, including active sportswear, fleece jackets, socks, seatbelts, tents and waterproof garments. The combination of warmth, durability and weatherproofing makes it a very versatile textile. It is also easy to clean and does not crease easily.

Neoprene is produced by the polymerisation of 2-chlorobutadiene. This process occurs in an aqueous emulsion. It is a versatile material that is durable, elastic and maintains flexibility over a wide temperature range. As such, it is used on wetsuits, laptop sleeves, orthopaedic braces, electrical insulation, padding for bicycle seats & saddles, and automotive fan belts. Kevlar is another synthetic textile which has multiple uses. It is an aramid fibre, that is, a fibre that is an aromatic polyamide. It can be manufactured in many different grades for a range of clothing, accessories & equipment that is safe and cut resistant. Like carbon fibre, which is commonly used on racing cars, Kevlar is both lightweight and strong. Other properties of Kevlar include low density, good impact resistance, good abrasion resistance, good chemical resistance & good resistance to thermal degradation. It is usually yellow in appearance but is frequently coloured, depending on its application, such as camouflaging colours for bullet-proof vests. While the manufacture of bullet-proof vests is probably the best known application for Kevlar, it is a very versatile material with uses that transcend textile uses. Other applications include protective apparel (such as gloves, motorcycle protective clothing, hunting gaiters and chaps; sails for boats), belts and hosing for industrial & automotive applications, aircraft body parts, boat hulls, fibre optic and electromechanical cables, friction linings for clutch plates and brake pads, gaskets for high temperature & pressure applications, and adhesives & sealants.

As well as producing fabrics, the petrochemicals industry also has a hand in creating products that are related to the textiles industry. Fabric softener, dye fix, dyestuffs, stain removers and fabric brighteners all rely on petrochemicals.

There has been concern that while these are desirable qualities in fabrics, the fabrics are nonbiodegradable so they do not break: down in soil if they are discarded. The synthetic textile industry has also come under criticism for the effects on the environment, wildlife and health, especially for those who work to produce these fabrics. Factories have been accused of consuming excessive amounts of water and energy.

USES OF PETROCHEMICALS IN SPORTS

Petrochemicals are especially relevant in the manufacturing process of many items which have become integral to modern sports and leisure activities. For example, plastics are widely used in high-tech sports equipment, which ranges from fairly common items, such as tennis racquets, skis, footballs and running shoes, as well as the kind of sporting equipment that is not quite so commonplace, such as Formula-1 race cars and yachts.

Making sporting equipment from synthetics has made mass production of such items easier and made sporting equipment cheaper and more easily accessible to more people. As well as the economic and social benefits, many synthetic materials used in sports and leisure equipment are also more durable than naturally sourced materials, such as leather for footballs.

Hard, shock-absorbing surfaces

Petrochemical-derived synthetics also play an important role in the manufacture of athletics tracks, tennis courts, netball and basketball courts, and artificial turf for football and rugby fields. An example of this is the development of Rebound Ace, a surface that is essentially a high density, high shock absorption rubber cushion mat that has been combined with multi-layered acrylic surfacing. It has become a popular surface for tennis and netball courts. The rubber is a multilayered polyurethane combined with fiberglass & EPDM (Ethylene Propylene Diene Monomer) rubber granule cushioning. This is laid on top of an asphalt or reinforced concrete base. The main benefit of a shock-absorbing surface, such as Rebound Ace, is that it responds to body impact to reduce fatigue on the legs, ankles and feet of players. Cushioned courts have kept players playing for as many as 10 years beyond what is usually the norm on hard courts. It was also the surface for the Australian Open tennis courts from 1988 until 2007, when it was replaced with another petrochemical-derived surface, Plexicushion. This is another synthetic rubber surface, a blend of latex, synthetic rubber and plastic particles.

Artificial Turf

Artificial turf has been refined and developed over the years as a low-maintenance, durable, affordable alternative to grass. It is commonly used on playing fields and tennis courts. Astro Turf use became widespread in the US in the 1970s, predominantly for baseball and American football. At the 1976 Olympics in Montreal, hockey matches were played on a nylon synthetic turf – these days most top-level hockey matches are played on such surfaces and it has changed the sport by making the game faster and adding new playing techniques.

While it is a registered trademark, Astro Turf became a colloquialism for any artificial turf. Originally, artificial turf was made from short-pile synthetic fibres without infill while second-generation artificial turf used sand as an infill from the 1980s. The latest artificial turfs, developed in the 1990s, tend to use infills made from a blend of sand and recycled rubber.

The basic material for the short-pile turf is polypropylene and polyethylene granules which are melted and extruded to create long yarns or a film that is then cut into individual yarns. The yarns are then tufted into a polypropylene fabric, in a similar technique to that used for domestic carpet – hundreds of needles are positioned on a beam and every needle pulls a yarn through a base fabric and form a loop. Each loop is then cut so the fibres stand upright.

As well as sporting uses, artificial turf has been used at airports instead of natural grass. It is easier for vehicles to drive over, it doesn't attract wildlife so the risk of wildlife colliding with planes is reduced, it stays green in all weathers so it is easily spotted by pilots and it is easier to maintain.

Athletics Tracks

One of the most prominent sporting uses for petrochemical products is on sporting tracks. Leading chemical companies have created sporting surfaces using petrochemicals. Their tracks are approved by the International Association of Athletics Federations. BASF's Conica surfaces, for example, have been created to be not too hard or too soft and are made from a three-layer solid plastic. A polyurethane coating is applied before the track is laid and then synthetic rubber granules (EPDM) are used before the top most polyurethane layer is applied. This surface is also weather-proof.

Versalis, Italy's leading petrochemical company, have also developed a polymer athletics track made from polyethylene and synthetic rubber. The blend of polymers create a durable track that is UV-resistant.

Other Sporting Equipments

Whether it's a more durable material than leather for footballs and rugby balls or wood for tennis racquets, or replacing heavier steels in bicycles, petrochemicals play a major role in the manufacture of many items of modern sporting equipment. For example, China, manufactures carbon fibre bicycles for the global market. Carbon fibre is strong, rust resistant and light so it is an obvious choice for improving bicycles. DuPont has used ionomer (a polymer with ethylene as the major component and contains covalent and ionic bonds) resins to improve the performance and durability of its golf balls.

Plastics have revolutionised the world of sporting equipment, often replacing traditional materials, such as leather and wood, to create items that are durable, easy to mass-produce and are generally more affordable. Some companies, supplies moulded plastics which are then turned into a wide range of products for different sports by other manufacturers. Sporting products which can be made from plastics include helmets, archery bows, canoes, tennis racquets, caps & nose plugs for swimmers & divers, bats & balls (especially for children), baskets on ball collecting carts for golf driving ranges and the seats at stadiums.

Appendix

C

ALL ABOUT INDIAN PETROCHEMICAL INDUSTRY

Petrochemical industry in India is growing steadily. The locations of petrochemical units are given in Table below along with feedstock. The major feedstock in Indian petrochemical units is naphtha and natural gas. From feedstock of natural gas, olefins compounds are produced. In refinery, generally aromatic compounds are produced. The major intermediate products produced in the country are ethylene, propylene, butadiene, benzene, toluene and xylene. The major petrochemical complexes are located in the State of Maharashtra and Gujarat. The first naphtha cracker unit (Olefins) of United Carbide India Limited (UCIL) went into production in Mumbai in 1960. National Organic Chemical Industry Limited (NOCIL), Thane (near Mumbai) went into production in the year 1966. The third petrochemical complex *i.e.* Indian Petrochemical Corporation Limited (IPCL), Vadodara, using naphtha with facilities of production of primary, intermediate & downstream petrochemical products started during 1978-79. IPCL commissioned their gas-cracking unit in 1989 at Nagothane in Maharashtra. Reliance India Limited (RIL) commissioned their plant in 1997 at Hazira, Gujarat. In the year 1999-2000, three petrochemical complexes came into existence, these are Haldia Petrochemical, West Bengal, Gas Authority of India Limited (GAIL), Auriya, Uttar Pradesh, IPCL, Gandhar in Gujarat.

Table-List of Indian Petrochemical Industries with their feedstock

S. N.	Name of Industry	Feedstock
1.	Bongaigaon Refineries, Assam	Naphtha
2.	Bharat Petroleum Corporation Limited, Mumbai, Maharashtra	Naphtha
3.	Cochin Refinery Limited, Ernakulam, Kerala	Naphtha
4.	Gas Authority of India Limited, Auriya, Uttar Pradesh	Natural gas
5.	Hindustan Organic Chemical Limited, Cochin, Kerala	Gas
6.	Hindustan Petroleum Corporation Limited, Vizag, Andhra Pradesh	Gas
7.	Haldia Petrochemical Limited, Haldia, West Bengal	Naphtha
8.	Indian Oil Corporation, Mathura, Uttar Pradesh	Gas
9.	Indian Oil Corporation, Koyali, Vadodara, Gujarat	Naphtha
10.	Indian Petrochemical Corporation Limited, Vadodara, Gujarat	Naphtha
11.	Indian Petrochemical Corporation Limited, Gandhar, Gujarat	Natural gas
12.	Indian Petrochemical Corporation Limited, Nagothane, Maharastra	Natural gas

13.	Mangalore Refinery Limited, Mangalore, Karnataka	Gas
14.	National Organic Chemical Industries Limited, Thane, Maharashtra	Naphtha
15.	Oswal Agro, Mumbai, Maharashtra	Naphtha
16.	Reliance India Limited, Hazira, Gujarat	Naphtha/ Natural gas
17.	Reliance India Limited, Jamnagar, Gujarat	Gas
18.	Reliance India Limited, Patalganga, Maharashtra	Naphtha
19.	Rashtriya Ispat Nigam Limited, Vizag, Andhra Pradesh	Naphtha
20.	Steel Authority of India Limited, Bhilai, Chhattisgarh	Coke oven gas
21.	Steel Authority of India Limited, Bokaro, Jharkhand	Coke oven gas
22.	Steel Authority of India Limited, Durgapur, West Bengal	Coke oven gas
23.	Steel Authority of India Limited, Rourkela, Orissa	Coke oven gas
24.	IISCO Steel Plant, Burnpur, West Bengal	Coke oven gas

Major petrochemical producing Indian states are as follows:

Gujrat - 60%
Maharashtra - 20%
West Bengal - 10%
Utter Pradesh - 5%
Rest of India - 5% (including 2.5% in Tamilnadu)

With installation of so many units, the installed capacities of major intermediate products such as ethylene, propylene, butadiene, benzene, toluene and xylene has substantially increased.

Product-wise installed capacity & production of major petrochemicals inIndia during 14-15 is tabulated below: (In thousand MT)

Major Groups/Products	Installed Capacity 2014-15	Production 2014-15
A : Basic Major Petrochemicals		
I: Synthetic Fibres/Yarn		
1. Polyester Filament Yarm (PFY) **($)**	2791	2179
2. Bylon Filament Yam(NFY) **($$)**	20	32
3. Nylon Industrial Yam (NIY) **($$)**	61	101
4. Polypropylene Filament Yarn (PPFY) **($$)**	8	5
Sub Total Yarn (1+2+3+4)	**2880**	**2317**
5. Acrylic Fibre (Inc. Dry Spun) (AF)	107	90
6. Polyester Staple Fibre (PSF)	1170	1021
7. Polypropylene Staple Fibre (PPSF)	32	25
8. Polyester Staple Fibrefil (PSFF)	87	57
9. Polyester Industrial Yarn (PIY)	22	17
Total Synthetic, Fibre/Yarn	**4298**	**3527**

II:Polymers		
1. Linear Low Density Polyethylene (LLDPE)	No Separate Capacity	910
2. High Density Polyethylene (HDPE)	No separate Capacity	1156
LLDPE/HDPE (Combined) ($$$	**2735**	**2066**
3. Low Density Polyethylene (LDPE)	160	184
4. Polyestyrene (PS)	462	281
5. Polypropylene(PP)	3116	2590
6. Poly Vinyl Chloride(PVC)	1423	1330
7. Expandable Polystyrene (Ex-PS)	109	81
Total Polymers	**8005**	**6533**
III. Synthetic Rubber		
1. Styrene Butadiene Rubber (SBR)	271	57
2. Poly Butadiene Rubber (PER)	114	108
3. Nitrile Butadiene Rubber	25	0
4. Ethyl Vinyl Acetate (EVA)	15	6
Total Synthetic Rubber	**425**	**172**
IV: Synthetic Detergent Intermediates		
1. Linear Alkyl Benzene (LAB)	547	411
2. Ethylene Oxide (EO)	140	185
Total Synthetic Detergent Intermediaties	**687**	**596**
V: Performance Plastics		
1. ABS Resin	128	107
2. Nylon-6 & Nylon 66	23	21
3. Polymethyl Methacrylate (PMMA)	4	1
4. Styrene Acrylonitrile (SAN)	136	89
5. PET Chips/Polyester Chips	1118	547
6. PTFE (TEFLON)	20	2
Total Performance Plastics	**1428**	**766**
Total Basic Major Petrochemicals (I+II+III+IV+V)	**14843**	**11594**
B: Intermediates		
I: Fibre Intermediates		
1. Acrylonitrile (ACN)	41	34
2. Caprolactum	120	87

3. Mono Ethylene Glycol (MEG)	1153	1001
4. Purified Terephthalic Acid (PTA)	3753	3755
Total FibreIntermediates	**5067**	**4877**
II: Building Blocks		
Olefins		
1. Ethylene	3783	3192
2. Propylene	3368	2845
3. Butadiene	433	239
Total Olefins	**7584**	**6276**
Aromatics		
1. Benzene	1566	1094
2. Tolune	258	108
3. Mixed Xylene	898	215
4. Ortho-xylene	420	462
5. Paraxylene	3132	2758
Total Aromatics	**6274**	**4638**
C : Other Petro-Based Chemicals		
1. Butanol	26	4
2. C4-Raffinate	292	365
3. Di-Ethylene Glycol	85	101
4. Diacetone Alcohol	10	0
5. Ethylene Dichloride (By Product)	593	285
6. 2-Ethyl Hexanol**	55	14
7. Epichlorohydrine	0	0
8. Iso-Butanol	3	1
9. Isopropanol (IPA)	70	75
10. Methyl Methacrylate (MMA)	4	3
11. Phthalic Anhydride (PAN)	349	292
12. Propylene Oxide (PO)	36	37
13. Propylene Glycol (PG)	20	16
14. Polyvinyl Acetate Resin	17	0
15. Vinyl Acetate Monomer (VAM)	30	0
16. Vinyl Chloride Monomer (VCM) (By Product)	541	718
17. Polyol	114	52
Total Other Petro-based Chemicals	**2244**	**1962**

($) : Includes capacity of all the units producing PFY, NFY, NIY and PPFY under broadbanding as Synthetic Filament Yarn.
($$) : Independent capacity of units producing only NFY, NIY and PPFY.
As the capacities of these products are also included in Synthetic Filament yarn, capacity utilisation can not be worked out.
($$$) : Combined capacity to produce both LLDPE and HDPE and hence capacity utilisation can not be worked out.However production is independent.

ETHYLENE

Ethylene is the basic petrochemical. Ethylene is the basic building block of polyethylene (PE), polyvinyl chloride (PVC) and ethylene glycol. Ethylene is a prime raw material for downstream petrochemicals. Usage ratio of ethylene in the manufacture of various downstream petrochemicals is given below:

S.N.	Products from Ethylene	Ethylene required per unit
1.	LDPE/LLDPE (Linear low density polythene)	1.03
2.	HDPE (High density polythene)	1.04
3.	PVC (Poly vinyl chloride)	0.54
4.	EO (Ethylene oxide)	0.80
5.	MEG (Mono ethylene glycol)	0.62
6.	Styrene	0.33
7.	AO (Alpha olefin)	1.03

Polyethylene like LDPE/LLDPE, HDPE are the major ethylene based down streams products. End use pattern for ethylene according to sector of usage is presented in Fig. below.

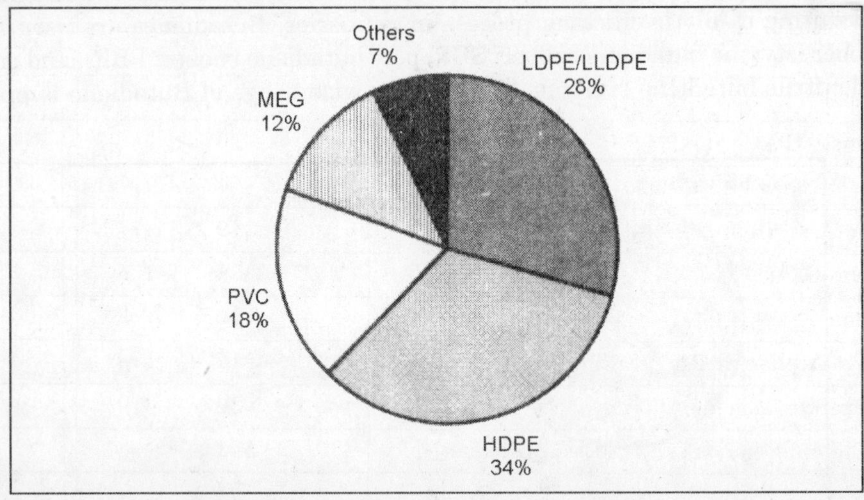

Fig. Sectoral usage pattern of Ethylene.

Propylene

Propylene is a co-product of ethylene from a cracker complex. It can also be produced by refineries, which set up a propylene recovery unit. Its most important end-uses are polypropylene, acrylonitrile, (used to make acrylic fibre, elastomers like acrylonitrile butadiene rubber: (ABR), and speciality polymers like acrylonitrile butadiene styrene: (ABS) and cumene, which is further processed into co-products – acetone and phenol. Usage ratio of propylene is given in Table below and sector-wise usage pattern is presented in Fig. below.

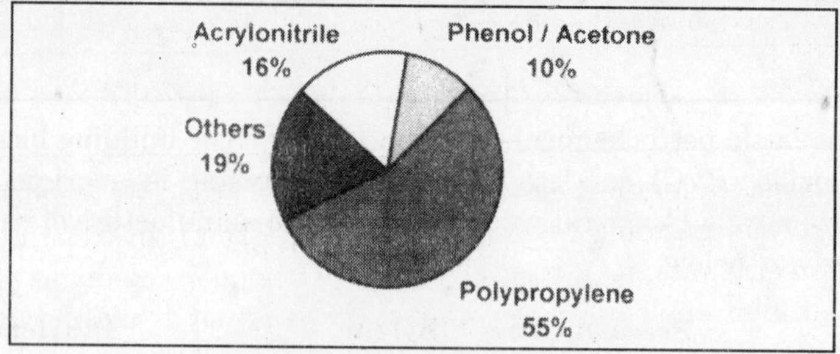

Fig. Sectoral Usage of Propylene

Table: Propylene: Usage Ratio

S.N.	Downstream Chemical from propylene	Propylene required per unit
1.	Polypropylene	1.02
2.	Arcylonitrile	1.10
3.	Phenol/Acetone	0.54

C4-Fractions-Butylenes/Butadienes

C4-fractions are co-produced during manufacture of ethylene and propylene in a cracker complex and during catalytic cracking process in refineries. Butadienes are used mainly for synthetic rubber (styrene butadiene rubber: SBR, poly butadiene rubber: PBR), and engineering plastic (acrylonitrile butadiene styrene: ABS). Sector-wise usage of Butadiene is presented in Fig below.

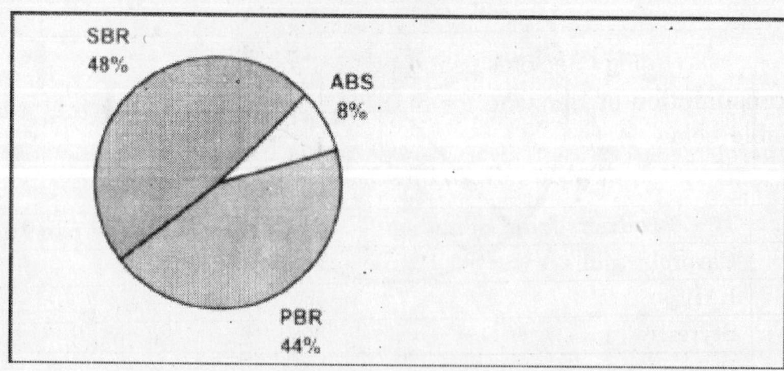

Fig. Sectoral Usage of Butadiene

Usage ratio of Butadiene for the manufacture of various products is presented in Table below.

Table: Butadiene Usage Ratio

S.N.	Product from butadiene	Butadiene required per unit
1.	Styrene Butadiene Rubber (SBR)	0.52
2.	Poly Butadiene Rubber (PBR)	0.70
3.	Acrylonitrile Butadiene Styrene (ABS)	0.54

BENZENE

Benzene is a basic aromatic chemical. In India, it is produced from a variety of sources: recovered from pyrolysis gasoline (during naphtha cracking), by reforming of naphtha (in refineries), and in steel plants, as a recovery product from coke oven gas obtained during the carbonisation of coal. Benzene is used as a raw material for several important products. These include caprolactum (used for making nylon filament yarn: NFY), linear alkyl benzene: LAB (which is used in detergents), styrene (used in polystyrene and Styrene butadiene rubber), phenol (used for laminates), nitro-benzene and chloro-benzenes (dye intermediates), and pesticides (DDT and malathion). Sector-wise usage pattern for benzene is depicted in Fig below.

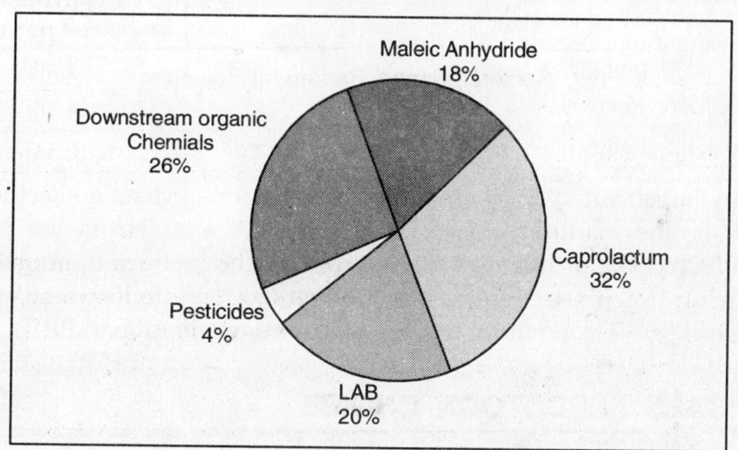

Fig. Sectoral usage pattern of benzene.

Unit wise consumption of Benzene for manufacture of various down stream products is indidicated in Table below.

Table : Benzene Usage Ratios

S.N.	Products from benzene	Benzene required per unit
1.	Carprolactum	1.00
2.	LAB	0.37
3.	Styrene	0.75
4.	Phenol	1.16
5.	Others	1.33

Toluene

Toluene is a basic aromatic chemical produced in reformer along with benzene and xylene.It is also produced as a by-product of naphtha cracking. It is mainly used as a solvent in a wide range of end-use sectors. The other major end uses are nitrotolunes, toluene sulphonamide, dyes, pesticides,chlorinated derivatives and drugs. Sectoral usage of Toluene is presented in Fig. below. Besides, toluene & xylene are also produced by the distillation of benzol recovered from coke oven gas in steel plant.

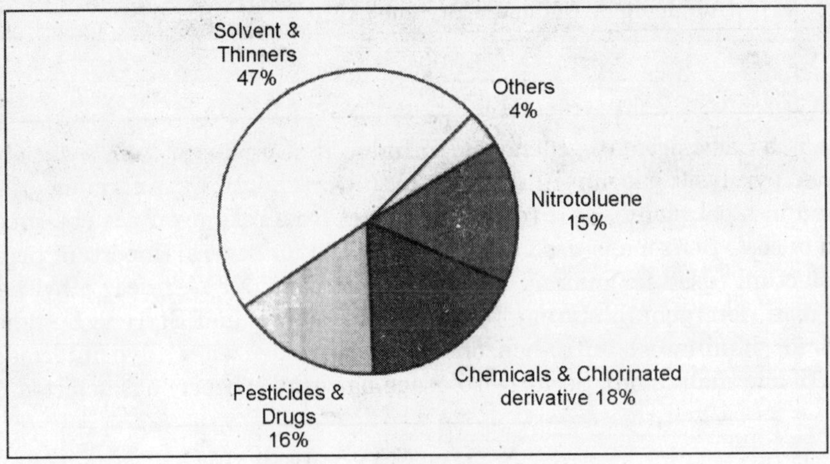

Fig. Sectoral Usage Pattern of Toluene.

Xylene

Commercially important xylenes are of two kinds: para - xylene and ortho - xylene. Para - xylene is used for the manufacture of DMT and PTA and this is the backbone of the synthetic fibre industry. Ortho xylene is mainly used in the production of phthalic anhydride (used for manufacture of plasticisers/paints /thinners). Ortho xylene and para xylene are produced in a reformer with naphtha or C5 - reformate as inputs. *

PETROCHEMICALS FEEDSTOCK CHOICE

A wide range of alternative feed stocks such as naphtha, ethane/propane, alcohol, LPG, NGL and gas oil can be used for production of Petrochemicals. In India, naphtha and C_2/C_3 fractions from natural gas are the main feedstock used. LPG is normally used as domestic fuel, while gas oil is not used because it is heavier fraction and needs complex processing. In India, some refineries crack LPG in their fluid catalytic cracking units to produce propylene. Summary of various feedstock used by Indian petrochemicals majors is presented in Table below.

From the table, it is apparent that about 60% of India's cracking capacity is based on natural gas, whereas 40% is based on Naphtha feedstock. Industrial alcohol which was an attractive feedstock in the days of alcohol price control is no longer an important feedstock and accounts for only 0.8% of the total ethylene production in the country. The major factors, which affect the choice of feedstock, are the relative yields of olefins and aromatics desired, energy costs, investment levels, availability and relative pricing. Natural gas and NGL yield a much higher proportion of ethylene. Hence, they are preferred when the polyolefins output of a cracker is sought to be maximised. On the other hand naphtha is preferred when a wider range of output products (including propylene and butadiene derivatives) is desired.

Table : Plant-wise petrochemicals feedstock requirement

S.N.	Petrochemicals Complex	Ethylene Capacity (TPA)	Feed stocks
1.	IPCL	1,30,000	Naphtha
2.	IPCL-MGCS, Nagothane	4,00,000	Gas (C_2/C_3 7:3), ethane-propane fraction
3.	IPCL, Gandhar	3,00,000	Gas (C_2/C_3 7:3), ethane-propane fraction
4.	RIL, Hazira	7,50,000	Naphtha/Natural Gas Liquid(NGL)
5.	NOCIL, Thane	75,000	Naphtha
6.	HPL,Haldia	4,20,000	Naphtha
7.	GAIL, Auriya	4,00,000	Gas (C_2/C_3 9:1)
8.	Oswal Agro, Mumbai	22,000	Alcohol

TPA – Tons per annum

DOWNSTREAM PETROCHEMICALS

Downstream Petrochemicals are the products made from basic petrochemicals and thus are derivatives of naphtha or gas. The downstream petrochemicals can be further classified into the Synthetic Organic Chemical Manufacturing Industry (comprising chemicals/ intermediates made from basic petrochemicals) and polymers (such as polyethylene/PVC etc.). The detail of downstream petrochemicals production in India is described below.

Downstream Products Based on Ethylene:

Ethylene Oxide (EO) and Ethylene Glycol (EG)

Ethylene oxide is an intermediate product during the manufacture of mono ethylene glycol (MEG). In India, EO is produced through the Petrochemical and alcohol routes. Ethylene glycol is used in the polyester fibre/filament industry (@70%) with minor usage in explosive and anti-freeze coolants.

Installed capacities for ethylene oxide/ethylene glycols are given in Table below.

Table : Installed capacities of ethylene oxide/ethylene glycol

S.N.	Unit	Location	Capacity (TPA)	Feedstock
1.	IPCL	Vadodara, Gujarat	20,000	Naphtha ethylene
2.	NOCIL	Thane, Maharashtra	24,000	Naphtha ethylene
3.	RIL	Hazira, Gujarat	3,40,000	Naphtha / NGL ethylene

4.	SM Dyechem	Pune, Maharashtra	14,000	Alcohol ethylene
5.	India Glycols		13,000	Alcohol ethylene
6.	IPCL-MGCC	Nagothane, Maharashtra	55,000	Gas ethylene
7.	IPCL	Gandhar, Gujarat	1,20,000	Gas ethylene

Ethylene oxide finds other uses in surfactants (50%), dye & amine derivatives and glycol ethers.

Ethylene Dichloride (EDC)/Vinyl Chloride (VCM)

Ethylene dichloride is an intermediate for vinyl chloride monomer, which polymerises to polyvinyl chloride (PVC). Installed capacities of major units are presented in Table below.

Table : Installed capacity of EDC/VCM

S.N.	Manufacturer	Year of Establishment	Technology Supplier	Design Capacity (TPA)
1.	IPCL, Vadodara	1983-84	Stauffer Chemicals, USA	57,300
2.	IPCL, Dahej	1997-98	Innovyl Belgium	1,70,000
3.	NOCIL, TTC	1961	Shell, Netherlands	30,000
4.	RIL, Hazira	1996	Geon, USA	2,70,000
5.	Finolex Pipes Limited, Ratnagiri	1992/2003	UHDE, Germany	2,40,000
Installed Capacity Data – Basis VCM				

Ethyl Benzene/Styrene

Major use of Ethyl Benzene is for production of styrene. Ethyl Benzene is made by alkylation of Benzene in presence of catalyst.

Propylene Based downstream Petrochemicals:

Propylene Oxide and Propylene Glycol

Propylene Oxide is used for the manufacture of propylene glycol which finds application in Polyester resins, cellophane and food & drug industries. Propylene oxide is produced by the direct oxidation of propylene.

Acrylonitrile (CH_2=CHCN)

Acrylonitrile (ACN) is the basic input for the production of acrylic fibre (AF). It is also used to produce acrylonitrile butadiene styrene rubber (ABS) and acrylates. Acrylonitrile is produced by air oxidation of propylene and ammonia mixture. IPCL is the sole producer of ACN in India producing about 30,000 TPA at its Vadodara Complex.

Isopropyl Alcohol [$(CH_3)_2$. CHOH]

Isopropanol finds its largest use as multipurpose industrial solvent and for manufacture of various drugs and fine chemicals. NOCIL (25,000 TPA), Herdellia in Thane district and IOC in Raigad, Maharashtra manufacture Isopropanol from propylene.

Cumene [Isopropyl Benzene - C_6H_5 . CH . $(CH_3)_2$]

Cumene is produced by propylene alkylation of Benzene. Most of the cumene produced is used for captive production of phenol/acetone. HOC - Ernakulum unit extracts propylene from LPG supplied by Cochin refineries limited and uses this for the production of cumene. Its installed capacity is 54,000 TPA. Herdellia chemicals, Thane gets propylene from NOCIL/MRL and uses this for production of cumene. Its installed capacity is 40,000 TPA.

Phenol/Acetone

Phenol is used to produce phenol - formaldehyde resins and derivatives such as bisphenol - A, salicylic acid & alkyl phenols. Acetone is used as a solvent, for manufacture of chemicals such as acetone cyanohydrin, di-acetone alcohol, in pesticides, pharmaceuticals, explosives *etc.* Phenol and acetone are co-produced from air oxidation of cumene and benzene (cumene in turn is produced from propylene). Hence, phenol and acetone economics are closely interlinked. Acetone is also produced through the alcohol route. The current installed capacity for phenol is about 65,000 TPA. Acetone capacity is about the same, with 50,000 TPA produced as co-product of phenol and remaining through the alcohol route.

Butanol and 2-Ethyl Hexanol

An oxo process involves substitution of HCHO across a double bond using a reaction of olefin and CO + H_2 synthesis gas. The primary product is an aldehyde (CHO group), which can be hydrogenated to an alcohol. NOCIL employs the oxo process for manufacture of Butanol (normal & Iso) and 2 Ethyl hexanol. The total capacity for chemicals manufactured by the process is 20,000 TPA.

Downstream Petrochemicals from Benzene:

Cyclohexane and Caprolactum

Cyclohexane is used as an intermediate for production of caprolactum and as a solvent for manufacture of HDPE. It is produced by hydrogenation of Benzene. Cyclohexane is produced in the country by GSFC, Gujarat and FACT, Cochin; both manufacture caprolactum. Caprolactum is the main input for the production of nylon filament yarn, nylon tyre cord and nylon industrial yarns. Caprolactum is made from cyclohexanone oxime, which is obtained by treating cyclohexanone with hydroxylamine.

Maleic Anhydride (MAN-$C_4H_2O_3$)

Maleic anhydride (MAN) is used to produce agrochemical, unsaturated polyester resins, alkyd resins and food acids. These in turn are used in a wide range of end use sectors including engineering plastics, helmets, table top lamination, pharmaceuticals, varnishes, paints etc. MAN is manufactured by air oxidation of Benzene or n-Butadiene in a process similar to phthalic anhydride. Major producers include Thirumalai Chemicals, Tamilnadu (10,OOOTPA) and Ganesh Anhydride, Tarapur, Maharashtra (12,000 TPA).

Downstream Petrochemicals from Toluene:

Toluene Di Isocyanate (TDI)

TDI is used in the manufacture of polyurethane plastics. It is manufactured by phosgenation of toluene diamine. The producer of TDI in the country is Narmada Chemateur India Ltd. Production capacity is 10,000 TPA.

Downstream Chemicals from Xylene:

Purified Terephthalic Acid (PTA) and Dimethyl Terephthalate (DMT)

Purified terephthalic acid (PTA) and dimethyl terephthalate (DMT) are substitute raw materials for production of polyesters (PSF, PFY, PET, polyester films and chips). There are four manufacturers of DMT/PTA in India. Reliance is the producer of PTA at its Patalganga & Hazira plants. Bombay Dyeing (Patalganga), IPCL (Vadodara) and BRPL, Assam produce DMT.

Phthalic Anhydride (PAN)

Phthalic anhydride is used to produce unsaturated polyester resins, esters, alkyd resins, specific dyes and pigments. These in turn are used in a wide range of applications of which the most important are the PVC processing industry (esters) and the paint industry (alkyd resins). Ortho xylene is the main raw material of PAN. The installed capacity of some plants for phthalic anhydride are presented in Table below.

Table: Installed Capacity of some Phthalic Anhydride (PA) plants

S.N.	Company	Capacity, TPA
1.	IGPL, Taloja	1,20,000
2.	MPCL, Karnataka	12,000
3.	TCL, Thirumalai	90,000
4.	Asian Paints, Ankleshwar	25,000
5.	Herdillia, Mumbai	10,000
6.	Ambuja, Andhra Pradesh	9,000

Amongst the basic petrochemicals, the olefins (viz. ethylene, propylene and butadiene) are made by cracking of natural gas, C_2/C_3 fraction of natural gas and naphtha.

Use of alcohol as feedstock for ethylene manufacture is declining fast and many units are facing closure. However, alcohol is used to manufacture number of downstream petrochemicals like aldehydes/ketones.

The aromatics (benzene, toluene and xylene) on the other hand are produced by number of processes: recovered from pyrolysis gasoline (during naphtha cracking), as recovery products from coke oven gas and by steam reforming naphtha. The refineries account for 48%, whereas crackers account for 42% of total Benzene production. There have been rapid increases in p-xylene production capacity due to capacity enhancement of synthetic fibres.

Natural gas results in higher yields of ethylene compared to other olefins. Naphtha on the other hand gives a wider range of output products (including propylene, butadiene derivatives).

The downstream petrochemical industry is widely dispersed, manufacturing wide range of products of varying capacities and employing differing technologies.

Among the various fractions produced by distillation of crude oil, petroleum gases, naphtha, kerosene and gas oil are the main feed-stocks for the petrochemical industry. Ethane, propane, butane and natural gas liquids obtained from natural gas are the other important feed stocks used in the petrochemical industry. Olefins (ethylene, propylene & butadiene) and Aromatics (benzene, toluene & xylenes) are the major building blocks from which most Chemicals and Petrochemicals are produced.

Petrochemical manufacturing involves manufacture of building blocks by cracking or reforming operation; conversion of building blocks into intermediates such as fibre intermediates (acrylonitrile, caprolactum, dimethyl terephthalate/purified terephthalic acid, mono ethylene glycol); precursors (styrene, ethylene dichloride, vinyl chloride monomer etc.) and other chemical intermediates; production of synthetic fibres, plastics, elastomers, other chemicals and processing of plastics to produce consumer & industrial products. In addition building blocks (benzene, toluene, OX, PX, etc) are basic raw materials for manufacture of pesticides, dyes & dyes intermediate and pharmaceutical bulk drugs, etc.

Petrochemical products namely synthetic fibres cater to the clothing needs of mankind and are used in both apparel and non-apparel applications. Polymers find major applications in packaging for preservation of food articles, moulded industrial and home appliances, furniture, extruded pipes, electronics, telecommunication, aeroplanes, etc. Synthetic rubbers are used for making various types of tyres & non-tyre rubber goods and supplement natural rubber. Surfactant intermediate are used in detergents and surfactants.

Petrochemicals industry consists of upstream integrated naphtha/gas cracker complexes which are technology intensive and operating with state-of-the-art technology and economic scales of operation. Downstream plastic processing industry is fragmented and spread throughout the country.

The competitiveness lies in the strong domestic demand from the downstream consuming sectors, which are widely spread out in sectors and geographic. Plastic processing industry is dominated by SMEs with substantial portion with first generation entrepreneur. The rising labour costs in developed markets in plastic converting sector offers an opportunity for India to expand capacities and export.

The planned capacity additions in upstream petrochemical industry are by the crude oil refineries for integrating refinery petrochemical complexes to increase the hydrocarbon value addition and to remain cost competitive. In case of downstream plastic processing industry the technology is in-built in the processing machinery. Major Plastic Processing international companies have put up fabrication shops. There are domestic plastic processing machinery manufactures that meets the international standards and are also exporting plastic processing machinery. Multinational companies are entering Indian markets through mergers & acquisition

Downstream plastic processing industry is dominated by tiny, small and medium scale industry. Large number of first generation entrepreneurs has entered the plastic processing sector. The increase in investment limits in the small scale industry up to ₹ 5 crore in plant and machinery has contributed significantly to the upgradation of this industry.

The trade in polymers over a period of time has become relatively easy. Even small parcel of 20-50 Mts can be imported. Large number of trader are present in India in polymer trade. India is already short in PVC, LDPE, Metalocene PE and Engineering Plastics. Large quantities are being imported.

Market prices are determined based on SEA (South Eastern Asian) prices plus import duties (Presently 5% for polymers). As imports are relatively free, domestic producers have to price their products in line with the prices prevailing in SEA region. As end product prices are market driven, most producers irrespective of their cost of production have to maintain the selling prices in line with market prices. This results in varying margins for producers having different feedstocks.

In Middle East, one striking feature is that the gas is supplied at highly subsidized price. Also, in the Middle East crackers, the cracking of gas into ethylene has huge cost advantage and these crackers are integrated to add value to ethylene to yield polymers, mono ethylene glycol etc. These Middle-East producers sell their products at the market driven prices which are essentially in line with naphtha-based product price band. This enables them to reap higher profit margins. This creates significant challenge to the naphtha-based cracker elsewhere in the world.

As far the profitability of Indian companies is concerned, they will be somewhere in the middle of the cost curve and margins will be moderate compared to MidEast producers. Major impact will be felt by companies who are at the top of cost curve like some of the European producers.

POLYETHYLENE

Polyethylene is broadly classified into High Density Polyethylene (HDPE), Linear Low Density Polyethylene (LLDPE) & Low Density Polyethylene (LDPE). Historically ICI Scientist produced polyethylene by accident while studying the relation between Ethylene & Benzaldehyde.

High Density Polyethylene (HDPE)

High Density Polyethylene is predominantly linear in structure, closely packed with density of 0.960 – 0.980 gram/cm^3, crystalline as high as 95% and a melting point as high as 138.5°C.

Currently High Density Polyethylene is showing annual growth of 10-12%. It finds its major application in Raffia, Blow Moulding, Film, Pipes and Injection Moulding. Consumption wise it is in order of HD-Film GP *Blow Moulding, Raffia, Pipe, have the major share.

Large and Medium Blow Moulding containers, Pipe, General Purpose BM containers are currently the leading growth sectors.

Global capacity of HDPE (2010) was close to 40.8 MMTPA (Million metric ton per annum), with consumption close to 33 MMTPA. The capacity is growing at 4.5% & demand is seeing a growth of 2.9%.

The domestic HDPE capacity during the end of 10th plan was 1177 KT (Kiloton) and by the end of 11th plan it is 1885 KT & an increase of 788 KT. The growth in capacity has been 9.9% & demand has increased by 10.9%.

A capacity growth of 940 KT is expected during 12th Five Year Plan & an increase in consumption from 1657 to 2573 KT i.e. by 916 KT. The capacity growth in PE (HD/LL) will mainly be from OPAL (510 KT) RIL – 550 KT (1460 KT from existing 910 KT), GAIL 620 KT (1120 KT from 500 KT), though the major share of it will be in High Density.

HDPE because of its unique properties and increased availability offers abundant opportunities for growth. There is a huge untapped potential for the sector like for e.g. by the dilution of JPMA, loose to packed packaging in Blow moulding containers, automotive sector, infrastructure, and in installing drinking water & sewerage system etc.

LLDPE - Linear Low Density Polyethylene

LLDPE is linear in structure. The branches are formed due to presence of co monomer. The co-monomer used is Butene (C4), Hexene (C6) & Octene (C8). The low pressure used in manufacture of LLDPE does not allow the formation of large branches.

LLDPE finds its application in mono and multi-layered films, roto-moulding, wire & cables and high flow LL is used in master batch compounds etc.

Currently 59% of LL share comes from Butene film, followed by HAO & Metallocene LL film, roto-moulding, wire & cables, high flow LL grades.

Currently the fast growing sector in LL is Metallocene, followed by wire & cable & high flow grades. Metallocene LL grades are fast moving due to its properties like better strength, high dart, higher line speed, better sealing, downsizing of the film etc.

The global capacity of LL in (2010) was 27.7 MMT though the demand is 20.8 MMT. The capacity has been increasing by 6.3% & demand by 3.2% in last five years.

Domestic capacity has increased by 601 KT in last five years i.e. from 591 KT (2006-07) to 1198 KT (2011-12). The growth in capacity has been 14.9%.

Demand has grown from 631 KT (2006-07) to 1198 KT (2011-12) during the period i.e. an increase of 13.7%.

Demand is likely to increase in 12th Five Year Plan from 1319 KT to 2076 KT i.e. a growth of 11.6%. It will be met by coming capacities of OPAL & expansion plan of RIL/GAIL.

LLD is the major constituents of flexible packaging growing at more than 17%. Key end uses of flexible packaging being lamination, edible oil and milk.

Low Density Polyethylene – LDPE

LDPE is predominantly a branched polymer. The properties depend upon the frequency and the length of branches.

LDPE helps in processibility, has excellent transparency & good sealing properties & low glass transition temp. Due to its inherent strength, LD is extensively used as a blend with other polymers.

LDPE finds its application in general purpose film, heavy duty film, liquid packaging, injection moulding, extrusion coating, wire & cables, adhesive lamination.

The major application wise share is in the order of general purpose film, extrusion coating, milk packaging, heavy duty, injection moulding, wire & cables & others. Flexible Packaging constitutes 84% of the LD consumption.

The growing sectors are general purpose film, heavy duty, milk packaging, injection moulding sectors, adhesive laminates.

The global capacity of LDPE is 22 MMT and the demand is 18.7 MMT. The capacity & demand for last five years have grown 1.9% & 1.1% respectively.

Domestically there has been an increase in demand of 9.9% from 2006-07 to 2011-12 from 253 KT to 405 KT. The capacity has remained the same i.e. 205 KT. Excess demand over & above capacity is being met by Imports. The major imports are from Basell, Qapco, Sasool, Tasnee.

It is expected that demand will grow at 8.1% during 12th plan period i.e. from 438 KT to 597 KT. M/s RIL has been contemplating to put up the plant of LD (400 KT) at Jamnagar.

EVA (ETHYLENE VINYL ACETATE)

Ethylene Vinyl Acetate is manufactured from polymerisation of Ethylene and Co-monomer (Vinyl Acetate/Ethyl Acrylate/Acrylic Acid/Methyl Acrylate). EVA is the most popular ethylene copolymer. Vinyl Acetate (VA) content in EVA varies from 9% to 40%. The increase in VA content improves the properties like flexibility, ESCR, compatibility, filler acceptability, X-likability adhesion, etc.

EVA due to its flexibility for various processing techniques like (Foaming, Moulding, Coating, blending, extrusion) makes it useful for broader application range.

Resin properties which makes it unique and useful are: very high low temperature impact strength, good elasticity & residence, high resistance to flex cracking & high ESCR, excellent clarity & gloss and good resistance to weathering.

Global demand for EVA copolymers was 2.8 MMT in 2009 and is expected to grow @ 5.6% from 2009 to 2015. The growth of EVA was 9% as against the 0.7% growth of LDPE (2004-2009).

Asian economies is the key driver behind an increased global demand for EVA with the Chinese economy being the leader, followed by India and South Korea. Shift in manufacturing of footwear to Asian countries like China has created a demand growth in the Asia pacific region.

EVA consumption grew in India @ 18% (2004-2010). Market is expected to grow from 116 KT (2010-11) to 200 KT (2016-17). Significant growth of EVA has been possible due to tremendous technological developments in shoe manufacturing technology by IM process. Apart from footwear (shoes, slippers, beach sandals, etc) market, other applications like film; HMA are also witnessing a healthy growth. Emerging sector "Solar Film" is highly technical and quality sensitive film produced from high VA is growing at a very healthy pace.

High growth in India gives immense opportunity for capacity expansion.

POLYPROPYLENE (PP)

Polypropylene today is one of the youngest members of the polymer family after being invented some six decades ago in the year 1957. This wonder polymer became one of the most invented inventions touching our lives in every possible way.

Globally PP has seen phenomenal growth over the last six decades to become a 48 million tonnes product with an overall capacity of 61 million tonnes with over 150 prominent Producers. Polypropylene Global demand witnessed a robust growth of 7.5% in 2010 and is expected to grow at a healthy AAGR of 5.7% up to 2015. Asia in line with its growing global stature has taken a dominant position in the PP Industry both in terms of demand and capacities. For all additions up to 2015, about 98% of incremental capacities are to come up in Asia reflecting the regions increasing global prominence.

The Indian Polypropylene story though a late starter has caught up with the global industry and over the past decade has grown at a phenomenal pace riding on India's growth story. At the start of the century, the PP industry in India was at 839KT and was a small player in the global market. In a span of ten years, the industry has now grown up over 3 times to 2638 KT. The future is expected to be no different with the industry expected to cross the 4 million tonnes mark by 2014-15 growing at a healthy CAGR (Compound annual growth rate) of almost 11%.Tho average hiotorical. The current per capita global consumption of Polypropylene stands at 7 kg per person vis a vis 2.2 Kg per person in India. This presents a wonderful opportunity for Indian Plastic industry in the coming years.

The Current domestic capacity stands at 3.7 million tonnes which is well over the current consumption of 2.6 million Tonnes. The new facilities will take the total Indian capacity to over 4500 KT by 2012-13 which is going to create surplus in the Domestic industry with Indian PP Exports reaching a peak of over 0.9 million tonnes by 2012-13. At the same time India has ambitious plans to significantly augment its petrochemical capabilities.

With this, stage is ready for Polypropylene upstream and downstream investment, by both Public & Private sector players. Indian Polypropylene Market is poised to grow at healthy double digits rates far surpassing the global growth rates.

PVC (POLY VINYL CHLORIDE)

The world's most versatile plastic had a rather humble beginning: A rubber scientist during the early 1920s stumbled onto a new material with fantastic properties during his search for a synthetic adhesive. Waldo Semon was intrigued with his finding, and experimented by making golf balls and shoe heels out of the versatile material called polyvinyl chloride, or Vinyl or PVC.

Soon after his discovery, PVC-based products such as raincoats, shower curtains and insulated wire hit the market. PVC today is recognized as 'an infrastructure plastic' finding applications in pipes, conduits, ducts, wire & cables, flooring, windows & doors and roofing etc. It is also used in many other sectors including automobiles, medical & healthcare, packaging, sports & leisure.

Indian PVC manufacturing is now 65 years old. In the year 1951, Calico started the first PVC plant in Mumbai with a production capacity of 6 KTA. Today the capacity has increased many folds to 1.3 MMTA with Chemplast Sanmar, DCM Shriram Consolidated, DCW, Finolex and Reliance Industries contributing to the same. Vivanta group has recently announced their entry in PVC manufacturing with a capacity of 240 KTA which is supposed to be operational by 2013.

However the domestic consumption at 1.9 MMTA is resulting in almost 600 KTA being imported from different parts of the world. There are 2000+ processors converting this PVC into various end use applications in the country.

Pipes continues to be the biggest end use sector with a share of ~ 70%, other sectors being calendered products, wire & cables, films & sheets, profiles, footwear and medical & healthcare.

PVC demand in the country is increasing @ ~10% which is more than twice compared to the world average of 4%. Heavy investments by the Government in infrastructure projects and focus on increase in irrigational land are the main growth drivers for the increased consumption of PVC. Funding by various International Agencies such as World Bank, Asian Development Bank for water management projects in the country is also creating huge potentials for PVC.

Green Building concept is increasingly being accepted in the country. Indian Green building Council, TERI Griha, and Bureau of Energy Efficiency are working towards green construction activities with usage of materials that save energy at all stages. PVC is known to save energy during the entire life cycle. Besides; products like windows, doors, partition, wall paneling false ceiling, and furniture when made in PVC, save natural resource like wood which would help in saving fast depleting forests.

Knowing the fact that the development focus has now shifted to Asia, world renowned window manufacturers have set up their fabrication units in India with future plans for backward integration to profile extrusion. Also special technologies in pipe manufacturing are being introduced in India. Indian processors have shown willingness to adopt new technologies and are keen on development of new products for widening the spectrum of applications of PVC products.

With this kind of demands, it is expected that the PVC consumption in the country shall be close to 30 MMTA by 2016-17.

POLYSTYRENE (PS) – GPS/HIPS AND EPS

We have three types of polystyrenes with different application - General purpose Polystyrene (GP/S), High Impact Polystyrene (HIPS) and Expandable Polystyrene (EPS).

General Purpose polystyrene has excellent transparency, processability and its grades are used to cover a wide variety of applications such as disposable containers, cutlery, CD storage cases, co-extrusion to obtain gloss layer, toys, stationery & refrigeration components.

High Impact polystyrenes resins are specifically designed to meet the mechanical properties and productivity requirements of a wide variety of applications both in injection and extrusion/ thermoforming processes. Typical applications include Electronics, household Appliances, disposable containers, toys and refrigeration components.

Polystyrene foam also known as Expandable Polystyrene is prepared by suspension polymerization of styrene, and then adding foaming agent and white bead-like particles, with relative density of 1.05 are obtained. Thermal conductivity is low. EPS is a rigid foam honeycomb material with good insulation & vibration absorption, high compressive strength, very light weight and resistance to moisture; its uses include building insulation & sound insulation, sidewall, & within the walls of the Cover, packaging materials and disposable containers.

Polystyrene current capacity is of 472 KT that can swing between GP and HIPS production. There are three producers in India i.e. Supreme PC (272 KT), BASF (100 KT) and LGP (100 KT). Production has been about 305 KT.

Domestic supplies are 275 KT, plus there are imports of (10 KT). The current domestic demand is 285 KT and exports are in the vicinity of 64 KT.

By the end of 12th five year plan, the demand is expected to be in the vicinity of 453 KT with a growth rate of 11.9%.

PS end use is mainly into the "Electronics/Appliances" sector, which accounts for almost 60 percent of total demand. This sector is forecast to have growth rate of 8-9% per year. Export of PS destinations have been West Europe, Africa, and the Middle East.

One reason why PS demand in the Indian Subcontinent has been below expectations is the competition from other polymers, mainly polypropylene in both packaging and durables, primarily due to cost reasons.

Expandable Polystyrene current capacity is in the vicinity of 168 KT. There are three major producers in India i.e. Supreme, BASF and LGP. M/s Supreme has been expanding by takeover and adding new capacities.

Current supplies from domestic route are 81 KT, plus there are imports of (11 KT). The current domestic demand is 92 KT and exports are in the vicinity of 9 KT.

By the end of 12th five year plan, the same is expected to be in the vicinity of 145 KT with a growth rate of 11.5%.

Future demand growth is expected with strong performance in the packaging sector. India's low-cost labor pool has been touted as one of the causes for manufacturing capabilities moving to India, which has supported the largest polystyrene end use of "Electronics / Appliances." This large PS end use consumes EPS through the packaging sector, making packaging the largest sector for EPS.

The table below summarises list of consumer end products produced from five commodity plastics.

Table: Polymers & Corresponding Plastic Products

LD/LLDPE	HDPE	PP	PVC	PS
Cost & stretch cling films	Raffia Mono filaments	Raffia/FIBC	Irrigation pipes & fittings	Thermocol
Mono/Multi layer films	Crates	Fish nets	Wire and cables	Foam
Wires & Cables	Multi layer films	Buckets, Tiffin boxes	Floorings/Tiles	Refrigerator trays
Water storage tanks	Water Conveyance pipes	Automobile bumpers	Shower curtains	CD/DVD cases
House wares	Pesticide bottles	BOPP films	Blister films	Structural insulation
Carrier bags	Woven sacks	Washing machine tubs	Profiles & Builders' ware	Bottles
Furniture	Caps & closures	Leno bags	Footwear	Disposable cups/cutleries
Milk Packaging	Pallets	Fibres & Filaments	IV bottles & blood bags	Smoke detectors
Detergent Packaging	Thin wall moldings	Syringes	Bag liners	Films

PET (POLYTHYLENE TEREPHTHALATE)

Polyethylene Terephthalate (PET) is a thermo plastic polyester. PET, the material used for the manufacturing of the plastic beverage bottles is of high strength and high barrier material. Pepsi-cola was the first company to commercially introduce its beverages products in PET.

PET containers when properly manufactured, offer excellent clarity & gloss, toughness & impact resistance, chemical resistance, low gas permeation, good process ability and good dimensional stability.

Features of PET is shown below in figure.

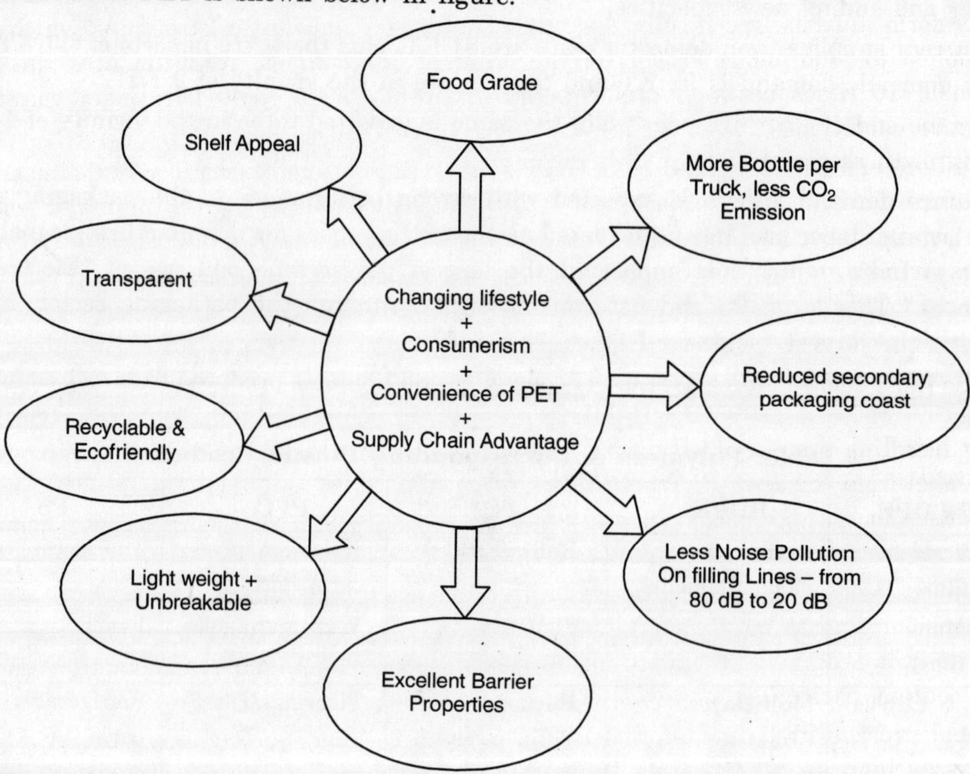

PET is used for the production of bottles, sheet, strapping and injection moulded products. PET bottles are used for packaging wide range of consumer products such as all types of Beverages including alcoholic, Edible Oil, Pharmaceutical formulations, Agro chemicals, Personal and Home Care etc. PET is also used for non-bottle applications such as thin/thick thermoformable sheet, dual-oven able crystallised PET (CPET) containers, injection moulded components, fastening straps etc.

Building blocks for PET is granular form of Purified Terephthalic acid (PTA) & mono ethylene glycol (MEG) are some of these building blocks used for manufacturing PET. It is a condensation polymer made by continuous melt-phase polymerization process (CP) followed by Solid State Polymerization (SSP). In the past, it was manufactured either by DMT or PTA route:

- DMT Route (Di Methyl Terepthalate): DMT + MEG = PET Polymer
- PTA Route (Purified Terephthalic acid): PTA + MEG = PET Polymer

DMT is seldom used for PET now a day. It is manufactured using PTA to help achieve better quality and consistency.

Injection Moulding of PET is becoming more prevalent in today's market place. PET can be best injection moulded into parts with wall thickness of up to 4.0 mm.

The Stretch Blow Moulding process is used to convert the injection moulded preform which is of amorphous structure to a specific bottle design of bi-axially oriented structure. To achieve the best bottle performance from a preform, it must be stretched by its natural amount both in longitudinal and transverse direction. This means for any desired bottle shape, preform must be specifically designed. The natural amount of stretching which need to be applied to amorphous PET is of the order of 10-12 times, resulting in a thickness reduction of 10 times between preform and bottle blown at 35-40 bar under controlled temperature conditions..

The concern for Recycled PET (RPET) is continuously growing in recent times. PET bottles, which form the major market of PET packaging resin (94%), are the major concern from the point of recycling. More than 90% of PET is consumed in food packaging with drinks/ beverages forming almost 80%. Since drinks and beverages are consumed mostly in residential houses, restaurants, stadia, railway stations, airports, entertainment venues and other public places, the importance of organized collection & recycling of post-consumer PET bottles need not be over-emphasized.

The PET bottle is 100% recyclable and most eco-friendly in nature. The need of the hour is proper handling system at all stages. The end-consumers should take care to remove the cap and label from the used PET bottle before disposing it off and also should not throw the empty bottles in a public place. In order to ease the handling at different stages, the PET bottle should be 'flattened' and baled before sending it to the intended area of usage.

Recycled PET is becoming alternative for PET packaging options today and is getting serious consideration by the packaging world. There is a growing awareness among consumers towards their environment obligation in individual capacity and a marked preference thus for recycled content packaging.

Global average PET consumption is 2.13 kg/capita. While consumption in USA, Mexico & Italy is as high as 8.0 kg/capita, Indonesia, India & Philippines are still in the lowest quadrant with consumption less than < 0.5 kg/capita.

Global Demand of virgin PET has observed a phenomenal growth rate. Demand for 2011-12 is estimated as 16.5 MMT at CARG 7.7% over last year. For 2016-17 demand is projected as 23.2 MMT with CARG of 7.1% in next 5 yrs. And further down the line demand is estimated to reach 30 MMT by 2021-22 with CAGR of 6.5%.

Capacity Growth - Capacity for 2010-11 exit rate is 19.3 MMT. Projected capacity for FY'11-12: 20.6 MMT; FY'16-17: 29.7 MMT; FY'21-22: 34 MMT.

PET consumption in India is growing rapidly. Strong fundamentals for growth lies within robust GDP growth rate, steadily increasing urbanization, improving affluence among middle class population, change in youth life style, boom in organized retail and inorganic consumption growth through non- conversion of non-PET.

Predominantly PET consumption in India is driven by Beverages, Health Care, Personal Care, Agro Chemicals, Cooking oil, Confectioneries and Non-bottle segments.

In domestic market, beverages mostly covered Mineral water, Fortified water, flavored water, Sparkling water, Juices, Coke & Pepsi Cola and other alcoholic drinks. Projected consumption of PET in this segment for 2011 is 280 KT and expected to grow by 30%. For FMCG, Pharma & Agrochemical sector, current consumption is 190 KT with a vigorous growth of 25%. In non-bottle segment e.g sheets, straps lamipack, punnets & blisters presently 100KT PET is consumed with a steady growth rate of 25%.

India's domestic demand is growing with CARG 25.5% in last 5 years, increased from 174 KT to 542 KT, and projected as 1319 KT by FY 2016-17 with a CARG of 19%. By 2021-22, domestic consumption is expected to touch 2614KT.

Petrochemical Products from Panipat Refinery Complex is shown below in figure.

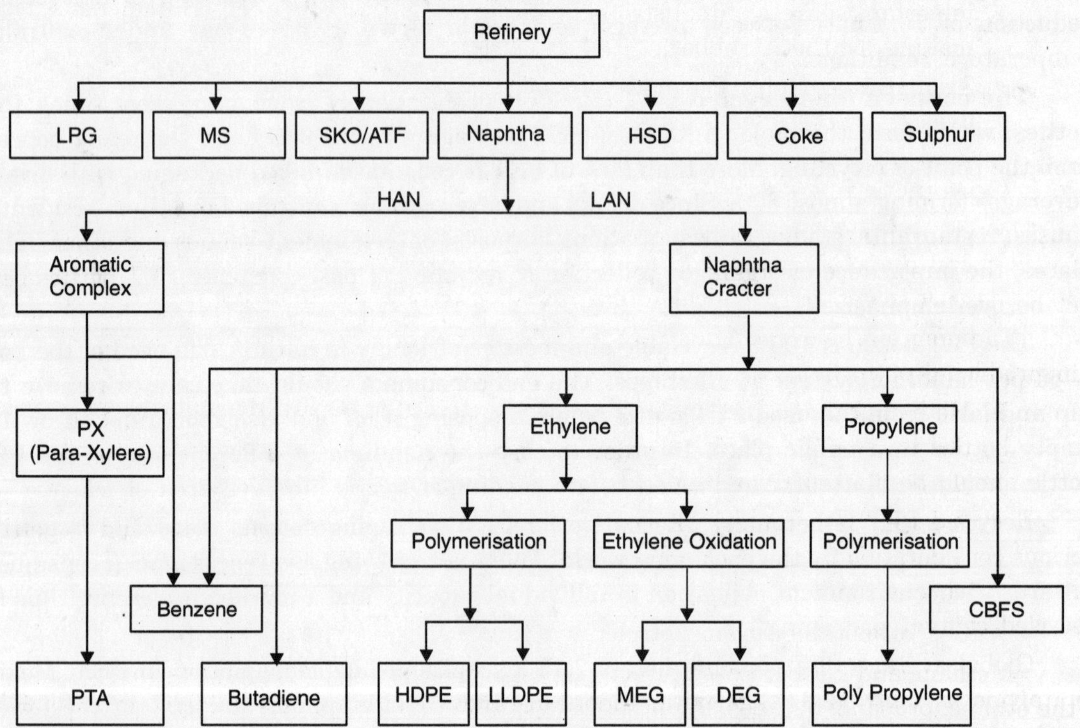

SKO-Superior KerosineOil, ATF-Aviation Turbine Fuel, HSD-High Speed Diesel, HAN-HighAromatic Naphtha, LAN-Low Aromatic Naphtha, PTA-Purified Terephthalic Acid, MEG-Mono Ethylene Glycol,DEG-Di-ethylene glycol.

Note:

Petrochemicals are chemical products derived from petroleum. Some chemical compounds made from petroleum are also obtained from other fossil fuels, such as coal or natural gas, or renewable sources such as corn or sugar cane.

The two most common petrochemical classes are olefins (including ethylene and propylene) and aroinatics (including benzene, toluene and xylene isomers). Oil refineries produce olefins and aromatics by fluid catalytic cracking of petroleum fractions. Chemical plants produce olefins by steam cracking of natural gas liquids like ethane and propane. Aromatics are produced by catalytic reforming of naphtha. Olefins and aromatics are the building-blocks for a wide range of materials such as solvents, detergents, and adhesives. Olefins are the basis for polymers and oligomers used in plastics, resins, fibers, elastomers, lubricants, and gels.

Global ethylene and propylene production are about 115 million tonnes and 70 million tonnes per annum, respectively. Aromatics production is approximately 70 million tonnes. The largest petrochemical industries are located in the USA and Western Europe; however, major growth in new production capacity is in the Middle East and Asia.

Primary petrochemicals are divided into three groups depending on their chemical structure:

- Olefins includes ethylene. propylene, and butadiene. Ethylene and propylene are important sources of industrial chemicals and plastics products. Butadiene is used in making synthetic rubber.
- Aromatics includes benzene, toluene, and xylenes. Benzene is a raw material for dyes & synthetic detergents, and benzene & toluene for isocyanates MDI and TDI used in making polyurethanes. Manufacturers use xylenes to produce plastics and synthetic fibers.
- Synthesis gas is a mixture of carbon monoxide and hydrogen used to make ammonia and methanol. Ammonia is used to make the fertilizer urea and methanol is used as a solvent & chemical intermediate.

The diagram below schematically depicts the major hydrocarbon feedstock sources used in producing petrochemicals which are:

- Methane, ethane, propane and butanes: Obtained primarily from natural gas processing plants.
- Naphtha obtained from petroleum refineries.
- Benzene, toluene and xylenes, as a whole referred to as BTX and primarily obtained from petroleum refineries by extraction from the reformate produced in catalytic reformers.
- Gas oil obtained from petroleum refineries.

Methane and BTX are used directly as feedstocks for producing petrochemicals. However, the ethane, propane, butanes, naphtha and gas oils serve as the optional feedstocks for steam-assisted thermal cracking plants referred to as *steam crackers* that produce following intermediate petrochemical feedstocks:

- Ethylene
- Propylene
- Butenes and butadiene
- Benzene

Steam crackers are not to be confused with steam reforming plants used to produce hydrogen and ammonia.

Like commodity chemicals, petrochemicals are made on a very large scale. Petrochemical manufacturing units differ from commodity chemical plants in that they often produce a number of related products. Compare this with specialty chemical and fine chemical manufacture where products are made in discrete batch processes.

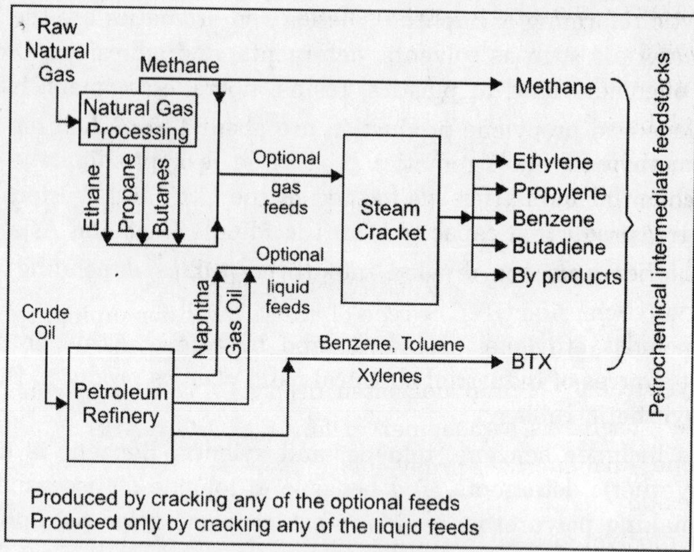

Fig. Petrochemical feedstock sources

The large scale petrochemical manufacturing locations have clusters of manufacturing units that share utilities and large scale infrastructure such as power stations, storage tanks, port facilities, road and rail terminals.

LIST OF SIGNIFICANT PETROCHEMICALS AND THEIR DERIVATIVES

The following is a partial list of the major commercial petrochemicals and their derivatives:

- ethylene - the simplest olefin; used as a chemical feedstock and ripening stimulant.

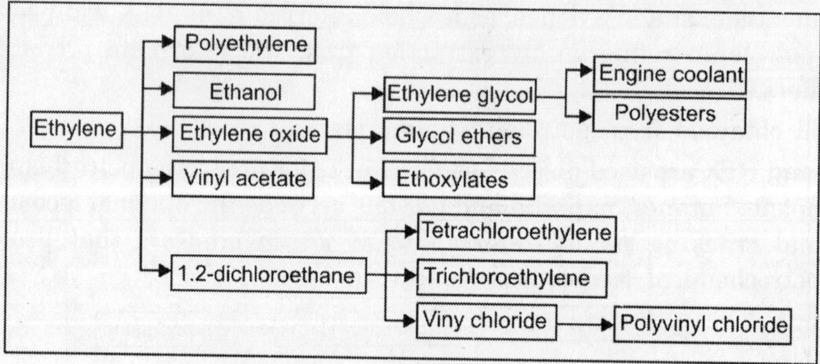

Fig. Chemicals produced from ethylene.

- polyethylene - polymerized ethylene
- ethanol - via ethylene hydration (chemical reaction adding water) of ethylene
- ethylene oxide - via ethylene oxidation
  - ethylene glycol - via ethylene oxide hydration
  - engine coolant - ethylene glycol, water and inhibitor mixture
  - polyesters - any of several polymers with ester linkages in the backbone chain.

  - glycol ethers - via glycol condensation
  - ethoxylates
- vinyl acetate
- 1,2-dichloroethane
  - trichloroethylene
  - tetrachloroethylene - also called perchloroethylene; used as a dry cleaning solvent and degreaser
  - vinyl chloride - monomer for polyvinyl chloride
    - polyvinyl chloride (PVC) - type of plastic used for piping, tubing, other things
- propylene - used as a monomer and a chemical feedstock
  - isopropyl alcohol - 2-propanol; often used as a solvent or rubbing alcohol
- acrylonitrile - useful as a monomer in forming Orion, ABS
- polypropylene - polymerized propylene
- propylene oxide
  - polyether polyol - used in the production of polyurethanes
  - propylene glycol - used in engine coolant and aircraft deicer fluid
  - glycol ethers - from condensation of glycols
- acrylic acid
  - acrylic polymers
- allyl chloride -
  - epichlorohydrin - chloro-oxirane; used in epoxy resin formation
    - epoxy resins - a type of polymerizing glue from bisphenol A, epichlorohydrin, and some amine.

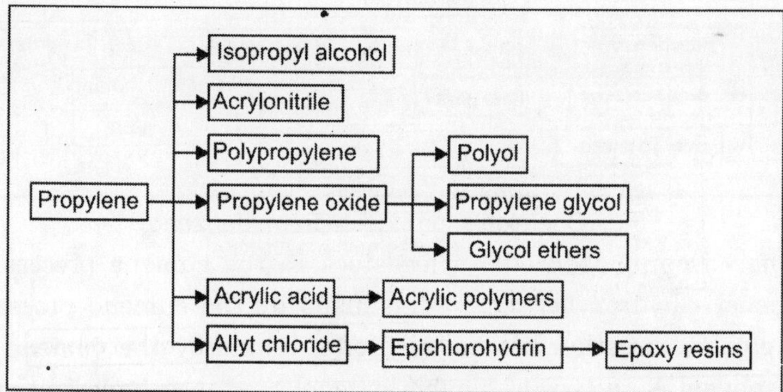

Fig. Chemicals produced from propylene.

- C4 hydrocarbons - a mixture consisting of butanes, butylenes and butadienes
  - isomers of butylene - useful as monomers or co-monomers
    - isobutylene - feed for making methyl *tert*-butyl ether (MTBE) or monomer for copolymerization with a low percentage of isoprene to make butyl rubber.
- 1,3-butadiene (or buta-1.3-diene) - a diene often used as a monomer or co-monomer for polymerization to elastomers such as polybutadiene, styrene-butadiene rubber, or a plastic such as acrylonitrile-butadiene-styrene (ABS)

    - synthetic rubbers - synthetic elastomers made of any one or more of several petrochemical (usually) monomers such as 1,3-butadiene, styrene, isobutylene, isoprene, chloroprene; elastomeric polymers are often made with a high percentage of conjugated diene monomers such as 1,3-butadiene, isoprene, or chloroprene.
- higher olefins:
    - polyolefins such as poly-alpha-olefins, which are used as lubricants.
    - alpha-olefins - used as monomers, co-monomers, and other chemical precursors. For example, a small amount of 1-hexene can be copolymerized with ethylene into a more flexible form of polyethylene.
    - other higher olefins
    - detergent alcohols
- benzene - the simplest aromatic hydrocarbon
    - ethylbenzene - made from benzene and ethylene
        - styrene made by dehydrogenation of ethylbenzene; used as a monomer
        - polystyrenes-polymers with styrene as a monomer.

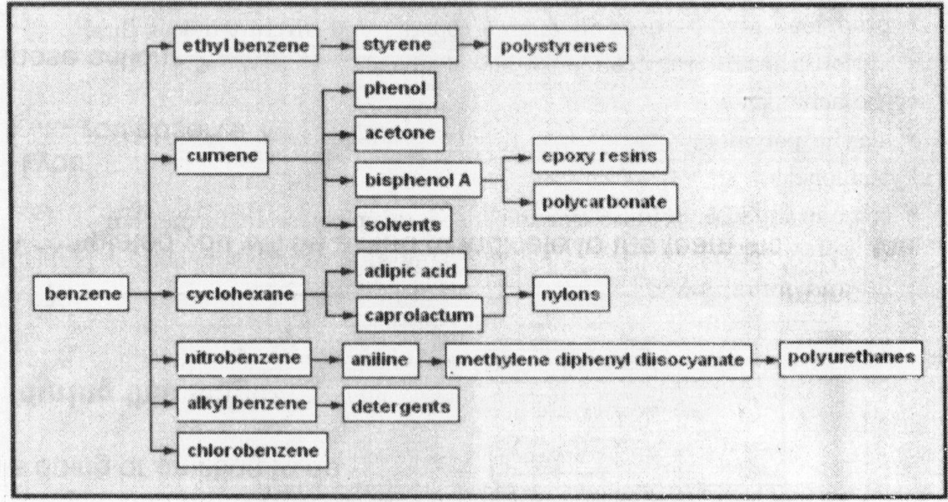

Fig. Chemicals produced from benzene.

    - cumene - isopropylbenzene; a feedstock in the cumene process.
        - phenol - hydroxybenzene; often made by the cumene process.
        - acetone - dimethyl ketone; also often made by the cumene process.
        - bisphenol A - a type of "double" phenol used in polymerization in epoxy resins and making a common type of polycarbonate.
            - epoxy resins - a type of polymerizing glue from bisphenol - A, epichlorohydrin, and some amine.
            - polycarbonate - a plastic polymer made from bisphenol - A and phosgene (carbonyl dichloride).
        - solvents - liquids used for dissolving materials; examples often made from petrochemicals include ethanol, isopropyl alcohol, acetone, benzene, toluene, xylenes.

- cyclohexane - a 6-carbon aliphatic cyclic hydrocarbon sometimes used as a non-polar solvent.
  - adipic acid - a 6-carbon dicarboxylic acid, which can be a precursor used as a co-monomer together with a diamine to form an alternating copolymer form of nylon.
    - nylons - types of polyamides, some are alternating copolymers formed from copolymerizing dicarboxylic acid or derivatives with diamines.
  - caprolactam - a 6-carbon cyclic amide.
    - nylons - types of polyamides, some are from polymerizing caprolactam
- nitrobenzene - can be made by single nitration of benzene.
  - aniline - aminobenzene
    - methylene diphenyl di-isocyanate (MDI) - used as a co-monomer with diols or polyols to form polyurethanes or with di- or polyamines to form poly ureas.
- alkylbenzene - a general type of aromatic hydrocarbon, which can be used as a presursor for a sultbnate surfactant (detergent).
  - detergents - often include surfactants types such as alkyl benzene sulfonates (ABS) and nonyl phenol ethoxylates.
- chlorobenzene

- toluene - methylbenzene; can be a solvent or precursor for other chemicals.
  - benzene
  - toluene di-isocyanate (TDI) - used as co-monomers with polyether polyols to form polyurethanes or with di- or polyamines to form polyureas polyurethanes.
  - benzoic acid - carboxybenzene
    - caprolactam
      - nylon

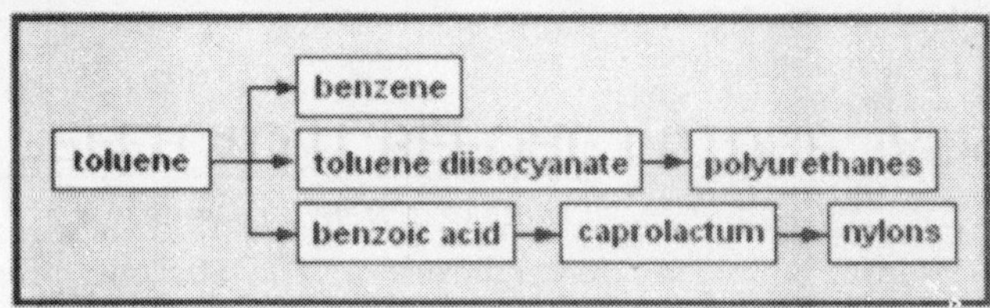

Fig. Chemicals produced from tolune.

- mixed xylenes - any of three dimethylbenzene isomers, could be a solvent but more often precursor chemicals.
  - *ortho-xylene* - both methyl groups can be oxidized to form (*ortho-*)phthalic acid
    - phthalic anhydride
  - para-xylene - both methyl groups can be oxidized to form terephthalic acid
    - dimethyl terephthalate - can be copolymerized to form certain polyesters.
      - polyesters - although there can be many types, polyethylene terephthalate is made from petrochemical products and is very widely used.
    - purified terephthalic acid - often copolymerized to form polyethylene terephthalate.

    - polyesters
- *mela-xykne*
  - isophthajic acid
    - alkyd resins
    - Polyamide Resins
    - Unsaturated Polyesters

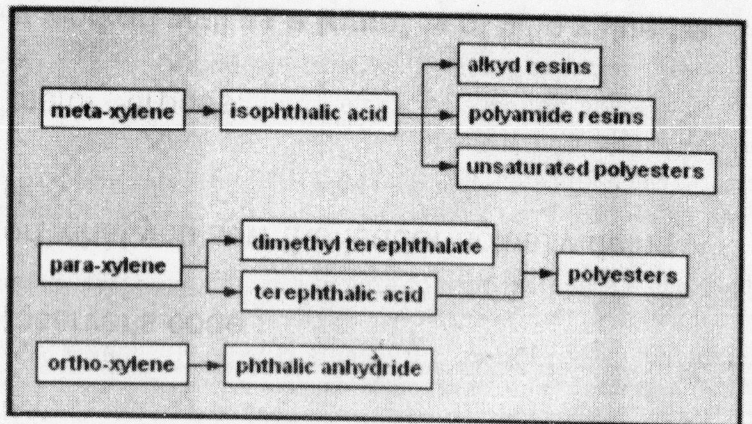

Fig. Chemicals produced from xylenes.

Petrochemicals products are tabulated below:

Petrochemicals	Fibers	Petroleum	Chemicals
Basic Feedstock Benzene Butadiene Ethylene *p*-Xylene Propylene			
Intermediates 2-Ethylhexanol (2-EH) Acetic acid Acrylonitrile (AN) Ammonia Bis(2-ethylhexyl) phthalate (dioctyl phthalate) *n*-Butene Cyclobexane Dimethyl terephthalate (DMT) Dodecylbenzene Ethanol Ethanolamine	Acrylic fiber Acrylonitrile butadiene styrene (ABS) Acrylonitrile styrene (AS) Polybutadiene (PBR) Polyvinyl chloride (PVC) Polyethylene (PE) Polyethylene	Lubricants Additives Catalysts Marine fuel oil Petroleum refining	Adhesives and sealants Agrochemicals Construction chemicals Corrosion control chemicals Cosmetics raw materials Electronic chemicals and materials Flavourings,

Ethoxylate 1,2-Dichloroethane (ethylene dichloride or EDC) Ethylene glycol (EG) Ethylene oxide (EO) Formaldehyde Moulding Compound (FMC) *n*-Hexene Linear alkyl benzene (LAB) Methanol Methyl tert-butyl ether (MTBE) . Phenol Propylene oxide Purified terephthalic acid (PTA) Styrene monomer (SM) Thermosetting Resin (Urea/Melamine) Vinyl acetate monomer (VAM) Vinyl chloride monomer (VCM)	terephthalate (PET) Polyol Polypropylene (PP) Polystyrene (PS) Styrene butadiene (SBR) Aery lie-formaldehgde (AF)		fragrances, food additives Pharmaceutical drugs Specialty and industrial chemicals Specialty and industrial gases Inks, dyes and printing supplies Packaging, bottles, and containers Paint, coatings, and resins Polymer additives Speciallty and life sciences chemicals Surfactants and cleaning agents

Petrochemicals are backbone of chemical industry. Although the origin of petrochemical industry in world was in 1920, however, sixties marks the era of petrochemicals in India when Union Carbide set up first ethylene complex with capacity 20,000 TPA in 1963 in Mumbai. It was followed by NOCIL with 60,000 TPA ethylene complexes in 1968 in Thane near Mumbai and PSF plant of Chemical and Fibers India Ltd.(CAFI) at Thane - Belapur road. Indian Petrochemical Corp. Ltd. (IPCL) set up first integrated petrochemical complex in 1970 in public sector at Vadodara. With the modest beginning in 1960 by setting up of 20,000 tonnes naphtha cracker by Union Carbide in Mumbai, Indian petrochemical industry has sustained a high and steady rate of growth during last six decades and has entered the world market. The petrochemical industry has been the fastest growing sectors in India and become a major segment of chemical industry, which is growing faster than industries overall and within chemicals. It has posed serious threat to chemical industry based on natural feedstock - Biomass and Coal. The«petrochemical industry is major supplier of chemical inputs to a large and growing number of downstream.

Basic Petrochemicals

C1 group : Methane, CO - H_2 synthesis, synthesis gas derivatives

C2 group : Ethane, ethylene, ethylene derivatives, acetylene

C3 group : Propane, propylene and propylene derivatives

C4, C5 group : Butadiene, Butanes, Butenes, Pentane, Pentene, Isoprene, Cyclopentadiene

Aromatic : Benzene, Toluene, Xylenes, Naphthalene, BTX derivatives

Major End Products

Polymer, Synthetic fibre, synthetic rubber, synthetic detergent, Chemical intermediate, dyes and intermediates chemical intermediates, pesticides.

Basic Building Block Processes

Petrochemical manufacturing involves following building blocks processes for the manufacture of building blocks and intermediates:

Cracking: Steam cracking, Catalytic cracking for olefins pyrolysis gasoline by product

Steam reforming and Partial oxidation: Synthesis gas

Catalytic Reforming: Aromatic production

Aromatic conversion processes: Aromatic production

Alkylation: Linear alkyl benzene (LAB)

OXO Process: Oxo-alcohol

Polymerisation Process: Polymer, elastomers and synthetic fibre

Petrochemicals Feed Stocks

One of the major issues, which have directed the worldwide growth of petrochemical industry, has been the availability of feedstock, which has led to replacement of the natural resources coal, molasses, fats, etc. Basic feedstock used in petrochemical industry for manufacture of olefins and aromatics are derived either from natural gas or petroleum fractions and includes natural gas condensate, naphtha, kerosene, catalytic cracking & reformer gases, waxes, pyrolysis gasoline. Natural Gas and Petroleum Fractions as Petrochemicals Feedstock is given in Table below. Alternative routes to principal petrochemicals is given in next Table. Some of the alternative feed stock choice for petrochemical industry are:

- Naphtha from methane from natural gas to liquid (NGL) process.
- Naphtha from coal via direct liquification or indirect liquification by FT process.
- Plastic waste to naphtha and other hydrocarbons through liquefaction, pyrolysis and separation processes.
- FT naphtha from biomass.
- Methanol routes: Synthesis gas from methane, coal & biomass; conversion of synthesis gas to methanol and production of olefin by methnol to olefin technology.
- Conversion of methanol to dimethyl ether.
- Product recovery & separation, Recovery of C_4&C_5 stream from FCC and steam cracker.
- Oxidative coupling of methane.
- Ethanol from biomass: direct fermentation of sugar rich biomass, hydrolysis of lingo-cellulosic biomass

- Gasification of lingo-cellulosic biomass followed by fermentation or chemical catalysis to ethanol.
- Carbon dioxide to liquid fuel by engineered bacteria.
- Gasification of petrocoke to hydrogen.

Table : Natural Gas and Petroleum Fractions as Petrochemicals Feedstocks

S. N.	Petroleum Fractions and Natural Gases	Source	Composition	Intermediate Processes	Intermediate Feedstock
1.	Refinery Gases	Distillation, catalytic cracking, catalytic reforming	Methane, ethane, propane, butane, BP upto 25°C	Liquefaction, cracking	LPG, ethylene, propylene, butane, butadiene.
2.	Naphtha	Distillation and thermal & catalytic cracking, visbreaking	C_4-C_{12} hydrocarbon, BP 70 - 200°C	Cracking, reforming, alkylation, disproportionation, isomerisation	Ethylene, propylene, butane, butadiene, benzene, toluene, xylene
3.	Kerosene	Distillation and secondary conversion processes	C_9-C_{10} hydrocarbon, BP 175-275 °C	Fractionation to obtain C_{10}-C_{14} range hydrocarbons	Linear n C_{10} - n C_{14} alkanes
4.	Gas Oil	Distillation of crude oil and cracking	C_{10}-C_{25} hydrocarbons, BP 200-400 °C	Cracking	Ethylene, propylene, butadiene. butylenes
5.	Wax	Dewaxing of lubricating oil	C_8-C_{56} hydrocarbon	Cracking	C_6-C_{20} alkanes
6.	Pyrolysis Gasoline	Ethylene cracker	Aromatic, alkenes, dienes, alkanes, cycloalkane	Hydrogenation distillation, extraction, crystallisation, adsorption	Aromatics
7.	Natural Gases & Natural Gas Condensate	Gas fields and crude oil stabilisation	Hydrogen, methane, ethane, propane, pentane, aromatics	Cracking, reforming, separation	Ethylene, propylene, EPG, aromatics, etc.
8.	Petroleum coke	Crude oil	Carbon	Residue upgradation processes, gasification	Carbon electrode, acetylene, fuel

Table : Alternative Routes to Principal Petrochemicals

S.N.	Chemieuls	Petroleum Source	Alternate Source (Europe, except where stated)
1.	Methane	Natural gas, Refinery light gases (demethaniser overheads)	Coal, as byproduct of separation of coke oven gases (1920-30) or of coal hydrogenation (1930-40)
2.	Ammonia	Methane, Light liquid hydrocarbons	From coal via water gas (1910-20)
3.	Methyl alcohol	Methane, Light liquid hydrocarbons	From coal via water-gas (1920-30); from methane (from coal) by methane-stream and methane oxygen processes (1930-40)
4.	Ethylene	Pyrolysis of gaseous & liquid hydrocarbons	Dehydration of ethyl alcohol (original route). By-product in fractional distillation of coke oven gas (1925-35). Hydrogenation of acetylene (1940-45)
5.	Acetylene	Ethylene	Calcium carbide (original process), methane from coal by partial combustion and by arc process (1935-45)
6.	Ethylene glycol	Ethylene	From ethylene made as above (1925). In America, from coal via carbon-monoxide and formaldehyde (1935-40)
7.	Ethyl alcohol	Synthetic ethyl alcohol Co-product of paraffin gas oxidation.	Fermentation of molasses (original route)
8.	Acetaldehyde	Direct oxidation of ethylene	Fermentation of ethyl alcohol, or acetylene from carbide (1900-10)
9.	Acetone	Propylene	Wood distillation (original process). Pyrolysis of acetic acid (1920-30) or by acetylene-stream reaction (1930-40)
10.	Glycerol	Propylene	By-product of soap manufacture (original process)
11.	Butadiene	1- and 2-Butenes, Butane, Synthetic ethyl alcohol, By-product of ethylene by pyrolysis of liquid hydrocarbons	Ethyl alcohol (1915); acetaldehyde via 1:3-butanediol (1920-30); acetylene and formaldehyde from coal via 1:4-butanediol (1940-45); from 2:3-Butanediol by fermentation (1940-45)
12.	Aromatic hydrocarbons	Aromatic-rich and naphthenic-rich fractions by catalytic reforming and direct extraction or by hydroalkylation.	By-products of coal-tar distillation

Figure: Refinery Streams for Petrochemicals/Chemicals

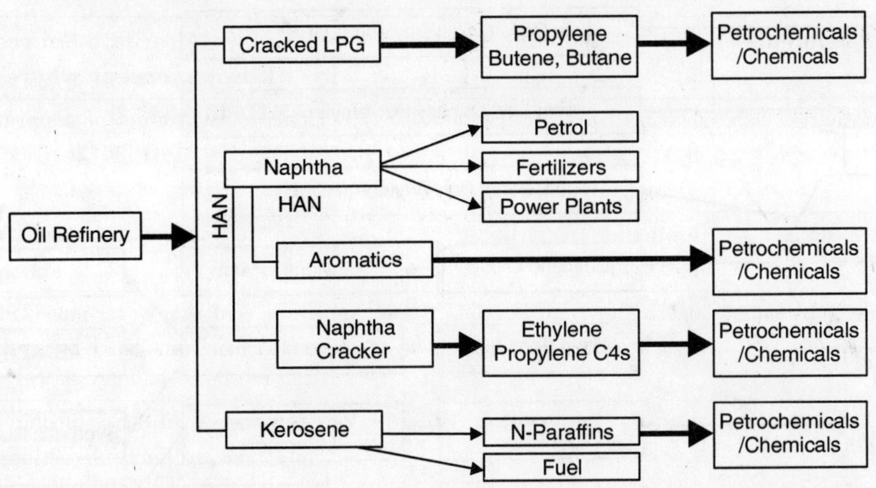

HAN-High Aromatic Naphtha, LAN-Low Aromatic Naphtha, LPG-Liquefied Petroleum Gas.

Figure: Olefin Cracking Process Applications

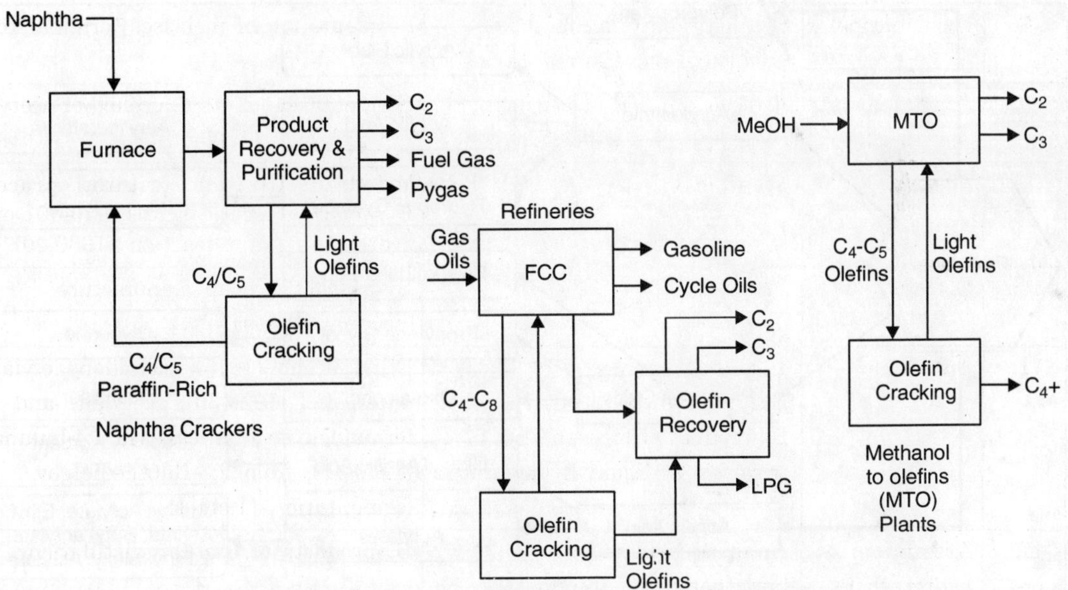

FCC-Fluid Catalytic Cracker, MeOH-Methyl alecohal

Fig. Olefin Cracking can be integrated with a variety of other processes.

Figure: Ethylene Products tree is shown below:

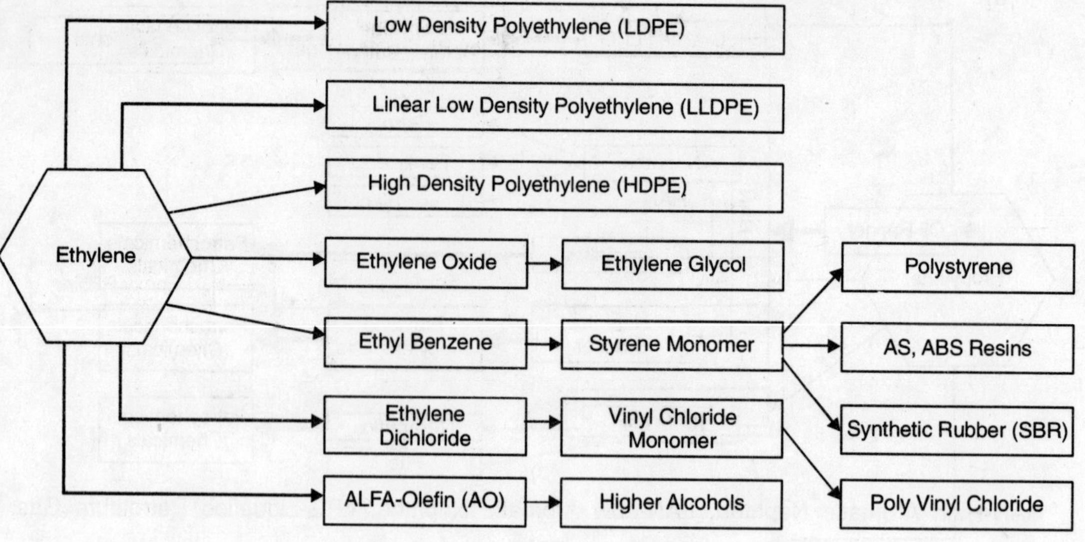

Figure: Propylene Products tree is shown below:

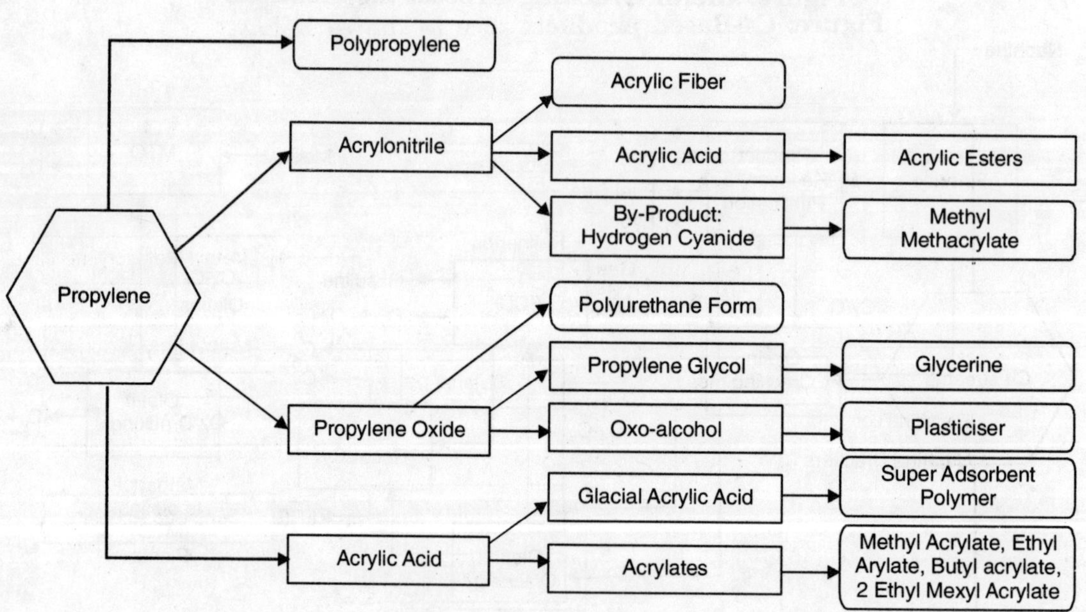

Propylene contd...

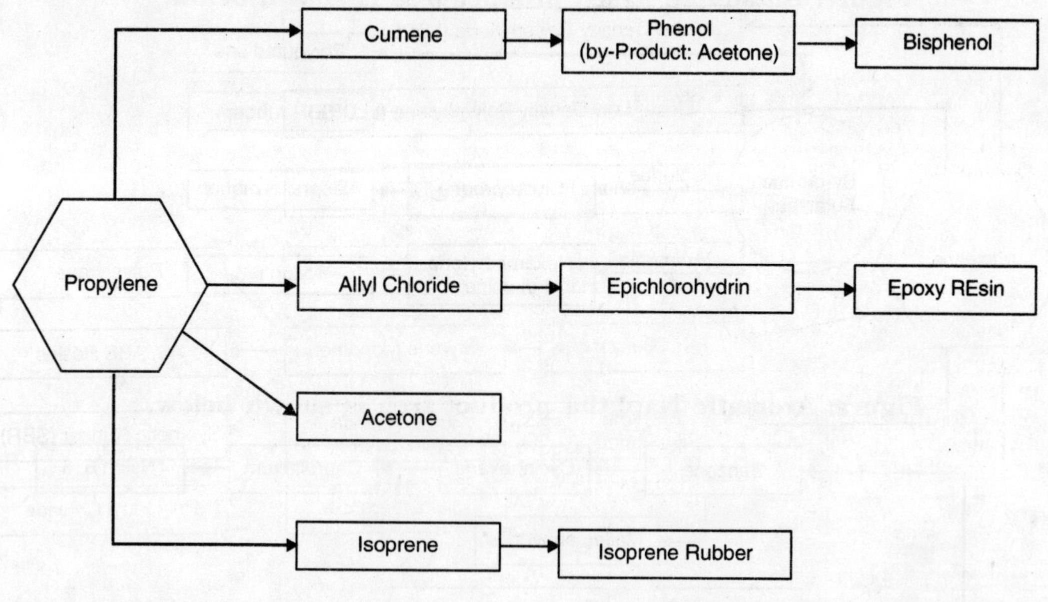

Figure: C4-Based products tree is shown below.

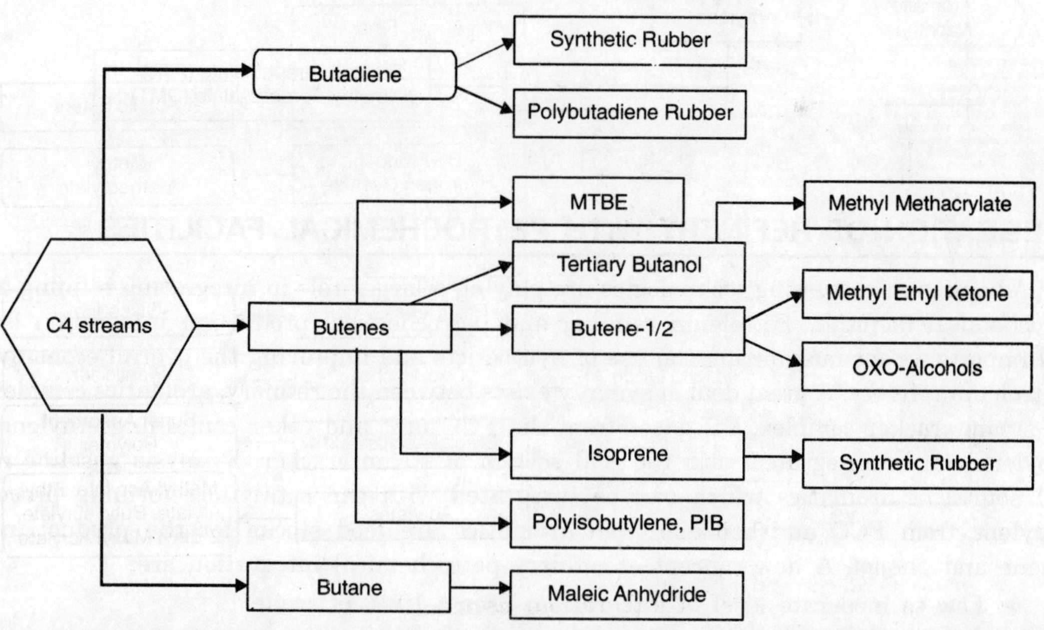

Figure: Butadiene Based product free is shown below:

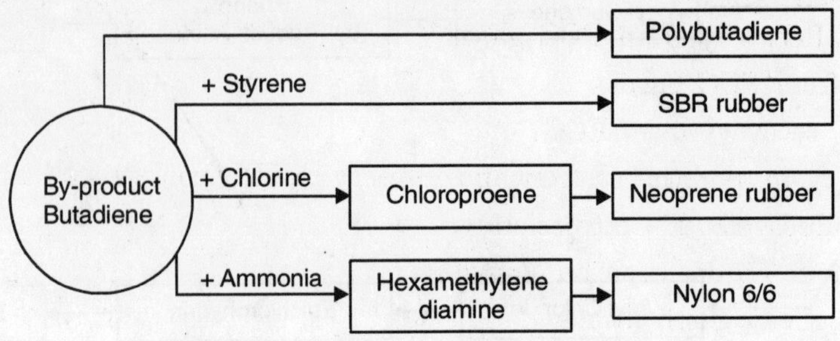

Figure: Aromatic Naphtha product tree is shown below.

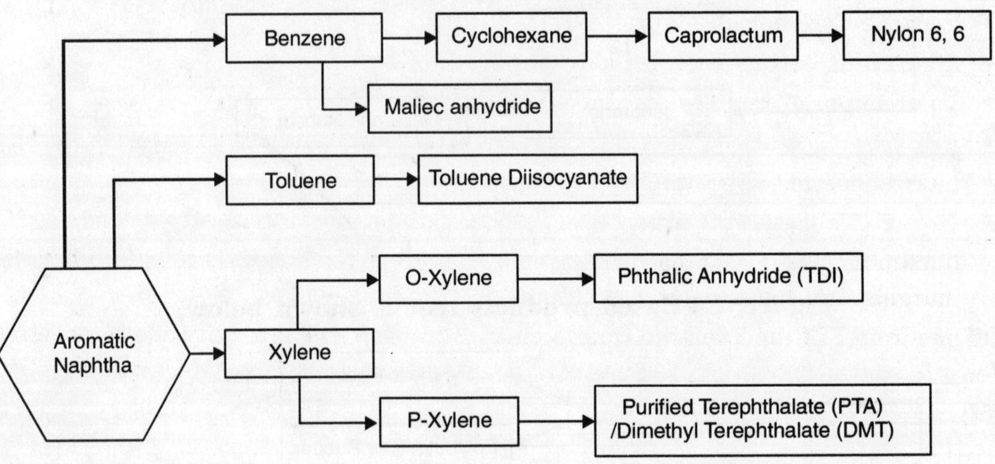

INTEGRATION OF REFINERY WITH PETROCHEMICAL FACILITIES

Advances in processing technologies are playing a larger role in integrating refining and petrochemical facilities. Petroleum refining and petrochemical production integration is of vital importance for maximizing the use of byproducts and improving the overall economy of a petroleum refinery. A great deal of synergy exists between the refinery, aromatics complexes and steam cracker complex. Off gases from the FCC unit and coker containing ethylene & propylene can be integrated with the cold section of steam cracker. Pyrolysis gasoline is a good source of aromatics which can be integrated with the catalytic reforming process. Propylene from FCC and benzene from aromatics are feed stocks for the production of cumene and phenol. A new concept of refinery petrochemical integration are:

- Low to moderate level of integration: uses 5-10 % of crude.
- High level integration: these complexes convert 10-25% of the crude oil.
- Petrochemical refinery: these complexes produce a significant amount of petrochemicals as compared to fuels.

Petrochemical Processes within refinery which will help in integration of refinery and petrochemicals include:

- Propylene Recovery from FCC gases
- Ethylene from FCC gases
- C_4 and C_5 recovery from FCC
- C_4 and C_5 from naphtha cracker and refinery to LPG pool as well as feed to cracker.
- Aromatic Recovery & Conversions.
- Light ends & Light Naphtha Conversion.
- Residue & Coke gasification.
- Hydrogen Production.
- Butane to Maleic anhydride & Derivative.
- Benzene-Cumene-Phenol-Acetone.
- Benzene-Cyclohexane-Caprolactum.
- n-Paraffins extraction from kerosene for LAB.
- Valorization of refinery streams- LCO, LCGO, HCGO.
- Recovery of Valuable Chemicals cyclopentadiene, diclopentadiene, isoprene, propolyene.
- Isobutylene for alkylation.
- Use of C7-C8 stream from benzene extortion for separation of p-xylene for PTA.
- Maximizing the use of natural gas in a refinery-petrochemicals complex offers higher margins and lower carbon emissions.

Off gas from FCC and delayed coking units contains a good quantity of ethane, ethylene, propylene & some propane and recovery of these hydrocarbons may be economical in the refinery.

INDALIN PROCESS

Indalin is a versatile indigenous technology adding value to upstream and downstream oil industry. Indalin is a catalytic cracking process for upgradation of low value naphtha to very high yield of LPG, containing high olefins such as propylene, ethylene, butylenes etc. Surplus kerosene and gas oil range fraction can also be processed along with naphtha. Indalin can integrate a refinery with petrochemicals complex and therefore offers a tremendous opportunity for value addition through upgradation of low value streams to petrochemical feed stock.

Figure: Value Chain from Refinery Streams is shown below.

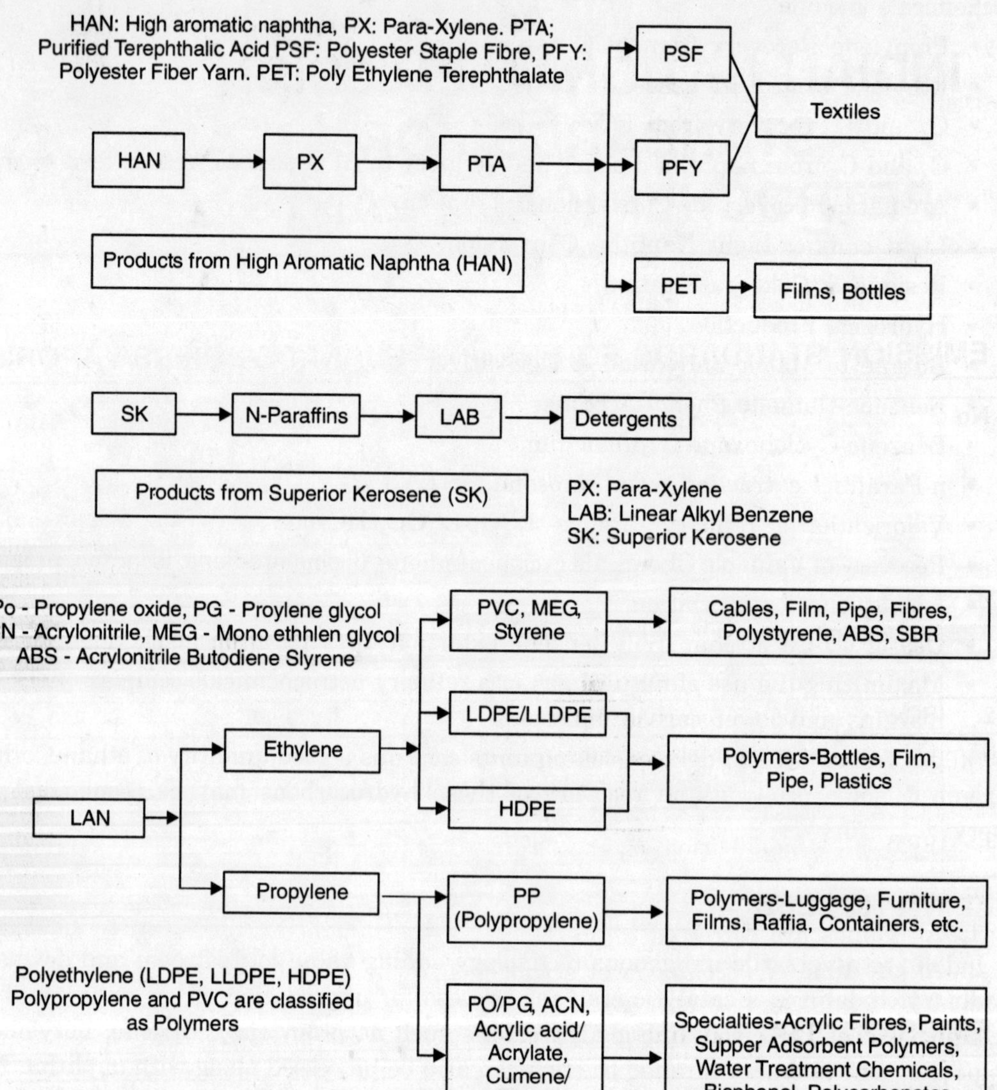

Appendix D

INDIAN NATIONAL EMISSION STANDARDS FOR PETROCHEMICAL PLANTS

1.0 EMISSION STANDARDS FOR HEATER/FURNACE/BOIERS/VAPORISERS

S. No.	Parameter		Maximum emission limit (mg/Nm^3)	
			Existing plants	New plants/ Expansion (commissioned after January 01, 2007
1.	NOx	Gas firing	350	150
		Liquid firing	400	250
2.	SOx	Liquid firing	1700	850
3.	CO	Carbon mono oxide limit in case of partial oxidation in PA, MA, PTA and DMT plant	150	150
4.	SPM	Liquid firing	150	100

Notes:

1. All values are corrected to 3% O_2.
2. At the time of decoking, wet scrubber shall be operated.

2.0 EMISSION STANDARDS FOR ORGANIC PARTICULATES

S.No.	Petrochemical compound	Maximum Emission limit (mg/Nm^3)		Mass flow limit (gm/hr)
		Existing plant	New plants/ Expansion (commissioned after January 01, 2007	
1.	Phthalic anhydride (PA), Maleic anhydride (MA), Toluene Di-isocyanate (TDI)	50	25	100*

Note:

* - Mass flow limit (gm/hr) is applicable for new plants and expansion plants.

3.6 EMISSION STANDARDS FOR PROCESS EMISSION (SPECIFIC ORGANIC POLLUTANTS

S. No.	Parameter	Source	Maximum emission Limit (mg/Nm³)
1.	Chlorine	EDC/VCM plant and incinerator	10
2.	HCl	EDC/VCM plant and incinerator	30
3.	Ammonia	Process vent (wastewater stripper) acrylonitrile plant, caprolactum plant	75
4.	H_2S	Naphtha pre-treatment plant, olefin plant	05
5.	Phosgene	Generated in TDI and MDI plants	01
6.	HCN	Acrylonitrile plant	10

4.0 EMISSION STANDARDS FOR VOC-HAPS FROM PROCESS VENTS

S.No.	Parameters	Maximum emission Limit		
		Existing plant (ma/Nm³)	New plants/ Expansion	
			(mg/Nm³)	(gm/hr)
1.	(Toluene Di-isocyanate) TDI, Methylenediphenyl Di-isocyanate (MDI)	0.1	0.1	0.5
2.	Benzene, Butadiene	5.0	5.0	25.0
3.	EO, VCM, EDC, ACN, PO	20.0	10.0	50.0

5.0 EMISSION STANDARDS FOR VOCS (GENERAL) FROM PROCESS VENTS

S.No.	Petrochemical process / compounds	Maximum emission Limit (mg/Nm³), dry basis
1.	MA, PA, Phenol	20
2.	Ethyl benzene (EB), Styrene, Toluene, Xylene, Aromatics, EG, PG	100
3.	Non-methane HC (paraffin), Acetone, olefins	150

6.0 STANDARDS FOR STORAGE AMD TRANSFER POINT (LOADING AND UNLOADING)

6.1 Standards for Atmospheric Storage Tanks of Petrochemical Products

S. No.	True Vapour Pressure (TVP), kPa at 20°C	Storage Tank Capacity (M^3)
1.	> 10	4 – 75
2.	10 – 76	75 – 500
3.	10 – 76	> 500
4.	> 76	> 75

Notes:

1. Requirement for seals in Floating Roof Tanks:
   (*i*) IFRT & EFRT are to be provided with secondary seal with minimum vapour recovery of 96%.
   (*ii*) Primary seal will be liquid or shoe mounted for EFRT and vapour mounted for IFRT. Maximum seal gap width will be 4 cm and maximum gap area will be 200 cm^2/m of tank diameter.
   (*iii*) Secondary seal will be rim mounted. Maximum seal gap width will be 1.3 cm and maximum gap area will be 20 cm^2/m of tank diameter.
   (*iv*) Material of seal and construction should ensure high performance and durability.
2. Fixed Roof Tanks will have vapour control efficiency of 95% or vapour recovery/ balancing efficiency of 90%.
3. Inspection and maintenance of storage tanks should be carried out under strict control. For the inspection, API RP 575 may be adopted. In-service inspection with regard to seal gap should be carried out once in every six months and repair to be implemented in short time. In future, possibility of on-stream repair of both seals will be examined.
4. Tanks shall have paint with white colour shade, except for derogation of visually sensitive area.

6.2 Storage of benzene, VCM and ACN

For storage of benzene, VCM and ACN, following shall be followed:

1. FRT with vapour to incineration with 99.9% of removal efficiency for volatile organic compounds (VOC's).

   (OR)

2. EFRT with double seals, emission-reducing roof fitting and fitted with fixed roof with vapor removal efficiency of at least 99%.

   (OR)

3. Internal floating roof and nitrogen blanketing in between fixed and floating roofs.

7.0 STANDARDS FOR EMISSION FROM LOADING OF VOLATILE PRODUCTS

S.No.	Item	(Standards) Maximum emission limit
1.	Naphtha: (i) VOC reduction, % (or) (ii) Emission, gm/m^3	 (i) 99.5% (or) (ii) 5 gm/m^3
2.	Benzene and Butadiene: (i) VOC reduction, % (or) (ii) Emission, gm/m^3	 (i) 99.5% (or) (ii) 20 gm/m^3
3.	Toluene/Xylene: (i) VOC reduction, % (or) (ii) Emission, gm/m^3	 (i) 99.98% (or) (ii) 150 gm/m^3

GUIDELINES

1.0 GUIDELINES FOR ATMOSPHERIC STORAGE TANK PRACTICES

(i) For true vapour pressure up to 10 kPa with tank capacity in the range of 4-75 m^3, Fixed Roof Tank (FRT) with pressure valve vent may be provided,

(ii) For true vapour pressure of 10-76 kPa with tank capacity in the range of 75-500 m^3, Internal Floating Roof Tank (IFRT) or External Floating Roof Tank (EFRT) or Fixed Roof Tank with vapour control or vapour balancing system may be provided.

(iii) For true vapour pressure more than 10-76 kPa with tank capacity more than 500 m^3, Internal Floating Roof Tank or External Floating Roof Tank or Fixed Roof Tank with vapour control system may be provided.

(iv) For true vapour pressure more than 76 kPa with tank capacity more than 75 m^3, Fixed Roof Tank with vapour control system may be provided.

2.0 LDAR AND MONITORING PROTOCOL

Leak detection and repair (LDAR) programme include (i) Block valves; (ii) Control valves; (iii) Pump seals; (iv) Compressor seals; (v) Pressure relief valves; (vi) Flanges – Heat Exchangers; (vii) Flanges – Piping; (viii) Connectors – Piping; (ix) Open ended lines; and (x) Sampling connections. Equipment and line sizes more than 2.54 cm are to be covered.

LDAR programme would be applicable to components (given at 2 above) for following products/compounds: (i) hydrocarbon gases; (ii) Light liquid with vapour pressure @ 20°C > 1.0 kPa; and (iii) Heavy liquid with vapour pressure @ 20°C between 0.3 to 1.0 kPa.

LDAR programme would not be applicable for (i) heavy liquids with vapour pressure < 0.3 kPa, it will be desirable to check for liquid dripping as indication of leak (ii) Equipment and line sizes less than 2.54 cm, less than 300 h service and in vacuum service, (iii) Equipments and piping during start up and shut down. (iv) Pumps (Canned, diaphragm, magnetic), Valves (Diaphragm, bellow) and close loop Sampling points and (v) Non-access able points to the extent of 5% of total plant.

A leak is defined as the detection of VOC concentration more than the values (in ppm) specified below at the emission source using a hydrocarbon analyser according to measurement protocol (US EPA – 40 CFR part 60 Appendix-A, method 21 for determination of VOC leaks may be referred):

S. No.	Component	HAP (General) in ppm		Volatile HAP* in ppm	
		w.e.f. 1.1.07	w.e.f. 1.1.10	w.e.f. 1.1.07	w.e.f. 1.1.10
1.	Pump/Compressor	10000	5000	3000	2000
2.	Valves / Flanges	10000	3000	2000	1000
3.	Other components	10000	3000	2000	1000

Note: * - Benzene, butadiene, VCM, EDC, ACN, EO, PO

In addition, any component observed to be leaking by sight, sound or smell, regardless of concentration (liquid dripping, visible vapour leak) or presence of bubbles using soap solution should be considered as leak.

Following frequency of monitoring of leaks and schedule for repair of leaks shall be followed:

S. No.	Component	Frequency of monitoring	Repair schedule
1.	Valves/Flanges	Quarterly (semi-annual after two consecutive periods with < 2% leaks and annual after 5 periods with < 2% leaks)	Repair will be started within 5 working days and shall be completed within 15 working days after detection of leak for general hydrocarbons. In case of benzene, the leak shall be attended immediately for repair.
2.	Pump seals	Quarterly	
3.	Compressor seals	Quarterly	
4.	Pressure relief devices	Quarterly	
5.	Pressure relief devices (after venting)	Within 24 hours	
6.	Heat Exchangers	Quarterly	
7.	Process drains	Annually	
8.	Components that are difficult to monitor	Annually	
9.	Pump seals with visible liquid dripping	Weekly	Immediately
10.	Any component with visible leaks	Weekly	Immediately
11.	Any component after repair/replacement	Within a week	

Following types of monitoring methods may be judiciously employed for detection of leaks: (i) Photo ionisation detector (PID) or flame ionisation detector (FID) Instrumental method of measurement of leaks; (ii) Audio, visual and olfactory (AVO) leak detection; and (Hi) Soap bubble method.

Data on time of measurement & concentration value for leak detection; time of repair of leak; and time of measurement & concentration value after repair of leak should be documented for all the components.

Pressure relief and blow down systems should discharge to a vapour collection and recovery system or to flare.

Open-ended lines should be closed by a blind flange or plugged.

Totally closed-loop should be used in all routine samples.

Low emission packing should be used for valves.

High integrity sealing materials should be used for flanges.

3.0 GENERAL NOTES

1. Emission monitoring shall be carried out as per the Emission Regulations – Part III, published by Central Pollution Control Board.
2. Following methods may be used for measurement of pollutant concentrations in the emissions:

S.No.	Parameter	Method of measurement
1.	Sulphur Dioxide (SO_2)	USEPA CFR – 40 Part 60 Appendix A Method 6
2.	Oxides of Nitrogen (NO_X)	USEPA CFR – 40 Part 60 Appendix A Method 7
3.	Particulate Matter (PM)	USEPA CFR – 40 Part 60 Appendix A Method 5
4.	Carbon Monoxide (CO)	USEPA CFR – 40 Part 60 Appendix A Method IOA/Combustion analyzer with electro chemical detector/NDIR detector

NOTE: Common Emission Sources in Petrochemical Plants

Petrochemicals production processes are specific to the feedstock product and so are the process emission sources & the air pollutants. Broadly petrochemical production unit producing a specific intermediates or final product generally involves distinct processing modules viz.:

- Feed raw material supply and pre-treatment.
- Synthetics
- Production separation, refinement & co-product recovery

Besides, the specific unit may involve intermediates product storage & handling and also dedicated disposal system for gaseous streams like incinerators etc. The emission sources from the petrochemical complex are classified also as "point sources" and "non-point sources" as given below:

Point sources

- Combustion sources (power and steam generation)
- Process/intermittent vents

- Emission due to transfer process
- Storage emission
- Emission control and disposal modules (Viz. flare, incinerator etc.)

Non Point sources (Fugitive emission)

- Equipment leak.
- Cooling water, wastewater collection and treatment system.

Emission from Combustion sources (power and steam generation)

Combustion gases may originate from primary sources such as process furnaces, steam boilers, turbines & engines, but also from pollutant abatement facilities (e.g. incinerators & flares). Whilst process furnaces are usually dedicated to one process, steam & electricity producing units serve a complete petrochemical complex and their emissions cannot be allocated easily to one process.

Combustion units will generate emissions to air that are related to combustion conditions (e.g. CO_2, H_2O, NOx, CxHy, CO, soot) and fuel composition (e.g. SO_2, fuel-NOx, metals, soot).

Generally, gaseous fuels are predominantly used in the petrochemical industry, these are the low-boiling gaseous fractions from the processes (e.g. hydrogen, C1-C4 hydrocarbons). In general, gaseous fuels combust cleanly and result in the lowest emissions. Gaseous fuels are normally low in sulphur and have a low content of bound nitrogen, and so the SOx and fuel NOx emissions from gas firing are relatively low. Emissions may be increased by air pre-heating (higher thermal-NOx emissions) and sulphur or nitrogen compounds in the fuel. The high temperatures in 'high temperature process furnaces' may also increase thermal-NOx emissions.

Light liquid fuels viz. Naphtha are occasionally used in the petrochemical industry. Further liquid fuels can also be residual higher boiling fractions from the process and industrial gas oil or fuel oil. 'Heavy' liquid fuels may cause emissions of dust & heavy metals (due to ash content), emissions of NOx & SO_2 (due to nitrogen & sulphur content) and have an increased potential for soot formation.

Nitrogen oxides (NOx) is generally the critical pollutant from Petrochemical Complex generated at the combustion devices located within the process plant (viz. fired heater, incinerator) or at the utility facilities (viz. Steam & Power generation).

(A) Emission Profile from Olefin Complex

• *Plant boundary definition and the degree of integration*

A number of operations are directly associated with olefins production, including feed pre-treatment, butadiene recovery or hydrogenation, gasoline heat soaking or hydro treatment benzene concentration or extraction and tar (residue from heavy gas Oil fields) handling.

Olefin plants require the principal utilities (stream, power & cooling water) as well as provision for wastewater treatments. In addition, they require the capability to flare waste gases during an upset condition and certain intermittent operations.

Olefin plants pre frequently used to recover vent & purge streams from other units (e.g. polymer plants), on an integraated site, and thus eliminates the requirement for off gas disposal. Residues from the cracking of heavy feedstock such as gas oil can also be recovered as fuel oil for subsequent steam & power generation.

• *Emission related to feed stock*

Emission levels are normally reported on a per ton ethylene basis.

On a per ton ethylene basis, emission will tend to be lower for those plants using gas feed stocks than those using naphtha or gas oil. As general rule, the percentage conversion of hydrocarbon to lower olefins is reduced as the molecular weight of feed stock increases. For illustration, approximately 80% of ethane is converted to ethylene in the gas cracking process, where as ethylene yield from naphtha is typically 30-35%. However, virtually no propylene (the next most important olefin) is produced from ethane gas cracking, whereas propylene production from naphtha can be as high as 70% ethylene production.

• *Air emission/emission factors and inventory*

The principal pollutants and their sources & the emission factor (kg pollutant/ton of ethylene product) for olefin (cracker) unit is given below.

S.N.	Source of emission from olefin complex	Air pollutants				
		NO_x	SO_2	CO	VOC's	Particulates
1.	Gas-fired Furnaces/heaters	X	X	X	X	—
2.	Gas-fired turbines	X	X	X	X	—
3.	Liquid Fuel Fired Boilers	X	X	X	X	X
4.	Point Sources	X	X	X	X	X
5.	Maintenance activities	—	—	—	X	—
6.	Fugitive Emissions	—	—	—	X	—
7.	Decoking Operations	X	—	X	X	X
8.	Sour Gas Disposal	—	X	—	—	—
9.	Regeneration Furnaces	X	X	X	X	—
10.	Emission Factor (kg/t ethylene)	1.0-2.8	0.01-3.3	0.2-1.0	0.03-6.0	0.05-1.5

X Indicates emission of pollutants; — indicates no emission of pollutants.

(B) Emission from cracker furnace

The furnace area is defined as that part of the process comprising the pyrolsis heaters, complete with heat exchange equipment for generating high-pressure steam, and any separately fired steam super heaters. It excludes auxiliary boilers & regeneration furnaces.

The most significant emissions to air result from the combustion of fuels in the pyrolsis cracking furnaces.

The cracker furnaces generally use the recovered off-gas as fuel. This off-gas is very lean & clean fuel (rich in C1 & H_2) and generally having no sulphur. The Nitrogen Oxides (NOx) and Carbon Monoxide (CO) are the major pollutanto in the cracker flue gas.

Data on emissions of carbon monoxide (CO) and nitrogen oxides (NOx) are given below.

	CO (mg/Nm³)		NOx (mg/Nm³)	
	Typical	Range	Typical	Range
Full range operation	35	0.2 – 620	39	61 – 250

Sulphur Dioxide Emissions: International survey reported levels of SO_2 emissions in tbe range 20 – 100 mg/Nm^3 (as 30-60 minute averages at normal temperature & pressure, 3% oxygen, dry gas). The SO_2 emissions are not normally considered to be significant for steam crackers. However, the SO_2 emission will be governed specially by the fuel used in the furnace.

Particulates Emissions: A general survey world wide reflects particulate emissions in the range 0.2-25 mg/Nm^3. Which is again governed by the type of fuel used and the furnace design and the burner design/efficiency. Besides, polymer production and polymer dust handling can contribute but not very significantly. Nevertheless, the normal particulate emission is not of critical consequence for an Olefin Cracker Unit.

The NO_X generation depends on the flame temperature (which is a function of burner technology, fuel gas composition and furnace geometry) and quantity of excess air.

• *Emission from cracker furnace (de-coke operations)*

All cracking furnaces require periodic de-coking to remove carbon build-up on the radiant coils. The carbon layer acts as an insulator, and requires the use of higher tube metal temperatures to maintain the desired feed stock conversion. The furnace must be de-coked to restore its performance and the carbon is burned to carbon dioxide. Cycle times vary significantly for different feed stocks, coil configurations & the operating severity, but are typically in the range 14-100 days.

The particulate matter (SPM) and Carbon Monoxide (CO) are the critical pollutants emitted during decoking operation. Particulate emissions are generally observed in the range 80 – 600 mg/Nm^3, although the highest figures relate to peak values observed during dry de-dusting. De-coking emissions are not monitored by dedicated equipment (e.g. on-line analysers) since furnace de-coking is an infrequent operational mode (typically only 3% of the time). During the de-coking phase, process control is important to minimize particulate emissions and there is typically visual inspection of the emission point and close supervision of the process parameters (e.g. temperatures).

CO emissions are reported in the range 1-2700 mg/Nm^3 (as 30 – 60 minute averages at normal temperature & pressure, 3% oxygen, dry gas). The highest figures are encountered at the start of the operation.

• *VOC emission from cracking process*

During normal operation, there are very few VOC's (Volatile Organic Compounds) emissions from the cracking process because they are recycled into the process, used as a fuel or routed to associated processes on an integrated site. Higher VOC emission from ethylene plants are rather intermittent, but occur during plant start-up & shutdown, process upsets and emergencies. Volatile organic compounds may be emitted from pressure relief devices, intentional venting of off-specification materials or depressurising & purging of equipment for maintenance. Crack gas compressor & refrigeration compressor outages are potential sources of short-term, high rate VOC emissions. The chief source of benzene emissions during normal operation is the crack gas compressor lubricating oil vent. In general, intermittent emissions, all pressure relief devices, and emergency vents are routed to flare through the main process vent, which is usually controlled. The relief valve from the de-methaniser usually vents to atmosphere, but the valve is operated very infrequently and emits mainly hydrogen & methane. Volatile organic compounds from such sources as cracking furnace flue gas, pyrolysis furnace decoking, acid gas removal and hydrogenation catalyst regeneration are not generally significant.

In typical olefin plant, hydrocarbons are mostly emitted due to leakage and flaring of the residual gases. Generally, VOC emissions are attributed to 72% from leakage losses from appendages, pumps, etc., 18% from flaring & disruption, 1% from losses due to storage & handling, 5% from combustion and 4% from other process emissions.

• *Emission inventory from olefin complex*

The emission inventory presents the typical emission factor and the annual emission load expected from an olefin complex having a capacity of 500,000 TPA Ethylene is given below in Table.

Table: Defin Complex (0.5 MMTPA Ethylene) Typical Emission Invertory

S.N.	Parameter	Concentration at the source (mg/Nm^3)	Emission Factor, Kg/T Ethylene	Typical Emission Inventory (Annual Load in TPA)
1.	CO	25 – 50	0.5 – 1.0	250 – 500
2.	SOx	40 – 100	0.04 – 0.1	20 – 50
3.	NOx	200 – 250	2.0 – 2.5	1000 – 1250
4.	Flare Load (Normal) (Max)	– –	10 – 15 35 – 45	5000 – 7500 17500 – 22500
5.	Total VOC Load (uncontrolled)	–	3.5 – 5.0	1750 – 2500
6.	Fugitive VOC	–	3.2 – 3.5	160 – 1750

(c) Emission from Aromatic Plants

Emissions from aromatic plants are contributed to a large extent due to the utilities needed by the aromatic separation processes.

The principal pollutants and emission sources from various aromatic plants are given in Table below.

Table: Emission sources and pollutants from various aromatic plants

Source of emission from aromatic plants	Air pollutant				
	NO_x	SO_2	CO/CO_2	VOC's	Particulates
Gas Fired Heater	X	X	X	X	—
Point Source	X	X	X	X	X
Maintenance activities	—	—	—	X	—
Non Point emission (Fugitive)	—	—	—	X	—
Regeneration furnaces	X	X	X	X	—

X Indicates emission of pollutants; — indicates no emission of pollutants.

• *Volatile organic compounds (VOCs) from aromatic plants*

There are generally no continuous VOC emission point sources in aromatics plants, although some plants may use vacuum systems that have a continuous air emission. Most VOC emissions are normally from non-point i.e. fugitives (e.g. valve, flange & pump seal leaks) and from non-routine operations (maintenance & inspection). Due to lower operating temperatures and pressures, the fugitive emissions from some aromatics processes are considerably less than in other Petrochemical processes where higher temperatures & pressures are employed.

The typical emission factor for aromatics plant is given below.

Aromatic plant	Emission Source	Emission source type	Pollutant	Emission Factor (kg/T of BTX product)
Process A *(BTX from aromatic mixture-reformate)*	Flare	Point	SO_2 NO_x	0.53 0.86
	Tank	Point	VOC	0.05
	Various others	Fugitive	VOC Methane	0.15 0.09
Process B *(benzene from pygas i.e. Naphtha Cracker)*	Desulphurising	Point	SO_2	0
	Furnaces	Combustion	NOX CO VOC	0.013 0 0.0008
	Process	Point	VOC	0
		Fugitive	Benzene Toluene Pentane VOC saturated C_1-C_4 VOC aliphatics C_2-C_{10} Total aromatics Total VOC	0.010 0.004 0.004 0.0005 0.0018 0.017 0.030

A release of 50 t/year of hydrocarbons (including benzene) is a plausible order of magnitude for non-routine emissions from a typical aromatics installation with proper leak detection &

repair programme in place and also use of high integrity (i.e., inherently low leak) equipment. VOCs may arise from small leaks in the cooling unit as ethylene, propylene and/or propane can be used as coolant fluids in the p-xylene crystallization unit.

VOCs may also result from storage tank breathing losses and displacement of tanks for raw materials, intermediate products and final products. The VOCs can be aromatics (benzene, toluene), saturated aliphatics (C_1-C_4) or other aliphatics (C_2-C_{10}).

(D) Emission from Butadiene Plant

• *Process vent*

This process vent emission comprises:

- Hydrocarbons from extractive distillation to fuel gas system.
- Hydrocarbons from Butadiene purification (propynes) & solvent recovery vent (C_4 acetylene).

The C_4 vapours (raffinate) from extractive distillation steps contain primarily butanes and some quantity of butanes are either sold as product or taken to catalytic hydrogenation unit for blending in LPG.

The C4 vapours vented during solvent recovery (C_4 acetylene) and butadiene purification (propynes) are either flared or taken to fuel gas system.

• Fugitive Emissions

The fugitive emissions sources of butadiene plants include pumps, valves, flanges, sampling systems etc. and emissions during plant decommissioning for maintenance. The fugitive emission profile of two butadiene plants of India is given in Table below.

Table: Fugitive emissions of Butadiene plant

S.N.	Plant	Equipment	Fugitive Emission kg/yr
1	Unit-I	Pumps	1081.48
2	(Capacity 36000 TPA)	Valves	7008.41
3		Flanges	13246.08
4		Screwed fitting	367.97
5		PRV's	380.42
6		Open ended lines	138.58
	Total		**22222.94**
A	Unit-II	Pumps	1351.85
B	(capacity 11000 TPA)	Valves	9034.27
C		Flanges	9000.45
D		Screwed fitting	459.96
E		PRV's	475.52
F		Open Ended Lines	173.26
	Total		**20495.31**

Work place monitoring results recorded at two units are reported in Table below.

Table: Work Place Monitoring Data of butadiene plant

Component	Range of Values, ppm (Threshold limit value–time weighted average)	OSHA Standard, ppm
Unit-I	0 – 7.0	1
Unit-II	0 – 10.0	1

(E) Emission from Xylene Plants

The gaseous emissions occur from xylene plants as follows:

- Heating furnace stacks
- Process vents
- Fugitive emissions
- Storage emissions

Heating furnace stacks: Heating furnaces (in hydro-treating/reformers /transalkylation etc) normally use process off gas separated during separatiun/stabilisation processes and other lighters separated during distillation. In case of shortage of off gas and lighters other conventional fuels such as furnace oil/LSHS etc are used. The parameters of significance are as follows:

Gas fired furnace	Oil/Dual fired furnace
NO_x	SO_2
SO_2 (*)	NO_X
	TPM

(*) Sometimes H_2S generated in hydro-treating section is recycled to heating furnaces and hence some SO_2 gets liberated in the furnaces.

Process Vents: The vents opening directly into the atmosphere are Continuous Catalyst Regeneration.

Fugitive Emissions: Xylene have high boiling point and thus lower volatility as also health hazards are not severe.

Flue gases. are monitored in process heaters and continuous catalyst generation unit. The typical data is presented in Table below.

Table: Typical Monitoring of flue gases in Xylene plants

S.N.	Parameter	Process Heater			CCR (*)
		D_1	D_2	D_3	
1.	Temperature (°C)	305	246	296	—
	Fuel:	Process off gas			
2.	Flue gas quantity (Nm^3/hr)	2583	21091	13988	—
3.	SO_2 (mg/Nm^3)	42.66	17.38	14.22	—
4.	NOx (mg/Nm^3)	124.1	97.86	175.2	—
5.	Carbon Monoxide (mg/Nm^3)	—	—	—	19.5
6.	TPM (mg/Nm^3)	—	—	—	37.4

* (+) with low NOx burners

(*) CCR - Continuous catalyst Regeneration

(F) Emission from Ethylene Oxide (EO) and Ethylene Glycol (EG) Plant

The Process emissions from the EO/EG plant include:

- Re-absorber vent
- CO_2 Purge
- Inerts Vent
- Residual EO Absorber

Re-absorber Vent

A recycle gas compressor is generally provided to recycle re-absorber vent to oxidation reactor. In one unit, the vent has to be sent to atmosphere. Other plants have recycle gas compressor.

CO_2 Purge

The CO_2 purge is removed from the recycle gas in oxidation section to prevent CO_2 build-up. The CO_2 is absorbed in K_2CO_3 solution and off-gas sent back to oxidation reactor. The K_2CO_3 solution is stripped off and K_2CO_3 recycled back to absorber. The stripper vent is released to atmosphere.

New plants have an inbuilt flasher, which removes organics from the K_2CO_3 solution and recycles them to oxidation reactor before stripping off CO_2. This ensures that maximum amount of organics are re-used and reduces organics concentration in vent.

Inerts Vent

The inerts vent in nearly, all plants is taken to fuel gas system/flare tip (by separate line) and hence is not monitored.

Residual EO Absorber

This is provided to scrub EO from vent gases originating from various sources such as releases of Pressure Relief Valves/emissions during tanker loading/equipment depressurising for maintenance/pure EO distillation column etc.

The emission data in respect of process vent sources is presented in Table below. As observed, the installation of flasher has reduced volatile organic compounds in the vent at a new unit, drastically, as compared to that at old unit.

Table: Process Vent Analysis (EO/EG)

S.N.	Parameter	CO_2 Purge Stream		EO Re-absorber		Residual EO Absorber	
		Unit-II	Unit-III	Unit-I		Unit-IV	Unit-II
				Normal Operation	Abnormal Operation		
1.	Stack temperature (°C)	108	—	45	51	Ambient	Ambient
2.	Flow rate (m^3/hr)	3478	8196	4891	5054	—	—
	Pollutant Parameters						
3.	Carbon di oxide (% wt)	37.35	66.81	60.11	59.86	—	—
4.	Ethylene (mg/m^3) (% weitht)	15 (1.2)	0.25 (0.02)	162 (13)	187 (15)	—	—
5.	Ethylene oxide (mg/m^3) (% wt)	373.2 (00.10)	Traces	392.8 (200)	982.14 (500)	13.72	19.6
6.	Ethylene Glycol (mg/m^3) (% wt)	3875 (0.14)	2214 (0.08)	—	—	—	—

Fugitive emissions occur primarily from pump seals, compressors, valves, flanges, pressure relief valves and sample connections particularly because large quantities of volatile organic compounds are handled at high pressure. Fugitive emissicns can be estimated by applying average USEPA emission factors to the number of pumps, valves, flanges etc.

Inspection and Maintenance (I/M) programs in which equipment is routinely monitored & leaks corrected will substantially reduce these emissions.

Work room environmental monitoring

Monitoring of workroom atmosphere typically shows following ranges of Ethylene Oxide concentration.

Work Room Monitoring Results – Ethylene Oxide	
Plant	**Monitored Range (ppm)**
UNIT-I	0 - 3
UNIT-II	0 - 2

(G) Emissions from EDC (Ethylene Direct Chlorination) and VCM (Vinyl Chloride Monomer) Plants

These include (i) vent from direct chlorination section and (ii) vent from oxy-chlorination. These two vents along with other releases are scrubbed with caustic before discharging to atmosphere.

The vent from direct chlorination and oxy-chlorination are taken to low temperature chlorinator, where they are reacted again with chlorine to utilize maximum amount of ethylene. The vent from low temperature chlorinator are passed via refrigerated condenser, scrubbed and then discharged to atmosphere. This results in much reduced concentration of organics as can be seen in Table below.

Process emissions of EDC also result from the release of gases from the column vents. Column vents include vents from the wastewater steam stripper, the drying column, the heads column and the EDC/VCM finishing column.

Table: Process Vent Analysis Results - EDC/VCM

Stack Connected to	Caustic scrubber	
	Unit-I	Unit-III*
(a) Stack height (m)	21.5	50
(b) Stack Diameter (m)	1.36	1.7
(c) Temperature (°C)	40	40
(d) Velocity (m/s)	-	-
(e) Flow rate (m^3/hr)	4,183.69	1634.3
Pollutant Concentration		
(a) Oxygen (%)	1.85	1.5
(b) Chlorine (kg/hr)	Nil	Nil
(c) Ethylene g/m^3 (%)	150 (12.04)	309 (24.72)
(d) Ethyl Chloride mg/m^3 (%)	4422 (0.165)	14285 (0.05)
(e) Vinyl Chloride Monomer mg/m^3 (%)	650 (0.025)	48714 (1.76)
(f) Ethylene dichloride mg/m^3 (%)	20961 (0.51)	15026 (0.34)

Note: * Data obtained from manufacturer.

In addition to the above, modern EDC/VCM plants have a dry vent header (DVH) and wet vent header (WVH) to capture any releases of chlorinated hydrocarbons due to leaking equipment/during maintenance/process operation such as equipment depressurising and SRV/ PRV releaser. The DVH is provided to collect hydrocarbon emissions from process areas where moisture content is low (<20 ppm); this is normally the area up to EDC cracker. The WVH is provided to collect hydrocarbon emissions where appreciable moisture may be present; this is the area near oxy-chlorination and near EDC waste water stripper.

Stack connected to fuel burning sources

These include:

- Stack connected to EDC cracker furnace
- Stack connected to Incinerator

The fuel for the above equipment is the fuel gas generated in cracker/or the natural gas feedstock. The emissions from EDC cracker will mostly comprise Nitrogen Oxides, while Incinerator emissions will contain Nitrogen Oxide and other traces of chlorinated hydrocarbons. Possibility of dioxin generation in the Incineration cannot be ruled out considering high temperature and presence of chlorinated hydrocarbons.

The typical analysis results of two incinerators system have been given in Table below.

Table Typical Analysis of Incineration off-Gases

S.N.	Parameters	Analysis Results	
		Unit-IV	Unit-Ill (+)
1.	Temperature (°C)	20	Ambient
2.	Fuel	Off-gas	Off-gas
3.	Flow-rate (Nm^3/hr)	11,000	2,500
4.	Vinyl Chloride (ppmv)	Nil (+)	10
5.	Chlorine (mg/Nm^3)	3.2	9
6.	Nitrogen oxides (ppm)	125	150
7.	Hydrogen chloride (mg/Nm^3)	30.4	20
8.	Ethylene dichloride (ppm)	0.333	—
9.	Sulphur di oxide (mg/Nm^3)	17.2	40
	(+) Data provided by company.		

Fugitive Emission

Fugitive emissions of EDC/VCM and other volatile organics result from leaks through valves, pumps, compressors, and pressure relief valves.

The component emissions are computed from theUSEPA average emission factors and component inventory and typical results are presented in Table below.

Table: Estimation of Fugative Emissions from A Typical plant

S.N.	Equipment	Nos.	Fugitive emission, TPA
1.	Rotary compressor	2	3.097
2.	Centrifugal pump	68	9.195
3.	Valves	2176	59.57
4.	Flanges	9180	114.12
5.	Screwed fitting	252	3.087
6.	PRV's	68	3.233
7.	Open ended lines	102	1.177
		Total	**193.48**

As can be seen, the total fugitive emissions are estimated at 193.48 TPA from the said unit.

(H) Emission from acrylonitrile plant

Air emission may occur from several sources such as process vents, storages, transport & loading facilities, and incinerator/flare. Fugitive emissions occur from leaks in pumps compressors & valves. These emission sources are described below:

• *Process Vents*

Process emissions occur from the absorber vent, the distillation column purge vents and the Hot Alkali Digester (provided to reduce Cyanide in effluents to 1 ppm level). The emissions from absorber vent and Hot Alkali Digester vent during normal operation (@ capacity of 85 TPD) at one acrylonitrile plant is presented in Table below.

Table: Acrylonitrile Plant Process Vents Plant operating load : 85 TPD

S.N.	Parameters	Stack Connected to HCN Absorber	(Hot Alkali Digester) Vent
1.	Stack height (m)	65	67
2.	Stack diameter (m)	1.6	3.0
3.	Fuel used	—	—
4.	Fuel quantity (Kg/hr)	—	—
5.	Temperature (°C)	40-50	—
6.	Velocity (m/s)	2.7	—
7.	Flow rate (m^3/hr)	19,380	—
	Parameter concentration		
(a)	NOx ppm	345.0	—
(b)	NH_3, mg/Nm^3	—	7.20
(c)	HCN TPD (mg/m^3)	0.0079 (17.1)	—
(d)	SO_2, (mg/m^3)	—	—
(e)	Acetonitrile (mg/m^3)	0.26 (558.56)	—
(f)	Acrylonitrile (mg/m^3)	0.13 (276.64)	—

VOCs emission during start-up is substantially higher than during normal operation. During start-up the reactor is heated to operating temperature before the reactants (propylene & ammonia) are introduced. Thus, the reactor product stream is initially oxygen-rich. As the start-up progresses and the reactants are introduced, the acrylonitrile & VOC content of the reactor effluent increases until the acrylonitrile-rich composition indicative of normal operation is reached. During part of this start-up process, the composition of the reactor product stream is within its explosive limits and must, therefore, be vented to the atmosphere to prevent explosions in the lines to the absorber. Emission of acrylonitrile from a single reactor during startup has been reported to be as high as 4500 kg/hr. However, emissions associated with start-up occur rather infrequently, with each reactor having about four start-ups of 1-hour duration per year. Even so, acrylonitrile emissions during start-up totalled over an entire year are higher than emissions from the absorber vent during normal operation. In one unit, the HCN vapour stream is connected to both flare and incinerator, so that during start-up there is 100% standby to destroy the cyanides before discharging to atmosphere. The vent gases from the recovery, acetonitrile, light-ends, and product columns are flared.

• *Combustion Emissions*

A multi-purpose Incinerator is provided in the plant to treat following liquid/hazardous/ gaseous stream:

- Quench column bottoms (flow 1.5 m^3/hr) containing ACN/Hydrogen cyanide/Acetonitrile & Acroliene.
- Ammonium sulphate stream (flow: 2 - 5 m^3/hr) ammonium sulphate & aqueous polymers.
- Treating kettle bottoms.
- Hydrogen cyanide vapours (during start-up & when no off-take by downstream plants)

Gas (Hydrogen & methane) is used as fuel for incinerator. LSHS is used a standby fuel. The temperature is maintained at 1600°C. The analysis result for incinerator off-gases is presented in Table below.

• Fugitive Emission

The process operating pressure is generally low and hence fugitive emission are generally low. Fugitive emission sources include leaks from pumps, compressors and valves. Considering the high toxicity of HCN/Acrylonitrile, numbers of measures have been taken to control fugitive emissions. These measures are summarised in next table below. In addition to the above, online monitors are provided at strategic locations for ACN and Hydrogen Cyanide. However, there are still number of pumps/valves in organic liquid service. The fugitive emissions from these sources are computed using the average emission factors of USEPA and component inventory data are presented in Table below.

Table: Typical Acrylonitrile Plant Stack Emissions

		Plant Operating Load: 85 TPD
S.N.	**Parameters**	**Stack Connected to Wastewater Incinerator**
1.	Stack height (m)	52
2.	Stack diameter (m)	2.0
3.	Fuel used	Gas
4.	Fuel quantity (Kg/hr)	465
5.	Temperature (°C)	327
6.	Velocity (m/s)	22
7.	Flow rate (m^3/hr)	2,48,400
Parameter concentration		
(a)	NOx ppm	41.60
(b)	NH_3, mg/Nm^3	—
(c)	HCN, TPD (mg/cum)	0.023 (3.915)
(d)	SO_2, (mg/m^3)	31.2
(e)	Acetonitrile, TPD (mg/m^3)	Traces
(f)	Acrylonitrile, TPD (mg/m^3)	Traces

Table: Fugitive Emission Prevention Measure from Acrylonitrile Plant

S.N.	Equipment	Type	Service
1.	Pumps	Use seal less pumps; costs 3 times ordinary pump	HCN/ACN
2.	Valves	Inside Opening Stem Valves with Teflon Packing	HCN
3.	Flanges	Plane Faced with Teflon packing	—
4.	Compressors	Tandem Seal	—

Table: Component Fugitive Emission from a typical Acrylonitrile Plant

S.N.	Components	Nos.	Service	Emission Rate		
				(kg/hr)	TPD	TPA
1.	Valves	512	Light liquid Service (LL)	3.635	0.0872	28.72
2.	Pump Seals	16	LL	0.7904	0.019	6.26
3.	Flanges/Screwed fitting	2219	LL	1.842	0.044	14.58
4.	Open ended lines	24	LL	0.0408	0.00097	0.32
5.	Pressure Relief Valves	16	LL	1.664	0.04	13.17
			Total	**7.972**	**0.1911**	**63.06**

TPD – Tons Per Day, TPA – Tons Per Annum

(I) Emission from Phthalic Anhydride Plant

• *Combustion Source*

The heat evolved during the reaction is sufficient to meet the entire electric power and steam requirements of the unit. The only source of combustion is the process heater provided to heat the thermic fluid. This uses both Furnace oil & heavier fraction from PA purification as fuel.

• *Process Vents*

The only source of air pollution from phthalic anhydride plant are the gases leaving from switch condenser since the gases evolved have organic acids (which can be removed by water scrubbing), three stage scrubber (packed bed type) is provided on the vent.

Monitoring data for phthalic anhydride scrubber vent are provided in Table below As can be seen, the scrubber is useful to control the organic acid concentration, but is of no use to control carbon monoxide emitted during the oxidation stage.

• *Fugitive Emissions*

The principal raw material (o-xylene) has a high boiling point (144°C) and poses no special health hazards. Phthalic anhydride also has a high boiling point (295°C). Thus, no special precautions are taken except at oxidation reactor where the heat evolved requires stringent control to prevent leakages. Generally, o-xylene pumps in large units are single mechanical seal type.

Table: Typical Analysis results of Phthalic Anhydride Absorber Vent

S.N.	Parameters	Values
1.	Temperature	51°C
2.	Flow	56520 Nm^3/hr
3.	Total Organic Carbon (TOC)	99 mg/Nm^3
4.	Maleic acid	70 mg/Nm^3
5.	Citraconic acid	15 mg/Nm^3
6.	Benzoic acid	10 mg/Nm^3
7.	o-Toluic acid	2 mg/Nm^3
8.	o-Toluic anhydride	2 mg/Nm^3
9.	CO	5750 mg/Nm^3

(J) Emission from Dimethyl Terephthalate (DMT) Plant

Process Vents

The main source of air pollution is the off-gas from oxidation stage, which is passed through carbon adsorbers to reduce the VOC emissions. The off-gases are analysed before/after adsorber to check the organics content. The results of a typical analysis are presented in Table below:

Table: Typical Analysis results of Adsorber off-gas Analysis

Temperature	51°C
Flow	45000 Nm^3/hr
Methanol	437 mg/Nm^3
Xylene	308 mg/Nm^3
CO	5612.5 mg/Nm^3

Note: All results corrected to 3% oxygen

(K) Emission Due to Transfer Operations

The principal method of transferring liquid product to tank, trucks & railcars is submerged loading, including submerged fill pipe loading and bottom loading. In submerged fill pipe loading, the fill pipe enters the vessel from the top but extends almost to the bottom to the vessel such that the fill pipe opening is completely submerged in the liquid product. In bottom loading, the fill pipe enters the vessel from the bottom, so that the fill pipe opening is positioned below the liquid level. Both submerged loading techniques significantly reduce turbulence and liquid surface area resulting in low vapour generation.

Top splash loading, rarely used in facilities, is another loading technique in which the fill pipe enters the vessel through the top but extends below the surface of the liquid. This type of loading results in higher vapour generation.

The loading rack is the equipment used to transfer materials into tank trucks and railcars. The loading rack and transfer vehicle are emission points during loading operations. A typical loading rack consists of loading arms, pumps, meters, shutoff valves, relief valves, and other associated piping necessary to perform either loading or unloading operations.

• *Emissions during tanker loading*

Benzene is a 'high' risk chemical & has very high toxicity. A large variation is seen in tanker loading practices adopted at individual units.

Monitoring of Benzene levels during Tanker Loading:

Monitoring of Benzene levels is carried out for following situations:

- Work Area Monitoring - (No vapour collection/control device provided)
- Monitoring to find control system efficiency and to find workroom monitoring levels during loading with system ON.

(a) **Work area Monitoring:** (No vapour collection/control device provided)

Samples are collected and analysed using short-term dragger tubes & using tenax tubes which are later thermally desorbed and analysed on gas chromotograph (GC). Range of values observed are given below:

Table: Work Area Benzene Monitoring Levels near Tanker Loading Bay (Situation: No vapour collection/Control device provided)

S.N.	Location	Value Range (ppm)	Permissible Standards (ppm)	
			TWA	STEL
1.	Tanker Top (*) (Breathing level)	20 – 80	1	5
2.	Tanker West side at Ground level (+)	0.5 – 5	1	5
3.	Tanker East side at Ground level (+)	1 – 3	1	5

Note : (*) values vary with height from open manhole/wind speed & direction (+) values vary with wind direction & speed & nature of enclosure. [TWA - Time Weighted Average, STEL - Short - Term Exposure Limit]

(b) **Monitoring with vapour collection system ON:**

Samples are also collected at inlet/outlet of scrubber system after connecting the tanker to vapour collection system.

Table: Monitoring at Tanker loading Bay Emission Control System

S.N.	Location	Sample Concentration (kg/hr)			Ambient levels at tanker top (ppm)
		Inlet to Scrubber	Outlet to Scrubber	Efficiency (%)	
1.	Day 1	4.69	1.02	78.2	2 – 4
2.	Day 2	3.86	0.18	95.3	0 – 0.5

As can be seen levels at tanker top reduced appreciably after connecting to the vapour collection system. The variations in efficiency of control system maybe because no blower is provided to suck the benzene emissions and benzene vapours are taken to the scrubber only due to displacement.

(L) Storage Emission

• Storage emission from ethylene and propylene plant

Ethylene, propylene & butylene are gases at normal temperature & pressure, hence are stored under pressure in spheres. Generally, these are used within the complex for manufacture of polyethylene and/or other downstream chemicals or supplied to neighbouring units manufacturing downstream chemicals. The storage spheres and road tankers are vapour balanced as also vapour balancing of storage tanks is done with each other in order to ensure minimum losses. The nitrogen vent and pressure releases from the tanks are taken to flare system. Some plants have made efforts to recover boil-offs' during loading/unloading into spheres to the process stream by providing separate compressor. Flare system for storage are generally provided seperately from the process flare.

The naphtha and other liquid hydrocarbon intermediates are stored in fixed/floating roof tanks and are sources for emission losses.

• Storage emission from butadine plant

Butadiene undergoes dimidiation as storage temperature increases. Hence it is essential to store butadiene at a low temperature. Also, butadiene reacts readily with oxygen in air to form polymeric peroxides & therefore following precautions are taken:

- ❑ Antioxidants are added
- ❑ Stored under N_2 pressure

Hence butadiene is generally stored under pressurized condition at low temperatures under N_2 blanket in spheres and no losses are expected from the storages. The Nitrogen vent and pressure releases from the tanks are generally taken to the flare. The storage tanks vapour balance with each other and with the road tankers & hence no loading/unloading emissions are expected to occur.

Losses of hydrocarbon vapours may take place however from the solvent storage system.

• Storage emission from benzene plants

The emissions from storage tanks of benzene are computed using TANKS software of USEPA. The emissions from vertical fixed roof tank due to breathing/filling are much higher than that for a internal floating roof tank.

• Monitoring of emission from benzene storage tanks

Three types of samples are taken:

- Underground tank during filling/breathing.
- Above Ground tank during filling/breathing.
- Internal floating roof tank during filling/breathing.

Samples during tank filling were collected using specially designed sampling train.

Benzene emitted from above ground/under ground storage tanks during filling is in the range of 5 – 7 kg/hr, whereas that emitted from IFR is only about 0.001 kg/hr. Similarly, emission expressed per Ton of Benzene filled also shows similar trend.

Emissions during breathing are much lower than that during filling. There are practically no breathing emissions from Underground tanks/Internal floating roof tanks.

• *Storage emissions from EO (Ethylene Oxide)/EG (Ethylene Glycol)*

EO is a gas at ambient temperature and is stored under pressure or by using chilling arrangements in liquefied form. Safety Relief valves (SRV's) with upstream rupture discs are provided on storage tanks, which vent to a water scrubber during emergencies. Losses can occur in handling EO during transfer from storage tank to road tanker. It is seen that most of the companies have provided arrangement to vent displaced exhaust gases from road tanker (during filling of road tanker) & these are directed to a water scrubber. Additional handling losses during transfer from EO storage to road tanker correspond to possible hold up in the transfer pipe. It is seen that emissions are very low due to high boiling point/low volatility of the glycols.

• *Storage emissions from ethylene dichloride (EDC)*

Ethylene dichloride emissions result from the storage of EDC during the process and final product stages. In some plants, EDC is stored in vertical fixed roof tanks at ambient temperature and pressure. Pure EDC and wet EDC is stored in dome-roof tanks in some unit. In new plants all tanks of Vertical Fixed Roof type provide with a closed loop recycle system, where cooled EDC is fed to the educator along with vapour EDC from strong tanks for absorbing the vapour. Liquid EDC is collected in storage tank and recycled. Uncondensed vapours are diverted to incinerator. The total storage losses (working & breathing) for storages are calculated to be 36.45 TPA.

Vinyl Chloride monomer (VCM) however occurs as a gas and is liquefied & stored under pressure & Nitrogen blanket. Alternately, the tanks may be refrigerated at abmbient pressure. The spheres are pressure balanced with each other and with road tankers. Hence emissions from VCM storage can occur only from safety valve discharge, which is routed, to incinerator in some plants.

There is generally no chlorine storage in the EDC/VCM plants as it is provided by a associated chlor-alkali plant. Storage of liquefied dry HCl is generally in closed system pressurized vessels at low tempeature.

• *Storage Emissions from ACN (Acrylonitrile & Acetonitrile)*

Acrylonitrile and Acetonitrile are stored in Vertical Fixed Roof tanks under Nitrogen blanketing. Due to its high toxicity, HCN is not stored in the facility. Typically, total acrylonitrile emissions from storage are estimated at 57.78 TPA and total acetonitrile emissions from storage at 1.1017 TPA.

The acrylonitrile produced is sent to downstream acrylic fibre plant directly by pipeline and hence no major emission during tanker loading take place.

• *Storage emissions from xylene*

o-Xylene is generally stored in vertical fixed roof tanks provided with breather valves. Due to the high boiling point & low hazard potential of o-xylene, no special precautions are taken for storage.

• *Storage emissions from DMT (Dimethyl Terephthalate)*

To control emission from storages, the following precautionary steps are taken:

- Day tanks for oxidate storage & methanol storage are connected to scrubber.
- Fixed roof storage tanks are provided for methanol storage tank.
- Pressure relief valves are connected to a Methanol recovery column through a common header.

Generally two types of storage vessels are of concern in a petrochemical facility *i.e.* fixed-roof storage vessels (*i.e.* with no internal floating roof) and floating roof storage vessels. They are exclusively above ground and cylindrical in shape with the axis perpendicular to the foundation. There are also horizontal tanks; these are generally smaller and not as widely used.

• *Fixed-roof storage vessel*

A typical fixed-roof vessel is a cylindrical steel shell with a cone or dome-shaped roof permanently affixed to it. Losses from fixed roof tanks are caused by changes in temperature, pressure and liquid level. Fixed roof tanks are either freely vented or equipped with a pressure/vacuum vent. The latter allows the tank to operate at a slight internal pressure or vacuum to prevent the release of vapours during very small changes in temperature, pressure or liquid level. Hatches/sample wells, float gauges, and roof manholes on the fixed roof, which provide access to these tanks, also are potential but less significant sources of emissions.

Two significant types of emissions from fixed roof (FR) tanks are storage & working losses. Storage loss/standing loss is the expulsion of vapour frorn a tank through vapour expansion and contraction, which are the result of changes in temperature & pressure. This loss occurs without any liquid level change in the tank. The combined loss during filling and emptying is called working loss.

Fixed roof tank emissions vary as a function of vessel capacity, vapour pressure of the stored liquid, utilisation rate of the tank and atmospheric condition at the tank location.

Emission control techniques available for fixed roof tanks include:

— Installing an internal floating roof and seals to minimise evaporation of the product being stored (efficiency: 60-99%).

— Vapour balancing (vapours expelled during filling of storage tank are directed to the emptying tanker truck. The truck then transports the vapour to a centralized station equipped with a vapour recovery/control system. (Control efficiency: 90-98%).

— Vapour Recovery System which (collects emissions from storage vessels and converts them to liquid product. Techniques used include vapour/liquid absorption, vapour compression, vapour cooling, vapour/solid adsorption or combinations. (Control efficiency: 90-98%).

• *Floating roof storage vessel*

A floating roof vessel is a cylindrical steel shell equipped with a disk-shaped deck with a diameter slightly less than the inside tank diameter. The floating deck floats freely on the surface of the stored liquid, rising and falling with the liquid level. The liquid surface is completely covered by the floating deck, except in the small annular space between the deck and the shell. A rim seal attached to the floating deck slides against the vessel wall as the deck is raised or lowered, covering the annular space where the deck is not covering the liquid. The primary function of the rim seal is to prevent emissions.

• *External floating roof (EFR) vessel*

An EFR vessel does not have a fixed roof; instead, its floating deck is the only barrier between the stored liquid & the atmosphere. An EFR vessel may have several types of rim seals & deck fittings that penetrate the deck & serve operational functions. The external floating roof design is such that evaporation losses from the stored liquids are limited to losses from the rim seal system and deck fittings (standing storage loss) and any exposed liquid on the tank walls (withdrawal loss).

• *Internal floating roof (IFR) vessel*

An IFR vessel is equipped with a permanently affixed roof above the floating deck. There are two basic types of internal floating roof tanks (i) tanks in which fixed roof is supported by vertical columns (ii) tanks with self supporting fixed roof. The deck in internal floating roof tanks rises and falls with the liquid level and either floats directly on the liquid surface (contact deck) or rests on pontoons several inches above the liquid surface (no contact deck). Evaporative losses from floating roofs may come from deck fittings, non-welded deck seams, the annular space between the deck and tank walls. The tanks are generally freely vented by circulation vents at the top of fixed roof, to minimise formation of explosive mixture in the vapour space.

• *Emissions from floating roof tanks*

Total emissions from floating roof tanks are the sum of withdrawal losses and standing storage losses. Withdrawal losses occur as the liquid level and thus the floating roof is lowered. Some liquid remains on the inner tank wall surface and evaporates. For an internal floating roof tank that has a column-supported roof, some liquid also clings to the columns & evaporates. Evaporative losses occur until the tank is filled and the exposed surfaces are again covered. Standing storage losses from floating roof tanks include rim seal, deck fitting losses and for internal floating roof tanks also include deck seam losses for constructions other than welded decks.

Other potential standing loss mechanism includes breathing losses as a result of temperature & pressure changes from the rim seal system and at the deck fittings. The majority of rim seal losses are wind induced. The rim seal system may consist of just a primary seal or a primary & a secondary seal, which is mounted above the primary seal.

The primary seal serves as a vapour conservation device by closing the annular space between the floating deck & the wall. Three basic types of primary seals are used on external floating roof tanks: Mechanical (metallic) shoe, resilient filled (non metallic) & flexible wiper seals. The primary seals are sometimes protected by a weather shield, which protect the primary seal fabric from weather, debris and sunlight. A secondary seal may be used to provide some additional evaporative loss control over that achieved by the primary seal. Secondary seals can be either flexible wiper seals or resilient foiled seals. The secondary seals maybe placed in two configurations viz. rim mounted and shoe mounted. Rim mounted secondary seals are more effective in reducing losses than shoe mounted secondary seals as they cover the entire rim vapour space.

(M) Flares

Flaring is a volatile organic compound (VOC) combustion control process in which the volatile organic compounds are piped to a remote, usually elevated location and burned in a flame in the open air using a specially designed burner tip, auxiliary fuel, and steam or air to promote mixing for nearly complete destruction of VOC (> 98%). Completeness of combustion in a flare is governed by flame temperature, residence time in the combustion zone, turbulent mixing of the components to complete the oxidation reaction and available oxygen for free radical formation. Combustion is complete if all volatile organic compounds are converted to carbon dioxide and water. Incomplete combustion results in some of the volatile organic compounds being unaltered or converted to other organic compounds such as aldehydes or acids.

• *Flare from cracker furnace*

All crackers are provided with flare gas systems to allow safe disposal of any hydrocarbons that cannot be recovered in the process. During startup, in a continuous process, it becomes a need of the process to vent the hydrocarbon holdup till the plant stabilises giving desired output. Also, during emergencies which maybe due to pressure or any other reason, it becomes necessary to vent the hydrocarbons. Considering the flammable nature of the hydrocarbons all such releases (start up & emergencies) are taken to the flare. Shut down operations can be classified into two categories i.e. normal shut down and emergency shutdown. During emergency shutdown it maybe necessary to flare the contents of units taken up for maintenance. Similarly, storage tanks of gaseous hydrocarbons are cryogenic in nature and a secondary flare (low pressure) is provided to take care of any refrigerent released in emergencies. The primary role of flare system is therefore to ensure safety, by preventing build-up of flammable hydrocarbon mixtures in the manufacturing area.

Generally, two flares are provided:

— One elevated flare for the fugitive losses & shutdown – startup losses.

— One low pressure flare for the cryogenic storages.

Generally, the flare monitoring is done by closed circuit television (CCTV) system having its output in the cracker control room. Motorised valves are provided on steam lines to have better control on steam mixing rate. The plants have an automatic trip arrangement in case the flare is put off.

(N) Fugitive Emissions from Equipment Leak

Fugitive emission from equipment leak (especially VOCs) i.e. diffused emissions (compressors, pumps) and fitting (flanges, valves) play a major role in petrochemical plants.

The chief source of VOC emission during normal operation is the charge gas compressor lubricating oil vent. Fugitive emission may also occur from cooling water contaminated with process stream through cooling tower and volatile organic compounds from wastewater treatment. The following is a brief description of the types of equipment from which fugitive emissions are generated, where the emissions occur, and what equipment changes can be made to reduce or eliminate these emissions.

(a) Connectors

The most numerous component in a plant is the connector, which is used to connect piping to other piping or to equipment. Flanges are connectors consisting of gasket-sealed junctions that are used on pipe with a diameter of $\geq$ 5 cms. Flanges may leak because of improperly selected gaskets or poor assembly. Other types of connectors are generally used on smaller-diameter pipe. Threaded connections, which leak when cross-threaded, are an example. Another type of small pipe connector is a nut-and-ferrule connection, which leaks when poorly assembled. All types of connectors are subject to thermal deformation and may leak as a result. Connectors and their associated emissions can be eliminated in some cases through the use of welded joints.

(b) Valves

In equipment with moving parts, releases generally occur around the moving part. Packing, which is subject to degradation, is often used around such parts to form a seal between the process fluid and the atmosphere. Valves with bent or nicked stems cannot be relied on to perform reliably in terms of leaks, and in these cases at least the stem must be replaced. Valve packing technologies that use rings to keep the packing from extruding and springs to maintain the packing under constant pressure (and thus in constant touch with the stem) have been developed. These systems can reduce leak rates and operate without maintenance for 10-50 times as long as valves with conventional packing.

There are two main types of "seal less" or "leak less" valves that have no emissions through the stem. They are bellows valves, which are expensive and used mostly in the nuclear power industry, and diaphragm valves, which separate the valve stem from the process fluid through the use of a diaphragm. The diaphragm in some designs, serves as the flow control device in addition to forming a barrier between the stem and the process fluid. If a diaphragm fails, emissions result, and packing is sometimes used as a backup for the diaphragm. Repair of a faulty diaphragm cannot be made without removing the valve from service. It can be difficult to replace conventional valves with leak less technology, if there are significant spaces constrains.

(c) Pumps

As with valves, emissions from pumps occur largely around the moving parts; releases occur where the pump shaft meets the stationery casing. Packed seals are used, but well maintained mechanical seals generally leak less. However, mechanical seals are costly and time-consuming to repair, and sudden failure of a mechanical seal can result in large emissions. Because of this, mechanical seals are often backed up by either more mechanical seals or packed seals. When dual mechanical seals are used, a barrier fluid may be circulated between the seals to further reduce fugitive emissions. This barrier fluid must be treated to remove process fluid. Mechanical seals for pumps have improved greatly in the last few years arid are a viable alternative to leak less technology in a wide variety of applications.

Seal less designs for pumps include the canned motor pump, where the pump bearings run in the process fluid; and the diaphragm pump, where a flexible diaphragm is used to drive the process fluid. There are also magnetic drive pumps in which the impeller is driven by magnets.

(d) Compressors

Compressors are similar to pumps in that they generally have rotating or reciprocating shafts. Like pumps, they move process fluid, but it is in the form of a gas instead of a liquid. Again packed and mechanical seals are used, but the use of packed seals is largely restricted to reciprocating compressors. Mechanical seals for compressors are not necessarily of the contact design used for pumps. Restrictive carbon rings and labyrinth-type seals that are composed of interlocking teeth are also used. Another type of seal used in compressors is a liquid film seal, in which an oil film is placed between the rotating shaft and a stationery gland.

(e) Pressure Relief Devices

Pressure relief devices are used to prevent operating pressure from exceeding the maximum allowable limit of the equipment. One type of pressure relief device is a valve that opens when the operating pressure exceeds a certain limit and closes when levels are safe again. These valves can leak because they did not reseal properly or because the operating pressure is near their limit and they are "simmering" (popping open and closed). Another type of pressure relief device is the rupture disk, which is leak less under normal operation. A rupture disk bursts when the operating pressure exceeds its limit, allowing process fluid to escape until a new disk is installed. Rupture disks can be mounted upstream of pressure-relief valves to eliminate emissions from poorly seated valves. Careful equipment design and proper process operation corrects the problem of simmering relief valves. Some pressure-relief valves have an improved "soft" seat that seals better on reseating.

(f) Open Ended Valves/Lines

Another class of components from which fugitive emissions originate is open-ended valves and lines. Drain valves, purge valves, and vent valves fall into this category. Process fluids leak when the valves are in poor repair or not fully closed. A pipe plug, cap, or blind flange can be installed over the open end to prevent emissions, or a second valve can be installed.

(g) Sampling Systems

Sampling systems are used to verify that a process unit is operating properly. They must be purged before sampling in order to obtain a representative sample. If the purge stream is drained to the ground or sewer, it is an episodic fugitive emission. The purge stream can be eliminated by modifying the sampling system so that the purge stream is round back to the process; such sampling systems are called closed-loop sampling system.

• *Estimating Fugitive Emission*

The EPA indicates that there are five methods for estimating emissions from equipment leak from a chemical processing unit.

- Average emission factor methods.
- Leak /no leak emission factor method.
- Three strata emission factor method.
- Application of ERA correlations.
- Development of new correlations.

All methods start with obtaining an accurate identification and count of equipment to be included in the emission estimate. The equipment count can simply be used with the EPA'S emission factors as given in table below.

- Liquid are classified based on most volatile components present @ > 20% weight.
- If components have total Vapour Pressure ≥ 0.04 psi @ 20°C, the material (containing ≥ 20% VOC) is classified as light liquid.
- All above to be checked at process conditions and not ambient conditions.

The next step in complexity and refinement is the use of a portable organic vapour analyser to find the number of leaking and non-leaking sources. (A leaking source is one whose screening concentration is greater than or equal to 10,000 ppmv.) The data on number of leaking/non leaking components can be used with the leaking and non-leaking emission factors previously developed by the EPA, given in next table.

Table: Average Emission Factors for Fugitive Emissions as per USEPA*

S.N.	Equipment	Service	Emission factor (kg/hr/source)
1.	Valves	Gas Light Liquid Heavy Liquid	0.0056 0.0071 0.00023
2.	Pump Seals	Light Liquid Heavy Liquid	0.0494 0.0214
3.	Compressor Seals	Gas/Vapour	0.228
4.	Pressure Relief Valves	Gas/Vapour	0.104
5.	Flanges	All	0.00083
6.	Open-Ended Lines	All	0.0071
7.	Sampling Connections	All	0.0150

* USEPA - United States Environmental Protection Agency

Table: Leaking and Non-Leaking Emission Factors for Fugitive Emissions (kg/hr/source)

S.N.	Equipment	Service	Leaking (≥10,000 ppm) Emission Factor	Non-leaking (<10,000 ppm) Emission Factor
1.	Valves	Gas	0.0451	0.00048
		LL[a]	0.0852	0.00171
		HL[b]	0.00023[c]	0.00023
2.	Pump Seals	LL	0.437	0.0120
		HL	0.3885	0.0135
3.	Compressor Seals[d]	Gas	1.608	0.0894
4.	Pressure Relief Valves	Gas	1.691	0.0447
5.	Flanges	All	0.0375	0.00006
6.	Open-Ended Lines	All	0.01195	0.00150

Note:

LL[a] : Light liquid service.

HL[b] : Heavy liquid service.

0.00023[c] : Leaking emission factors assumed equal to non-leaking emission factor since computed leaking emission factor (0.00005 kg/hr/source) was less than non-leaking emission factor.

Compressor Seals[d] : Emission factor reflects existing control level of 60 percent found in the industry; control is through the use of barrier fluid/degassing reservoir/vent-to-flare or other seal leakage capture system.

A final refinement is a method employing discrete emission factor that is applying emission factors to represent three different ranges of screening values. This has been called the stratified emission factor approach or the three strata approach. The stratified emission factors for equipment leaks are presented in Table below.

Table: Stratified Emission Factors for Equipment Leaks (Kg/Hr/Source) as per USEPA

S.N.	Sources of leakage	Service	Emission Factors (kg/hr/source) for Screening Value Ranges, ppmv		
			0-1,000	1,001-10,000	Over 10,000
1.	Compressor seals	Gas/vapour	0.01132	0.264	1.608
2.	Pump seals	Light liquid	0.00198	0.0335	0.437
		Heavy liquid	0.00380	0.0926	0.3885
3.	Valves	Gas/vapour	0.00014	0.0165	0.0451
		Light liquid	0.00028	0.00963	0.0852
		Heavy liquid	0.00023	0.00023	0.00023
4.	Flanges, connections	All	0.00002	0.00875	0.0375
5.	Pressure relief devices	Gas/vapour	0.0114	0.0279	1.691
6.	Open-ended lines	All	0.00013	0.00876	0.01195

USEPA - United States Environmental Protection Agency

Applying the stratified emission factors requires more rigorous measurement of organic vapour concentrations with a portable instrument because actual concentration reading must be recorded instead of noting whether a piece of equipment is classified as leaking or not leaking.

The remaining two methods make use of correlation equations relating mass emission to organic concentrations measured with a portable organic analyser. This is used in cases where the mass emission rates are statistically different from those represented by the EPA's emission factors; correlations are then developed specifically for that process unit.

GLOSSARY OF TERMS RELATED TO PETROCHEMICAL ENGINEERING

Appendix

E

Acetylene: Chemical compound with the formula C_2H_2. It is hydrocarbon that is a colourless gas, widely used as a fuel and chemical building block.

Acetyl: A functional group with the formula CH_3CO.

Acrylic fibres: Fibres where the major raw material is acrylonitrile, a derivative of propylene.

Aliphatic: Hydrocarbons characterised by a straight or branched chain. They do not have a ring structure. The simplest aliphatic is methane.

Alkane: Any of a group of hydrocarbons that have carbon atoms in changes linked by single bonds. They can be gaseous, liquid or solid and have the general chemical formula C_nH_{2n+2}. Alkanes occur naturally in petroleum and natural gas. They include methane, propane and butane.

Ammonia: A pungent, colourless gas with the formula NH_3, used to manufacture fertilisers and a range of nitrogen-containing organic & inorganic chemicals.

Aromatics: Primary petrochemicals which include benzene, toluene and xylenes. Aromatics are very stable and have multiple double bonds and a six-carbon ring structure.

Asphalt: A brown-black solid or semi-solid mixture of bitumens which are commonly used in roofing, paving & waterproofing. Derived from either native deposits or as a petroleum by-product.

Benzene: (C_6H_6) A colourless, volatile, flammable liquid used as a base structure to which different atoms and molecular structures can be attached. Common uses include medicine, crop protection chemicals and as a solvent and component in automotive fuels.

BTX: A collective abbreviation for benzene, toluene and xylenes.

Butadiene: A colourless, flammable hydrocarbon obtained from petroleum with the chemical formula, C_4H_6. Often used to make synthetic rubber.

Butane: Either of two isomers of a gaseous hydrocarbon with the chemical formula C_4H_{10}. It is produced synthetically from petroleum. Its uses include household fuel, as a refrigerant, aerosol propellant and in the manufacture of synthetic rubber.

Butylene: A colourless, flammable, liquid gas with a detectable odour. Butylenes have a chemical formula of C_4H_8 and are formed during the cracking of petroleum fractions. Butylene is used in the production of high-octane gasoline, butyl alcohols and synthetic rubber.

Catalytic conversion: The catalytic (of or relating to a catalyst) oxidation of carbon monoxide and hydrocarbons, especially in automotive exhaust gas to carbon dioxide and water.

Catalytic reforming: Catalytic reforming is the chemical process which is used to convert low-octane petroleum refinery naphthas into high-octane liquid products. These products are called reformates and they are components of high-octane petrol.

Coking: The process of deriving petroleum coke, a carbonaceous solid, from petroleum using oil refinery coker units or other cracking processes.

Cracking: The breaking down of large molecules as part of the refining process.

Dehydrogenation: Any process that involves the removal of hydrogen. Frequently used in catalytic reforming processes to convert non-aromatic hydrocarbons to aromatic. Cyclisation – the formation of one or more rings in a hydrocarbon – is also part of the catalytic reforming process.

Derivative products: Petrochemical derivative products which can be made in a number of different ways: via intermediates which still contain only carbon and hydrogen; through intermediates that incorporate chlorine, nitrogen or oxygen in the finished derivative. Some derivatives are finished products while further steps are required for others to arrive at the desired composition.

Detergent: A cleansing agent, especially a surface-active chemical such as an alkyl benzene-sulphonate (ABS).

Distillation: A method of physically separating mixtures in a boiling liquid mixture.

Dye: Substance, either natural or chemical, used to colour materials.

Elastomer: A material that can resume its original shape when a deforming force is removed, such as natural or synthetic rubber.

Ethane: A colourless, odourless flammable gaseous alkane with the formula, C_2H_6. It is used as a fuel and also in the manufacture of organic chemicals.

Ethylene: A colourless, flammable gas containing only two carbons that are double-bonded to one another. It is an olefin that is used extensively in chemical synthesis and to make many different plastics, such as plastic used for water bottles.

Feedstock: The raw material that is needed for some industrial processes. In relation to petrochemicals, feedstocks derived from petroleum are mainly used for the manufacture of chemicals, synthetic rubber and a variety of plastics.

Fraction: A component of a mixture that has been separated by a fractional process, such as fractional distillation.

Fractional distillation: The process of separating the different parts of a liquid mixture by heating it and separately condensing the parts according to their boiling points.

Hydrocarbon: A broad term that refers to organic chemicals that are characterised by various carbon and hydrogen molecular structures.

Hydrorefining: A refining process for treating petroleum in the presence of catalysts and substantial quantities of hydrogen. The process includes desulfurisation and the removal of substances that deactivate catlysts (such as nitrogen compounds). The process is used in the conversion of olefins to paraffins to reduce gum formation in gasoline and in other processes to upgrade the quality of a fraction.

Hydrogen: A flammable, colourless gas with the chemical symbol H and H_2 as a molecule of gas. It is the lightest and most abundant element in the universe. As petrochemicals are produced from hydrogen-containing hydrocarbons, hydrogen is involved in nearly all petrochemical processes. The most common application of hydrogen is as a reducing agent in catalytic hydrogenation and hydrorefining.

Inorganic: The class of chemicals that does not contain carbons and hydrogens bound together like those in organic compounds. These typically exist as salts, acids and alkaline, as well as some gases and elemental compounds. An inorganic petrochemical is one that does not contain carbon atoms, ammonia being the most common.

Intermediates: Petrochemical intermediates are generally produced by chemical conversion of primary petrochemicals to form more complicated derivative products. Common petrochemical intermediate products include vinyl acetate for paint, paper and textile coatings, vinyl chloride for polyvinyl chloride (PVC) resin manufacturing, ethylene glycol for polyester textile fibres and styrene which is used in rubber and plastic manufacturing.

Isobutylene: A four-carbon branched olefin, one of the four isomers of butane, with the chemical formula C_4H_8.

Isomer: Any of two or more substances that are composed of the same elements in the same pro-portions but have differing properties because of differences in how the atoms are arranged.

Isomerisation: The conversion of a compound into an isomer of itself. Different processes can be used to achieve isomerisation. For example, in petroleum refining, there are two processes commonly used: a process that uses an aluminium chloride catalyst plus hydrogen chloride; and a process where feedstock is mixed with organic chloride and recycled hydrogen, heated & transferred to a reactor.

Methane: A colourless, odourless, flammable gas with the chemical formula CH_4. It is the simplest alkane and the main constituent of natural gas.

Methanol: A colourless, volatile, poisonous liquid compound with the chemical formula CH_3OH. It is commonly used as a solvent and a fuel. Also known as methyl alcohol or wood alcohol.

Monomer: A molecule that can combine with other molecules of the same kind to form a polymer.

Naphtha: The collective name for any of the highly volatile flammable liquid mixtures of hydrocarbons which are distilled from petroleum, coal tar and natural gas. Naphtha mixtures are commonly used as solvents, fuel and to make various chemicals.

Natural gas: A fossil fuel. Natural gas is a mixture of naturally occurring hydrocarbon gases and it is primarily used a fuel and for making organic compounds. Deposits are found beneath the Earth's surface. Methane is the primary component of natural gas but it also contains varying quantities of ethane, propane, butane and nitrogen.

Olefins: Primary petrochemicals which include ethylene, propylene and butadiene. These are unsaturated molecules of carbon and hydrogen that appear as short chains of two, three or four carbons in length. Olefins are produced by steam-cracking of natural gas liquids. Also known as alkenes.

Organic: Chemical compounds that contain carbon atoms bonded to other carbon atoms, hydrogen atoms or other substitutes for hydrogen, such as halogens, sulphur and nitrogen.

Paraffin: An alkane in liquid or wax form. It consists mainly of alkane hydrocarbons with boiling points in the range of 150-300°C Used as an aircraft fuel, in domestic heaters and as a solvent.

Petrochemicals: Any substance obtained from petroleum or natural gas.

Petroleum: A thick, flammable mixture of gaseous liquid and solid hydrocarbons occurring naturally beneath the Earth's surface. The origins of petroleum are believed to come from accumulated remains of fossilised plants and animals. Petroleum can be separated into fractions including natural gas, gasoline, lubricating oils, naptha, kerosense, paraffin wax and asphalt. It can also be used as raw material for a range of derivative products.

Petroleum coke: Often abbreviated as "petcoke"or"pet coke", this is the carbonaceous solid derived from petroleum during a refining process.

Petroleum oil: Any light hydrocarbon oil that is a distillate of petroleum. It has a wide range of uses including cosmetics, preservatives, medicines, cleaning and industrial lubrication. Also known as mineral oil.

Plastic: A broad term describing any of a large number of synthetic (usually organic) materials that have a polymeric structure and can be moulded while soft and then set. In its finished state, a plastic might contain plasticiser, stabiliser, filler or pigments.

Plasticiser: A substance added to plastic to make it more pliable.

Polymer: A compound, either naturally occurring or synthetic, that has large molecules made up of many relatively simple repeated units. All plastics are examples of polymers. A common example of a synthetic polymer is Perspex. Naturally occurring polymers include DNA, tortoise shell and cellulose in trees. Major polymer products include PVC and polystyrene (derived from ethylene) and polypropylene (derived from propylene).

Polymerisation: The chemical process that combines several monomers to form a polymer or polymeric compound.

Propane: A colourless gas which is found in natural gas and petroleum. Widely used as fuel. Its chemical formula is C_3H_8.

Propylene: (C_3H_6) A three-carbon, flammable gaseous molecule containing a double-bond. Propylene is an olefin frequently used in organic synthesis. It is a base chemical used to make polypropylene fibres which are commonly used in carpets and clothing.

Refinery: An industrial plant where a crude substance, such as crude oil, natural gas or coal, is purified so it can then be turned into more useful products.

Solvent: A substance, usually liquid, which is dissolved with another substance to form a solution or is capable of dissolving another substance.

Solvent extraction: A method used to separate compounds based on their relative solubilities in two different immiscible liquids, usually water and an organic solvent. Also known as "liquid-liquid extraction" and "partitioning".

Steam cracking: The high-temperature, cracking of petroleum hydrocarbons in the presence of steam as part of the petroleum-refining process. Steam cracking is the main industrial method used for producing olefins. Also referred to as steam-assisted thermal cracking.

Sulfur recovery: The process whereby gas is desulfurised. Sulfur can be recovered from raw natural gas and by-product gases containing hydrogen sulphide derived from refining crude oil and other industrial processes. The Claus process is the most significant gas desulfurisation process. It involves a series of high-temperature reheating processes.

Surfactant: A surface-active agent, the basic cleaning agent of petrochemistry.

Synthesis gas: A mixture of carbon monoxide and hydrogen. This is mainly used in chemical synthesis to make hydrocarbons.

Synthetic: Produced by synthesis, not of natural origin.

Synthetic fibres: Manufactured fibres, the most common of which are polyester, a combination of ethylene glycol and terephtalic acid (made from xylene); and nylon, which has benzene as its most important raw material.

Thermal cracking: A petroleum refining process that decomposes, rearranges or combines hydrocarbon molecules using the application of heat, without the use of catalysts.

Thermoplastic: A synthetic plastic or resin, such as polystyrene, that becomes soft when heated and re-hardens on cooling without any appreciable change of properties.

Thermosetting plastic: A plastic that hardens permanently after one application of heat and pressure. Once hardened, a thermosetting plastic cannot be remoulded.

Toluene: A liquid aromatic hydrocarbon with a benzene-like structure. However, it is less flammable, toxic and volatile than benzene. It is used in organic synthesis as a solvent and an anti-knock agent in gasoline.

Wax: A natural, oily or greasy substance consisting of hydrocarbons or esters of fatty acids that are insoluble in water but soluble in non-polar organic solvents, such as benzene.

Xylene: A major aromatic feedstock usually obtained from petroleum or natural gas distillates. Used in the manufacture of plastics and synthetic fibres as a solvent, and in the blending of gasoline.

Appendix

F

BIBLIOGRAPHY/REFERENCES/ FURTHER READING

1. R.N. Shreev, Chemical Process Industries, McGraw Hill, Newyork.
2. Introduction to petrochemicals, Prof. S. Maiti.
3. W.L. Nelson, Petroleum Refinning Engineering, McGraw Hill, Newyork.
4. C.E. Dryden, Outlines of Chemical Technology, Affiliated East-West Press, New Delhi.
5. Ullman F., Encyclopedia of Industrial Chemistry, Wiley-VCH.
6. Kirk R.E, Othmer D.F; Encyclopedia of Chemical Technology, John Wiley and Sons.
7. Lecture notes from IPCL, Vadodara.
8. Lecture notes from IOCL.
9. Lecture notes from M/S Reliance Industries.

OBJECTIVE TYPE QUESTIONS AND ANSWERS

Appendix

G

1. Petrochemicals are chemicals
(a) obtained from petroleum and natural gas and made on large scale
(b) used for tertiary oil recovery
(c) added to crude petroleum for processing before it is fed to the distillation column for refining
(d) added to gasoline and other fuels in order to improve their properties

2. Dry natural gas contains mainly
(a) methane (plus small amount other components)
(b) ethane
(c) propane
(d) butane

3. Wet natural gas contains substantial amount of
(a) ethane (b) propane
(c) butane (d) all of the above

4. In petrochemical industry, any petroleum fraction having an approximate boiling point range between 20 and 200 ^{0}C which is used as a feedstock is called
(a) kerosene (b) gas oil
(c) naphtha
(d) none of the foregoing

5. Naphtha having an approximate boiling point range of 20 to 200°C contain compounds having ———— carbon atoms.
(a) 1 to 6 (b) 4 to 12
(c) 8 to 16 (d) 16 and higher

6. In petrochemical industry, thermal cracking or steam cracking process is used to produce mainly
(a) ethane (b) ethylene
(c) propane (d) Propylene

7. The steam reforming process employed in petrochemical industry produces mainly
(a) benzene, toluene and xylene
(b) n-butenes, butadiene and styrene
(c) carbon monoxide and hydrogen
(d) carbon dioxide and ethylene

8. Which of the following processes is employed in petrochemical industry to produce benzene, toluene and xylene (BTX)?
(a) thermal cracking,
(b) catalytic reforming,
(c) steam reforming,
(d) catalytic cracking,

9. Main feed stocks used for the manufacture of ethylene by thermal cracking process are
(a) ethane, propane
(b) naphtha and gas oil
(c) butane to a minor extent,
(d) all of the above

ANSWER

1. (a) **2.** (a) **3.** (d) **4.** (c) **5.** (b)
6. (b) **7.** (c) **8.** (b) **9.** (d)

10. In petrochemical industry thermal cracking of hydrocarbons to produce ethylene is accomplished in cracking furnace and the feed to the cracking furnace is a mixture of
(a) hydrocarbon and nitrogen which acts as an inert diluent
(b) hydrocarbon and carbon dioxide which acts as an inert diluent
(c) hydrocarbon and steam which acts as an inert diluent
(d) hydrocarbon and air which acts as an inert diluent

11. Under otherwise uniform conditions yield of ethylene from a thermal cracker will be
(a) lower than that obtained from a catalytic cracker
(b) equal than that obtained from a catalytic cracker
(c) higher than that obtained from a catalytic cracker
(d) none of the above

12. Under otherwise uniform conditions, branched chain alkanes and cycloalkanes, on thermal cracking, will yield
(a) higher amount of ethylene compared to the straight chain alkanes
(b) almost the same amount of ethylene compared to the straight chain alkanes
(c) lower amount of ethylene compared to the straight chain alkanes
(d) none of these

13. The most important outlet for ethylene is the manufacture of
(a) ethylene oxide (b) styrene
(c) vinyl chloride (d) polyethylene

14. Which of the following polymers of ethylene was introduced into the market in the late 1970s?
(a) low density polyethylene (LDPE)
(b) high density polyethylene (HDPE)
(c) linear low density polyethylene (LLDPE)
(d) none of the above

15. LDPE is manufactured by free radical polymerization of ethylene : reaction temperatures are in the range of 80 to 300^{0}C and pressures are in the range of 1000 to 3000 atm. Initiator used in the polymerization reaction is
(a) dodecyl mercaptans
(b) ethylene oxide or fluorine
(c) oxygen or an organic peroxide
(d) none of these

16. Linear low density polyethylene is made by co-polymerizing ethylene with small amount of
(a) propylene (b) isobutene
(c) isoamylene
(d) 1-alkenes such as 1-butene or 1-hexene

17. High density polyethylene (HDPE) is produced by a process
(a) similar to that used for the production of LDPE
(b) Called co-ordination or Ziegler polymerization
(c) called Clans process
(d) called Halcon process

18. Catalyst systems introduced by Ziegler consist of
(a) HF – BF_3 combination
(b) HF – $AlCl_3$ mixture
(c) complexes formed from aluminum

ANSWER

10. (c)	**11.** (c)	**12.** (c)	**13.** (d)	**14.** (a)
15. (c)	**16.** (d)	**17.** (b)	**18.** (c)	

alkyls and transition metal halides, typically $(C_2H_5)_3$ Al and $TiCl_4$

(d) none of these

19. Which of the following ethylene polymers is recommended for making buckets & washing up bowls?

(a) LDPE (b) HDPE

(c) Linear LDPE (d) none of these

20. Ethylene oxide is produced by oxidation of ethylene by air or oxygen at temperatures in the range of 250 to 330°C in presence of a catalyst which is

(a) silver on a solid support such as alumina or silicon carbide

(b) copper on r – alumina

(c) finely divided nickel

(d) platinum on carbon

21. The most important use of ethylene oxide (EO) is in the manufacture of

(a) Ethanol amines by reaction of EO with NH_3.

(b) Ethylene glycol by reaction of EO with water.

(c) Non-ionic surface-active agents by reaction of EO with alcohols, phenols, and/or amines,

(d) None of the above

22. The major uses of ethylene glycol are

(a) as antifreeze agents in automobiles & as a raw material in the manufacture of poly (ethyleneterephthalate)

(b) as a plasticizer

(c) as an antioxidant

(d) all of the above

23. Raw materials for the production of styrene are

(a) benzene and ethylene

(b) toluene and propylene

(c) toluene and ethanol

(d) m-xylene and ethylene

24. Since, styrene readily polymerizes; now in order to inhibit polymerization of styrene which of the following chemicals is usually added to it and at what concentration level?

(a) dioctylphthalate; 50 ppm

(b) p-tert-butylcatechol; 10 ppm

(c) acetic acid; 5 ppm

(d) hydroquinone; 100 ppm

25. Ethylation of benzene (to produce ethylbenzene) is conducted either in liquid phase or in vapor phase. The liquid phase process is conducted at temperatures ranging from about 80 to 130^0C and pressures from normal atmospheric to about 9 atm. The catalyst employed in the liquid phase ethylation of benzene is

(a) zeolite

(b) solid phosphoric acid

(c) super-acidic Nafion NR 50

(d) aluminum chloride promoted by either hydrogen chloride or ethylchloride

26. The gas phase process for ethylation of benzene, called Mobil – Badger process, carried out at about 420°C and 12 to 20 atm pressure employs as catalyst

(a) r – alumina

(b) macroporous cation exchange resin

(c) zeolites

(d) dispersed platinum on a porous carbon support

27. The main problem in the liquid phase process for ethylation of benzene is that

(a) the yield of ethyl-benzene is very low

(b) the quality of the product is poor

(c) the reaction mixture is very

ANSWER

19. (b) **20.** (a) **21.** (b) **22.** (a) **23.** (a)

24. (c) **25.** (d) **26.** (c) **27.** (c)

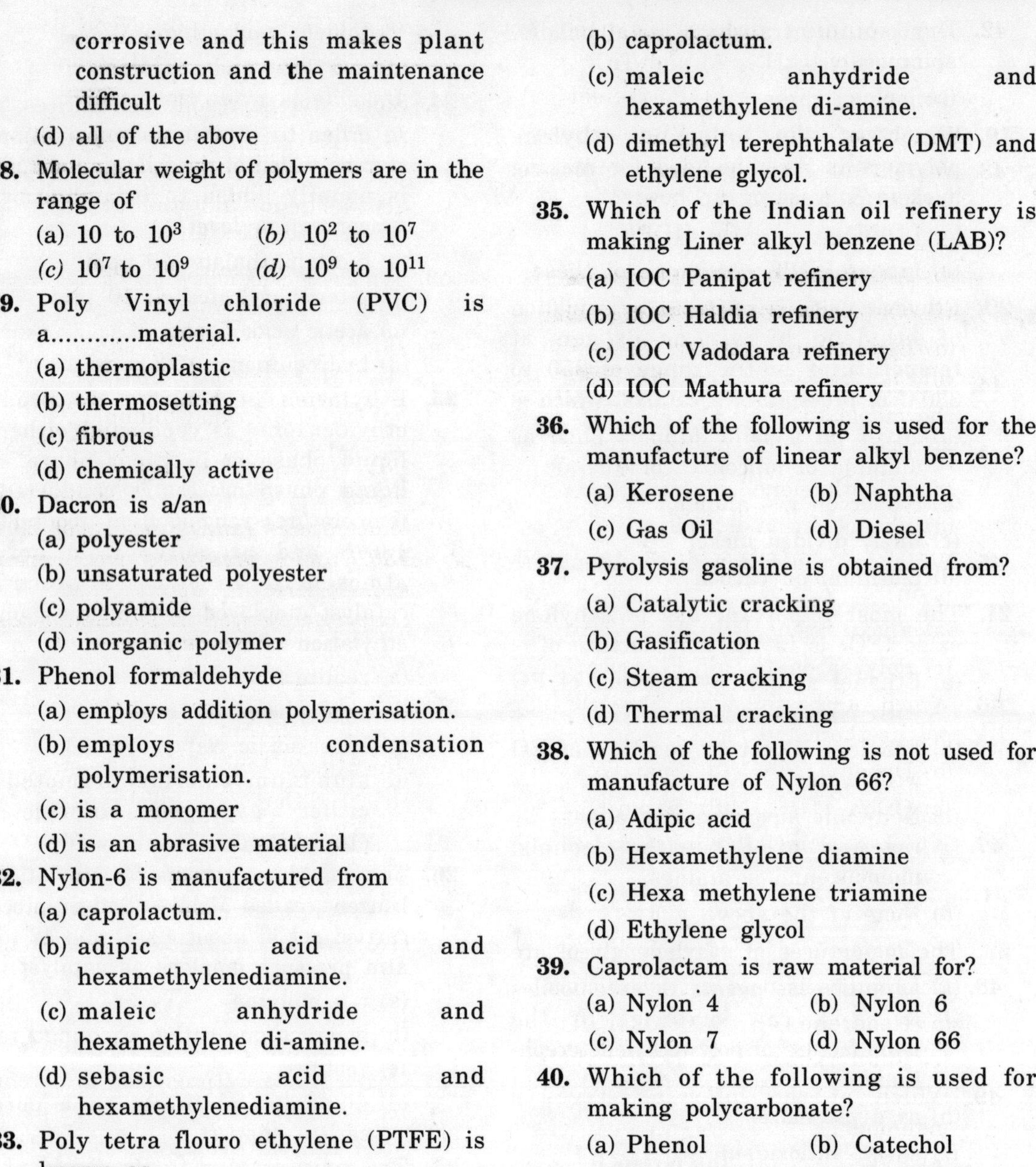

corrosive and this makes plant construction and the maintenance difficult

(d) all of the above

28. Molecular weight of polymers are in the range of

(a) 10 to 10^3 *(b)* 10^2 to 10^7

(c) 10^7 to 10^9 *(d)* 10^9 to 10^{11}

29. Poly Vinyl chloride (PVC) is a...........material.

(a) thermoplastic

(b) thermosetting

(c) fibrous

(d) chemically active

30. Dacron is a/an

(a) polyester

(b) unsaturated polyester

(c) polyamide

(d) inorganic polymer

31. Phenol formaldehyde

(a) employs addition polymerisation.

(b) employs condensation polymerisation.

(c) is a monomer

(d) is an abrasive material

32. Nylon-6 is manufactured from

(a) caprolactum.

(b) adipic acid and hexamethylenediamine.

(c) maleic anhydride and hexamethylene di-amine.

(d) sebasic acid and hexamethylenediamine.

33. Poly tetra flouro ethylene (PTFE) is known as

(a) dacron (b) teflon

(c) bakelite *(d)* celluloid

34. Nylon-66 is manufactured from

(a) adipic acid and hexamenthylene diamine.

(b) caprolactum.

(c) maleic anhydride and hexamethylene di-amine.

(d) dimethyl terephthalate (DMT) and ethylene glycol.

35. Which of the Indian oil refinery is making Liner alkyl benzene (LAB)?

(a) IOC Panipat refinery

(b) IOC Haldia refinery

(c) IOC Vadodara refinery

(d) IOC Mathura refinery

36. Which of the following is used for the manufacture of linear alkyl benzene?

(a) Kerosene (b) Naphtha

(c) Gas Oil (d) Diesel

37. Pyrolysis gasoline is obtained from?

(a) Catalytic cracking

(b) Gasification

(c) Steam cracking

(d) Thermal cracking

38. Which of the following is not used for manufacture of Nylon 66?

(a) Adipic acid

(b) Hexamethylene diamine

(c) Hexa methylene triamine

(d) Ethylene glycol

39. Caprolactam is raw material for?

(a) Nylon 4 (b) Nylon 6

(c) Nylon 7 (d) Nylon 66

40. Which of the following is used for making polycarbonate?

(a) Phenol (b) Catechol

(c) Resorcinol (d) Bis Phenol

41. Which of the following is used for manufacture of azo dyes?

(a) Beta Naphthol (b) Aniline

(c) Tobias Acid (d) Anthaquinone

ANSWER

28. (b)	**29.** (a)	**30.** (a)	**31.** (b)	**32.** (a)
33. (b)	**34.** (a)	**35.** (c)	**36.** (a)	**37.** (c)
38. (d)	**39.** (b)	**40.** (d)	**41.** (a)	

42. Dry spinning process is used for spinning of
(a) Nylon (b) Polysetser
(c) Acrylic fibre (d) Viscose rayon

43. Which of the refinery produces terephthalic acid?
(a) Jam Nagar refinery
(b) IOC Panipat refinery
(c) IOC Vadodara refinery
(d) BPCL Guna refinery

44. Which of the following is not thermoplastic?
(a) Polyester (b) PVC
(c) Polyethyelene
(d) Phenol formaldehyde

45. Which of the following is thermoset resin?
(a) Epoxy resin (b) PET resin
(c) Polycarbonate (d) Fluorocarbon

46. Which of the following fibre contains polyurethane?
(a) Dynel (b) Spandex
(c) Orlon (d) Saran

47. Which of the following involve poly condensation?
(a) Acrylic (b) Polyethylene
(c) Polyester (d) Nylon

48. Chloroprene is used in manufacture of?
(a) Neoprene (b) Thikol
(c) Butyl rubber d. Nitrile rubber

49. Which of the following have co-monomer?
(a) PVC (b) Nylon 6
(c) Nylon 66 (d) Polybutadiene

50. Which of the following is made from hexamethylene diamine and adipic acid?
(a) Nylon 66 (b) Nyon 6
(c) Terylene (d) Dacron

51. Bakelite is made from
(a) Phenol & formaldehyde
(b) Urea & formaldehyde
(c) Melamine & Formaldehyde
(d) None of these

52. Which of the following is not used in case of Polyurethane?
(a) Polyols
(b) Hexa methylene dissocynate
(c) Hexamethylene diamine
(d) Adipic acid

53. Major source of propylene in refinery is?
(a) Catalytic reforming
(b) Vis breaking
(c) Fluidised bed catalytic cracking
(d) Iso craking process

54. Steam consumption in steam cracker plant is high in case of?
(a) Ethane (b) Gas oil
(c) Naphtha (d) Propane

55. First gas cracker plant is located at?
(a) Digboi (b) Haldia
(c) Jam Nagar (d) Nagothane

56. For LAB which fraction of the kerosene is used?
(a) Paraffins (b) Olefins
(c) Benzene (d) Xylene

57. Major source of propylene in a refinery is
(a) Catalytic reforming
(b) Visbreaking
(c) Fluidised bed catalytic cracking
(d) Iso craking process

ANSWER

42. (c)	**43.** (b)	**44.** (d)	**45.** (a)	**46.** (b)
47. (c)	**48.** (a)	**49.** (c)	**50.** (a)	**51.** (a)
52. (d)	**53.** (c)	**54.** (a)	**55.** (d)	**56.** (a)
57. (c)				

58. Thermosetting plastic materials
(a) can be repeatedly melted.
(b) is useful for melt casting.
(c) can not be melted after forming.
(d) is useful for spinning.

59. Phthalic anhydride is used
(a) in making PVC
(b) as plasticisers
(c) in insecticides manufacture
(d) for making nylon-6

60. Bakelite is a/an
(a) addition polymer (b) elastomer
(c) thermoplastic (d) none of these

61. Synthetic polymer produced by using terephthalic acid and ethylene glycol is
(a) terylene (b) nylon-66
(c) PVC (d) polyesterene

62. Nylon-66 is a polyamide of
(a) hexamethylene diamine and adipic acid.
(b) adipic acid and methyl amine.
(c) vinyl chloride and formaldeyde.
(d) none of these.

63. Polycaprolactum is
(a) nylon-6 (b) nylon-66
(c) dacron (d) rayon

64. Acrylonitrile is mainly used in the................industry.
(a) polymer (b) printing
(c) dyeing (d) photographic

65. The main use of butadiene is
(a) as a plasticiser for unsaturated polyester.
(b) in the manufacture of synthetic rubber.
(c) as an anti-skimming agent in paint.
(d) none of these.

66. Flexible foam (for mattresses) are usually made of
(a) PVC (b) silicone rubber
(c) polyurethanes (d) polyamides

67. Which of the following is not an elastomer ?
(a) Polyisoprene (b) Neoprene
(c) Nitrile-butadiene (d) None of these

68. Nylon-6 as compared to nylon-66 is
(a) harder
(b) more abrasion resistant.
(c) having higher melting point.
(d) None of these

69. Plasticisers are added to synthetic plastics to
(a) impart flexibility.
(b) improve workability during fabrication.
(c) develop new unproved properties not present in the original resin.
(d) all (a), (b) and (c)

70. Polythene a/an
(a) addition polymerisation product.
(b) condensation polymerisation product.
(c) thermosetting material
(d) none of these.

71. The process involved in converting rubber into a thin sheet or coating it on fabric is called
(a) extrusion (b) mastication
(c) calendering (d) vulcanization

72. Main constituent of natural rubber is
(a) polystyrene (b) polyisoprene
(c) polybutadiene
(d) poly chloroprene

ANSWER

58. (c)	**59.** (b)	**60.** (d)	**61.** (a)	**62.** (a)
63. (a)	**64.** (a)	**65.** (b)	**66.** (c)	**67.** (d)
68. (d)	**69.** (d)	**70.** (a)	**71.** (c)	**72.** (b)

73. Caprolactum, a raw material for the manufacture of nylon-6, is produced from
(a) phenol (b) naphthalene
(c) benzene (d) pyridine

74. Polymethyl methacrylate (PMMA) is known as
(a) bakelite (b) teflon
(c) perspex (d) nylon "6

75. ....is a thermosetting plastic.
(a) PVC (b) Polythene
(c) Bakelite (d) Polystyrene

76. Phenol formaldehyde is produced by condensation polymerisation. It is also known as
(a) teflon (b) bakelite
(c) polyester (d) nylon-66

77. Which of the following is not a thermoplastic material?
(a) Epoxy polymer (b) PVC
(c) Polystyrene (d) Polythene

78. Polymerisation product of C_2F_4 (carbon tetraflouride) is called P.T.F.E. (poly chloro tetraflouro ethylene). It is also called
(a) polyurethane (b) silicone rubber
(c) teflon (d) epoxy resin

79. Starting material for the production of butadiene in India is
(a) naphthalene (b) benzol
(c) ethyl alcohol
(d) phthalic anhydride

80. Styrene which is a monomer for the production of polystyrene, is commercially produced by the
(a) catalytic dehydrogenation of ethyl benzene.
(b) dehydration of ethyl alcohol followed by hydrogenation.
(c) reacting ethylene oxide with acetaldehyde.
(d) fermentation of starch.

81. Reaction of dimethyl terephthalate (DMT) and ethylene glycol produces
(a) nylon-6 (b) dacron
(c) polyester (d) PVC

82. Thermal pyrolysis of ethylene dichloride produces
(a) trichlorethylene(b) vinyl chloride
(c) ethanol amine (d) ethylene oxide

83. Pick out the wrong statement.
(a) Addition polymers are generally formed bychain growth polymerisation.
(b) Condensation polymers are generally formed by step growth polymerisation.
(c) Teflon is formed by step growth polymerisation.
(d) Bakelite is formed by step growth polymerisation.

84. Polymers are classified into four categories namely thermosetting, thermoplastic, elastomer and fibre depending upon their
(a) molecular sizes.
(b) magnitude of intermolecular forces
(c) resistance to heat.
(d) polymerisation mechanism.

85. Low density polythene as compared to high density polythene is
(a) harder (b) tougher
(c) chemically inert(d) more flexible

86. Pick out the wrong statement.
(a) Polymers made of only one monomer is called homopolymer.

ANSWER

73. (c) **74.** (c) **75.** (c) **76.** (b) **77.** (a)
78. (c) **79.** (c) **80.** (a) **81.** (b) **82.** (b)
83. (c) **84.** (b) **85.** (d) **86.** (a)

(b) Polymers made of more than one monomer is called copolymer or mixed polymer.

(c) Vulcanised rubber is more elastic the natural rubber.

(d) The intermoleculer forces in thermosplastic polymers are intermediate to that of elastomers & fibres.

87. Polypropylene compared to polythene is

(a) harder (b) stronger

(c) lighter (d) all (a), (b) and (c)

88. Polystyrene is a light, transparent, thermoplastic material used for making

(a) toys and combs (b) packaging bags

(c) non-sticking utensils

(d) electrical insulation

89. Rain coats are made of

(a) neoprene (b) PVC

(c) polyurathane (d) SBR

90. ....tubes are good substitude for human blood vessels in heart by-pass operation.

(a) PVC (b) Polythene

(c) Teflon/dacron (d) Polystyrene

91. The starting material used for the manufacture of caprolactum is,

(a) ethyl benzene (b) cyclohexane

(c) ethylene glycol (d) DMT

92. Trinitro-toluene (TNT) is

(a) used in glycerine manufacture.

(b) an explosive.

(c) used in dye manufacture.

(d) none of these.

93. Plasticisers are added to paints to

(a) make it corrosion resistant.

(b) make glossy surface.

(c) give elasticity & prevent cracking of the film.

(d) increase atmospheric oxidation.

94. Acrylonitrile is mainly used in the.............industry.

(a) polymer (b) printing

(c) dyeing (d) photographic

95. Phenol is mainly used

(a) to produce benzene.

(b) to produce phenol formaldehyde.

(c) to produce polyester resin.

(d) as a plasticiser for unsaturated polyster.

96. Phthalic anhydride is made by the

(a) oxidation of naphthalene.

(b) oxidation of benzene.

(c) dehydrogenation of ethyl benzene.

(d) none of these.

97. Trinitrotoluene (TNT), an explosive, is made by the nitration of

(a) nitrobenzene (b) toluene

(c) nitrotoluene (d) benzene

98. Zeigler process

(a) produces high density polyethylene.

(b) produces low density polyethylene.

(c) uses no catalyst.

(d) employs very high pressure.

99. Poly Vinyl Chloride (P.V.C.) is a..........material.

(a) thermosetting (b) thermoplastic

(c) fibrous

(d) chemically active

100. Poly tetraflouro ethylene (P.T.F.E.) is known as

(a) bakelite (b) teflon

(c) celluloid (d) dacron

101. Epoxy resin

(a) is a good adhesive.

(b) is an elastomer.

ANSWER

87. (d)	**88.** (a)	**89.** (b)	**90.** (c)	**91.** (b)
92. (b)	**93.** (c)	**94.** (a)	**95.** (b)	**96.** (a)
97. (b)	**98.** (a)	**99.** (b)	**100.** (b)	**101.** (a)

(c) cannot be used for surface coatings.
(d) is a polyester.

102. Vulcanisation of rubber
(a) decreases its tensile strength.
(b) increases its ozone & oxygen reactivity,
(c) increases its oil & solvent resistance.
(d) converts its plasticity into elasticity.

103. Most commonly used rubber vulcanising agent is
(a) sulphur (b) bromine
(c) platinum (d) alumina

104. Thermoplastic materials
(a) do not soften on application of heat.
(b) are heavily branched molecules.
(c) are solvent insoluble.
(d) none of these.

105. Thermosetting materials
(a) are cross-linked molecules.
(b)soften on application of heat.
(c) are solvent soluble.
(d) none of these.

106. Cumene (isopropyl benzene) is made by
(a) oxidation of napthalene.
(b) propylene alkylation of benzene.
(c) polymerisation of a mixture of benzene & propylene.
(d) none of these.

107. Cumene is the starting material for the production of
(a) benzoic acid (b) phenol and acetone
(c) isoprene (d) styrene

108. The amount of benzene present in pure benzol is about...............percent.
(a) 30 (b) 50
(c) 70 (d) 90

109. Plasticisers are added to synthetic plastics to
(a) impart flexibility.
(b) improve workability during fabrication.
(c) develop new improved properties not present in the original resins.
(d) all (a), (b) and (c).

110. Bakelite is
(a) same as poly tetrafluoroethylene (PTFE).
(b) an inorganic polymer.
(c) same as thermosetting phenol formaldehyde.
(d) not a polymer.

111. Styrene-butadiene-rubber (SBR) as compared to natural rubber has
(a) Poorer tensile strength.
(b) poorer resistance to oxidation.
(c) greater amount of heat build-up under heavy loading.
(d) all (a), (b) and (c).

112. Nylon-6 is a
(a) polyamide (b) thermosetting resin
(c) polyester (d) none of these

113. Dacron (or terylene) fibres as compared to nylon fibres have
(a) better heat & acid resistant properties.
(b) poorer resistance to alkalis.
(c) poorer dyeability.
(d) all (a), (b) and (c).

114. Teflon is
(a) phenol formaldehyde.
(b)an inorganic polymer.
(c) poly tetrafluoroethylene (P.T.F.E.).
(d) a monomer.

ANSWER

102. (a)	**103.** (a)	**104.** (d)	**105.** (a)	**106.** (b)
107. (b)	**108.** (c)	**109.** (d)	**110.** (c)	**111.** (d)
112. (a)	**113.** (d)	**114.** (c)		

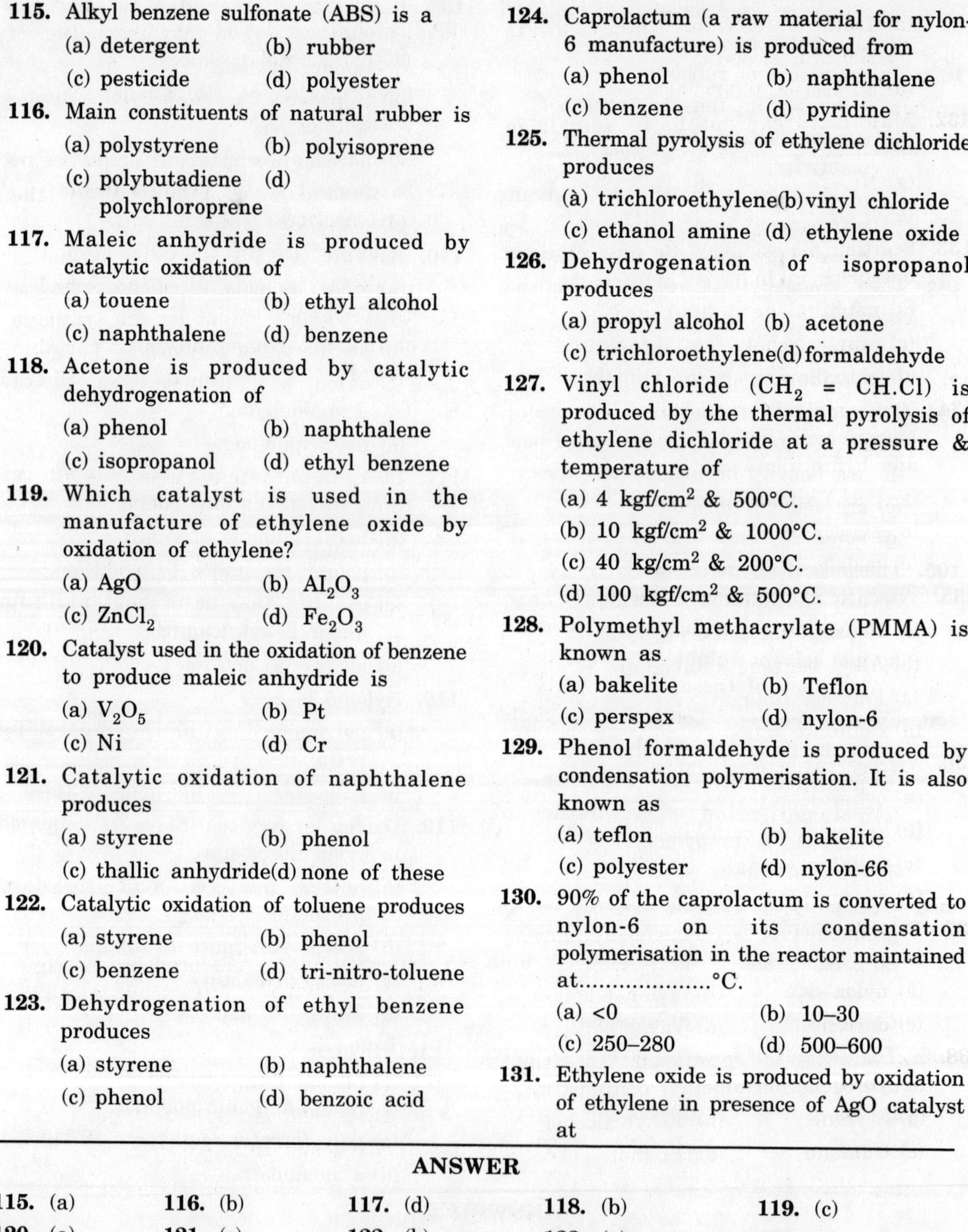

115. Alkyl benzene sulfonate (ABS) is a
(a) detergent (b) rubber
(c) pesticide (d) polyester

116. Main constituents of natural rubber is
(a) polystyrene (b) polyisoprene
(c) polybutadiene (d) polychloroprene

117. Maleic anhydride is produced by catalytic oxidation of
(a) touene (b) ethyl alcohol
(c) naphthalene (d) benzene

118. Acetone is produced by catalytic dehydrogenation of
(a) phenol (b) naphthalene
(c) isopropanol (d) ethyl benzene

119. Which catalyst is used in the manufacture of ethylene oxide by oxidation of ethylene?
(a) AgO (b) Al_2O_3
(c) $ZnCl_2$ (d) Fe_2O_3

120. Catalyst used in the oxidation of benzene to produce maleic anhydride is
(a) V_2O_5 (b) Pt
(c) Ni (d) Cr

121. Catalytic oxidation of naphthalene produces
(a) styrene (b) phenol
(c) thallic anhydride (d) none of these

122. Catalytic oxidation of toluene produces
(a) styrene (b) phenol
(c) benzene (d) tri-nitro-toluene

123. Dehydrogenation of ethyl benzene produces
(a) styrene (b) naphthalene
(c) phenol (d) benzoic acid

124. Caprolactum (a raw material for nylon-6 manufacture) is produced from
(a) phenol (b) naphthalene
(c) benzene (d) pyridine

125. Thermal pyrolysis of ethylene dichloride produces
(a) trichloroethylene (b) vinyl chloride
(c) ethanol amine (d) ethylene oxide

126. Dehydrogenation of isopropanol produces
(a) propyl alcohol (b) acetone
(c) trichloroethylene (d) formaldehyde

127. Vinyl chloride (CH_2 = CH.Cl) is produced by the thermal pyrolysis of ethylene dichloride at a pressure & temperature of
(a) 4 kgf/cm^2 & 500°C.
(b) 10 kgf/cm^2 & 1000°C.
(c) 40 kg/cm^2 & 200°C.
(d) 100 kgf/cm^2 & 500°C.

128. Polymethyl methacrylate (PMMA) is known as
(a) bakelite (b) Teflon
(c) perspex (d) nylon-6

129. Phenol formaldehyde is produced by condensation polymerisation. It is also known as
(a) teflon (b) bakelite
(c) polyester (d) nylon-66

130. 90% of the caprolactum is converted to nylon-6 on its condensation polymerisation in the reactor maintained at.................... °C.
(a) <0 (b) 10–30
(c) 250–280 (d) 500–600

131. Ethylene oxide is produced by oxidation of ethylene in presence of AgO catalyst at

ANSWER

115. (a)	**116.** (b)	**117.** (d)	**118.** (b)	**119.** (c)
120. (a)	**121.** (c)	**122.** (b)	**123.** (a)	**124.** (c)
125. (b)	**126.** (b)	**127.** (a)	**128.** (c)	**129.** (b)
130. (b)	**131.** (b)			

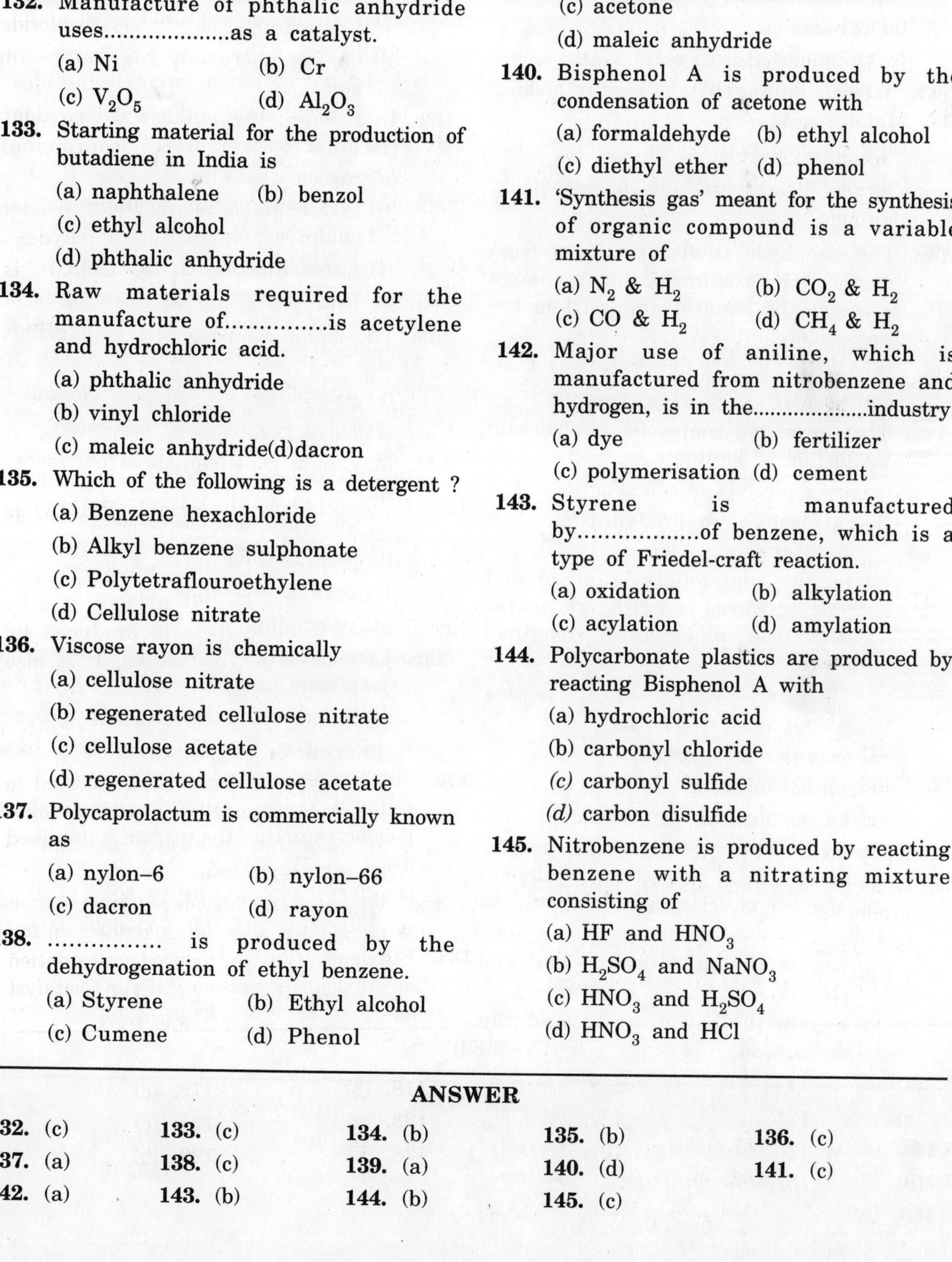

(a) 1 atm & 100°C (b) 5 atm & 275°C
(c) 100 atm & 500°C
(d) 50 atm & 1000°C

132. Manufacture of phthalic anhydride uses..................as a catalyst.
(a) Ni (b) Cr
(c) V_2O_5 (d) Al_2O_3

133. Starting material for the production of butadiene in India is
(a) naphthalene (b) benzol
(c) ethyl alcohol
(d) phthalic anhydride

134. Raw materials required for the manufacture of...............is acetylene and hydrochloric acid.
(a) phthalic anhydride
(b) vinyl chloride
(c) maleic anhydride(d)dacron

135. Which of the following is a detergent ?
(a) Benzene hexachloride
(b) Alkyl benzene sulphonate
(c) Polytetraflouroethylene
(d) Cellulose nitrate

136. Viscose rayon is chemically
(a) cellulose nitrate
(b) regenerated cellulose nitrate
(c) cellulose acetate
(d) regenerated cellulose acetate

137. Polycaprolactum is commercially known as
(a) nylon–6 (b) nylon–66
(c) dacron (d) rayon

138. .... is produced by the dehydrogenation of ethyl benzene.
(a) Styrene (b) Ethyl alcohol
(c) Cumene (d) Phenol

139. Catalytic oxidation-dehydrogenation of methyl alcohol produces
(a) formaldehyde (b) phenol
(c) acetone
(d) maleic anhydride

140. Bisphenol A is produced by the condensation of acetone with
(a) formaldehyde (b) ethyl alcohol
(c) diethyl ether (d) phenol

141. 'Synthesis gas' meant for the synthesis of organic compound is a variable mixture of
(a) N_2 & H_2 (b) CO_2 & H_2
(c) CO & H_2 (d) CH_4 & H_2

142. Major use of aniline, which is manufactured from nitrobenzene and hydrogen, is in the..................industry.
(a) dye (b) fertilizer
(c) polymerisation (d) cement

143. Styrene is manufactured by.................of benzene, which is a type of Friedel-craft reaction.
(a) oxidation (b) alkylation
(c) acylation (d) amylation

144. Polycarbonate plastics are produced by reacting Bisphenol A with
(a) hydrochloric acid
(b) carbonyl chloride
(c) carbonyl sulfide
(d) carbon disulfide

145. Nitrobenzene is produced by reacting benzene with a nitrating mixture consisting of
(a) HF and HNO_3
(b) H_2SO_4 and $NaNO_3$
(c) HNO_3 and H_2SO_4
(d) HNO_3 and HCl

ANSWER

132. (c)	**133.** (c)	**134.** (b)	**135.** (b)	**136.** (c)
137. (a)	**138.** (c)	**139.** (a)	**140.** (d)	**141.** (c)
142. (a)	**143.** (b)	**144.** (b)	**145.** (c)	

146. Which of the following compounds is made on a fairly large scale starting from toluene as a raw material?
(a) Benzoic acid (b) p–Nitrotoluene
(c) Phenol
(d) Tolylene di-isocyanate (TDI)

147. Tolylene di-isocyanate is used in making
(a) dyestuffs
(b) pharmaceutical products
(c) flexible polyurethane foams
(d) all a, b & c

148. Formaldehyde is also produced from methanol by oxidation dehydrogenation process. The catalyst employed in the process is
(a) Pt–Rh (b) ZrO_2
(c) Pd – C (d) Ag –gauze

149. There are two routes for commercial production of aniline :
(i) ammoniolysis of phenol,
(ii) catalytic hydrogenation of nitrobenzene
In the ammoniolysis of phenol process, phenol and NH_3 are reacted in vapour phase in a fixed-bed reactor.
The catalyst employed in the process is
(a) granular ion exchange resin
(b) silica – alumina
(c) Pd on alumina (d) Thoria

150. In nylon-66, the first and second numbers (i.e., 6) respectively designate the number of carbon atoms present in the
(a) hexamethylene diamine and the ring.
(b) hexamethylene diamine and the adipic acid.
(c) adipic acid and the ring.
(d) none of these.

151. Linear polymers are normally
(a) thermosetting. (b) thermosplastic.
(c) elastometric.
(d) having extremely high softening point

152. In nylon-6, the number 6 represents the total number of
(a) carbon atoms in the ring.
(b) carbon atoms in the linear polymer chain.
(c) nitrogen atoms in the ring.
(d) hydrogen atoms in the ring.

153. The major component of acrylic fibres is
(a) polyamides (b) poly olefins
(c) polyacrylonitrile(d) polyesters

154. Maximum consumption of polymers is in
(a) electrical insulation
(b) toys making
(c) coating and films
(d) packaging

155. Low pressure Zeigler process of polythene manufacture
(a) employs a pressure of 30 kgf/cm^2.
(b) achieves an yield of 95-98% based on ethylene.
(c) produces very low density polythene.
(d) does not use any catalyst for polymerization

156. Which of the following polymers is used for making a non stick coating on frying pans?
(a) Bakelite (b) Teflon
(c) Perspex (d) PVC

ANSWER

146. (d)	**147.** (c)	**148.** (d)	**149.** (b)	**150.** (b)
151. (b)	**152.** (a)	**153.** (c)	**154.** (c)	**155.** (b)
156. (b)				

157. Which of the following is a copolymer?
(a) PVC (b) Bakelite
(c) Polythene (d) Teflon

158. Which of the following is not a condensation polymer ?
(a) Bakelite
(b) Melamine polymer
(c) Polymethylmethacrylate (PMMA)
(d) None of these

159. Vulcanisation of rubber does not increase its
(a) elasticity (b) plasticity
(c) ductility (d) none of these

160. Polypropylene is preferred to polythene because the former is
(a) non-inflammable(b) harder
(c) stronger (d) both (b) & (c)

161. Branched chain polymers compared to linear polymers have higher
(a) density.
(b) tensile strength.
(c) melting point.
(d) degree of irregularity in atomic packing.

162. Which of the following is not a polyolefin ?
(a) Polystyrene (b) Polypropylene
(c) Neoprene (d) None of these

163. Contact lenses for eyes are made of perspex, which is nothing but
(a) polymethylmethacrylate
(b) polystyrene
(c) unsaturated polyester
(d) polypropylene

164. The monomer for the production of neoprenerubber is
(a) acetylene (b) chloroprene
(c) isoprene (d) none of these

165. Size range of polymer molecules varies from...................metre.
(a) 10^2 to10^5 (b) 10^4 to 10^7
(c) 10^1 to 10^2 (d) 10^8 to 10^{-10}

166. Vulcanisation of raw rubber makes it
(a) soft (b) less elastic
(c) plastic (d) tacky

167. Thermocol is a spongy, porous, rigid or flexible foamed plastic, obtained by blowing gas/air through molten
(a) saturated polyester
(b) polyurathane
(c) polystyrene (d) either (b) or (c)

168. Thermocol (expanded polystyrene) is not used for
(a) low temperature thermal insulation as in refrigerator and air conditioners.
(b) accoustic control and ceiling for building.
(c) high temperature thermal insulation infurnaces.
(d) packing of delicate electronic gadgets.

169. Rexin (also called artificial leather), which is used for making table cover, automobile seat cover, shoes etc. is made by coating thick cloth with molten
(a) teflon (b) bakelite
(c) SBR (d) PVC

170. Which of the following low molecular weight ($<10^4$), soft & waxy polymer is used in chewing gum?
(a) Cellulose acetate
(b) Polyvinyl acetate
(c) Thiokol (d) PVC

ANSWER

157. (b)	**158.** (c)	**159.** (b)	**160.** (d)	**161.** (d)
162. (c)	**163.** (a)	**164.** (b)	**165.** (a)	**166.** (b)
167. (d)	**168.** (c)	**169.** (d)	**170.** (b)	

171. Thermosetting resins/polymers as compared to thermoplastic ones are

(a) soluble in all organic solvents.

(b) more brittle.

(c) formed by addition polymerisation only.

(d) easily reshaped & reused.

172. ....moulding is used for shaping of thermosetting plastics exclusively.

(a) Compression (b) Injection

(c) Transfer (d) Extrusion

173. Automobile steering wheels are normally made of

(a) cellulose acetate

(b) cellulose nitrate

(c) PVC

(d) high density polythene

174. Density of low density polythene is about..................gm/c.c.

(a) 0.38 (b) 0.56

(c) 0.81 (d) 0.91

175. Density of high density polythene is about..................gm/c.c.

(a) 1.18 (b) 1.05

(c) 0.95 (d) 0.99

176. Zeigler – Natta catalyst (AIR_3 – $AlCl_3$) is used in the polymerisation of

(a) vinyl acetate (b) vinyl chloride

(c) propylene (d) styrene

177. Softening point of high density polythene is about..................°C.

(a) 85 (b) 135

(c) 165 (d) 205

178. ....are produced by reacting polybasic acid (e.g. phthalic anhydride) with polyhydric alcohol (e.g., glycerol).

(a) Unsaturated polyester

(b) Alkyd resins

(c) Saturated polyester

(d) Amino resins

179. Tubeless tyres are made of .... rubber, which is a co-polymer of isoprene & isobutylene.

(a) nitrile (b) silicone

(c) neoprene (d) butyl

180. Which of the following rubbers has the widest service temperature range (–75 to 275°C)?

(a) Butyl rubber (b) Silicon rubber

(c) Nitrile rubber (d) Silicone rubber

181. Cellulose is the main contituent of most................fibres.

(a) acrylic (b) spandex

(c) synthetic (d) natural

182. Tyres are made by

(a) injection moulding

(b) extrusion

(c) rotational moulding

(d) compression moulding

183. Plastic tubes & pipes are generally made by..................moulding.

(a) injection (b) transfer

(c) extrusion (d) compression

184. Which of the following is stretched into fibres ?

(a) Saturated polyester

(b) Unsaturated polyester

(c) Isoprene (d) Bakelite

ANSWER

171. (b)	**172.** (c)	**173.** (a)	**174.** (d)	**175.** (c)
176. (c)	**177.** (b)	**178.** (b)	**179.** (d)	**180.** (b)
181. (d)	**182.** (c)	**183.** (c)	**184.** (a)	

185. Mastication of rubber means
(a) its softening.
(b) a treatment to retard its deterioration due to oxidation.
(c) improving its curing rate.
(d) depression of its freezing point.

186. Polyvinyl acetate is never used for making
(a) moulded articles(b) fibres
(c) adhesives (d) all (a), (b) &(c)

187. Synthetic rubber
(a) deforms, if stretched to double of its original dimension.
(b) is brittle at low temperature.
(c) is softer at higher temperature.
(d) is highly permeable to air & water and is readily attacked by chemicals & atmospheric gases.

188. Styrene butadiene rubber (SBR) is never used for making
(a) coated fabrics
(b) front wheel tyres of aeroplanes (i.e., heavy duty tyres)
(c) gaskets
(d) soles of shoes

189. Which of the following is the lowest cost plastic commercially available ?
(a) Polythene (b) Teflon
(c) Bakelite (d) PVC

190. Paper like thin plastic articles can be produced by
(a) blow moulding
(b) vacuum thermo forming
(c) injection moulding
(d) none of these

191. Most of the plastics are safe to be used upto a maximum temperature of....................°C.
(a) 100 (b) 150
(c) 350 (d) 450

192. Which of the following polymers has the tendency of decomposing before melting?
(a) Polystyrene (b) Nylon
(c) PVC (d) None of these

193. Polycondensation of saturated dicarboxylic acid with polyhydric alcohol produces
(a) epoxy resin (b) polyamide
(c) alkyd resin (d) phenolic resin

194. Out of all the elastomers, natural rubber has the longest elongation range & flexibility of the order of................ percent.
(a) 1-1000 (b) 1000-1500
(c) 1500-2000 (d) 2000-2500

195. ....polythene is most prone to stresscracking.
(a) High density (b) Low density
(c) Cross-linked
(d) Linear low density

196. Commercial production of polypropylene employs..................polymerisation.
(a) emulsion (b) suspension
(c) solution (d) bulk

197. Addition of stabiliser during PVC manufacture is done to
(a) improve its impact strength.
(b) improve its elasticity.
(c) reduce the melt viscosity & glass transition temperature.
(d) prevent its thermal degradation

ANSWER

185. (a)	**186.** (a)	**187.** (d)	**188.** (b)	**189.** (a)
190. (b)	**191.** (b)	**192.** (c)	**193.** (c)	**194.** (a)
195. (a)	**196.** (c)	**197.** (d)		

198. Nylon 6 as compared to nylon 6,6 has lower

(a) abrasion resistance.

(b) thermal stability.

(c) adhesion to rubber.

(d) hardness.

199. Nylon 6, 10 which is used for bristles making is superior to nylon 6, 6 due to its lower-water absorption capacity, is a/an

(a) polyester (b) polyamide

(c) polyisoprene (d) polystyrene

200. Cation exchange resins (0.3 to 1 mm size) used in water treatment is prepared from..............resins.

(a) epoxy

(b) phenol formaldehyde

(c) urea formaldehyde

(d) melamine formaldehyde

201. Alkyd resin e.g., glyptal resin formed byphthalic anhydride and glycerene is not used

(a) for surface coating of automobiles & aircrafts.

(b) for fibre making.

(c) as plasticiser for PVC & nitrocellulose.

(d) for film forming materials

202. Scouring is a finishing operation during the manufacture of fibre, which aims at .... of fibres.

(a) improving the stretchability

(b) dyeing/colouring

(c) detergent solution washing

(d) glycol soaking

203. Most of the fibre forming polymers are crystalline in nature, one of the exceptions being.............which is amorphous in nature.

(a) nylon

(b) polyacrylonitrile

(c) polypropylene (d) polyester

204. All thermoplastic, thermosetting & elastic materials can be processed in a extrusion machine, however it can not be used for the production of plastic

(a) filaments (b) pipes

(c) buckets (d) tubings

205. ....scrap can be recycled & reutilised.

(a) Bakelite (b) Epoxy resin

(c) Polythene (d) None of these

206. Lavatory cisterns are normally made of

(a) expanded polystyrene

(b) saturated polyester

(c) Perspex (d) PVC

207. ....resins are produced by the condensation polymerisation of formaldehyde with urea or melamine.

(a) Epoxy (b) Amino

(c) Alkyd (d) Phenolic

208. Thermoplastic resins usually

(a) remain hard as long as they are hot.

(b) can not be reclaimed from waste.

(c) permanent setting resins.

(d) less brittle than thermosetting resins.

209. In a co-polymer, the repeating units contain two different monomers. Which of the following is a copolymer ?

(a) PTFE (b) Buna-S

(c) PMMA (d) Polycaprolactum

ANSWER

198. (a) **199.** (b) **200.** (c) **201.** (b) **202.** (c)
203. (b) **204.** (c) **205.** (c) **206.** (a) **207.** (b)
208. (d) **209.** (b)

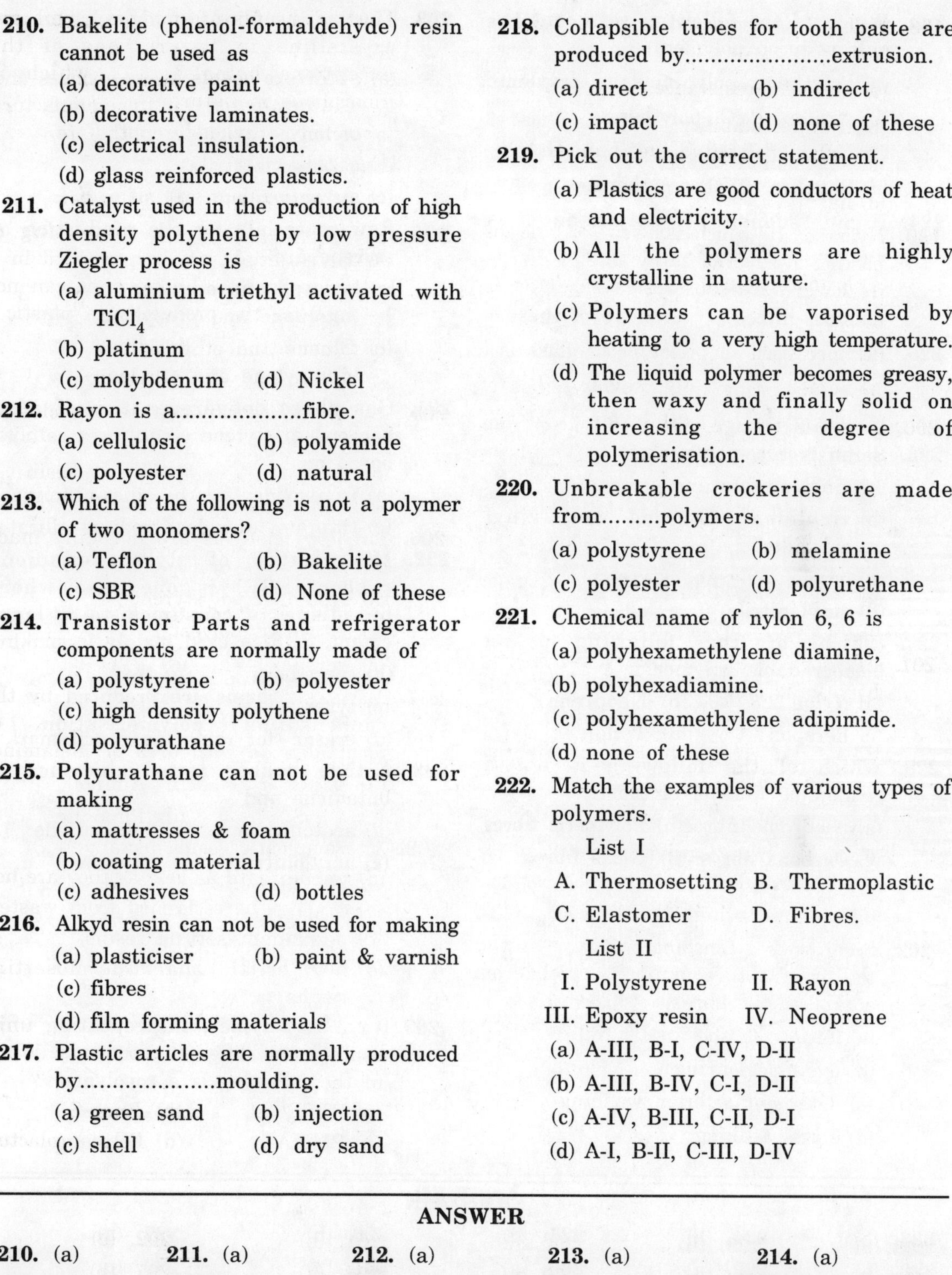

210. Bakelite (phenol-formaldehyde) resin cannot be used as
(a) decorative paint
(b) decorative laminates.
(c) electrical insulation.
(d) glass reinforced plastics.

211. Catalyst used in the production of high density polythene by low pressure Ziegler process is
(a) aluminium triethyl activated with $TiCl_4$
(b) platinum
(c) molybdenum (d) Nickel

212. Rayon is a....................fibre.
(a) cellulosic (b) polyamide
(c) polyester (d) natural

213. Which of the following is not a polymer of two monomers?
(a) Teflon (b) Bakelite
(c) SBR (d) None of these

214. Transistor Parts and refrigerator components are normally made of
(a) polystyrene (b) polyester
(c) high density polythene
(d) polyurathane

215. Polyurathane can not be used for making
(a) mattresses & foam
(b) coating material
(c) adhesives (d) bottles

216. Alkyd resin can not be used for making
(a) plasticiser (b) paint & varnish
(c) fibres
(d) film forming materials

217. Plastic articles are normally produced by.....................moulding.
(a) green sand (b) injection
(c) shell (d) dry sand

218. Collapsible tubes for tooth paste are produced by.......................extrusion.
(a) direct (b) indirect
(c) impact (d) none of these

219. Pick out the correct statement.
(a) Plastics are good conductors of heat and electricity.
(b) All the polymers are highly crystalline in nature.
(c) Polymers can be vaporised by heating to a very high temperature.
(d) The liquid polymer becomes greasy, then waxy and finally solid on increasing the degree of polymerisation.

220. Unbreakable crockeries are made from.........polymers.
(a) polystyrene (b) melamine
(c) polyester (d) polyurethane

221. Chemical name of nylon 6, 6 is
(a) polyhexamethylene diamine,
(b) polyhexadiamine.
(c) polyhexamethylene adipimide.
(d) none of these

222. Match the examples of various types of polymers.
List I
A. Thermosetting B. Thermoplastic
C. Elastomer D. Fibres.
List II
I. Polystyrene II. Rayon
III. Epoxy resin IV. Neoprene
(a) A-III, B-I, C-IV, D-II
(b) A-III, B-IV, C-I, D-II
(c) A-IV, B-III, C-II, D-I
(d) A-I, B-II, C-III, D-IV

ANSWER

210. (a)	**211.** (a)	**212.** (a)	**213.** (a)	**214.** (a)
215. (d)	**216.** (c)	**217.** (b)	**218.** (c)	**219.** (d)
220. (b)	**221.** (c)	**222.** (a)		

223. Which of the following sets contains only addition polymers ?

(a) Polythene, polypropylene, terylene.

(b) Acrilon, PVC, polythene.

(c) Nylon, polybutadiene, buna-S.

(d) Bakelite, PVC, polythene.

224. High density polythene is a..............polymer.

(a) branch-chain (b) linear

(c) cross-linked (d) none of these

225. Polymer used in bullet proof glass is

(a) nomex (b) Kevlar

(c) plexiglass (d) lexan

226. Saran is a co-polymer of

(a) vinyl acetate and styrene.

(b) vinylidene chloride and vinyl chloride.

(c) acrylonitrile and styrene.

(d) acrylonitrile.

227. Which of the following is a biodegradable polymer?

(a) Nylon 2,6 (b) Neoprene

(c) Perspex (d) Teflon

228. Which of the following is not a reinforcing fibre of or plastic?

(a) Nichrome fibres(b) Polymeric fibres

(c) Carbons fibres (d) Glass fibres

229. The major uses of ethylene glycol are as

(a) antifreeze agents in automobiles and as a raw material in the manufacture of poly ethyleneterephthalate.

(b) a plasticizer

(c) an antioxidant (d) all a, b & c

230. Raw materials for the production of styrene are

(a) benzene and ethylene

(b)toluene and propylene

(c) toluene and ethanol

(d)m-xylene and ethylene

231. Gas phase dehydrogenation of ethyl benzene to styrene occurs over catalyst based on

(a) iron oxide. (b) silica-alumina.

(c) titanium dioxide.(d) sodium silicate.

232. Manufacture of styrene-butadiene rubber (SBR) is one case where butadiene is consumed to a large extent. SBR is used mainly in making

(a) carry-bags (b) toys

(c) tyres

(d) eraser (for removing pencil marks)

233. Nitrile rubber is a copolymer of butadiene and

(a) acetonitrile (b) adiponitrile

(c) acrylonitrile (d) styrene

ANSWER

223. (b) **224.** (b) **225.** (d) **226.** (b) **227.** (a)
228. (a) **229.** (a) **230.** (a) **231.** (a) **232.** (c)
233. (c)

Appendix

H

SHORT QUESTIONS AND ANSWERS

Q.1. *What are petrochemicals?*

Ans. Petrochemicals are chemicals derived from petroleum products. Examples of petrochemicals are plastics, rubbers, fibres, paints, solvents, and detergents. In fact, petroleum products are mixtures of hydrocarbons, whereas the raw materials for petrochemicals are pure hydrocarbons separated & converted to desirable products, such as polymers, solvents&detergents , usually in several stages and may be grouped as (1) feedstocks (first-generation petrochemicals), (2) intermediates (second-generation petrochemicals), and (3) finished products (third-generation petrochemicals). Products similar to petrochemicals derived from non-petroleum sources are not strictly petrochemicals. For example, cellulose, natural rubber, natural resins, nylon 11, and ethanol of plant origin are strictly non-petrochemicals. Coal distillation is also a source of varieties of coal chemicals, e.g., benzene, toluene, xylene , naphthalene etc. Many of the chemicals from non-petroleumsourcesareco processed with petrochemicals to the finished product. Non-hydrocarbons obtained from petroleum, e.g., hydrogen, carbon monoxide, carbon dioxide, sulfur, and carbon, are also loosely called petrochemicals. Hydrogen, nitrogen and oxides of carbon manufactured from steam reforming and partialoxidation of naphtha are also petrochemicals. These are used for production of ammonia, urea, melamine, fertilizer, etc.

Q2. *What are basic petrochemicals and their major end products?*

Ans. Basic petrochemicals are as follows :

- C_1 group hydrocarbons : Methane, CO + H_2 synthesis gas, synthesis gas derivatives
- C_2 grouphydrocarbons : Ethane, ethylene, ethylene derivatives, acetylene
- C_3 grouphydrocarbons : Propane, propylene and propylene derivatives
- C_4, C_5 group hydrocarbons :Butadiene, Butanes, Butenes, Pentane, Pentene, Isoprene, Cyclopentadiene
- Aromatics : Benzene, Toluene, Xylenes, Naphthalene, BTX (Benzene, Toluene, Xylene) derivatives

Major end products of petrochemicals include, Polymer, Synthetic fibre, synthetic rubber, synthetic detergent, chemical intermediate, dyes & intermediates, chemical intermediates, pesticides etc.

Q.3. *What are the basic feedstock choices for Petrochemical industry?*

Ans. Basic feedstock used in petrochemical industry for manufacture of olefins and aromatics are derived either from natural gas or petroleum fractions and includes natural gas, condensate, naphtha, kerosene, catalytic cracking &reformer gases, waxes, pyrolysis gasoline. Natural Gas and Petroleum Fractions as Petrochemicals Feedstock is given below in table.

Table-Natural Gas & Petroleum Fractions as Petrochemicals Feedstock

Petroleum Fractions and Natural Gases	Source	Composition	Intermediate Processes	Intermediate Feedstock
Refinery Gases	Distillation, catalytic cracking, catalytic reforming	Methane, ethane, propane, butane, BP upto 25 °C	Liquefaction, cracking	LPG, ethylene, propylene, butane, butadiene.
Naphtha	Distillation and thermal & catalytic cracking, visbreaking	C4-C12 hydrocarbon, BP 70 - 200 °C	Cracking, reforming, alkylation, disproportionation , isomerisation	Ethylene, propylene, butane, butadiene, benzene, toluene, xylene
Kerosene	Distillation and secondary conversion processes	C9-C10 hydrocarbon, BP 175-275 °C	Fractionation to obtain C10-C14 range hydrocarbon	Linear n C10 - n C14 alkanes
Gas Oil	Distillation of crude oil and cracking	C10-C25 hydrocarbons BP 200-400 °C	Cracking	Ethylene, propylene, butadiene, butylenes
Wax	Dewaxing of lubricating oil	C8-C56 hydrocarbon	Cracking	C6-C20 alkanes
Pyrolysis Gasoline	Ethylene cracker	Aromatic, alkenes, dienes, alkanes, cycloalkane	Hydrogenation distillation, extraction, crystallisation, adsorption	Aromatics
Natural Gases & Natural Gas Condensate	Gas fields and crude oil stabilisation	Hydrogen, methane, ethane, propane, pentane, aromatics	Cracking, reforming, separation	Ethylene, propylene, LPG, aromatics, etc.
Petroleum coke	Crude oil	Carbon	Residue upgradation processes, gasification	Carbon electrode, acetylene, fuel

Q.4. *What are the alternative feed stock choices for petrochemical industry?*

Ans. Some of the alternative feed stock choice for petrochemical industryare:

- Naphtha from methane from natural gas to liquidprocess
- Naphtha from coal via direct liquification or indirect liquification by FTprocess
- Plastic waste to naphtha and other hydrocarbons through liquefaction, pyrolysis and separationprocesses

- FT (Fischer Tropsch) naphtha from biomass
- Methanol routes: Synthesis gas from methane, coal and biomass; conversion of synthesis gas to methanol and production of olefin by methnol to olefintechnology.
- Conversion of methanol to dimethylether
- Product recovery and separation Recovery of C4&C5 stream from FCC and steamcracker
- Oxidative coupling of methane
- Ethanol from biomass: direct fermentation of sugar rich biomass, hydrolysis of lingo-cellulosic biomass
- Gasification of lingo-cellulosic biomass followed by fermentation or chemical catalysis toethanol.
- Carbon dioxide to liquid fuel by engineered bacteria
- Gasification of petrocoke to hydrogen

Q.5. *List out the alternative routes/sources of production of Principal Petrochemicals.*

Ans. Alternative Routes to Principal Petrochemicals are tabulated below.

Table : Alternative Routes to Principal Petrochemicals

Chemicals	Petroleum Source	Alternate Source
Methane	Natural gas Refinery light gases (demethaniser overheads)	Coal, as byproduct of separation of coke oven gases (1920-30) or of coal hydrogenation (1930-40)
Ammonia	Methane Light liquid hydrocarbons	From coal via water gas (1910-20)
Methyl alcohol	Methane Light liquid hydrocarbons	From coal via water-gas (1920-30); from methane (from coal) by methane-stream and methane oxygen processes (1930-40)
Ethylene	Pyrolysis of gaseous liquid hydrocarbons	Dehydration of ethyl alcohol (original route). By-product in fractional distillation of coke oven gas (1925-35). Hydrogenation of acetylene (1940-45)
Acetylene	Ethylene	Calcium carbide (original process). methane from coal by partial combustion and by arc process (1935-45)
Ethylene glycol	Ethylene	From ethylene made as above (1925). In America, from coal via carbon-monoxide and formaldehyde (1935-40)
Ethyl alcohol	Synthetic ethyl alcohol Co-product	Fermentation of molasses (original route)
Acetaldehyde	of paraffin gas oxidation. Direct oxidation of ethylene	Fermentation of ethyl alcohol, or acetylene from carbide (1900-10)
Acetone	Propylene	Wood distillation (original process). Pyrolysis of acetic acid (1920-30) or by acetylene-stream reaction (1930-40)
Glycerol	Propylene	By-product of soap manufacture (original process)

Butadiene	1- and 2-Butenes Butane Synthetic ethyl alcohol By-product of ethylene by pyrolysis of liquid hydrocarbons	Ethyl alcohol (1915); acetaldehyde via 1:3-butanediol (1920-30); acetylene and formaldehyde from coal via 1:4-butanediol (1940-45); from 2:3-Butanediol by fermentation (1940-45)
Aromatic hydrocarbons	Aromatic-rich and naphthenic-rich fractions by catalytic reforming and direct extraction	By-products of coal-tar distillation

Q.6. ***What are various building block processes for the manufacture of building blocks & intermediates in Petrochemical manufacturing?***

Ans. Petrochemical manufacturing involves building blocks processes for the manufacture of building blocks and intermediates which are as follows :

- **Cracking:** Steam cracking, Catalytic cracking for olefins pyrolysis gasoline by product
- **Steam reforming and Partial oxidation:** Synthesis gas
- **Catalytic Reforming:** Aromatic production
- **Aromatic conversion processes:** Aromatic production
- **Alkylation:** Linear alkyl benzene
- **OXO Process:** Oxo-alcohol
- **Polymerisation Process:** Polymer, elastomers and synthetic fibre

Q.7. ***List out the Major Petrochemicals obtained from Petroleum Products as feedstocks***

Ans. Feedstocks are the raw hydrocarbons obtained from crude oil refining by distillation and thermal & catalytic processes. For instance, hydrocarbon gases & naphtha are available from atmospheric distillation of crude oil; similarly, benzene, toluene & xylene (BTX) , obtained by catalytic reforming & catalytic cracking processes, are the major raw materials for the manufacture of second generation petrochemicals. Benzene, toluene, xylene, and heavier aromatics are also generated as by-products from petrochemical plants. Thus, the feedstocks for petrochemical plants are either directly obtained from refineries or are further processed to generate them in the petrochemical plant itself.

Natural gas and refinery products are the major source of feedstocks for petrochemicals.

Thermal cracking of ethane, propane, butane, & naphtha produces cracked gasesor olefins (ethylene, propylene, butylenes, acetylene, etc.) andliquids (benzene, toluene, xylene, etc.). Olefins are thestartingmaterial (monomers) for polyolefin plants. Olefins are also reacted with other hydrocarbons or non-hydrocarbon chemicals to generate vinyl chloride, ethylene glycol, neoprene, ethylene oxide, etc. and these are used as thes tarting materials (monomers) forth emanu facture of avariety of polymers.

A list of the major value added petrochemicals is given below.

Products from refinery/oil field Value added Petrochemicals

Natural gas or methane: Carbon black, methanol, acetic acid, formaldehyde, synthesis gas, chlorinatedmethanes, ethylene glycol, MTBE, acetylene, chloroprene, vinyl chloride, vinyl acetate, etc.

Ethane: Ethylene, polyethylene, ethyl alcohol, ethylene oxide, ethylene glycol, styrene, vinylchloride, acrylonitrile, etc.

Propane: Propylene, polypropylene, propylene oxide, isopropanol, acrylonitrile, acetone, etc.

Butanes: Isobutanol, butadiene, acrylonitrile, polybutadiene, etc.

Benzene: Phenol, styrene and polystyrene, dodecyl benzene, cyclohexane, caprolactum fibres, nylons, etc.

Toluene: Nitrotoluene, phenol

m-Xylene: p-Xylene

p-Xylene: Terephthalic acid, PET

o-Xylene: Phthalic anhydride

Naphtha: Olefins for PE, PP, LPG, CBFS, pyrolysis gasoline, butadiene, linear alkyl benzene

Sulfur: Sulfuric acid

Wax: Micro crystalline wax, food grade coating wax, cosmetics, medicinal wax, etc.

Coke: Metallurgical coke, electrode grade coke, etc

CBFS: Carbon black

Q.8. ***What are petrochemicals and their major building blocks? What are the main petrochemical products?***

Ans. Petrochemicals are derived from various chemical compounds,mainlyhydrocarbons. These hydrocarbons are derived from crude oil and natural gas. Among the various fractions produced by distillation of crude oil, petroleum gases, naphtha, kerosene and gas oil are the main feed-stocks for the petrochemical industry. Ethane, propane, butane and natural gas liquids obtained from natural gas are the other important feed stocks used in the petrochemical industry.

Olefins (ethylene, propylene & butadiene) and Aromatics (benzene, toluene &xylenes) are the major building blocks from which most Chemicals &Petrochemicals are produced. Petrochemical manufacturing involves manufacture of building blocks by cracking or reforming operation; conversion of building blocks into intermediates such as fibre intermediates (acrylonitrile, caprolactum, dimethyl terephthalate/purified terephthalic acid, mono ethylene glycol); precursors (styrene, ethylene dichloride, vinyl chloride monomer etc.) and other chemical intermediates; production of synthetic fibres, plastics, elastomers, other chemicals and processing of plastics to produce consumer and industrial products. In addition building blocks (benzene, toluene, xylene setc) are basic raw materials for the manufacture of pesticides, dyes & dyes intermediate and pharmaceutical bulk drugs, etc.

Petrochemical products namely synthetic fibres cater to the clothing needs of mankind and are used in both apparel & non-apparel applications. Polymers find applications in packaging for preservation of food articles, moulded industrial & home appliances, furniture, extruded pipes, electronics, telecommunication, aeroplanes, etc. Synthetic rubbers are used for making various types of tyres & non-tyre rubber goods and supplement natural rubber. Surfactant intermediates are used in detergents & surfactants.

Q.9. ***Describe the importance of methane and synthesis gas as petrochemical feedstock for manufacture of chemicals, which are used directly or as intermediates.***

Ans. Methane and synthesis gas are important petrochemical feedstock for manufacture of a large number of chemicals, which are used directly or as intermediates . These products find use in plastic, synthetic fiber, rubber, pharmaceutical and other industries. 'Synthesis gas' is commonly used to describe two basic gas mixtures - synthesis gas containing carbon monoxide (CO) & hydrogen(H_2) and synthesis gas containing hydrogen(H_2) & nitrogen (N_2) for the production of ammonia(NH_3).

The technologies in utilization of synthesis gas & methane for the production of petrochemicals include Fischer Tropschsynthesis for manufacture of olefins from synthesis gas, oxidative coupling of methane with chlorine to yield ethane & ethylene, methanol to olefin technology (MTO) etc. CO that can be separated from synthesis gas either by cryogenic or by pressure swing adsorption is a promising feedstock for production of a variety of products.

Petrochemical derivatives based on synthesis gas and carbon monoxide have experienced steady growth due to large scale utilization of methanol and development of a carbonylation process for acetic acid and Oxo synthesis process for detergents, plasticizers & alcohols. Methanol is the largest consumer of synthesis gas.

Q.10. ***What are the raw material for synthesis gas production? How is synthesis gas produced?***

Ans. Various raw materials for synthesis gas production are natural gas, refinery gases, naphtha, fuel oil/residual heavy hydrocarbons and coal. Although coal was earlier used for production of synthesis gas, it has now been replaced by petroleum fractions and natural gas. Petrocoke is the emerging source for Synthesis gas production.

Various synthesis gas production technologies are steam methane reforming, naphtha reforming, auto thermal reforming, oxygen secondary reforming, and partial oxidation of heavy hydrocarbons, petroleum coke & coal.

Various steps involved in synthesis gas production through steam reforming ofare:

- Desulphurization of gas
- Steam reforming and compression
- Separation of CO_2

Various available synthesis gas generation schemes are:

- Conventional steam reforming
- Partial oxidation
- Combined, Parallel or Gas heated reforming

Q.11. ***Show various routes for Synthesis gas, Ammonia and Methanol (Methyl Alcohol) manufacture.***

Ans. Various routes for Synthesis gas, Ammonia and Methanol manufacture are shown below diagrammatically.

Q.12. ***List out the process steps, chemical reactions & process conditions (Catalysts, temperature etc.) involved in the manufacture of synthesis gas by Steam reforming and Partial oxidation.***

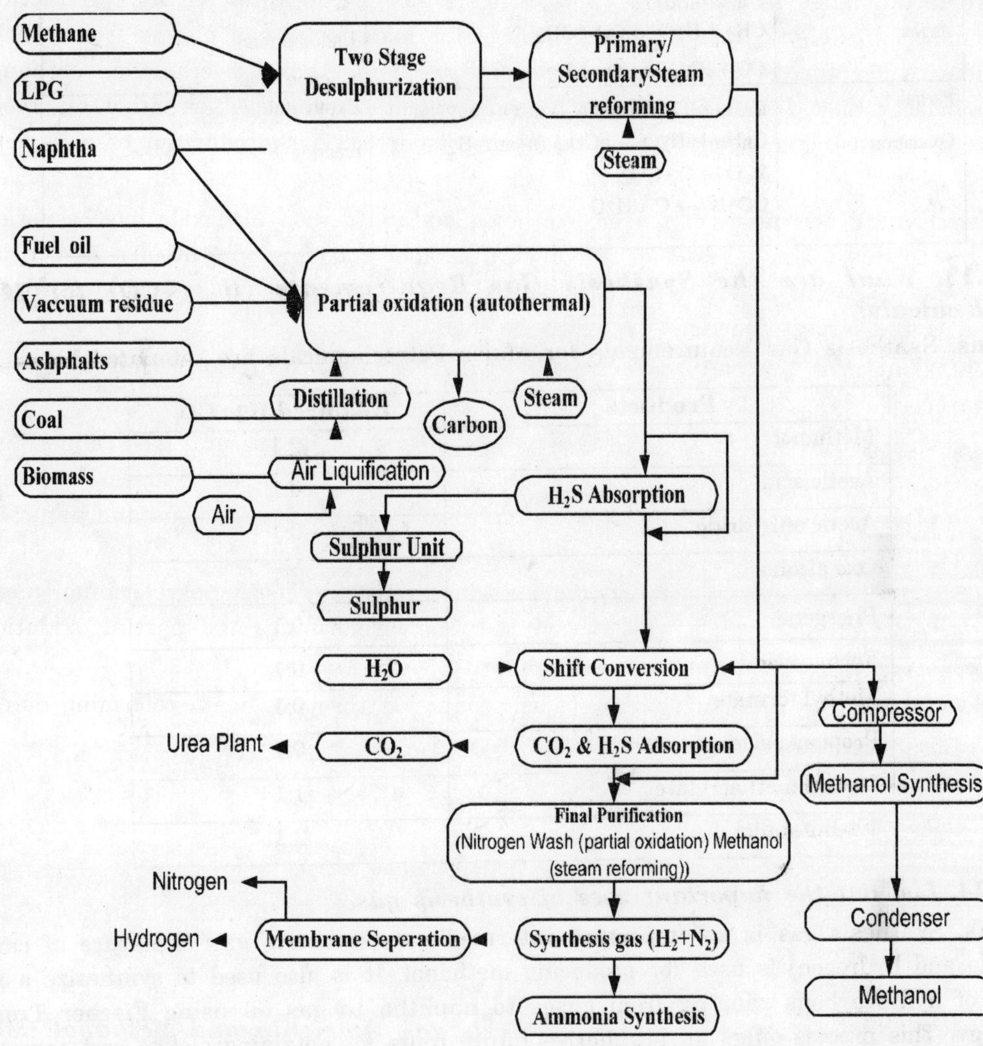

Ans. The process steps, chemical reactions & process conditions involved in the manufacture of synthesis gas by Steam reforming and Partial oxidation are tabulated below.

Process steps	Chemical Reactions	Process Conditions
Desulphurisat--ion: 1st Stage	$C2H5SH + H2 \rightarrow H2S + C2H6$ $C6H5SH + H2 \rightarrow H2S + C6H6$ $C4H4SH + 3H2 \rightarrow H2S + C4H9$ $CS2 + 4H2 \rightarrow 2H2S + CH4$	Al-Co-Mo Al-Ni-Mo Catalyst 350-400 ^{0}C
2nd Stage	$COS + H2 \rightarrow H2S + CO$ $CH3SC2H5 + H2 \rightarrow H2S + CH4 + C2H4$ $H2S + ZnO \rightarrow ZnS + H2O$	Zinc oxide absorbent 200-500 ^{0}C

Steam reforming two stages	$C_nH_m + 1/4(4n-m)H_2O \rightarrow 1/8(4n+m)CH_4$ $1/8(4n-m)CO_2$ $CH_4 + H_2O \rightarrow CO + 3H_2$ $CO + H_2O \rightarrow CO_2 + H_2$	Nickel catalyst 800 ^oC Endothermic reaction
Partial Oxidation	$C_nH_m + [(2n+m)/4]O_2 \rightarrow nCO + m/2H_2O$ $C_nH_m + nH_2O \rightarrow nCO + (n+m/2)H_2$ $2CO \rightarrow C + CO_2$ $CO + H_2 \rightarrow C + H_2O$	Exothermic reaction

Q.13. ***What are the Synthesis Gas Requirements (H_2 : CO) for Major Petrochemicals?.***

Ans. Synthesis Gas Requirements for Major Petrochemicals are tabulated below.

Products	Required H_2 : CO
Methanol	2:1
Acetic acid	0:1
Acetic anhydride	0:1
Oxo alcohol	2:1
Phosgene	0:1
Formic acid	0:1
Methyl formate	0:1
Propionic acid	0:1
Methyl methacrylate	1:1
1,4-Butandiol	2:1

Q.14. ***List out the important uses of synthesis gas.***

Ans. Synthesis gas is an important intermediate. Synthesis gas(amixture of carbon monoxide and hydrogen) is used for producing methanol. It is also used to synthesize a wide variety of hydrocarbons ranging from gases to naphtha to gas oil using Fischer Tropsch technology. This process offers an alternative future route for obtaining olefins and chemicals. The hydroformylation reaction (Oxo synthesis) is based on the reaction of synthesis gas with olefins for the production of Oxo aldehydes and alcohols.

Synthesis gas is a major source of hydrogen, which is used for producing ammonia. Ammonia is the host of many chemicals such as urea, ammonium nitrate, and hydrazine. Carbon dioxide, a by-product from synthesis gas, reacts with ammonia to produce urea.

Q.15. *List out the product profile of methane, synthesis gas and CO based building blocks.*

Ans. The products profile of methane, synthesis gas and CO based building blocks are shown below in figure.

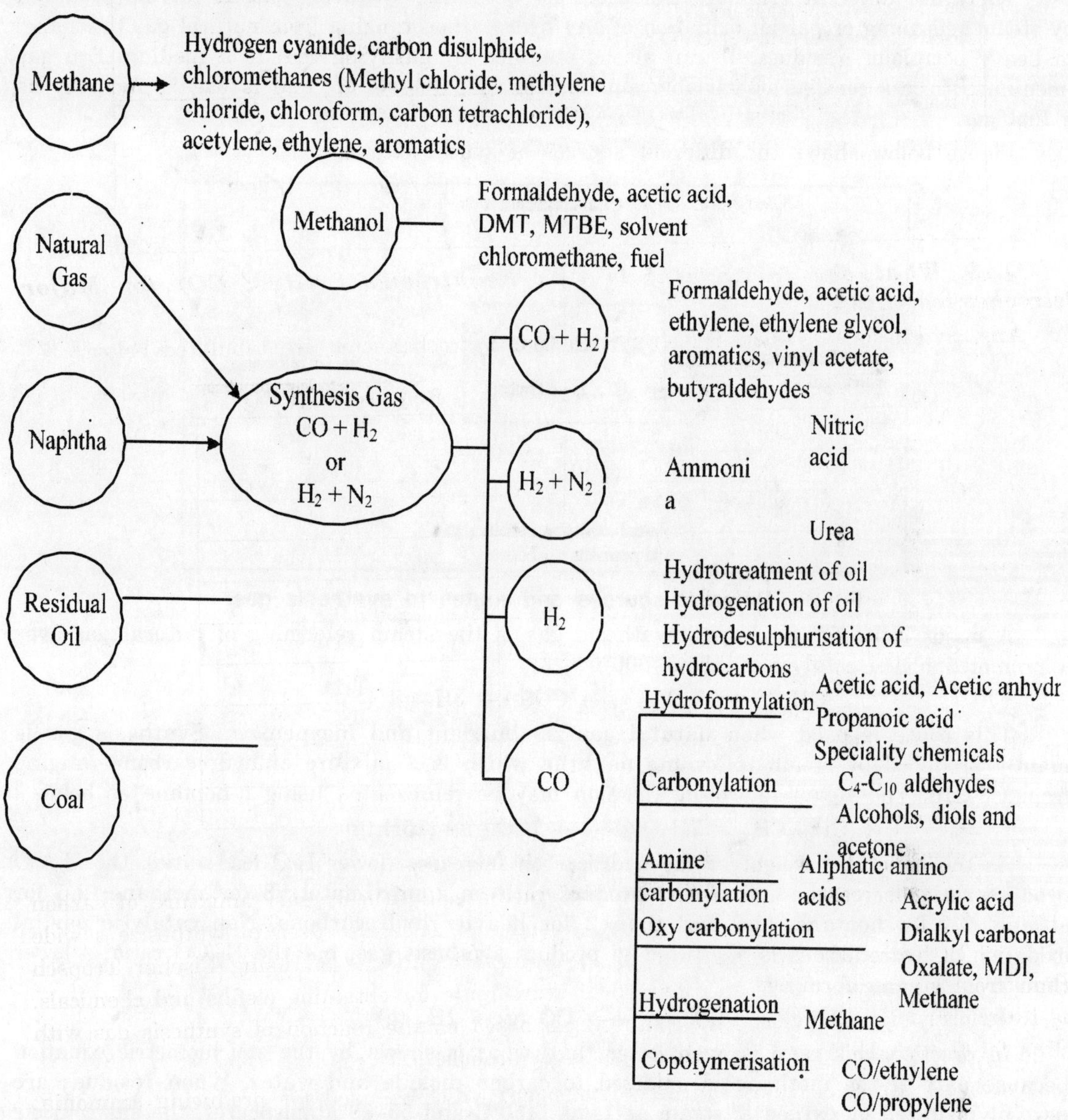

Figure: Product Profile of Methane, synthesis gas and CO building blocks

Q.16. ***Describe the various sources and routes of Synthesis gas production.***

Ans. Synthesis gas generally refers to a mixture of carbon monoxide and hydrogen. The ratio of hydrogen to carbon monoxide varies according to the type of feed, the method of production, and the end use of the synthesis gas.

During World War II, the Germans obtained synthesis gas by gasifying coal. The mixture was used for producing a liquid hydrocarbon mixture in the gasoline boiling range using Fischer-Tropsch technology. Although this route was abandoned after the war due to the high production cost of these hydrocarbons, it is currently being used in South Africa (SASOL) where coal is cheap.

There are different commercial sources for obtaining synthesis gas. It can be produced by steam reforming or partial oxidation of any hydrocarbon ranging from natural gas (methane) to heavy petroleum residues. It can alsobe obtained by gasifying coal to a medium Btu gas (medium Btu gas consists of variable amounts of CO, CO_2, & H_2 and is used principally as a fuel gas.

Figure below shows the different sources of synthesis gas.

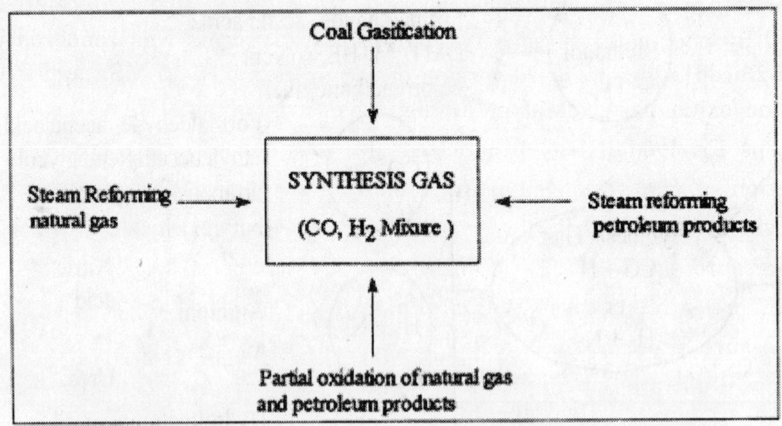

Figure: Different sources and routes to synthesis gas.

A major route for producing synthesis gas is the steam reforming of natural gas over a promoted nickel catalyst at about 800°C:

$$[CH_4(g) + H_2O(g) \longrightarrow CO(g) + 3H_2(g)]$$

This route is used when natural gas is abundant and inexpensive. Synthesis gas is mainly produced by steam reforming naphtha which is a mixture of hydrocarbons ranging from C5-C10 .The steam reforming reaction may be represented using n-heptane as below :

$$CH_3(CH_2)_5CH_3 + 7H_2O(g) \longrightarrow 7CO(g) + 15H_2(g)$$

As the molecular weight of the hydrocarbon increases (lower H/C feed ratio), the H_2/CO product ratio decreases. The H_2/CO product ratio is approximately 3 for methane, 2.5 for ethane, 2.1 for heptane, and less than 2 for heavier hydrocarbons. Non catalytic partial oxidation of hydrocarbons is also used to produce synthesis gas, but the H_2/CO ratio is lower than from steamreforming:

$$CH_4(g) + 1/2O_2\ (g) \longrightarrow CO\ (g) + 2H_2\ (g)$$

In practice, this ratio is even lower than what is shown by the stoichiometric equation because part of the methane is oxidized to carbon dioxide and water. When residues are partially oxidized by oxygen & steam at 1400–1450°C and 55–60 atmospheres, the gas consists of equal parts of hydrogen & carbon monoxide.

Table below compares products from steam reforming natural gas with products from partial oxidation of heavy fueloil.

Partial oxidation of fuel oil					
	Volume % dry sulfur free				
Process	**CO**	**H2**	**CO2**	**N2+A**	**CH4**
Steam reforming natural gas	15.5	75.	8.1	0.2	0.5
Partial oxidation-heavy fuel	47.5	46.	4.3	1.4	0.3

Q.17. ***Describe the manufacturing process of methanol (Methyl Alcohol) by steam reforming of natural gas/naphtha with a neat sketch.***

Ans. The most common industrialmethod for the production of methanol was first developed by BASF in 1923 in Germany from synthesis gas utilising high pressure process using zinc-chromic oxide catalyst. However, due to high capital and compression energy costscompounded by poor catalyst activity, high-pressure process was rendered obsolete when ICI in 1966 introduced a low-pressure version of the process at 5-10 MPa and 210-270 ºC, with a new copper-zinc oxide based catalyst of high selectivity andstability.

Process steps involved in the production of methanol (CH_3OH) are:

- Production of synthesis gas using steam reforming or partialoxidation
- Synthesis of methanol
- High-pressure process (25 – 30MPa)
- Medium pressure process (10-25 MPa)
- Low-pressure process (5-10 MPa)

Figure below illustrate the production of methanol from steam reforming of natural gas and naphtha.

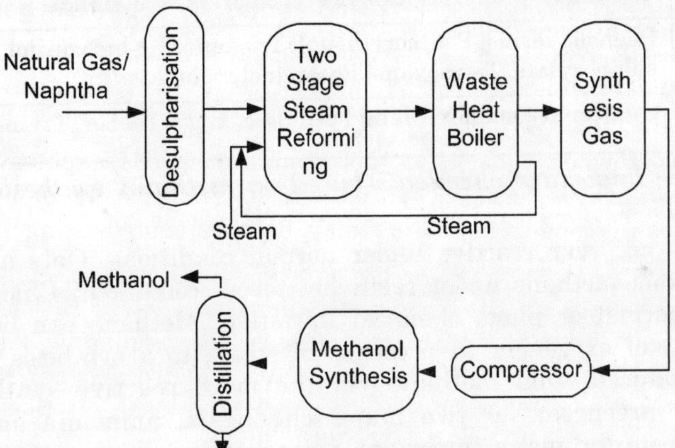

Figure: Methanol from steam reforming of Natural gas and Naphtha

The major reactions during methanol synthesis converter is as follows :

$$CO + 2H_2 \rightarrow CH_3OH$$

$$CO_2 + 3\,H_2 \rightarrow CH_3OH + H_2O$$

$$H_2 \rightarrow CO + H_2O$$

The first two reactions are exothermic and proceed with reduction in volume. In order to achieve a maximum yield of methanol and a maximum conversion of synthesis gas, the process must be carried out at low temperature& high pressure.

After cooling to ambient temperature, the synthesis gas is compressed to 5-10 MPa and is added to the synthesis loop which comprises of circulator, converters, heat exchanger, heat recovery exchanger, cooler, & separator. The catalyst used in methanol synthesis must be very selective towards the methanol reaction, i.e. give a reaction rate for methanol production which is faster than that of competing catalysts.

Q.18. *Describe the various products produced from methanol & their uses.*

Ans. Product Profile of Methanol & their uses are tabulated below.

Product	Uses
DMT/ Polyethylene terephthalate	Polyester fiber and film, Adhesives, Wire coating, Textile sizing, Herbicides
Methyl methacryla te (MMA)	Cast sheet, surface coating, molding resins, oil additives
MTBE	Oxygenate
Mono methanolami ne	Naphthyl-n-methyl carbamate, monoethyl hydrazine, Monomethylamine nitrate
Dimethylamine	Dimethyl acetamide, Dimethyl formamide, Dimethyl hydrazine, 2,4-Dichlorophenoxyacetic salt
Methylacetate	Paint remover
Dimethylaniline	Solvent, Flavoring, Dyes, Fragrance
Aceticacid	Vinyl acetate, Acetic anhydride, Chloro acetic acid, Ethyl acetate, Butyl acetate, Isopropyl acetate, Acetyl chloride, Acetanilide
Formaldehyde	Phenolic resins, Pentaerythritol, Trioxane, 1,4-butanediol, Formaldehyde, sulphoxylate, Tetraoxane, Resorcinol resin
Methylhalides	Quaternary amines, Methyl cellulose, Butyl rubber, Tri-methanol propene

Q.19. *List out the important chemicals based on methane, synthesis gas, ammonia, & methanol.*

Ans. Methane is not very reactive under normal conditions. Only a few chemicals can be produced directly from methane under relatively severe conditions. Chlorination of methane is only possible by thermal or photo chemical initiation. Methane can be partially oxidized with a limited amount of oxygen or in presence of steam to a synthesis gas mixture. Many chemicals can be produced from methane via the more reactive synthesis gas mixture. Synthesis gas is the precursor for two major chemicals, ammonia and methanol. Both compounds are the hosts for many important petrochemical products.

Figure below shows the important chemicals based on methane, synthesis gas, methanol & ammonia.

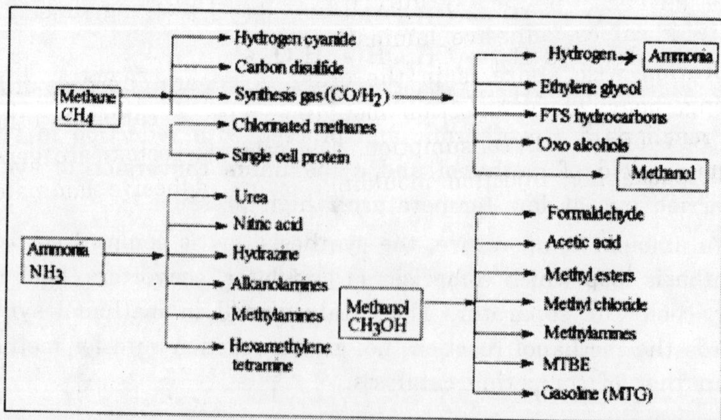

Figure: Important chemicals based on methane, synthesis gas, ammonia & methanol.

Q.20. ***What is the catalyst used in the Low Pressure Zeigler Process for the manufacture of polyethylene?***

Ans. Aluminiumtriethyl combined with titanium tetrachloride.

Q.21. ***What does the Ziegler-Natta catalyst consist of?***

Ans. The Ziegler Natta catalyst consists of a combination of HF and BF_3.

Q.22. ***What is the main criteria based upon which Polyethylene is classified?***

Ans. Polyethylene is classifiedbased upon its density and molecular bonding.

Q.23. ***Describe the classification and properties & applications of various types of polyethylene.***

Ans. Ethylene is the monomer for manufacturing Polyethylene. Polyethylene is broadly classified into High Density Polyethylene (HDPE), Linear Low Density Polyethylene (LLDPE) & Low Density Polyethylene (LDPE).

High Density Polyethylene (HDPE) is predominantly linear structure, closelypacked with density of 0.960 - 0.980 gram/cm^3, crystalline as high as 95% and a melting point as high as 138.5^{0}C. It finds its major application in Raffia, Blow Moulding, Film, Pipes and Injection Moulding.

HDPE because of its unique properties and increased availability offers abundant opportunities for growth. There is a huge untapped potential for the sector like loose to packed packaging in Blow moulding containers, automotive sector,infrastructure, and in installing drinking water & sewerage system etc.

Linear Low Density Polyethylene (LLDPE) is linear in structure. The branches are formed due topresence of co monomer. The co-monomer used is Butene (C_4), Hexene (C_6) & Octene (C_8). The low pressure used in manufacture of LLDPE does not allow the formation of large branches. LLDPE finds its application in mono & multi-layered films, rotomoulding, wire & cables and high flow LL is used in master batchcompounds etc.

Currently the fast growing sector in LL is Metallocene, followed by wire & cable & high flow grades. Metallocene LL grades are fast moving due to its properties like better strength, high dart, higher line speed, better sealing, downsizing of the film etc.

Low Density Polyethylene (LDPE) is predominantly a branched polymer. The properties depend upon the frequency and the length of branches. LDPE helps in processibility, has excellent transparency & good sealing properties & low glass transition temperature. Due to its inherent strength, LD is extensively used as a blend with other polymers. LDPE finds its application in general purpose film, heavy duty film,liquid packaging, injection moulding, extrusion coating, wire & cables, adhesive lamination.

The major application wise share is in the order of general purpose film, extrusion coating, milk packaging, heavy duty, injection moulding, wire & cables & others. Flexible Packaging constitutes 84% of the LD consumption. The growing sectors are general purpose film, heavy duty, milk packaging, injection moulding sectors, adhesive laminates etc.

Important properties of polyethylenes

Polyethylenes	Melting point range °C	Density g/cm^3	Degree of crystal-linity %	Stiffness modules psi × 10^3
Branched, Low density	107–121	0.92	60–65	25–30
Medium density	—	0.935	75	60–65
Linear, High density: Ziegler type	125–132	0.95	85	90–110
Phillips type	—	0.96	91	130–150

Q.24. *Discuss the salient features of Polyethylene (PE) Manufacturing Processes.*

Ans. Three classes of PE, namely, LDPE, HDPE, and LLDPE, are produced in different processes. LDPE is produced in a very high-pressure process, and HDPE & LLDPE are obtained in moderately low-pressure processes.

Salient features of PE manufacturing processes is given below in Table .

Table. Salient features of Polyethylene Manufacturing Processes

Reactant	Polymer Type of Polymerisation		Catalyst Temperature,°C Pressure, atm
Ethylene LDPE	Solution or bulk, tubular reactor, liquid phase	H_2O_2	200–240 3500
Ethylene HDPE	Suspension, stirred tank reactor, liquid phase	$TiCl_4$	60"80 20"35
Ethylene LLDPE	suspension, fluidised bed and 1-butenereactor, vapourphase	$TiCl_4$	1007–20

Q.25. *Describe the process of manufacture of Low Density Polyethylene (LDPE).*

Ans. In a tubular reactor, pure liquid ethylene is mixed with hydrogen peroxide and forced through the tubular reactor at very high pressure (3500 atm), surrounded by a cooling medium to extract the at of polymerisation reaction. Oxygen is used with hydrogen peroxide as the initiator.

There action takes place in solution. The heat of there action is given as 3650 kJ/kg.

$$nC_2H_4 = (-\ C_2H_4\ -)_n - 3650 \text{ kJ/kg.}$$

The temperature in the reactor is controlled above 200°C to avoid crystallisation of LDPE, which will damage the reactor tube. Conversion per pass is about 20%. Effluent from the reactor, consisting of the product and the unconverted monomer, is separated by high and low pressure separator vessels. Ethylene is recycled to the reactor. The overall conversion achieved by recycling is about 95"97%. Finally, the molten LDPE is withdrawn from the low pressure separator and extruded, followed by cooling, drying, and pilling. As tubular reactors are prone to plugging and poor heat transfer problems, a thick-walled, stirred tank vessel reactor is used. High speed agitation helps in good heat transfer. If the agitator stops, the temperature will rise to such an extent that explosion may takes place because of the presence of oxygen in the reaction mixture.

LDPE is used for making films, sheets, tubes, blocks, insulation, hoses etc. But, it cannot be used at temperatures above 80°C because deformation will occur. It has excellent dielectric properties, good elasticity up to "60°C, and it is anticorrosive. It is soft and waxy and is used to make films.

Figure below is a schematic flow diagram of an LDPE manufacturing process.

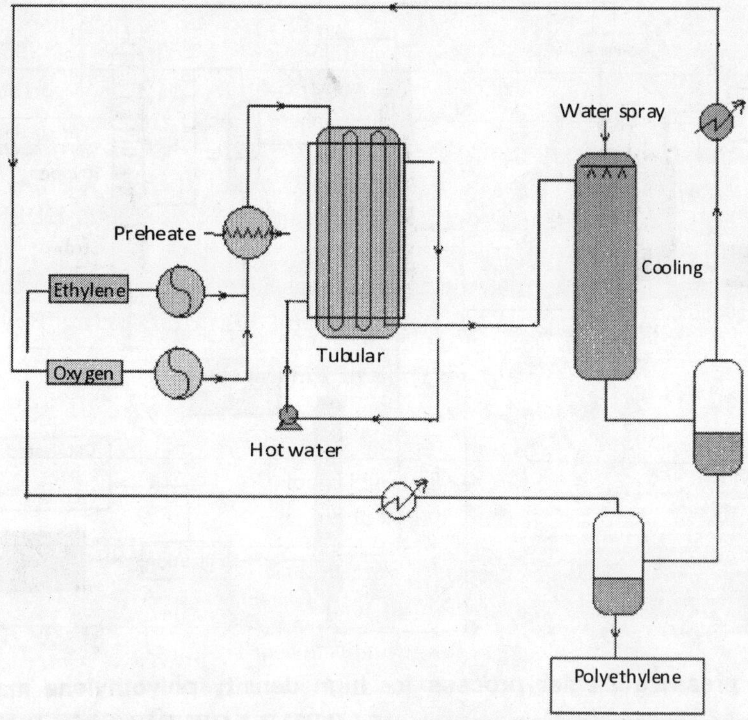

Figure: A high pressure low density polyethylene (LDPE) plant

Q.26. ***Describe the manufacturing process of High Density Polyethylene (HDPE).***

Ans. HDPE is manufactured by the suspension polymerisation method. In this method, high purity ethylene is introduced into the reactor vessel in which a catalyst (Ziegler-Nutta catalyst, TiCl4 in alkyl aluminium) is suspended in benzene at a pressure of 20"35 atm and at a temperature of 60"80°C. The ratio of alkyl aluminium and titanium chloride determines the size of the polymer. The greater the ratio, the greater the molecular weight of the polymer.

After the reaction, the polymer mixture is separated from ethylene and inerts in a flash drum. The polymer is water washed and filtered to recover the catalyst (water soluble) and reused. However, recovery of the catalyst from the polymer is not complete. Catalyst consumption is about 1 g of titanium (Ti) per 1500 kg of polymer. This is due to the high residual presence of the catalyst within the polymer. The presence of this impurity limits the HDPE's applications. A modern catalyst has been developed that can yield 6000 kg polymer per gram of Ti, where the contamination of Ti metal in the polymer is a few parts per million (ppm) only.

HDPE has a rigid and translucent property and is suitable for making electrical goods, bottles, ropes, etc. The main economic advantage of HDPE is that it can be manufactured at much lower pressure as compared to LDPE. HDPE has a higher melting temperature than LDPE and has almost the same glass transition temperature.

A flow diagram of a HDPE plant is shown below in Figure.

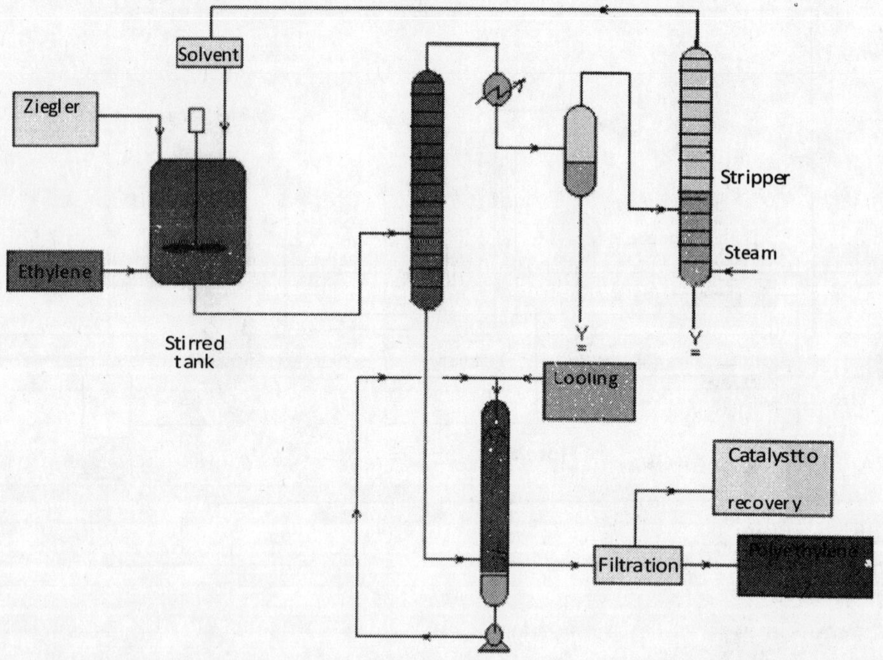

Figure: Low pressure Ziegler process for high density polyethylene manufacture.

Q.27. ***Describe the production process of LINEAR LOW DENSITY POLYETHYLENE (LLDPE).***

Ans. LLDPE is a copolymer of ethylene and 1-butene having linear structure. It is produced by a low pressure fluidised bed process where the temperature & pressure are 100°C & 7 to 20 atm. respectively. Gaseous monomers are used to fluidise the polymer granules previously prepared. During polymerisation, additional polymers are produced of specific sizes depending on the seed polymers present. Unreacted monomers are separated from the effluent and recycled to the reactor. Along residence time in the range of 3 to 5 hours is required for there action. Such a process is shown below in Figure. HDPE is also produced by a similar fluid used method in modern plants.

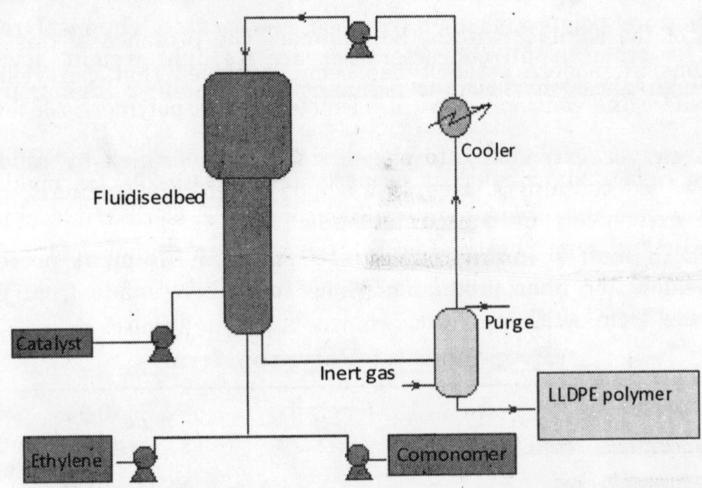

Figure: A fluidised bed LLDPE manufacturing unit.

Q28. ***Describe the manufacture, properties& applications of Ethylene Vinyl Acetate (EVA).***

Ans. Ethylene Vinyl Acetate is manufactured from polymerisation of Ethylene and Co-monomer (Vinyl Acetate/Ethyl Acrylate/Acrylic Acid/Methyl Acrylate). EVA is the most popular ethylene copolymer.

Vinyl Acetate (VA) content in EVA varies from 9 to 40%. The increase in VA content improves the properties like flexibility, ESCR, compatibility, filler acceptability, X-likability adhesion. etc.

EVA due to its flexibility for various processing techniques like Foaming, Moulding, Coating, blending, extrusion makes it useful for broader application range. Resin properties which makes it unique and useful are: very high low temperature impact strength, good elasticity & resilience, high resistance to flex cracking & high ESCR, excellent clarity & gloss, good resistance to weathering etc.

Significant growth of EVA has been possible due to technological developments in shoe manufacturing technology. Apart from footwear (shoes, slippers, beach sandals, etc) market other applications like film; HMA are also witnessing a healthy growth. Emerging sector "Solar Film" is highly technical & quality sensitive film produced from high VA and growing at a very healthy pace.

Q.29. ***How POLY PROPYLENE (PP) is produced?***

Ans. After polyethylene, polypropylene is a valuable polymer and is used as plastic for making pipes, ropes, fibres, etc. It is manufactured by catalytic reaction in a stirred tank reactor, where Ti & aluminium halides are used as catalysts at a temperature of 60"70 ^{0}C and a pressure of 1"2 MPa. An unreacted monomer is recycled after it is separated from the catalyst and polymer mixture in a flash chamber under vigorous stirring conditions. The mixture of polymer and catalyst is then passed to a centrifugal separator where a catalyst and polypropylene polymer is recovered. Further processing of the spent catalyst in the presence of alcohol is carried out to recover the active components of the catalyst for its reuse.

Q.30. ***Describe the properties and areas of applications of polypropylene (PP).***

Ans. The properties of polypropylene vary widely according to the percentage of crystalline isotactic polymer and the degree of polymerization.

Articles made from polypropylene have good electrical & chemical resistance and low water absorption. Its other useful characteristics are its light weight (lowest thermoplastic polymer density), high abrasion resistance, dimensional stability, high impact strength, and no toxicity.

Polypropylene can be extruded into sheets and thermoformed by solid- phase pressure forming into thin walled containers. Due to its light weight & toughness, polypropylene and its copolymers are extensively used in automobile parts.

Improvements in melt spinning techniques and film filament processes have made polypropylene accessible for fiber production. Low cost fibers made from polypropylene are replacing those made from sisal and jute.

Properties of Polypropylene

Density, g/cm^3	0.90–0.91
Fill temperature, max.°C	130
Tensile strength, psi	3,200–5,000
Water absorption, 24 hr.,%	0.01
Elongation,%	3–700
Melting point, Tm °C	176
Thermal expansion, 10–5in./in.°C	5.8–10
Specific volume, cm^3	30.4–30.8

Polypropylene today is one of the youngest members of the polymer which became one of the most invented inventions touching our lives in every possible way.

Main areas of applications of PP are in packaging, furniture, automobiles & domestic appliances sectors as shown below in figures.

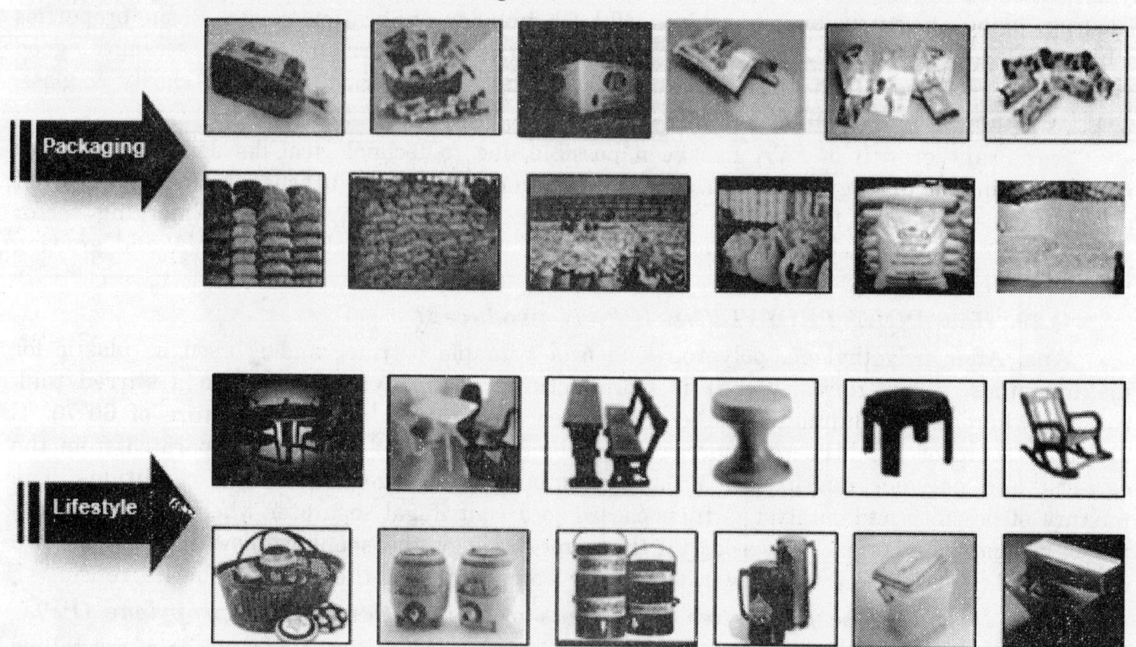

Figure: PP Applications - Homopolymers

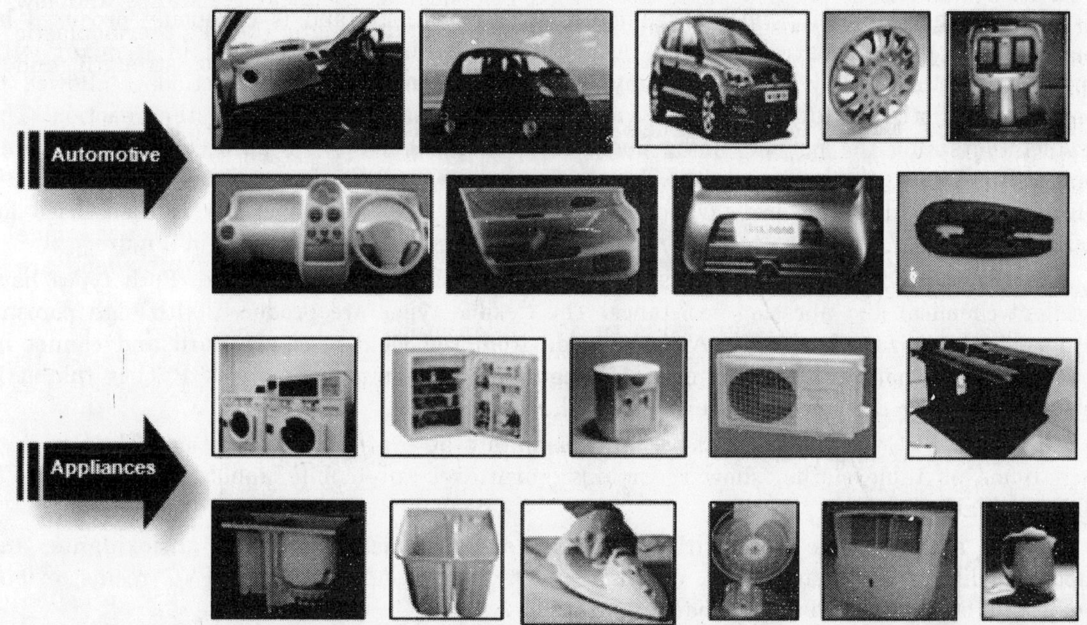

Figure: PP Applications – Copolymer

Q.31. ***Which are commonly used two polymerization methods for the manufacture of polyvinyl chloride? Mention some of the disadvantages of using PVC as a polymer.***

Ans. Two methods for manufacture of PVC are Emulsion Polymerization and Suspension Polymerization. Disadvantage of PVC is that it has low solvent solubility. In addition to this, the heat stability of PVC is poor.

Q.32. ***Describe the manufacturing process, Properties and Uses of Polyvinyl Chloride.***

Ans. PVC is a polymer of vinyl chloride, $n\text{CH}_2$ = CHCl= (–CH_2 – CHCl– CH_2–$)_n$.

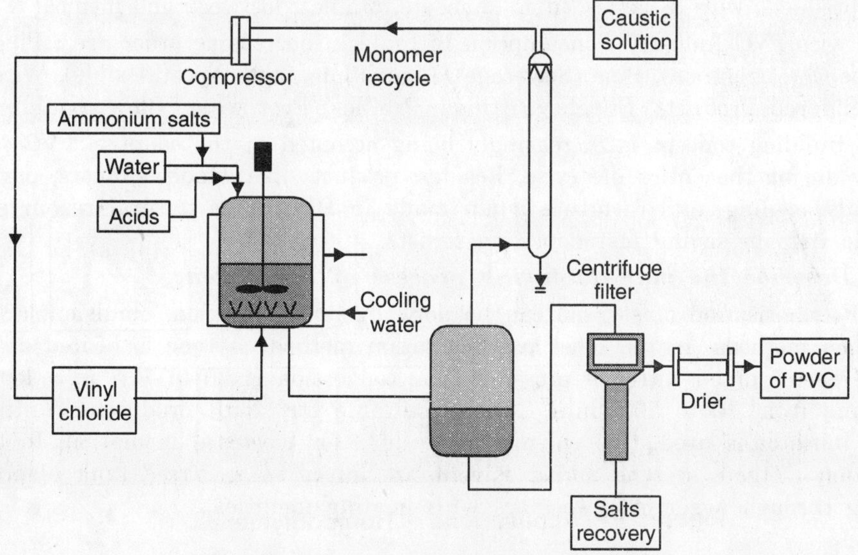

Figure: A PVC manufacturing plant.

It is a thermoplastic, widely used for many applications, and is commonly produced by the emulsion polymerisation method. An aqueous emulsion is produced in a mixer with ammonium salts of fatty acids & a vinyl chloride monomer and the reaction is allowed to continue in a stirred tank reactor. A very low pH is maintained during the reaction. The product containing the polymer latex and unreacted monomer are separated in a gas–liquid separator. These parated vapour of them on omerisre cycled after scrubbing with caustic solution to neutralise the acidic vapours. The polymer from the separator is then dried for making powder of polymer. A PVC manufacturing plant is shown below in Figure.

Two types of the homopolymer are available, the flexible and the rigid. Both types have excellent chemical and abrasion resistance. The flexible types are produced with high porosity to permit plasticizer absorption. Articles made from the rigid type are hard and cannot be stretched more than 40% of their original length. An important property of PVC is that it is self-extinguishing due to presence of the chlorine atom.

Flexible PVC grades account for approximately 50% of PVC production. They go into such items as table cloths, shower curtains, furniture, automobile upholstery, and wire & cable insulation.

Many additives are used with PVC polymers such as plasticizers, antioxidants, and impact modifiers. Heat stabilizers, which are particularly important with PVC resins, extend the useful life of the finished product.

Rigid PVC is used in many items such as pipes, fittings roofing, automobile parts, siding, and bottles.

Q.33. ***Describe the areas of application of PVC in India.***

Ans. The world's most versatile plastic, PVC had a humble beginning by making golf balls and shoe heels . Soon after this discovery, PVC-based products such as raincoats, shower curtains and insulated wire hit the market. PVC is recognized as 'an infrastructure plastic' finding applications in pipes, conduits, ducts, wire & cables, flooring, windows & doors and roofing etc. It is also used in many other sectors including automobiles, medical & healthcare, packaging, sports & leisure.

Pipes are the biggest end use sector of PVC with a share of 70%, other sectors being calendered products, wire & cables, films & sheets, profiles, footwear and medical & healthcare.

Sector wise PVC Industry Consumption in India in decreasing order are : Pipe (for water supply), Pipe (for Irrigation), Pipe (Sewerage), Pipe (Plumbing), Pipe (Flexible), Wire & Cables, Films, Calendered Products, Sheets, Fittings, Profiles, Foot wear, Blow moulding etc.

Green Building concept is increasingly being accepted in the country. PVC is known to save energy during the entire life cycle. Besides; products like windows, doors, partition, wall, paneling, false ceiling, and furniture when made in PVC, save natural resource like wood which would help in saving fast depleting forests.

Q.34. ***Describe the manufacturing process of Polystyrene.***

Ans. Polymerisation of styrene can be done by mass, solution, emulsion, or suspension polymerisation methods. In the mass polymerisation method, styrene is heated at 80 to 85°C in a batch reactor under nitrogen pressure. The conversion is 30 to 40% in a long residence time, ranging from 40 to 50 hours. This is suitable for small production units. Since no catalyst or initiator is used, this polymer is suitable for electrical insulation. In the solution polymerisation method, styrene and a solvent are mixed in a stirred tank blender followed by pumping through a series of reactors with heating facilities.

In a batch suspension polymerisation process, drops of styrene are dispersed in water in the presence of a benzoyl peroxide initiator. In the emulsion polymerisation method, a stirred tank autoclave is employed, in which styrene and water are emul-sified with an initiator. The product is then separated from unreacted monomers, which are then recycled.

Q.35. ***What are the areas of applications of different types of Polystyrene.***

Ans. There are three types of polystyrenes with different applications - General purpose Polystyrene (GPPS), High Impact Polystyrene (HIPS) and Expandable Polystyrene (EPS).

General Purpose polystyrene (GPPS) has excellent transparency, processability and its grades are used to cover a wide variety of applications like disposable containers, cutlery, CD storage cases, co-extrusion to obtain gloss layer, toys, stationery & refrigeration components.

High Impact polystyrenes (HIPS) resins are for a wide variety of applications both in injection & extrusion/thermoforming processes. Typical applications include Electronics, household appliances, disposable containers, toys and refrigeration components.

Expandable Polystyrene (EPS) also known as Polystyrene foam is prepared by suspension polymerization of styrene, and then adding foaming agent and white bead-like particles, with relative density of 1.05 are obtained. Thermal conductivity is low. EPS is polystyrene in a rigid foam honeycomb material with good insulation & vibration absorption, high compressive strength, very light weight and resistance to moisture. Its uses include building insulation & sound insulation, sidewall, and within the walls of the Cover, packaging materials and disposable containers.

Polystyrene (PS) end use is in India is mainly into the "Electronics/Appliances" sector, which accounts for almost 60 percent of total demand. PS demand in the India faces competition from other polymers, mainly polypropylene in both packaging and durables, primarily due to cost reasons.

Q.36. ***How are General purpose Polystyrene (GPPS) different from High impact Polystyrene (HIPS)?***

Ans. GPPS is also known as crystal polystyrene. It has good clarity and fairly good mechanical properties. Although GPPS can be processed into most types of objects, it is relatively brittle. It is used in applications where light transmission properties are important. On the other hand, HIPS contains an elastomer like polybutadiene rubber, which improves its mechanical properties. It has appreciable impact strength and hence, is milky or opaque in appearance. HIPS is tougher than GPPS, and also has good processing properties.

Q.37. *List out the important products made from five important commodity plastics.*

Ans. Important products made from five important commodity polymers/plastics namely LD/LLDPE, HDPE, PP, PVC and PS are tabulated below.

Table: Polymers/Plastics & Corresponding Products

LD/LLDPE	HDPE	PP	PVC	PS
Cast & stretch cling films	Raffia/ Monofilaments	Raffia/FIBC	Irrigation pipes & fittings	Thermocol
Mono/ Multi layer films	Crates	Fish nets	Wire and cables	Foam
Wires & Cables	Multi layer films	Buckets, Tiffin boxes	Floorings/Tiles	Refrigerator trays

Water storage tanks	Water Conveyance pipes	Automobile bumpers	Shower curtains	CD/DVD cases
House wares	Pesticide bottles	BOPP films	Blister films	Structural insulation
Carrier bags	Woven sacks	Washing machine tubs	Profiles & Builders' ware	Bottles
Furniture	Caps & closures	Leno bags	Footwear	Disposable cups/ cutleries
Milk Packaging	Pallets	Fibres Filaments	IV bottles & blood bags	Smoke detectors
Detergent Packaging	Thin wall moldings	Syringes	Bag liners	Films

Q.38. *Describe Polyethylene Terephthalate (PET) Production Processes.*

Ans. Polyethylene terephthalate (PET) is produced by esterifying terephthalic acid (TPA) & ethylene glycol or, more commonly, by the trans-esterification of dimethyl terephthalate (DMT) & ethylene glycol (EG). This route is favored because the free acid is not soluble in many organic solvents. The reaction occurs in two stages .Methanol is released in the first stage at approximately 200°C with the formation of bis (2-hydroxyethyl) terephthalate. In the second stage, polycondensation occurs, and excess ethylene glycol is driven away at approximately 280°C and at lower pressures ($\approx$ 0.01atm).

Using excess ethylene glycol is the usual practice because it drives the equilibrium to near completion and terminates the acid end groups. This results in a polymer with terminal -OH. When the free acid is used (esterification), the reaction is self catalyzed. However, an acid catalyst is used to compensate for the decrease in terephthalic acid as the esterification nears completion. In addition to the catalyst and terminator, other additives are used such as color improvers and dulling agents. For example, PET is delustred by the addition of titanium dioxide.

The molecular weight of the polymer is a function of the extent of polymerization and could be monitored through the melt viscosity. The final polymer may be directly extruded or transformed to chips, which are stored.

Batch polymerization is still used. However, most new processes use continuous polymerization and direct spinning.

An alternative route to PET is by the direct reaction of terephthalic acid (TPA) and ethylene oxide. The product bis (2-hydroxyethyl) terephthalate reacts in a second step with TPA to form a dimer and ethylene glycol, which is released under reduced pressure at approximately 300°C.

This process differs from the direct esterification and the transesterification routes in that only ethylene glycol is released. In the former two routes, water or methanol are co-produced and the excess glycol released.

PET is a condensation polymer made by continuous melt-phase polymerization process followed by Solid State Polymerization (SSP). In the past, it was manufactured either by DMT or PTA route:

- DMT Route (Di Methyl Terepthalate) DMT + MEG = PET Polymer
- PTA Route (Purified Terephthalic acid) PTA + MEG = PET Polymer

DMT is seldom used for PET now a day. It is manufactured using PTA to help achieve better quality and consistency.

Q.39. *Describe the properties and applications of Polyethylene Terephthalate (PET).*

Ans. Polyethylene Terephthalate (PET) is a thermo plastic polyester. PET, the material used for the manufacturing of the plastic beverage bottles is of high strength and high barrier material. Pepsi-cola was the first company to commercially introduce its beverages products in PET.

PET containers when properly manufactured, offer excellent clarity & gloss, toughness & impact resistance, chemical resistance, low gas permeation, good process ability and good dimensional stability.

Why PET is turning into a very useful material is shown below in figure.

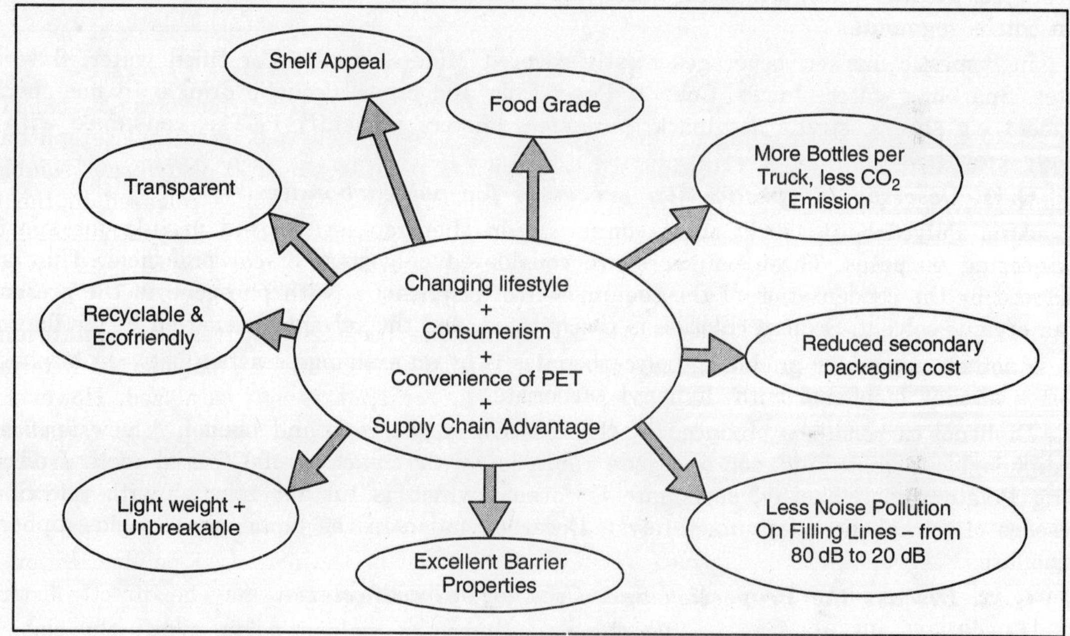

Figure: Why PET?

PET is used for the production of bottles, sheet, strapping and injection moulded products. PET bottles are used for packaging wide range of consumer products such as all types of Beverages including alcoholic, Edible Oil, Pharmaceutical formulations, Agro chemicals, Personal & Home Care etc. PET is also used for non-bottle applications such as thin/thick thermo formable sheet, dual-oven able ,crystallised PET (CPET) containers, injection moulded components, fastening straps etc.

Injection Moulding of PETis becoming more prevalent in today's market place. PET can be best injection moulded into parts with wall thickness of up to 4.0 mm.

Q.40. *Why there is a growing concern for the Recycling of PET?*

Ans. The concern for Recycled PETis continuously growing in recent times. PET bottles, which form the major market of PET packaging resin are the major concern from the point of recycling. More than 90% of PET is consumed in food packaging with drinks /beverages forming almost 80%. Since drinks & beverages are consumed mostly in residential houses, restaurants, stadia, railway stations, airports, entertainment venues & other public places, the organized collection & recycling of post-consumer PET bottles is important.

The PET bottle is 100% recyclable and most eco-friendly in nature. The need of the hour is proper handling system at all stages. The end consumers should take care to remove the cap & label from the used PET bottle before disposing it off and also should not throw the empty bottles in a public place. In order to ease the handling at different stages, the PET bottle should be 'flattened' and baled before sending it to the intended area of usage.

Recycled PET is becoming alternative for PET packaging options today and is getting serious consideration by the packaging world. There is a growing awareness among consumers towards their environment obligation in individual capacity and a marked preference thus for recycled content packaging.

PET consumption in India is growing rapidly. PET consumption in India is driven by Beverages, Health Care, Personal Care, Agro Chemicals, Cooking oil, Confectioneries and Non-bottle segments.

In domestic market, beverages mostly covered Mineral water, For tified water, flavored water, Sparkling water, Juices, Coke & Pepsi Cola and other alcoholic drinks. In non- bottle segment e.g sheets, straps ,lamipack, punnets& blisters presently PET is consumed with a steady growth .

Q.41. ***Describe the production processes for polycarbonates.***

Ans. Polycarbonates (PC) are condensation thermoplastics used mainly for special engineering purposes. These polymers are considered polyesters of carbonic acid. They are produced by the condensation of the sodium salt of bisphenol A with phosgene in the presence of an organic solvent. Sodium chloride is precipitated, and the solvent is removed by distillation.

Another method for producing polycarbonates is by an exchange reaction between bisphenol A or a similar bisphenol with diphenyl carbonate.

Diphenol carbonate is produced by the reaction of phosgene and phenol. A new approach to diphenyl carbonate and non-phosgene route is by the reaction of CO and methyl nitrite using Pd/alumina. Dimethyl carbonate is formed which is further reacted with phenol in presence of tetraphenox titanium catalyst. Decarbonylation in the liquid phase yields diphenyl carbonate.

Q.42. ***Discuss the Properties and Uses of Polycarbonates.***

Ans. Polycarbonates, known for their toughness in molded parts, typify the class of polymers known as engineering thermoplastics. These materials, designed to replace metals & glass in applications demanding strength & temperature resistance, offer advantages of light weight, low cost, and ease of fabrication.

Materials made from polycarbonates are transparent, strong, and heat & break resistant. However, these materials are subject to stress cracking and can be attacked by weak alkalies& acids.

Polycarbonates are used in a variety of articles such as laboratory safety shields, street light globes, and safety helmets. The maximum usage temperature for polycarbonate objects is 125°C.

Polycarbonate is a high performance engineering resin which demonstrated phenomenal growth during the 1990s averaging close to 20 percent per year largely due to its use in optical media applications - primarily the production of CDs and DVDs.

Polycarbonate finds its application in Sheets/film, Automotive (window and non window), optical media, Appliances, Housewares, Electronics/Electrical, medical, ophthalmic, construction, sports, recreational, packaging etc. As blends, it finds usage as auto body.

Q.43. ***Write a short note on manufacture, properties and applications of Acrylonitrile Butadiene Styrene (ABS).***

Ans. Acrylonitrile Butadiene Styrene (ABS) is a copolymer of acrylonitrile, butadiene, and styrene (ABS). In a typical batch suspension process styrene is suspended in water by agitation and use of a stabilizer. The polymer forms beads. The bead/water slurry is separated by centrifugation, dried, and blended with additives. It is derived from Acrylonitrile, butadiene, and styrene either as such as graft polymer or SAN (Styrene Acrilo Nitrile) on Polybutadiene or as a blend of ABS plus SAN. Composition of the constituent monomers can be varied in order to alter the properties of the final resin.

ABS resins are well-established, high volume, amorphous engineering thermoplastics, which offer an excellent balance of heat, chemical and impact resistance with superior processing versatility.

ABS has a specific gravity of 1.03 to 1.06 and a tensile strength in the range of 6 to 7.5 $\times 10^3$ psi. These polymers are tough plastics with outstanding mechanical properties. A wide variety of ABS modifications are available with heat resistance comparable to or better than polysulfones and polycarbonates. Another outstanding property of ABS is its ability to be alloyed with other thermoplastics for improved properties. For example, ABS is alloyed with rigid PVC for a product with better flame resistance.

The advantage of ABS is that this material combines the strength and rigidity of Acrylonitrile & styrene polymers with the toughness of the Poly-Butadiene rubber. Resistance and Toughness are the most important mechanical properties of this product.

Among the major applications of ABS are extruded pipes and pipe fittings, appliances such as refrigerator door liners, and in molded automobile bodies.

ABS also finds its application in Buildings/Construction, Transportation, Electronics/ Electrical, Appliances, Alloys and other application like medical, toys & recreational.

Q.44. ***Write a short note on Styrene Acrylonitrile (SAN).***

Ans. Styrene and Acrylonitrile monomers can be copolymerized to form SAN which is a random, amorphous copolymer that has improved weatherability, stress crack resistance, and barrier properties. The SAN copolymer generally contains 70 to 80% styrene and 30 to 20% Acrylonitrile. This combination provides higher strength, rigidity& chemical resistance than polystyrene, but it is not quite as clear as crystal polystyrene and its appearancet ends to yellow more quickly.

SAN (Tg = 105°C) is stiffer and has better chemical and heat resistance than the homopolymer. However, it is not as clear as polystyrene, and it is used in articles that do not require optical clarity, such as appliances and houseware materials. SAN has a high clarity with High Heat resistance & good chemical resistance besides good dimensional stability and food contact acceptability.

SAN finds its application in Electrical/Electronic, Automotive Applications, General Purpose Containers, Household Goods, Cosmetics and Compounding with ABS, Acrylics, etc.

Q.45. ***Write a short note on manufacture, properties & application of polyurathanes.***

Ans. Urathanes are thermosetting resins obtained by condensation polymerisation of polyol (diols or triols) and a diisocyanate. Examples of monomers are toluene diisocyanate $[(CH_3)(C_6H_4)(NCO)_2]$ and dihydric alcohol or glycol $[(CH_2OH)_2]$. The polymerisation is carried out in a stirred tank autoclave reactor at a temperature of around 200°C under nitrogen pressure.

Polyurethane polymers are either rigid or flexible, depending on the type of the polyol used. For example, triols derived from glycerol and propylene oxide are used for producing block slab foams.

Other polyhydric compounds with higher functionality than glycerol (three-OH) are commonly used. Examples are sorbitol (six-OH) and sucrose (eight-OH). Triethanolamine, with three OH groups, is also used. Diisocyanates generally employed with polyols to produce polyurethanes are 2, 4-and 2, 6-toluene diisocyanates prepared from dinitro-toluenes.

A different diisocyanate used in polyurethane synthesis is methylene diisocyanate (MDI), which is prepared from aniline and formaldehyde. The diamine product is reacted with phosgene to get MDI.

Improved polyurethane can be produced by copolymerization. Block copolymers of polyurethanes connected with segments of isobutylenes exhibit high temperature properties, hydrolytic stability, and barrier characteristics. The hard segments of polyurethane block polymers consist of –(RNHCOO)–n, where R usually contains an aromatic moiety.

The physical properties of polyurethanes vary with the ratio of the polyol to the diisocyanate. For example, tensile strength can be adjusted within a range of 1,200–600 psi; elongation, between 150–800%.

The major use of polyurethanes is to produce foam. The density as well as the mechanical strength of the rigid & the flexible types varies widely with polyol type and reaction conditions. For example, polyurethanes could have densities ranging between 1–6 lb/ft^3 for the flexible types and 1–50 lb/ft^3 for the rigid types. Polyurethane foams have good abrasion resistance, low thermal conductance and good load bearing characteristics. However, they have moderate resistance to organic solvents and are attacked by strong acids. Flame retardency of polyurethanes could be improved by using special additives, spraying a coating material such as magnesium oxychloride, or by grafting a halogen phosphorous moiety to the polyol. Trichlorobutylene oxide is sometimes copolymerized with ethylene and propylene oxides to produce the polyol.

Major markets for polyurethanes are furniture, transportation, and building & construction. Other uses include carpet underlay, textural laminates & coatings, footwear, packaging, toys, and fibers.

The largest use for rigid polyurethane is in construction and industrial insulation due to its high insulating property.

Molded urethanes are used in items such as bumpers, steering wheels, instrument panels, and body panels. Elastomers from polyurethanes are characterized by toughness and resistance to oils, oxidation, & abrasion. They are produced using short chain polyols such as polytetramethylene glycol from 1,4-butanediol. Polyurethanes are also used to produce fibers. Spandex (trade name) is a copolymer of polyurethane (85%) and polyesters.

Polyurethane networks based on tri-isocyante and diisocyanate connected by segments consisting of polyisobutylene are rubbery and exhibit high temperature properties, hydrolytic stability & barrier characteristics.

Q.46. *What are epoxy resin? Describe their properties and uses.*

Ans. Epoxy resins are produced by reacting epichlorohydrin and a diphenol. Bisphenol A is the diphenol generally used. The reaction, a ring opening polymerization of the epoxide ring, is catalyzed with strong bases such as sodium hydroxide.

The linear polymer formed is cured by cross-linking either with an acid anhydride, which reacts with the -OH groups, or by an amine, which opens the terminal epoxide rings. Cresols and other bisphenols are also used for producing epoxy resins.

Epoxy resins have a wide range of molecular weights (1,000–10,000). Those with higher molecular weights, termed phenoxy, are hydrolyzed to transparent resins that do not have the epoxide groups. These could be used in molding purposes, or crosslinked by diisocyanates or by cyclic anhydrides.

Important properties of epoxy resins include their ability to adhere strongly to metal surfaces, their resistance to chemicals, and their high dimensional stability. They can withstand temperatures up to 500°C. Epoxy resins with improved stress cracking properties can be made by using toughening agents, such as carboxyl-terminated butadiene-acrylonitrile liquid polymers. The carboxyl group reacts with the terminal epoxy ring to form an ester. The ester, with its pendant hydroxyl groups, reacts with the remaining epoxide rings, then more crosslinking occurs by forming ether linkages. This material is tougher than epoxy resins and suitable for encapsulating electrical units.

Major uses of epoxy resins are coatings for appliance finishes, auto primers, adhesive, and in coatings for cans & drums. Interior coatings of drums used for chemicals and solvents manifests its chemical resistance.

Q.46. *Write a short note on thermoplastic polyester.*

Ans. Thermoplastic polyesters are among the large volume engineering thermoplastics produced by condensation polymerization of terephthalic acid with ethylene glycol or 1,4-butanediol. These materials are used to produce film for magnetic tapes due to their abrasion& chemical resistance, low water absorption, and low gas permeability. Polyethylene terephthalate (PET) is also used to make plastic bottles (approximately 25% of plastic bottles are made from PET). Similar to nylons, the most important use of PET is for producing synthetic fibers. Polybutylene terephthalate (PBT) is another thermoplastic polyester produced by the condensation reaction of terephthalic acid and 1,4-butanediol.

The polymer is either produced in a bulk or a solution process. It is among the fastest growing engineering thermoplastics, and leads the market of reinforced plastics.

Q.47. *What are unsaturated polyester?*

Ans. Unsaturated polyesters are a group of polymers and resins used in coatings or for castings with styrene. These polymers normally have maleic anhydride moiety or an unsaturated fatty acid to impart the required unsaturation. A typical example is the reaction between maleic anhydride and ethylene glycol.

Phthalic anhydride, a polyol, and an unsaturated fatty acid are usually copolymerized to unsaturated polyesters for coating purposes. Many other combinations in variable ratios are possible for preparing these resins.

Q.48. *Describe the manufacturing process of PHENOL-FORMALDEHYDE RESINS.*

Ans. Phenol (C_6H_5OH) and formaldehyde (HCHO) in the presence of an aqueous ammoniacal medium condense to a phenol-formaldehyde resin. The reaction is carried out in a kettle-type reactor at a temperature of 160°C and at atmospheric pressure. The reaction time is less than 1 hour. The product from the reactor is usually blended with hexamethyl enetriamine and other fillers to make finished resins ready for moulding processes. Depending on the ratio of phenol to formaldehyde, a variety of grades of this resin may be made. This is a thermosetting type resin and is also termed as phenolic resin.

The reaction between phenol and formaldehyde is either base or acid catalyzed, and the polymers are termed resols (for the base catalyzed) and novalacs (for the acid catalyzed). The condensation reaction between the methylolphenols with the elimination of water forms the polymer. Crosslinking occurs by a reaction between the methylol groups and results in the formation of ether bridges. It occurs also by the reaction of the methylol groups and the aromatic ring, which forms methylene bridges. The formed polymer is a three-dimensional network thermoset.

Usually, the ratio of formaldehyde to phenol is kept less than unity to produce a linear fusible polymer in the first stage. Crosslinking of the formed polymer can occur by adding more formaldehyde and a small amount of hexamethylene tetramine [hexamine,$(CH_2)_6N_4$]. Hexamine decomposes in the presence of traces of moisture to formaldehyde and ammonia. This results in crosslinking and formation of a thermoset resin.

Q.49. *Discuss the Properties and Uses of Phenolic Resins.*

Ans. Important properties of phenolic resins are their hardness, corrosion resistance, rigidity, and resistance to water hydrolysis. They are also less expensive than many other polymers.

Many additives are used with phenolic resins such as wood flour, oils, asbestos, and fiberglass. Fiberglass piping made with phenolic resins can operate at 150°C and pressure up to 10 kg/cm^2.

Molding applications dominate the market of phenolic resins. Articles produced by injection molding have outstanding heat resistance and dimensional stability. Compression molded glass-filled phenolic resins disk brake pistons are replacing the steel ones in many automobiles because of their light weight and corrosion resistance. Phenolic resins are also used in a variety of other applications such as adhesives, paints, laminates for building, automobile parts, and ion exchange resins.

Q.50. *Describe the manufacturing process of Amino resins (Aminoplasts) such as urea formaldehyde and urea melamine.*

Ans. Amino resins are condensation thermosetting polymers of formaldehyde with either urea or melamine. Melamine is a condensation product of three urea molecules. It is also prepared from cyanamide at high pressures and temperatures.

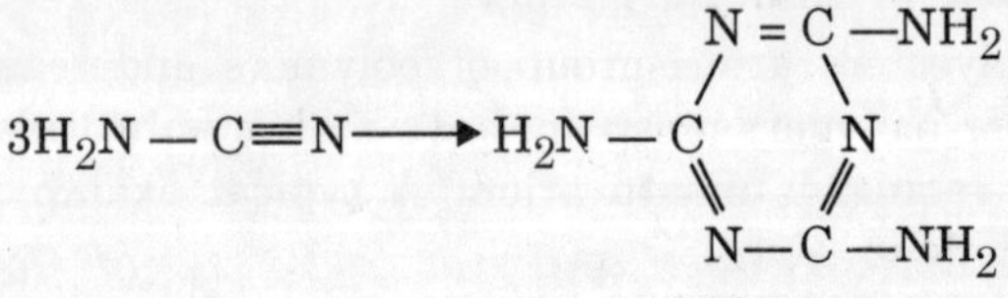

$$3H_2N-C{\equiv}N \longrightarrow H_2N-C\begin{matrix} \diagup N=C-NH_2 \diagdown \\ \\ \diagdown N-C-NH_2 \diagup \end{matrix}N$$

Melamine

Urea-Formaldehyderesin: The addition polymerization reaction of urea to formaldehyde produces mainly monomethylol urea and some dimethylol urea. When the mixture is heated in presence of an acid, condensation occurs, and water is released. This is accompanied by the formation of a crosslinked polymer:

Urea $(CONH_2)_2$ and formaldehyde (HCHO) readily react to polymerise into urea formaldehyde resin. Polymerisation is carried out in a glass-lined (or stainless steel) stirred tank reactor in an ammoniacal or weak alkaline medium. The polymerisation reaction takes place at room temperature and atmospheric pressure. This resin is a thermosetting resin.

Urea-Melamine Resins: A similar reaction occurs between melamine & formaldehyde and produces methylol melamines. A variety of methylols are possible due to the availability of six hydrogens in melamine. As with urea formaldehyde resins, polymerization occurs by a condensation reaction and the release of water.

Methylene Formaldehyde: Melamine [$C_3N_3(NH_2)_3$] and formaldehyde (HCHO) combine readily to form melamine-formaldehyde resin, which is also a thermosetting resin. The temperature of there action is 100°C in a neutral medium, i.e., at a pH of 7 in the aqueous medium at atmospheric pressure. Melamine formaldehyde resins are thermosets.

Q.51. ***Describe the Properties and Uses of Amino resins (Aminoplasts).***

Ans. Amino resins are characterized by being more clear and harder (tensile strength) than phenolic resins . However, their impact strength (breakability) and heat resistance are lower. Melamine resins have better heat & moisture resistance and better hardness than their urea analogs.

The most important use of amino resins is the production of adhesives for particle board and hardwood plywood. Compression and injection molding are used with amino resins to produce articles such as radio cabinets, buttons, and cover plates. Because melamine resins have lower water absorption and better chemical & heat resistance than urea resins, they are used to produce dinnerware and laminates used to cover furniture. Almost all molded objects use fillers such as cellulose, asbestos, glass, wood flour, glass fiber and paper.

Q.52. ***What is Poly Methyl Metha Acrylate (PMMA) polymer?***

Ans. This is a polymer of the methyl metha acrylate monomer, which is produced from acrylonitrile (CH_2CHCN) and sulfuric acid at a temperature below 90°C to yield methyl acrylate (CH_2CONH_2) followed by reacting with methanol to convert methyl acrylate to methyl methacrylate ($CH_2CHCOOCH_3$). This methyl metha acrylate monomer is then polymerised in a stirred tank autoclave under nitrogen pressure. Any of the solution, emulsion, massorsus pension polymerisation methods are used.

Q.53. ***What is Poly Tetra Fluoro Ethylene (PTFE) polymer? How it is prepared?***

Ans. Poly tetra fluoro ethylene (PTFE) or *teflon is* a polymer of the tetrafluoroethylene (C_2F_4) monomer. This monomer is prepared from methanol chlorination followed by fluorination. The reactions are carried out in three reactors. Trichloro methane produced in the chlorinator is reacted with hydrofluoric acid in the first reactor at about 65°C in the presence of $SbCl_5$ as the catalyst and is converted to produce mono-chloro-difluoro- methane (CHF_2Cl) in the presence of $AlCl_3$ as the catalyst in the second reactor. In the third reactor, CHF_2Cl is then converted to C_2F_4 by catalytic pyrolysis at a temperature of 650–800 °C in the presence of platinum (Pt) as the catalyst. Polymerisation of C_2F_4 is then carried out in a batch reactor preferably by the suspension polymerisation method at a temperature of 200°C & at a pressure of 70 kg/cm^2. The reaction time is about 1 hour.

Q.54. ***What are synthetic rubber and natural rubber? Compare their properties.***

Ans. Synthetic rubbers (elastomers) are long chain polymers with special chemical and physical as well as mechanical properties. These materials have chemical stability, high abrasion resistance, strength, and good dimensional stability. Many of these properties are imparted to the original polymer through crosslinking agents and additives. An important property of elastomeric materials is their ability to be stretched at least twice their original length and to return back to nearly their original length when released.

Synthetic rubbers include elastomers that could be crosslinked such as polybutadiene, polyisoprene, and ethylene-propylene-diene tere-polymer. It also includes thermoplastic elastomers that are not crosslinked and are adapted for special purposes such as automobile bumpers and wire & cable coatings. These materials could be scraped and reused. However, they cannot replace all traditional rubber since they do not have the wide temperature performance range of thermoset rubber.

Synthetic Rubber is broadly classified in two categories viz. General Purpose Rubber (GP) and Special Purpose Synthetic Rubber. GP includes Styrene Butadiene Rubber (SBR), Polybutadiene Rubber (PBR) and Polyisoprene Rubber (IR). Styrene Butadiene Rubber (SBR) is the largest tonnage synthetic rubber. Ethylene-Propylene Rubber (EPM and EDPM), Butyl Rubber (IIR) and Nitrile Rubber (NBR) are some of the other important types of synthetic rubbers.

Natural rubber is an elastomer constituted of isoprene units. These units are linked in a cis-1,4-configuration that gives natural rubber the outstanding properties of high resilience and strength. Natural rubber occurs as a latex (water emulsion) and is obtained from a tree. The latex can be vulcanized (crosslinked) by heating with sulfur or other agents. Vulcanization of rubber is a chemical reaction by which elastomer chains are linked together. The long chain molecules impart elasticity, and the crosslinks give load supporting strength.

The major use of rubber is for tire production. Non-tire consumption includes hoses, footwear, molded & extruded materials, and plasticizers.

Q.55. *What are various synthetic rubbers produced from 1, 3-Butadiene (BD). List out properties and uses of different rubbers.*

Ans. Butadiene is colourless easily liquefiable flammable gas used in the production of a wide variety of synthetic rubbers (SBR, BR, NBR etc.), Acrylonitrile-butadiene-styrene (ABS), polymer resins and few chemical intermediates. Chemical intermediates manufactured from butadiene are adiponitrile and chloroprene. Adiponitrile is used to make nylon fibres and polymers. Chloroprene is used to make polychloroprene (better known as Neoprene) which has uses such as wet suits, electrical insulation, car fan belts, gaskets, hoses, corrosion-resistant coatings and adhesives.

In India Butadiene (BD) is manufactured from C_4 hydrocarbons available from Naphtha & Gas Cracking units.

Selected properties of some elastomers

Elastomers	Durometer hardness range	Tensile strength at room temp, psi	Elongation at room temp, %	Temp. range of service, °C	Weather resis-tance
Natural rubber	20–100	1,000–4,000	100–700	–55–80	Fair
Styrene-butadiene rubber (SBR)	40–100	1,000–3,500	100–700	–55–110	Fair
Polybutadiene	30–100	1,000–3,000	100–700	–60–100	Fair
Polyisoprene	20–100	1,000–4,000	100–750	–55–80	Fair
Polychloroprene	20–90	1,000–4,000	100–700	–55–100	Very good
Polyurethane	62–95 A 40–80 D	1,000–8,000	100–700	–70–120	Excellent
Polyisobutylene	30–100	1,000–3,000	100–700	–55–100	Very good

Uses of various rubbers:

Styrene Butadiene Rubber (SBR) is used mainly in automobile tyres and also in adhesives, sealants, coatings & rubber articles such as shoesoles.

Butadiene Rubber (BR) is used mainly in tyres and also as an impact modifier of polymers such as high impact polystyrene (HIPS).

Nitrile rubbers are used mainly in the manufacture of hoses, gasket seals and fuel lines for the automobile industry as well as in gloves & footwear.

Acrylonitrile-butadiene-styrene (ABS) is used in business machines, computers, radios, televisions & telephone handsets and also in automobile industry.

Styrene-butadiene (SB) copolymer latex is used in paper coatings, carpet back coatings, foam mattresses and adhesives.

Q.56. ***Describe the manufacture, properties and applications of Polybutadiene Rubber (PBR).***

Ans. Butadiene (CH_2 = CH–CH=CH_2) is obtained as the by-product from a Naphtha Cracker Unit (NCU). By catalytic (Ziegler–Nutta) polymerisation of butadiene, a product PBR that resembles natural rubber is obtained. A solvent is used to keep the catalyst & the monomer in the solution during the reaction. The product is separated from its unconverted monomer, which is recycled. The solvent is evaporated, stripped, and recycled. The polymer from the evaporator is then dried to get the final product (PBR).

Butadiene can be polymerized using free radical initiators or ionic or coordination catalysts. When butadiene is polymerized in emulsion using a free radical initiator such as cumenehydroperoxide, a random polymer is obtained with three isomeric configurations, the 1,4-addition configuration dominating.

Polymerization of butadiene using anionic initiators (alkyllithium) in a nonpolar solvent produces a polymer with a high cis configuration. A high cis-polybutadiene is also obtained when coordination catalysts are used.

Manufacturing process of PBR from butadiene is shown below in figure.

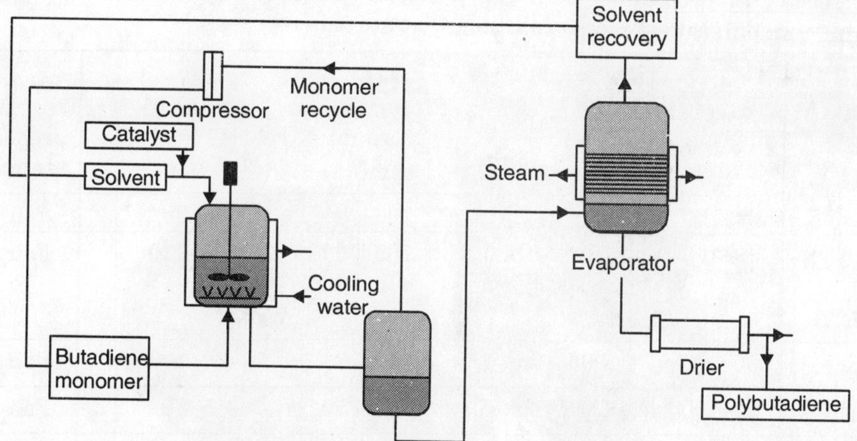

Figure: Manufacturing process of polybutadiene rubber from butadiene

cis-1, 4-Polybutadiene rubber has high elasticity, low heat buildup, high abrasion resistance, and resistance to oxidation. However, it has a relatively low mechanical strength. This is improved by incorporating a cis- trans block copolymer or 1, 2 (vinyl) block copolymer in the polybutadiene matrix. A small amount of natural rubber may be mixed with polybutadiene to improve its properties. trans 1,4-Polybuta- diene rubber has a higher glass transition temperature (Tg = –14°C) than the cis form (Tg = –108°C). The polymer has the toughness, resilience& abrasion resistance of natural rubber (Tg = –14°C).

PBR is used mainly in tyres and also as an impact modifier of polymers such as high impact polystyrene (HIPS).

Following figure shows the consumption pattern of PBR in India during 2010-2011.

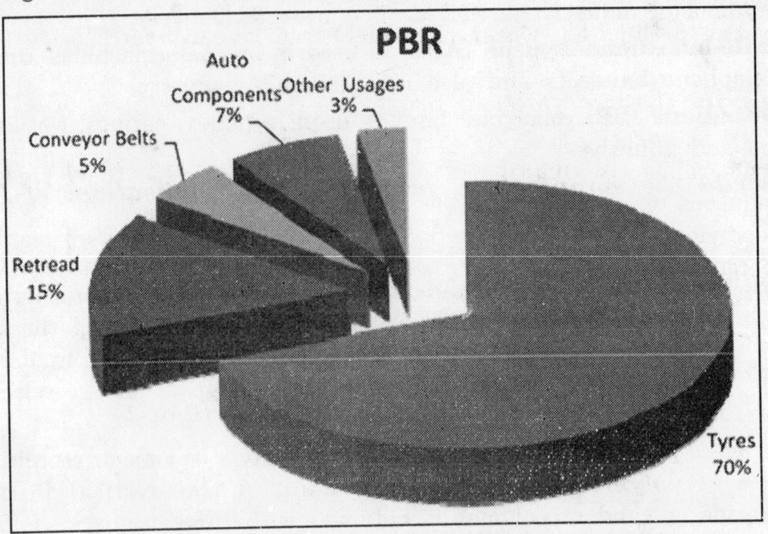

Figure: PBR – 2010-11 Consumption in India

Q.57. ***Describe the manufacture, properties and applications of STYRENE BUTADIENE RUBBER (SBR).***

Ans. SBR is produced by the copolymerization of butadiene (75%) & styrene (25%) using free radical initiators. A random copolymer is obtained. Polymerisation is carried out in a stirred tank reactor at around 50°C in the presence of free radical initiators, e.g., potassium sulfate, in an ammoniacal aqueous solution.

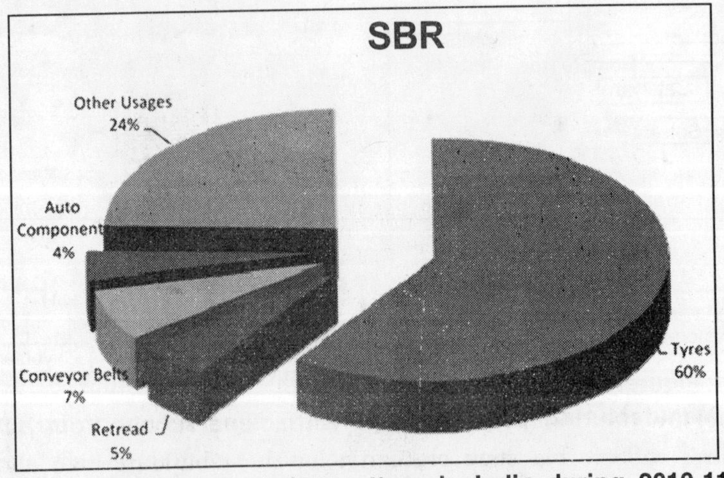

Figure: SBR Consumption pattern in India during 2010-11

Currently, more SBR is produced by copolymerizing the two monomers with anionic or coordination catalysts. The formed copolymer has better mechanical properties and a narrower molecular weight distribution. A random copolymer with ordered sequence can also be made in solution using butyllithium, provided that the two monomers are charged slowly. Block copolymers of butadiene and styrene may be produced in solution using coordination or anionic catalysts. Butadiene polymerizes first until it is consumed, then styrene starts to polymerize. SBR produced by coordination catalysts has better tensile strength than that produced by free radical initiators.

Styrenebutadiene rubber (SBR) is the most widely used synthetic rubber. The main use of SBR is for tire production. Other uses include footwear, coatings, carpet backing, and adhesives.

Q.58. ***Describe the manufacture, properties and applications of NITRILE RUBBER (NBR).***

Ans. Nitrile rubber is a copolymer of butadiene and acrylonitrile. The copolymerization occurs in an aqueous emulsion. When free radicals are used, a random copolymer is obtained. Alternating copolymers are produced when a Zieglar-Natta catalyst is employed. Molecular weight can be controlled by adding modifiers and inhibitors. When the polymerization reaches approximately 65%, the reaction mixture is vacuum distilled in presence of steam to recover the monomer.

The ratio of acrylonitrile/butadiene can be adjusted to obtain a polymer with specific properties. Increasing the acrylonitrile ratio increases oil resistance of the rubber, but decreases its plasticizer compatibility. NBR has the special property of being resistant to hydrocarbon liquids.

NBR is produced in different grades depending on the end use of the polymer. Low acrylonitrile rubber is flexible at low temperatures and is generally used in gaskets, O-rings, and adhesives. The medium type is used in less flexible articles such as kitchen mats and shoe soles. High acrylonitrile polymers are more rigid and highly resistant to hydrocarbons & oils and are used in fuel tanks& hoses, hydraulic equipment, and gaskets.

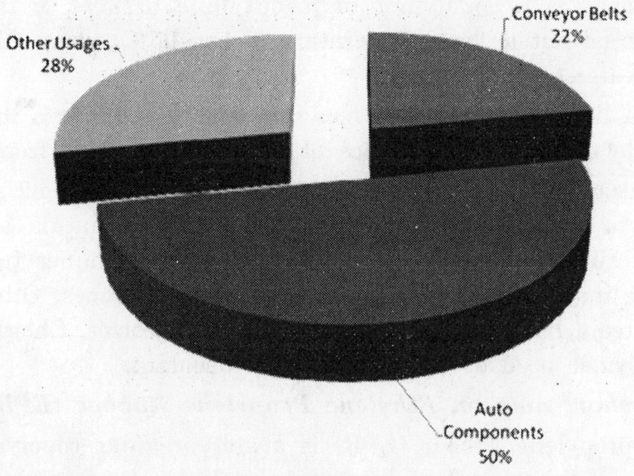

Figure: NBR Consumption pattern in India during 2010-11

Q.59. ***Describe the manufacture, properties and applications of polyisoprene.***

Ans. Natural rubber is a stereoregular polymer composed of isoprene units attached in a cis configuration. This arrangement gives the rubber high resilience and strength.

Isoprene can be polymerized to form polyisoprene using free radical initiators, but a random polymer is obtained. Stereo regularpoly is oprene is obtained when Zieglar-Natta catalysts or anionic initiators are used. The most important coordination catalyst is a- $TiCl_3$ cocatalyzed with aluminum alkyls. The polymerization rate and cis content depends upon

Al/Ti ratio, which should be greater than one. Lower ratio spread ominantly produced hetrans structure.

Polyisoprene is a synthetic polymer (elastomer) that can be vulcanized by the addition of sulfur. cis-Polyisoprene has properties similar to that of natural rubber. It is characterized by high tensile strength and insensitivity to temperature changes, but it has low abrasion resistance. It is attacked by oxygen and hydrocarbons.

trans-Polyisoprene is similar to Guttapercha, which is produced from the leaves and bark of the sapotacea tree. It has different properties from the cis form and cannot be vulcanized. Few commercial uses are based on trans-polyisoprene.

Important uses of cis-polyisoprene include the production of tires, specialized mechanical products, conveyor belts, footwear, and insulation.

Q.60. *Write a short note on POLYCHLOROPRENE (Neoprene Rubber).*

Ans. Polychloroprene is the oldest synthetic rubber. It is produced by the polymerization of 2-chloro-1, 3-butadiene in a water emulsion with potassium sulfate as acatalyst.

The product is a random polymer that is vulcanized with sulfur or with metal oxides (zinc oxide, magnesium oxide etc.). Vulcanization with sulfur is very slow, and an accelerator is usually required. Neoprene vulcanizates have a high tensile strength, excellent oil resistance (better than natural rubber), and heat resistance.

Neoprene rubber can be used for tire production, but it is expensive. Major uses include cable coatings, mechanical goods, gaskets, conveyor belts, & cables.

Q.61. *Write a short note on Butyl Rubber.*

Ans. Butyl rubber is a copolymer of isobutylene (97.5%) & isoprene (2.5%). The polymerization is carried out at low temperature (below–95°C) using $AlCl_3$ cocatalyzed with a small amount of water.

The product is a linear random copolymer that can be cured to a thermosetting polymer. This is made possible through the presence of some unsaturation from isoprene.

Butyl rubber vulcanizates have tensile strengths up to 140 kg/cm2 and are characterized by low permeability to air and a high resistance to many chemicals & to oxidation. These properties make it a suitable rubber for the production of tire inner tubes and inner liners of tubeless tires. The major use of butyl rubber is for inner tubes. Other uses include wire & cable insulation, steam hoses, mechanical goods, and adhesives. Chlorinated butyl is a low molecular weight polymer used as an adhesive and asealant.

Q.62. *Write a short note on Ethylene Propylene Rubber (EPR).*

Ans. Ethylene-propylene rubber (EPR) is a stereoregular copolymer of ethylene and propylene. Elastomers of this type do not possess the double bonds necessary for crosslinking. A third monomer, usually a monoconjugated diene, is used to provide the residual double bonds needed for cross linking. 1,4-Hexadiene and ethylidenenorbornene are examples of these dienes. The main polymer chain is completely saturated while the unsaturated part is pending from the main chain. The product elastomer, termed ethylene propylene terepolymer (EPT), can be crosslinked using sulfur. Crosslinking EPR is also possible without using a third component (a diene). This can be done with peroxides.

Important properties of vulcanized EPR and EPT include resistance to abrasion, oxidation, and heat & ozone; but they are susceptible to hydrocarbons attack.

The main use of ethylene propylene rubber is to produce automotive parts such as gaskets, mechanical goods, wire, and cable coating. It may also be used to produce tires.

Q.63. *Write a short note on THERMOPLASTIC ELASTOMERS (TPE's).*

Ans. Thermoplastic elastomers (TPE's), are plastic polymers with the physical properties of rubbers. They are soft, flexible, and possess the resilience needed of rubbers. However, they are processed like thermoplastics by extrusion and injection molding.

TPE's are more economical to produce than traditional thermoset materials because fewer steps are required to manufacture them than to manufacture and vulcanize thermoset rubber. An important property of these polymers is that they are recyclable.

Thermoplastic elastomers are multiphase composites, in which the phases are intimately depressed. In many cases, the phases are chemically bonded by block or graft copolymerization. At least one of the phases consists of a material that is hard at room temperature.

Important TPE's include blends of semicrystalline thermoplastic polyolefins such as propylene copolymers, with ethylene-propylene terepolymer elastomer. Block copolymers of styrene with other monomers such as butadiene, isoprene, and ethylene or ethylene/propylene are the most widely used TPE's. Styrene-butadiene-styrene (SBS) accounts for 70% of global styrene block copolymers (SBC). Polyurethane thermoplastic elastomers are relatively more expensive then other TPE's. However, they are noted for their flexibility, strength, toughness, and abrasion & chemical resistance. Blends of polyvinyl chloride with elastomers such as butyl are widely used in Japan.

Random block copolymers of polyesters (hard segments) and amorphous glycol soft segments, alloys of ethylene interpolymers, and chlorinated polyolefins are among the evolving thermoplastic elastomers.

Important properties of TPE's are their flexibility, softness, and resilience. However, compared to vulcanizable rubbers, they are inferior in resistance to deformation and solvents.

Important uses for TPE's include shoe soles, pressure sensitive adhesives, insulation, and recyclable bumpers.

Q.64. *Write a short note on synthetic fibres.*

Ans. Fibers are solid materials characterized by a high ratio of length to diameter. They can be manufactured from a natural origin such as silk, wool, and cotton, or derived from a natural fiber such as rayon. They may also be synthesized from certain monomers by polymerization (synthetic fibers). In general, polymers with high melting points, high crystallinity, and moderate thermal stability & tensile strengths are suitable for fiber production.

Man-made fibers include, in addition to synthetic fibers, those derived from cellulose (cotton, wood) but modified by chemical treatment such as rayon, cellophane, and cellulose acetate. These are sometimes termed "regenerated cellulose fibers." Rayon and cellophane have shorter chains than the original cellulose due to degradation by alkaline treatment.

Major fiber-making polymers are those of polyesters, polyamides (nylons), polyacrylics, and polyolefins. Polyesters and polyamides are produced by step polymerization reactions, while polyacrylics and poly-olefins are synthesized by chain-addition polymerization.

Q.65. ***Describe the manufacture, properties and applications of POLYESTER FIBERS.***

Ans. Polyesters are the most important class of synthetic fibers. In general, polyesters are produced by an esterification reaction of a diol and a diacid. First polyester polymer was prepared by using terephthalic acid (an aromatic diacid) and ethylene glycol.

Polyesters can be produced by an esterification of a dicarboxylic acid and a diol, a transesterification of an ester of a dicarboxylic acid and a diol, or by the reaction between an acid dichloride and a diol.

Polyethylene terephthalate (PET) is produced by esterifying terephthalic acid (TPA) and ethylene glycol or, more commonly, by the trans- esterification of dimethyl terephthalate and ethylene glycol. This route is favored because the free acid is not soluble in many organic solvents. The reaction occurs in two stages. Methanol is released in the first stage at approximately 200°C with the formation of bis (2-hydroxyethyl) terephthalate. In the second stage, polycondensation occurs, and excess ethylene glycol is driven away at approximately 280°C and at lower pressures (≈ 0.01 atm).

Batch polymerization is still used. However, most new processes use continuous polymerization and direct spinning.

An alternative route to PET is by the direct reaction of terephthalic acid and ethylene oxide. The product bis (2-hydroxyethyl) terephthalate reacts in a second step with TPA to form a dimer and ethylene glycol, which is released under reduced pressure at approximately 300°C.

This process differs from the direct esterification and the transesterification routes in that only ethylene glycol is released. In the former two routes, water or methanol are co-produced and the excess glycol released.

Polyethylene terephthalate is an important thermoplastic. However, most PET is consumed in the production of fibers. Polyester fibers contain crystalline as well as non-crystalline regions. The degree of crystallinity and molecular orientation are important in determining the tensile strength of the fiber (between 18–22 denier) and its shrinkage. The degree of crystallinity and molecular orientation can be determined by X-ray diffraction techniques.

Important properties of polyesters are the relatively high melting temperatures (≈ 265°C), high resistance to weather conditions & sunlight, and moderate tensile strength.

Melt spinning polyesters is preferred to solution spinning because of its lower cost. Due to the hydrophobic nature of the fiber, sulfonated terephthalic acid may be used as a comonomer to provide anionic sites for cationic dyes. Small amounts of aliphatic diacids such as adipic acid may also be used to increase the dyeability of the fibers by disturbing the fiber'scrystallinity.

Polyester fibers can be blended with natural fibers such as cotton and wool. The products have better qualities and are used for men's and women's wear, pillow cases, and bedspreads. Fiberfill, made from polyesters, is used in mattresses, pillows, and sleeping bags. High tenacity polymers for tire cord reinforcement are equivalent in strength to nylon tire cords and are superior because they do not "flatspot." V-belts and fire hoses made from industrial filaments are another market for polyesters.

Q.66. ***Write a short note on nylon resin.***

Ans. Nylon resins are important engineering thermoplastics. Nylons are produced by a condensation reaction of amino acids, a diacid and a diammine, or by ring opening lactams such as caprolactam. The polymers, however, are more important for producing synthetic fibers.

Important properties of nylons are toughness, abrasion & wear resistance, chemical resistance, and ease of processing. Reinforced nylons have higher tensile &impact strengths and lower expansion coefficients than metals. They are replacing metals in many of their applications. Objects made from nylons vary from extruded films used for pharmaceutical packaging to bearings & bushings, to cable and wire insulation.

Q.67. ***Write a short note on POLYAMIDES (Nylon Fibers).***

Ans. Polyamides/Nylons fibres are the second largest group of synthetic fibers after polyesters. Polymers of amine or polyamides are excellent for making fibres. These are well known as nylons.

Polyamides are produced by the reaction between a dicarboxylic acid and a diamine (e.g., nylon 66), ring openings of a lactam, (e.g., nylon 6) or by the polymerization of w-amino acids (e.g., nylon 11). Condensation polymerisation is carried out with the monomer/monomers in autoclaves or stirred tank batch reactors.

Numbers that follow the word "nylon" denote the number of carbons present within a repeating unit and whether one or two monomers are being used in polymer formation. For nylons using a single monomer such as nylon 6 or nylon 12, the numbers 6 and 12 denote the number of carbons in caprolactam and laurolactam, respectively. For nylons using two monomers such as nylon 610, the first number, 6, indicates the number of carbons in the hexamethylene diamine and the other number, 10, is for the second monomer sebacic acid.

Q.68. ***Describe the properties and uses of nylons.***

Ans. Nylons have relatively high melting points due to the presence of the amide linkage. They are also highly crystalline, and the degree of crystallinity depends upon factors such as the polymer structure & the distance between the amide linkages. An increase in polymer crystallinity increases its tensile strength, abrasion resistance, and modulus of elasticity.

Hydrogen bonding in polyamides is fairly strong and has a pronounced effect on the physical properties of the polymer such as the crystallinity, melting point, and water absorption. For example, nylon 6, with six carbons, has a melting point of 223°C, while it is only 190°C for nylon 11. This reflects the higher hydrogen bonding in nylon 6 than in nylon 11.

Moisture absorption of nylons differs according to the distancebetween the amide groups. For example, nylon 4 has a higher moisture absorption than most other nylons, and it is approximately similar to that of cotton. This is a result of the higher hydrophilic (water loving) character of nylon 4. Nylons, however, are to some extent subject to deterioration by light.

This has been explained on the basis of chain breaking and crosslinking. Nylons are liable to attack by mineral acids but are resistant to alkalies. They are difficult to ignite and are self-extinguishing.

In general, most nylons have remarkably similar properties, and the preference of using one nylon over the other is usually dictated by economic considerations except for specialized uses.

Melting points of various nylons and the monomer formula are tabulated below.

Monomer	Nylon type	Monomer Formula	Approximate M.P. °C
Caprolactam	6	CH_2-(CH_2)-$_4C$=0 (ring closed through NH)	223
Adipic Acid	66	HOOC-$(CH_2)_4$COOH	265
Hexamethylenediamine		H_2N-$(CH_2)_6$-NH_2	
Sebacic Acid	610	HOOC-$(CH_2)_8$COOH	215
Hexamethylenediamine		H_2N-$(CH_2)_6$-NH_2	
11-Aminoundecanoic Acid	11	H_2N-$(CH_2)_{10}$-COOH	190
Laurolactam	12	CH_2-$(CH_2)_{10}$-C=0 (ring closed through NH)	119
Acrylamide	3	CH_2=CHCONH_2	320
2-pyrrolidone	4	CH_2-$(CH_2)_2$-C=O (ring closed through NH)	265
Valerolactam	5	CH_2-$(CH_2)_3$-C=0 (ring closed through NH)	260
7-aminoheptanoic Acid or Ethyl 7-aminoheptanoate	7	N_2H-$(CH_2)_6$-COOH H_2H-$(CH_2)_6$-$COOC_2H_5$	233
Capryllactam	8	CH_2-$(CH_2)_6$-C=0 (ring closed through NH)	200
9-Aminopelargonic acid	9	H_2N-$(CH_2)_8$-COOH	209
10-Aminodecanoic acid	10	H_2N-$(CH_2)_9$-COOH	188

Nylons have a variety of uses ranging from tire cord to carpet to hosiery. The most important application is cord followed by apparel. Nylon staple and filaments are extensively used in the carpet industry. Nylon fiber is also used for a variety of other articles such as seat belts, monofilament finishes, and knitwear. Because of its high tenacity and elasticity, it is a valuable fiber for ropes, parachutes, & underwear.

Q.69. ***Describe the manufacturing process of Nylon 66 (Poly hexamethyleneadipate).***

Ans. *Nylon 6,6* is produced by polymerisingadipic acid and hexamethylene diamine in an autoclave in an aqueous solution at about 200°C and a pressure of 16 kg/cm^2 for a reaction time of 1 to 2 hour. The reaction of hexamethylenediamine and adipicacid produces hexamethylene diammoniumadipate salt. The product is a dilute salt solution which is concentrated to approximately 60% and charged with acetic acid to a reactor where water is continuously removed. The presence of a small amount of acetic acid limits the degree of polymerization to the desired level.

The temperature is then increased to 270–300°C and the pressure to 17 atmospheres, which favors the formation of the polymer. The pressure is finally reduced to atmospheric to permit further water removal. After a total of three hours, nylon 66 is extruded under nitrogen pressure.

Q.70. ***Describe the manufacture of Nylon 6 (Polycaproamide).***

Ans. Nylon 6 is produced by the polymerization of caprolactam. The monomer is first mixed with water, which opens the lactam ring and gives *w*-amino acid. The formed amino acid reacts with itself or with at approximately 250–280°C to form the polymer. The reaction time is about 8–12 hours .Temperature control is important, especially for depolymerization, which is directly proportional to reaction temperature and water content.

Q.71. Describe the production process of Nylon 12 (Polylaurylamide).

Ans. Nylon 12 is a product of the polymerisation of laurolactam at a temperature of about 260°C in an autoclave. Nylon 12 is produced in a similar way to nylon 6 by the ring opening polymerization of laurolactam. The polymer has a lower water capacity than nylon 6 due to its higher hydrophobic (water hating) properties. The polymerization reaction is slower than for caprolactam. Higher temperatures are used to increase the rate of the reaction.

The monomer (laurolactam) can be produced from 1,5,9-cyclododecatriene, a trimer of butadiene. The trimer is epoxidized with peracetic acid or acetaldehyde peracetate and then hydrogenated. The saturated epoxide is rearranged to the ketone with MgI_2 at 100°C. This is then changed to the oxime and rearranged to laurolactam.

Q.72. ***Describe the production process of Nylon 4 (Polybutyramide).***

Ans. Nylon 4 is produced by ring opening 2-pyrrolidone. Anionic polymerization is used to polymerize the lactam. Cocatalysts are used to increase the yield of the polymer. Carbon dioxide is an excellent polymerization activator.

Nylon 4 has a higher water absorption capacity than other nylons due to its lower hydrophobic property.

Q.73. ***Describe the production process of Nylon 11 (Polyundecanylamide).***

Ans. Nylon 11 is produced by the condensation reaction of 11- aminoundecanoic acid, which is obtained from castor oil, by polymerising in an aqueous medium at a temperature of 220°C in an autoclave. This is an example of the self-condensation of an amino acid where only one monomer is used. The monomer is first suspended in water, then heated to melt the monomer and to start the reaction. Water is continuously removed .The polymer is finally withdrawn for storage.

Many other nylons could be produced such as nylon nylon 5, nylon 7, nylon 610, and nylon 612. These have properties generally similar to those nylons described earlier.

Q.74. ***Describe the manufacturing processes for the various monomers used for producing different types of nylons.***

Ans. Manufacturing processes for the various monomers used for producing different types of nylons are described below:

- ***Adipicacid:*** Commercial a dipic acid is made from benzene. Benzene is catalytically hydrogenated to cyclohexane (C_6H_{12}) by Raney nickel as the catalyst at a temperature of 200–300°C under a hydrogen pressure of 3–4 MPa. The reaction takes place in a stirred tank reactor. Cyclohexane is then oxidised to a mixture of cyclohexanol ($C_6H_{11}OH$) and cyclohexanone ($C_6H_{10}O$) by cobalt naphthenate as the catalyst at a temperature of 145–175°C and at a pressure of 0.8–1 MPa in an oxygen atmosphere. Both the cyclo hexanol and cyclohexanone are then further oxidised to adipic acid in the presence of copper and ammonium metavan date as the catalyst in the presence of nitric acid. The oxidation is carried out in a stirred tank reactor at a temperature of 60–80°C and atmospheric pressure. There action time is about 5–7 minutes. The product adipic acid [$(CH_2)4(COOH)_2$] is then further purified by steam distillation.

- ***Hexamethylene diamine*:** This is obtained from adipic acid in two stages. In the first stage, adipicacidis converted to adiponitrile [$(CH_2)_4(CN)_2$] by reacting with gaseous ammonia in the presence of the catalyst, boron phosphate, and at a temperature of about 340°C in a fixed bed reactor. The product adiponitrile is then catalytically hydrogenated to hexamethylene diamine [$(CH_2)_6(NH_2)_2$] by a cobalt catalyst at a temperature of 100–135°C and at a hydrogen pressure of 60–65 MPa in a fixed bedreactor.
- ***Sebacic acid*:** This acid is commercially produced from castor oils a ponification with acaustic solution at about 250°C. In there actor, the castor oil (glyceride) is spliting to ricinoleic acid [$CH_3(CH_2)_{13}CHOH(CH)_2COOH$] and glycerine [$(CH_2OH)_2CHOH$]. Ricinoleic acid is separated from glycerine and further heated in a separate reactor to yield sebacic acid[$(CH_2)_8(COOH)_2$].
- ***11-Amino-undecanoic acid*:** This is also produced from castor oil. The oil is first esterified by methanol to yield methyl ester of ricinoleic acid, followed by pyrolysis at about 500°C. A mixture of products of heptyl alcohol, ester of undecylenic acid, heptaldehyde, etc., is obtained. Hydrolysis yields the undecylenic acid [$CH(CH_2)_9COOH$], which is then brominated by hydrobromic acid to bromo-undecanoic acid [$CHBr(CH_2)_{10}COOH$]. Treating bromo-undecanoic acid with ammonia yields 11-amino undecanoic acid [$NH_2(CH_2)_{10}COOH$].
- ***Caprolactam*:** Benzene is converted to cyclohexane followed by oxidation to cyclohexanoland cyclohexanone, as described in the production of adipic acid. Cyclohexanol from the mixture is further catalytically dehydrogenated to cyclohexanone in the presence of zinc at a temperature of 400°C at 1 atm. The cyclohexanone is then converted to cyclohexanone oxime [$C_6H_{10}NOH$] by reacting with hydroxylamine sulfate [$(NH_2OH)_2H_2SO_4$] at about 18–25°C. Cyclohexanone oxime is then further treated with sulfur trioxide in the presence of sulfuric acid to yield caprolactum [$(CH_2)_5(CO)(NH)$] by a reaction mechanism known as the Beckmann rearrangement. The temperature of this reaction is maintained at 140–150°C.
- ***Laurolacta mordo decyllactam*:** Butadiene is first catalytic allytrimerised in a liquid phase reaction by the Ziegler catalyst (TiCl4) in a stirred tank reactor. The product stream containing the desired trimer, 1, 5, 9-cyclo-dodecatriene, is separated from the by-products polybutadiene, cyclo-octadiene, and vinyl cyclohexene by distillation. This trimer is then catalytically hydrogenated using a nickel catalyst at 200°C and a hydrogen pressure of 1.4 MPa. The reaction is carried out in the liquid phase in a stirred tank autoclave. The hydrogenated product is cyclododecane[$(CH_2)_{12}$]. Cyclodode cane is then oxidised in the liquid phase at about 150°C in jacketed stirred tank reactors in the presence of boric acid as the catalyst. The oxidised product stream, containing cyclododecanol [$(CH_2)_{11}OH$] and cyclododecanone [$(CH_2)_{10}O$], is separated by distillation. Cyclodode canol is further dehydrogenated to yield cyclododecanone in a separate reactor at 200°C in the liquid phase and in the presence of a copper catalyst supported on alumina. Finally, cyclododecanone is oximated by hydroxylamine hydrogen sulfate [$(NH_2OH)_2H_2SO_4$] at a temperature of 100°C and converted to cyclododecanone oxime [$(CH_2)_{10}NOH$]. Finally, cyclododecanone oxime is converted to laurolactam [$(CH_2)_{11}(CO)(NH)$] ammonium sulfate and concentrated

sulfuric acid. The Beckmann rearrangement reaction takes place during the formation of laurolactam.

Q.75. ***What are ACRYLIC AND MODACRYLIC FIBERS? Describe the properties and uses of polyacrylic fibres.***

Ans. Modacrylic fibers are copolymers containing between 35–85% acrylonitrile. Acrylic fibers contain at least 85% acrylonitrile. Orlon is an acrylic fiber developed by DuPont in 1949; Dynel is a modacrylic fiber developed by Union Carbide in 1951.

Polyacrylicfibres (**ACRYLIC AND MODACRYLIC FIBERS**) are produced by copolymerizing acrylonitrile with other monomers such as vinyl acetate, vinyl chloride, and acrylamide. Solution polymerization may be used where water is the solvent in the presence of a redox catalyst. Free radical or anionic initiators may also be used. The produced polymer is insoluble in water and precipitates.

Copolymers of acrylonitrile are sensitive to heat, and melt spinning is not used. Solution spinning (wet or dry) is the preferred process where a polar solvent such as dimethyl formamide is used. In dry spinning the solvent is evaporated and recovered.

Dynel, a modacrylic fiber, is produced by copolymerizing vinyl chloride with acrylonitrile. Solution spinning is also used where the polymer is dissolved in a solvent such as acetone. After the solvent is evaporated, the fibers are washed and subjected to stretching, which extends the fiber 4–10 times of the original length.

Important properties of acrylics are resistance to solvents & sunlight, resistance to creasing, and quick drying. Acrylic fiber breaking strength ranges between 1500 & 2650 kg/cm^2 and they have a water absorption of approximately 5%. Dynel, due to the presence of chlorine, is less flammable than many other synthetic fibers.

Major uses of acrylic fibers are woven and knitted clothing fabrics (for apparel), carpets, and upholstery. Acrylic fibers are characterized by having properties similar to wool and have replaced wool in many uses such as blankets, carpets, & sweaters.

Q.76. ***Describe the important characteristics and uses of POLYPROPYLENE FIBERS.***

Ans. Polypropylene fibers represent a small percent of the total polypropylene production. (Most polypropylene is used as a thermoplastic.) The fibers are usually manufactured from isotactic polypropylene.

Important characteristics of polypropylene are high abrasion resistance, strength, low static buildup, and resistance to chemicals. Crystallinity of fiber grade polypropylene is moderate and when heated, it starts to soften at 145°C and then melts at 170°C. Melt spinning is normally used to produce the fibers.

Polypropylene fibers are used for face pile of needle felt, tufted carpets, upholstery fabrics, etc.

Q.77. ***Describe the merits and demerits of Emulsion Polymerization w.r.t. other polymerization processes.***

Ans. One of the major advantages of Emulsion Polymerization is that it has good heat control over the entire polymerization process. The heat control occurs by transfer to the aqueous phase, and this enables little change in the viscosity of the emulsion medium. Secondly, in this process, the reaction mass is emulsified with soap in a continuous aqueous medium. Soap has a stabilizing action, and hence, the emulsion obtained is stable and no agitation is necessary. Due to this, extra cost incurred in employing agitators is avoided. Moreover, Emulsion Polymerization overcomes the difficulty of low degree of polymerization, as encountered in Solution polymerization and Bulk Polymerization. Since the degree of polymerization directly depends upon the droplet sizes, and in this process, spherical monomer particles (1 – 10 μ) are yielded ; hence, the degree of polymerization is sufficiently higher than in the other two processes. However, Emulsion Polymerization has certain disadvantages. It blocks the emulsifiers used in the process. In addition, the process has poor capacity and low electrical resistance, which sometimes create major problems.

Q.78. ***What are the additives used in polymer products?***

Ans. Polymer additives include materials which make and modify polymers. These include surfactants, chain transfer agents, modifiers, plasticizers, stabilizers, initiators and cross linking agents.

Q.79. ***What is Glass Transition Temperature?***

Ans. Glass Transition Temperature of a non crystalline material is the temperature at which the material changes its behaviour from its glassy state to a rubbery state. Glassy state refers to the state at which a material is hard & brittle; whereas, rubbery state refers to the state at which the material is flexible & elastic.

Q.80. ***Mention the advantages of blow molding as compared to injection molding.***

Ans. Some of the advantages of blow molding as compared to injection molding are:

- Possibility of variable wall thickness.
- Low stresses and favorable cost factor.

Q.81. ***In molding process, how can we prevent the polymer material sticking to the wall of the mold (container)?***

Ans. The polymeric material can be prevented from sticking to the walls of the mold by changing the viscosity of the polymer. This can be achieved by selecting the MOC of the container (mold) in such a way that it will create substantial viscosity gradient between the walls and the polymeric material. It can also be done by heating the walls of the mold externally.

Q.82. ***What are the various commercial uses of formaldehyde?***

Ans. Formaldehyde is commercially available as aqueous solution with concentration ranging from 30-56 wt.% HCHO. It is also sold in solid form as paraformaldehyde or trioxane. Some major intermediates derived from formaldehyde are chelating agents, acetal resins, 1,4-butanediol, polyols, methylene diisocynate. It is also used for the manufacture of wide variety of chemicals, including sealant, herbicides, fertilisers, coating, and pharmaceutical. Product profile of formaldehyde is tabulated below.

Product	Uses
Formaldehyde	Thermosetting resin: Phenol, Urea Melamine, Formaldehyde resins Hexamethylenetetramine, Plastic & pharmaceuticals 1,4-Butadiol Methylenediisocyanate Fertiliser, Disinfectant, Biocide Preservative, Reducing agent, Corrosion inhibitor Polyaceta resin p-formaldehyde Pentaerythritol (Explosive-PETN), Alkyl resins

Q.83. ***Describe the industrial manufacturing process of formaldehyde.***

Ans. The main industrial route for producing formaldehyde (HCHO) is the catalyzed air oxidation of methanol (CH_3OH).

$$CH_3OH_{(g)} + 1/2\ O_{2(g)} \rightarrow H-\overset{\overset{\displaystyle O}{\|}}{C}-H_{(g)} + H_2O_{(g)}\ \Delta H° = -148.9\ KJ/mol$$

A silver-gauze catalyst is still used in some older processes that operate at a relatively higher temperature (about 500°C). New process reaction is exothermic and occurs at 400–425°C & atmospheric pressure. Excess air is used to keep the methanol air ratio below the explosion limits. Various industrial processes for manufacture of formaldehyde using processes use an iron-molybdenum oxide catalyst. Chromium or cobalt oxides are sometimes used to dope the catalyst. The oxidation reaceiver and iron-molybdenum catalyst are given below:

Silver catalyst processes: Bayer, Chemical construction, Ciba, DuPont, IG Farben, CdFChemie process , BASF process, ICI process,

Iron-molybdenum catalyst processes: Degussa process, Formox process, Fischer-Adler, Hiag-Lurgi, IFP-CdF ChimleLumus, Motedisous, Nikka Topsoe, Process

Process diagram for manufacture of formaldehyde using silver and iron- molybdenum catalyst is shown below in figure.

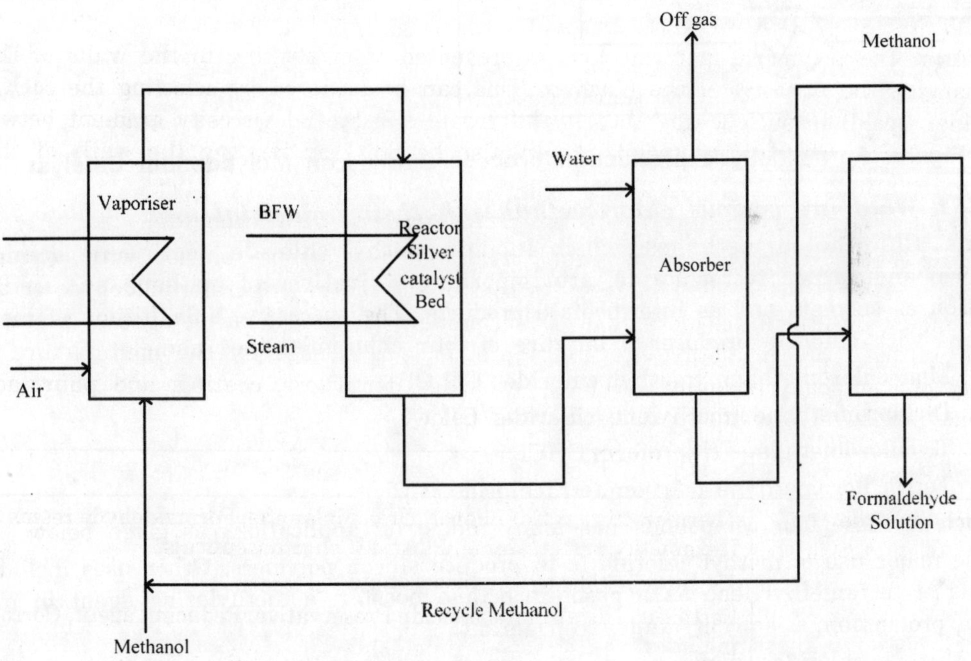

Figure: Formaldehyde production process Using Silver Catalyst

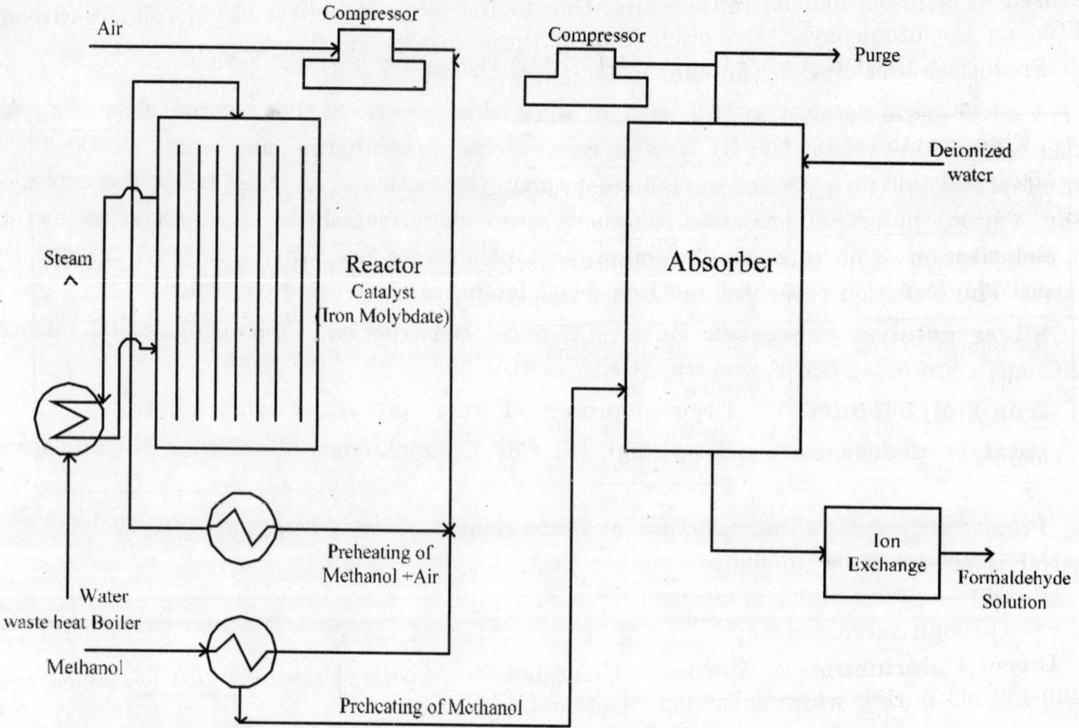

Figure: Formaldehyde production process using Iron Molybdenum Catalyst

Q.84. What are various chloromethanes & their industrial uses.

Ans. Chlorinated methanes, which include methyl chloride, methylene dichloride, chloroform and carbon tetrachloride, are important derivatives of methane and find wide application as solvents and as intermediate products. The successive substitution of methane hydrogens with chlorine produces a mixture of four chloromethanes namely:

- Monochloromethane (methyl chloride, CH_3Cl)
- Dichloromethane (methylene chloride, CH_2Cl_2)
- Trichloromethane (chloroform,$CHCl_3$)
- Tetrachloromethane (carbon tetrachloride, CCl_4)

Each of these four compounds has many industrial applications as given below :

The major use of methyl chloride is to produce silicon polymers. Other uses include the synthesis of tetramethyl lead as a gasoline octane booster, a methylating agent in methyl cellulose production, a solvent, and a refrigerant.

Methylene chloride's major use is as a paint remover. It is also used as a degreasing solvent, a blowing agent for polyurethane foams, and a solvent for cellulose acetate.

Chloroform is mainly used to produce chlorodifluoro methane (Fluoro- carbon 22) by the reaction with hydrogen fluoride . This compound is used as a refrigerant and as an aerosol propellent. It is also used to synthesize tetrafluoroethylene, which is polymerized to a heat resistant polymer Polytetrafluoroethylene (PTFE) which is commercially known as Teflon).

Carbon tetrachloride is used to produce chlorofluorocarbons by the reaction with hydrogen fluoride using an antimony pentachloride ($SbCl_5$) catalyst. The formed mixture is composed of trichlorofluoromethane (Freon-11) and dichlorodifluoromethane (Freon-12). These compounds are used as aerosols and as refrigerants. Due to the depleting effect of chlorofluoro-carbons (CFCs) on the ozone layer, the production of these compounds has been reduced.

Product profile of Chloromethanes is tabulated below.

Chloromethanes	Uses
Methyl chloride	Refrigerant, butyl rubber, silicones, solvent, tetramethyl lead, intermediates
Methylene dichloride	Solvent, Intermediates, Photographic film, Degreasing solvents, Aerosol, Propellants
Chloroform	Chlorodifluoro methane, (Refrigerants), Propellants, Pharmaceuticals
Carbon tetra chloride	Dichlorodifluoro methane, Trichlorofluoro methane, Solvent, Fire extinguishers

Q.85. *Describe the major routes for the manufacture of chloromethane.*

Ans. There are two major routes for the manufacture of chloromethane:

- Direct chlorination of methane
- Through methanolroute

Direct Chlorination of Methane: Chlorination of methane (natural gas) is carried out at 400-450 oC during which following reaction takes place:

$$CH_4 + Cl_2 \longrightarrow \underset{\text{Methyl chloride}}{CH_3Cl} + HCl$$

$$CH_3Cl + Cl_2 \longrightarrow \underset{\text{Methylene Dichloride}}{CH_2Cl_2} + HCl$$

$$CH_2Cl_2 + Cl_2 \longrightarrow \underset{\text{Chloroform}}{CHCl_3} + HCl$$

$$CHCl_3 + Cl_2 \longrightarrow \underset{\text{Carbon tetrachloride}}{CCl_4} + HCl$$

Production of Methyl Chloride (CH_3Cl) through methanol (methyl alcohol) route:

Methyl chloride is produced by the vapor phase reaction of methanol and hydrogen chloride:

$$CH_3OH + HCl \longrightarrow CH_3Cl + H_2O$$

Many catalysts are used to effect the above reaction, such as zinc chloride on pumice, cuprous chloride, and ignited alumina gel. The reaction conditions are 350°C at nearly atmospheric pressure. The yield is approximately 95%.

Zinc chloride is also a catalyst for a liquid-phase process using concentrated hydrochloric acid at 100–150°C. Hydrochloric acid may be generated in situ by reacting sodium chloride with sulfuric acid. Methyl chloride may also be produced directly from methane with other chloromethanes. However, methyl chloride from methanol may be further chlorinated to produce dichloromethane, chloroform, and carbon tetrachloride.

Q.86. *Describe the manufacture and uses of Acetic Acid (CH_3COOH).*

Ans. The carbonylation of methanol is currently one of the major routes for acetic acid production. The basic liquid-phase process uses a cobalt catalyst at 250°C and a high pressure of about 70 atmospheres. The newer process uses a rhodium complex catalyst in presence of CH_3I, which acts as a promoter. The reaction occurs at 150°C and atmospheric pressure. A 99% selectivity is achieved with this catalyst.

$$CH_3OH + CO \longrightarrow CH_3COOH$$

Acetic acid is also produced by the oxidation of acetaldehyde and the oxidation of n-butane. However, acetic acid from the carbonylation route has an advantage over the other commercial processes because both methanol and carbon monoxide come from synthesis gas, and the process conditions are quite mild.

Uses of Acetic Acid. The main use of acetic acid is to produce vinyl acetate, followed by acetic acid esters and acetic anhydride . Vinyl acetate is used for the production of adhesives, film, paper and textiles. Terephthalic acid production also consumes acetic acid. Acetic acid is also used to produce pharmaceuticals, dyes, and insecticides. Chloroacetic acid (from acetic acid) is are active intermediate used to manufacture many chemicals such as glycine and carboxy methyl cellulose (CMC).

Q.87. *Write a short note on Production and uses of Perchloroethylene.*

Ans. Chlorination of propane with chlorine at 480–640°C yields a mixture of perchloroethylene (Perchlor) and carbon tetrachloride:

$$\underset{\text{Perchlor}}{CH_3CH_2CH_3 + 8Cl_2} \longrightarrow CCl_2 = CCl_2 + CCl_4 + 8HCl$$

Carbon tetrachloride is usually recycled to produce more perchloroethylene:

$$2CCl_4 \longrightarrow CCl_2 = CCl_2 + 2Cl_2$$

Perchlor may also be produced from ethylene dichloride (1,2-dichloroethane) through an oxy chlorination-oxy hydrochlorination process. Trichloroethylene (trichlor) is co-produced.

Perchlor and trichlor are used as metal degreasing agents and as solvents in dry cleaning. Perchlor is also used as a cleaning & drying agent for electronic equipment and as a fumigant.

Q.88. *What are the various feedstocks & routes of ethylene production. What are various products made from ethylene?*

Ans. Ethylene, the key petrochemical building block is produced through multiple routes using different feedstocks like Naphtha, Natural Gas (Ethane), Propane, Butane, etc. However, Naphtha and Ethane are the most commonly used feedstocks, accounting for 82% of global ethylene production. Today, 50% of the ethylene produced in the world is through Naphtha cracking while one-third is produced through the Ethaneroute.

Ethylene is one of the most versatile petrochemicals and its production has steadily increased over the years. Ethylene is called as king of chemicals and surpasses all organic chemicals in production and in amount sold. Ethylene is the basic building block for petrochemicals.

Following Figure gives Shares of Feed Stocks in Global Ethylene Production in 2010.

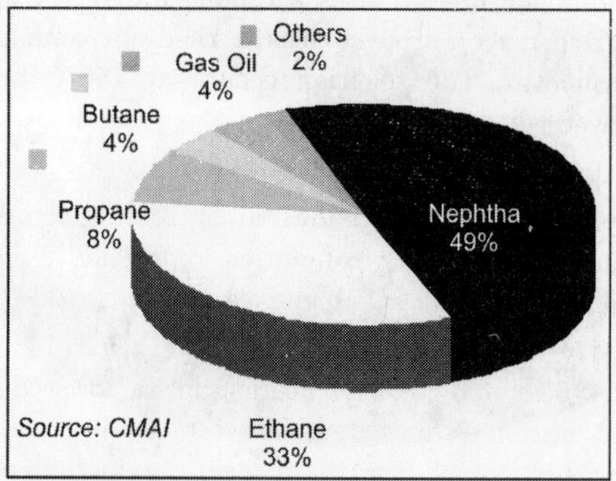

Figure: Shares of Feed Stocks in Global Ethylene Production in 2010

Because of its ready availability at low cost and high purity & reactivity, ethylene has become one of the important raw materials for large number of petrochemicals & products. Ethylene has replaced the earliest route of production of vinyl chloride, acetaldehyde, vinyl acetate and other chemicals through acetylene route. Large tonnage of ethylene is being used for the manufacture of polyethylene, ethylene oxide, ethylene glycol and styrene.

Product profile of ethylene is given below in Table .

Table: Product Profile of Ethylene

Product	Uses
Polyethylene LDPE, LLDPE, HDPE	Films, moldings, pipes, cable covering, netting
Ethylene oxide & Ethylene glycol	Antifreeze, polyester, solvents, detergent, textile, ethanol amine
Styrene	Synthetic rubber and polystyrene
Ethyl alcohol	Industrial solvent and Chemical intermediates
Acetaldehyde (from ethyl alcohol)	Acetic acid (Peracetic acid), Acetic anhydride, Cellulose acetate, Vinyl acetate, Pyridine, Butyraldehyde (ethyl hexanol)
Olefin	n-Butenes, Synthetic detergent, Oxo alcohols, Synthetic lubricants
Chlorinated Solvents	Trichloroethylene, Perchloroethylene
Ethyl Chloride	Tetraethyl lead, Chemical intermediates
Vinyl Acetate	Polyvinyl acetate, Polyvinyl alcohol
Vinyl Chloride	Polyvinyl chloride(PVC)

Over 60% of global Ethylene production is used for manufacturing Polyethylenes (PE), 13% for Ethylene oxide (EO), 11% for production of ethylene dichloride (EDC), and the residual Ethylene goes in producing a wide array of petrochemicals.

Following Figure gives the Global Ethylene Demand by End use in 2010.

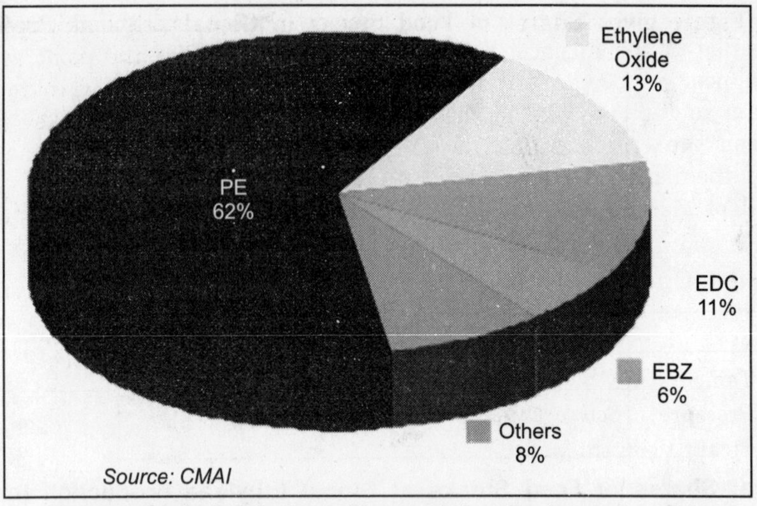

Figure: Global Ethylene Demand by End-use in 2010

Q.89. ***Describe the process of ethylene (Olefins) production by naphtha cracking.***

Ans. Due to the larger availability of naphtha & the low energy requirement for cracking as compared to gases (ethane, methane etc.), naphtha cracking is widely used for the manufacture of olefins. Naphtha is a mixture of hydrocarbons boiling in the range of the lowest boiling component (C_5) to 150°C which contains paraffins, naphthenes, and aromatics in various proportions. Paraffins, i.e., straight chain hydrocarbons, yield olefins when heated at a high temperature. The branched chain and aromatic hydrocarbons, on the other hand, become heavier or decomposed to carbon. Yield of olefins decreases with the increasing molecular weight (hence with boiling point) of the hydrocarbons. For example, if ethane is thermally cracked, it will yield 80% ethylene, while from propane, butane, naphtha, and gas oil as feedstocks, if cracked separately the yields will be 45%, 37%, 30%, and 25%, respectively. Hence, the desirable feedstock naphtha should have a very high content of paraffins and be in the lower boiling range. A suitable boiling range for feedstock naphtha for olefin production is below 100°C and should have a paraffin content of more than 75%.

Usually, naphtha in the boiling range of 90 to 150°C is catalytically reformed in a refinery either to produce gasoline or aromatics. Hence, in the refinery, C_5-90°C cut is separated in the naphtha redistillation unit and is sold to the petrochemical industry. Naphtha cracking is carried out in a tube still furnace at a temperature above 800°C. Light hydrocarbons obtained from the initial cracking reactions further crack into lighter olefins and propagate until the reaction temperature is brought down (quenched). If the reaction is continued indefinitely, branched & cyclised heavy hydrocarbons will be produced and coke will be generated as the ultimate product.

Therefore, the cracking reaction is carried out in a very short residence time, i.e., the naphtha feed passes the heater tubes at very high speed to avoid undesirable heavy end products and coke. Usually, residence time is maintained at <1 sec in the traditional cracker furnaces and it is of the order of a few milliseconds in the modern furnaces. Since a coke layer develops inside the tube surface, the heat transfer rate is rapidly reduced, causing reduced cracking and poor olefins yield. Steam is introduced with the feed to remove the coke layer on the tube surface by converting coke into carbon monoxide and hydrogen by water gas reaction : $C + H_2O$ (steam) $= CO + H_2$.

However, coke cannot be removed completely by steam and the thickness grows during the operating period of the furnace. When the coke layer reaches the point at which cracking operation shows poor yield, the furnace is taken out of service and decoking is carried out with air & steam to remove coke to the maximum extent. Thus, a cracker furnace operates cyclically between the cracking and decoking operations. Excess steam may also partially convert some of the naphtha/hydrocarbons components to carbon monoxide & hydrogen and reduce the yield of olefins.

Table below shows the operating parameters for a typical naphtha cracker.

Typical Operating Conditions in a Naphtha Cracking Furnace

Process Variable	Values
Temperature	800°C
Pressure	Atmospheric
Steam to naphtha ratio	0.6 kg/kg
Reaction time	<1 sec
Cycle time for cracker	20 days
Yield of ethylene	30% of the feed

Cracked liquids and gases are separated in a fractionating column the bottom product of which is the heavy cracked oil rich in high boiling aromatics. This heavy oil is also partly used as the quenching medium for the products from the furnace and partly sold as the carbon black feedstock (CBFS) due to its heavy aromatic contents.

Cracked gases containing hydrocarbons, both saturated & unsaturated, from methane to C_7 hydrocarbons emerge from the top of the column, which is then compressed and amine (or caustic) washed to remove hydrogen sulfide & carbon dioxide gases. Amine or caustic washed gases are then passed through a flash separator vessel, where liquid hydrocarbon gases are separated from hydrogen at high pressure & low temperature. Hydrogen from this vessel is used in the hydrogenating units, such as pyrolysis gasoline and butadiene hydrogenation, or as a fuel.

Liquified gases from a hydrogen separator are then separated from methane in a distillation column where methane (C_1) emerges from the top and is used as a fuel for the cracking furnace. The bottom of the column is then passed to a de-ethaniser which is also a distillation column that separates ethane and ethylene mixture (C_2 mixture) as the top product from the rest of the liquified gases containing propane, propylene, butane, butylenes, etc.

A C_2 mixture from the top of the de-ethaniser column is then passed through another distillation column that separates ethylene as the top product and ethane as the bottom product. Ethylene is sent to storage and is used up in the polyethylene (PE) synthesis plant. Ethane from this column is recycled to a small cracking furnace to yield additional ethylene.

The liquified gas mixture from the bottom of the de-ethaniser is separated from propane & propylene (C_3 mixture), which leaves from the top of the column and enters the propane–propylene fractionator. The bottom product contains the butanes, butenes, butadiene,& heavier components, which are then separated from the butane–butene mixture (C_4 mixture).

In this column, propylene is recovered as the top product and propane as the bottom product. Propylene is stored & used for manufacturing polypropylene, and propane is sold as a domestic fuel-liquified petroleum gas (LPG).

Butane, butenes & butadiene (C_4 mixture) are recovered as the top product and components heavier than the C_4 mixture, i.e., C_5 & heavier, are recovered as the pyrolysis gasoline (bottom product). Pyrolysis gasoline is catalytically hydro desulfurised before it is sold or blended as a gasoline component. A gasoline hydro desulfurisation unit is similar to a naphtha pretreatment unit.

Q.90. ***What are the major chemicals produced from ethylene?***

Ans. Major chemicals produced based on ethylene is shown below.

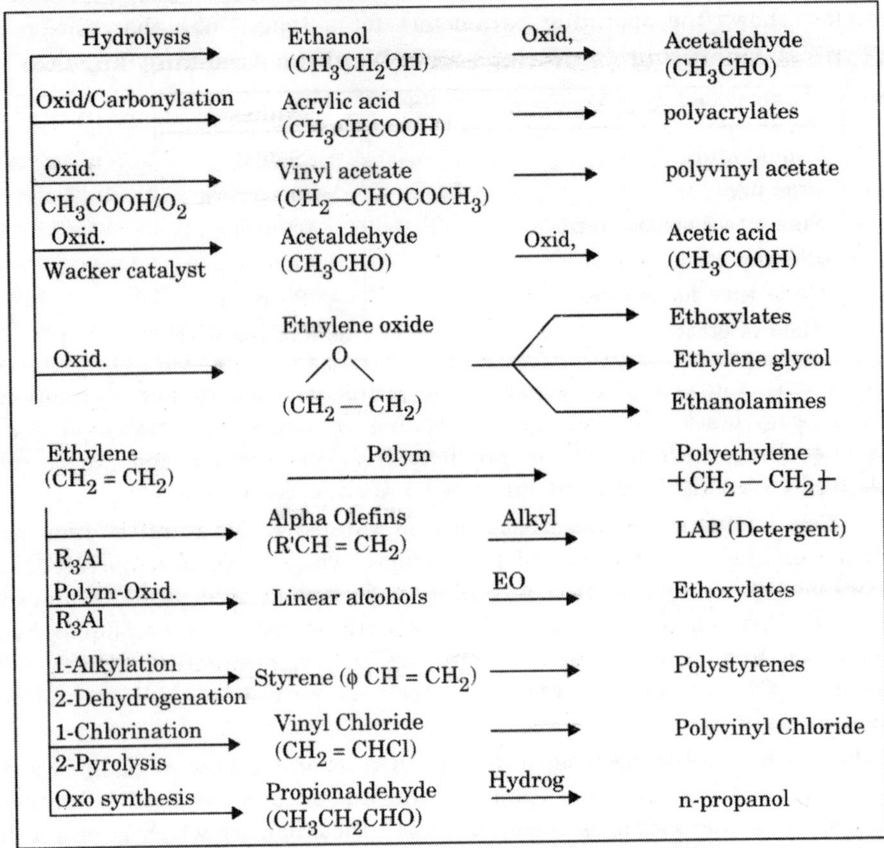

Figure: Major chemicals based on ethylene.

Q.91. ***How Ethylene Oxide is produced?***

Ans. Ethylene oxide (EO) is a colorless gas that liquefies when cooled below 12°C. It is highly soluble in water and in organic solvents. The main route to ethylene oxide is oxygen or air oxidation of ethylene over a silver catalyst. The reaction is exothermic, so heat control is important.

A concomitant reaction is the complete oxidation of ethylene to carbon dioxide and water:

$$CH_2 = CH_2 + 3O_2 \rightarrow 2CO_2 + 2H_2O \qquad \Delta H = -1{,}421 \text{ kJ/mol}$$

This reaction is highly exothermic; the excessive temperature increase reduces ethylene oxide yield and causes catalyst deterioration. Over oxidation can be minimized by using modifiers such as organic chlorides. It seems that silver is a unique epoxidation catalyst for ethylene.

In the process, compressed oxygen, ethylene, and recycled gas are fed to a multitubular reactor. The temperature of oxidation is controlled by boiling water in the shell side of the reactor. Effluent gases are cooled and passed to the scrubber where ethylene oxide is absorbed as a dilute aqueous solution. Unreacted gases are recycled. Epoxidation reaction occurs at 200–300°C with a short residence time of one second. A selectivity of 70–75% can be reached for the oxygen based process. Selectivity is the ratio of moles of ethylene oxide produced per mole of ethylene reacted. Ethylene oxide selectivity can be improved when the reaction temperature is lowered and the conversion of ethylene is decreased (higher recycle of unreacted gases).

Q.92. *What are the properties & uses of ethylene oxide?*

Ans. Ethylene oxide is a highly active intermediate. It reacts with all compounds that have a labile hydrogen such as water, alcohols, organic acids, and amines. The epoxide ring opens, and a new compound with a hydroxyethyl group is produced. The addition of a hydroxyethyl group increases the water solubility of the resulting compound. Further reaction of ethylene oxide produces polyethylene oxide derivatives with increased water solubility.

Ethylene oxide is a precursor for many chemicals of great commercial importance, including ethylene glycols, ethanol amines, and alcohol ethoxylates. Ethylene glycol is one of the monomers for polyesters, the most widely used synthetic fiber polymers.

Ethylene oxide is one of the important petrochemical intermediate used for the manufacture of large number of products; some of the major uses are in the manufacture monoethylene glycol, glycol ethers which is made by reaction of ethylne oxide and alcohols, ethanol amines. Surfactant industry is one of largest user of EO, both for industrial and house hold applications. Product profile of ethylene oxide is tabulated below .

Table: Product Profile of Ethylene Oxide (EO)

Product	Uses
Ethanol Amines	Detergent, Soap solvent, Cosmetics, Morpholine
Mono Ethylene Glycol	Polyester, Staple fiber yarn, PET Bottles, Film
Diethylene Glycol	Coolants, Pesticides, Rubber Compounding, Plasticizer, Polyurethane, Alkyl Resin
Tri ethylene Glycol	Natural gas conditioning agents plasticizer
Polyethylene Glycol	Pharmaceuticals, Brake Fluid, cosmetics
Non-ionic surfactants, Ethoxylates	Textile auxiliaries, Binders, Dyes, Pesticides, Pharmaceuticals, Cosmetics
Glycol Ethers	Brake Fluid and Protective Coating

Q.93. *Describe the manufacturing process of MONO-, DI- & TRI-ETHYLENE GLYCOLS (MEG, DEG, TEG).*

Ans. The main route for producing ethylene glycol is the hydration of ethylene oxide in presence of dilute sulfuric acid.

$$\overset{\displaystyle O}{CH_2\!\!-\!\!CH_2} + H_2O \xrightarrow{H^+} HO\!-\!CH_2\!-\!CH_2OH$$

The hydrolysis reaction occurs at about 200°C. Contact time is about 45 minutes. Di- and triethylene glycols are coproducts with the monoglycol. Increasing the water/ethylene oxide ratio and decreasing the contact time decreases the formation of higher glycols. A water/ethylene oxide ratio of 10 is normally used to get approximately 90% yield of the monoglycol. However, the di- and the triglycols are not an economic burden, because of their commercial uses. Ethylene oxide is preheated to 195°C. EO:H_2O ratio is maintained at 1:10 to maximize MEG production. By Products are DEG and TEG.

Chemical reactions involved are as follows:

$$C_2H_4 + H_2O \longrightarrow HO{-}\underset{H_2}{C}{-}C{-}OH$$

$$\underset{\text{MEG}}{HO{-}\underset{H_2}{C}{-}\overset{H_2}{C}{-}OH} + C_2H_4O \longrightarrow \underset{\text{DEG}}{HO(C_2H_4O)_2H}$$

$$\underset{\text{DEG}}{HO(C_2H_4O)_2H} + C_2H_4O \longrightarrow \underset{\text{TEG}}{RO(C_2H_4O)_3F}$$

Figure below gives detail manufacturing of MEG, DEG and TEG from Ethylene Oxide.

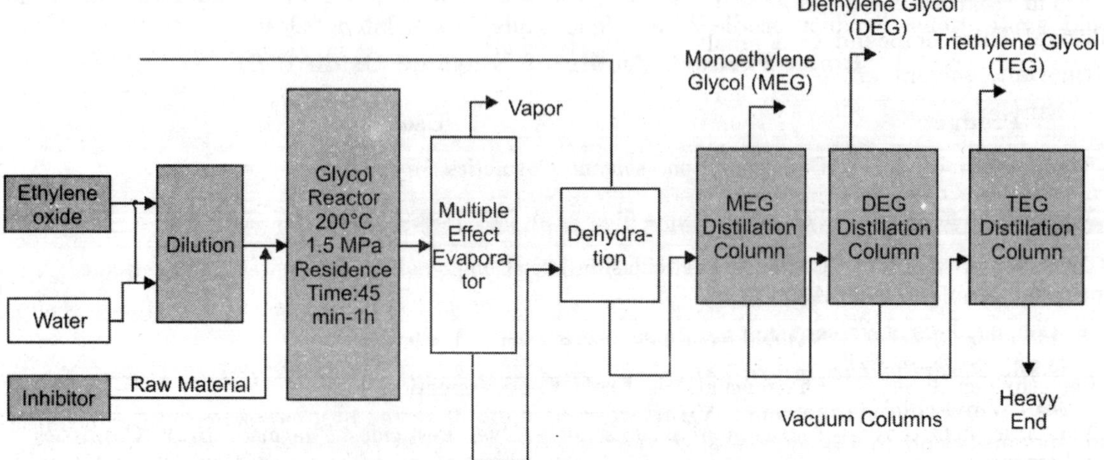

Figure: MEG, DEG and TEG from Ethylene Oxide

A new route to ethylene glycol from ethylene oxide via the intermediate formation of ethylene carbonate has recently been developed by Texaco. Ethylene carbonate may be formed by the reaction of carbon monoxide, ethylene oxide, and oxygen. Alternatively, it could be obtained by the reaction of phosgene and methanol.

Ethylene carbonate is a reactive chemical. It reacts smoothly with methanol and produces ethylene glycol in addition to dimethyl carbonate. The reaction occurs at approximately 80–130°C using the proper catalyst. Many catalysts have been tried for this reaction, and the best catalyst types are those of the tertiary amine and quaternary ammonium functionalized resins. This route produces ethylene glycol of a high purity and avoids selectivity problems associated with the hydrolysis of ethylene oxide.

The coproduct dimethyl carbonate is a liquid soluble in organic solvents. It is used as a specialty solvent, a methylating agent in organic synthesis, and a monomer for polycarbonate resins. It may also be used as a gasoline additive due to its high oxygen content and its high octane rating.

Q.94. *Write a short note on properties and uses of ethylene glycols.*

Ans. Ethylene glycol (EG) is colorless syrupy liquid, and is very soluble in water. The boiling & the freezing points of ethylene glycol are 197.2°C & –13.2°C, respectively.

Ethylene Glycol (CH_2OHCH_2OH) which finds major application in manufacture of polyester & as antifreeze compound account for about 70% of Ethylene oxide production.

Most of EG is used for producing polyethylene terephthalate (PET) resins (for fiber, film, bottles), antifreeze, and other products. Approximately 50% of the world EG is consumed in the manufacture of polyester fibers and another 25% for the antifreeze.

Q.95. *Describe the manufacturing process & uses of ethylene dichloride (EDC).*

Ans. The direct addition of chlorine to ethylene produces ethylene dichloride (1, 2-dichloroethane). Ethylene dichloride is the main precursor for vinyl chloride, which is an important monomer for polyvinyl chloride plastics & resins.

Other uses of ethylene dichloride include its formulation with tetraethyl and tetramethyl lead solutions as a lead scavenger, as a degreasing agent, and as an intermediate in the synthesis of many ethylene derivatives.

The reaction of ethylene with hydrogen chloride, on the other hand, produces ethyl chloride. This compound is a small volume chemical with diversified uses (alkylating agent, refrigerant, solvent etc).

Ethylene reacts also with hypochlorous acid, yielding ethylene chlorohydrin:

$$CH_2 = CH_2 + HOCl \rightarrow ClCH_2CH_2OH$$

Ethylene chlorohydrin via this route was previously used for producing ethylene oxide through an epoxidation step. Currently, the catalytic oxy-chlorination route is an alternative for producing ethylene glycol where ethylene chlorohydrin is an intermediate. In organic synthesis, ethylene chlorohydrin is a useful agent for introducing the ethyl hydroxy group. It is also used as a solvent for cellulose acetate.

Q.96. *Describe the manufacturing process of vinyl chloride ($CH_2 = CHCl$).*

Ans. Vinyl chloride monomer (VCM) was originally produced by the reaction of hydrochloric acid and acetylene in the presence of $HgCl_2$ catalyst at 100-180°C. The reaction is straight forward and proceeds with high conversion (96% on acetylene) .

However, ethylene as a cheap raw material from cracker plant has replaced acetylene for obtaining vinyl chloride. The production of vinyl chloride via ethylene is a three-step process. The first step is the direct chlorination of ethylene to produce ethylene dichloride. The exothermic reaction occurs at approximately 4 atmospheres and 50°C in the presence of $FeCl_3$, $CuCl_2$ or $SbCl_3$ catalysts. Ethylene bromide may also be used as a catalyst.

The second step is the dehydrochlorination of ethylene dichloride (EDC) to vinyl chloride and HCl. The pyrolysis reaction occurs at approximately 500°C and 25 atmospheres in the presence of pumice on charcoal.

The third step, the oxychlorination of ethylene, uses by-product HCl from the previous step to produce more ethylene dichloride.

$$CH_2 = CH_2 + 2HCl + 1/2O_2 \longrightarrow ClCH_2—CH_2Cl + H_2O$$

Ethylene dichloride from this step is combined with that produced from the chlorination of ethylene and introduced to the pyrolysis furnace.

The reaction conditions are approximately 225°C and 2–4 atmospheres. In practice the three steps, direct chlorination, oxychlorination, and dehydrochlorination, are integrated in one process so that no chlorine is lost.

Direct Chlorination of ethylene: The process of vinyl chloride manufacture takes place in two stages.

- **First stage:** Ethylene is reacted with chlorine in either liquid or vapor phase in presence of ferric chloride. However, the liquid phase process is more common and the reaction takes place at around 50⁰C and 4 atm pressure.
- **Second stage:** Vinyl chloride is produced by pyrolysis of vaporised ethylene dichloride in a set of tubular furnaces at temperature of about 500⁰C.

$$CH_2 = CH_2 + Cl_2 = ClCH_2–CH_2Cl$$

Oxychlorination

The original process of manufacture of vinyl chloride by ethylene chlorination and cracking of ethylene dichloride had been replaced by oxychlorination process in which no hydrochloric acid is formed as byproduct. The process involves production of ethylene dichloride by exothermic reaction of ethylene, hydrochloric acid and oxygen.

Liquid phase: at about 170-180°C in at 15-20 atm pressure in presence of copper chloride in either fixed or fluidised bed reactor.

Vapor phase reaction: The temperature and pressure are 200-220°C and 20-50 atm pressure.

- Direct chlorination: $CH_2 = CH_2 + Cl_2 \longrightarrow ClCH_2—CH_2Cl$
- Oxychlorination: $CH_2 = CH_2 + 2HCl + ½O_2 \longrightarrow ClCH_2—CH_2Cl + H_2O$
- Ethylene dichloride pyrolysis: $ClCH_2—CH_2Cl \longrightarrow CH_2 = CHCl + HCl$
- Overall reaction: $2\ CH_2 = CH_2 + Cl_2 + ½O_2 \longrightarrow 2\ CH_2 = CHCl + H_2O$

Process diagram of vinyl chloride production from oxychlorination process is shown below in Figure.

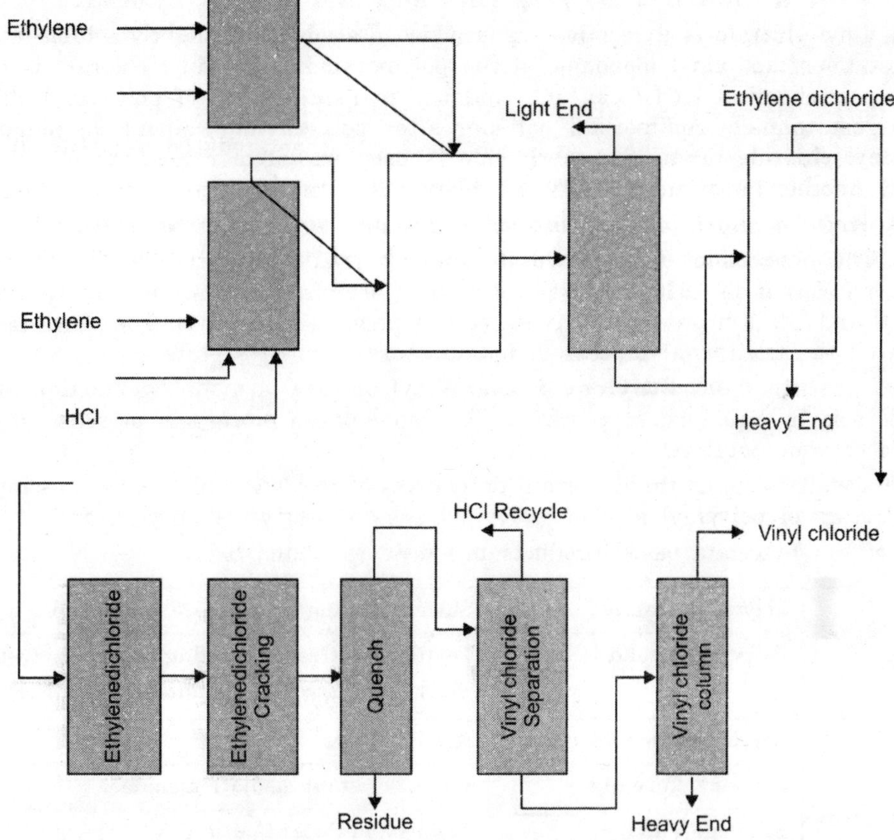

Figure: Vinyl Chloride production from Oxychlorination Process

Figure below shows the European Vinyls Corporation process for producing vinyl chloride.

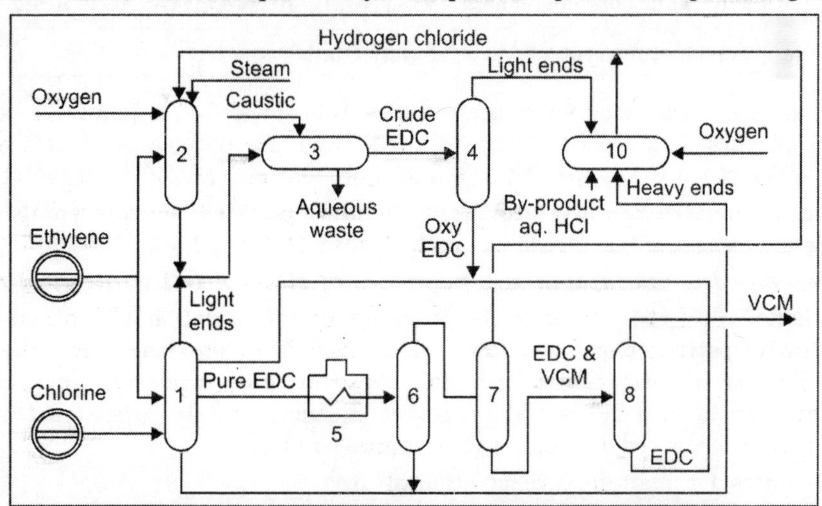

Figure: The European Vinyls Corporation process for producing vinyl chloride: (1) chlorination section, (2) oxychlorination reactor, (3) steam stripping and caustic treatment of water effluent, (4) EDC distillation, (5) pyrolysis furnace, (6,7,8) VCM and EDC separation, (10) by-product reactor.

Q.97. *Write a short note on properties and uses of vinyl chloride.*

Ans. Vinyl chloride is a reactive gas soluble in alcohol but slightly soluble in water. It is the most important vinyl monomer in the polymer industry. Vinyl chloride is one of the important petrochemical feedstocks and find use in manufacture of poly vinyl chloride the second largest tonnage commercial polymer after polyethylene. About 95 percent of the present vinyl chloride production worldwide is used in polymer production or co-polymer application. Another important use of vinyl chloride is in the production of vinylidiene chloride.

Q.98. *Write a short note on manufacture and uses of vinyl acetate.*

Ans. The production of vinyl acetate through traditional acetylene route, which was developed by Wacker in 1930, involves reaction of acetylene and acetic acid in liquid phase at 60-80 °C and 1-2 atm pressure in presence of mercury salt catalyst. The ethylene route has now replaced the traditional process of manufacture of vinyl acetate.

Vinyl acetate from Ethylene route: Vinyl acetate is made by reaction of ethylene with acetic acid by liquid phase process or by vapor phase process in presence of palladium and cupric chloride catalyst.

Vinyl acetate is one of the important derivatives of ethylene which is used as intermediate for manufacture of polyvinyl alcohol, polyvinyl acetate, polyvinyl butyral, etc.

Use of Vinyl Acetate based products has been tabulated below.

Vinyl Acetate	Polyvinyl Acetate	Surface coating adhesives, Textile resins
	Polyvinyl Alcohol	Textile size, Grease proofing paper, Vinyl emulsifier, Thickener, Viscosity regulators, Adhesives,
	Acrylo- nitrile Copolymer	Acrylic Fibers
	Polyvinyl Formate	Water resistant insulation enamel
	Ethylene Vinyl Acetate	Textile and Paper Coating
	Vinyl chloride comonomers	VC-VAC, LP Records, VC-VAC Coating
	Polyvinyl Butryaldehyde	Safety Glass

Global end use pattern of vinyl acetate is:- Adhesives (23%), paints & coating (29%), textiles (21%), plastics (17%), paper & board (10%). End use pattern of vinyl acetate in India is: Adhesives (35-40%), textiles (30-35%), paints and coating (15-20%), others (10-15%).

Consumption pattern of vinyl acetate in India is polyvinyl acetate emulsions & resins (50%), polyvinyl alcohol (25%), ethylene vinyl acetate (10%), others (15%).

Q.99. *Describe the manufacturing processes of ethanol (ethyl alcohol) and its uses.*

Ans. Ethanol (C_2H_5OH) is generally produced by fermentation of molasses, but due to the development of petrochemical industry and availability of ethylene, now ethylene provides another major route of formation of ethanol. However, still molasses are used to produce ethanol in India. In India some of the important chemical are still prepared through ethanol which were earlier prepared through petrochemical route.

Various routes for manufacture of ethanol are:

- Fermentation of molasses
- Catalytic hydration of ethylene
- Ethylene esterification and hydrolysis

Fermentation of Molasses:

Ethyl alcohol is prepared from molasses by fermentation process utilising yeast enzymes. Separation of 8-10% alcohol is achieved in a series of distillation columns, as alcohol and water at 95% concentration form a zeotropic mixture.

Ethanol by Esterification of ethylene and Hydrolysis:

- Ethylene and sulphuric acid are reacted at 80°C and 1.5 MPa to form a mixture of ethyl sulphates, which are then hydrolysed to ethylalcohol.
- Ethylene and sulphuric acid are reacted in absorber from which the mixture of ethylene sulphates thus formed is fed to hydrolyser from which the crude alcohol and sulphuric acid are fed to stripping section & caustic scrubbing section and finally to a series of two distillation columns for separation of ether and alcohol.

Ethanol by Vapor Phase Hydration of Ethylene:

- An ethylene rich gas is mixed with water and heated to about 300°C and passed on to fixed bed catalytic reactor where catalytic hydration of ethylene takes place.

$$C_2H_4 + H_2O \rightarrow C_2H_5OH$$

- The catalyst used is phosphoric acid deposited on silica gel. The reactor effluents are sent to separator for separation of vapor and liquid. The gases from the separator are cooled and scrubbed with water to recover traces of alcohol. The alcohol water mixture is sent to a series of distillation columns where ether is separated in the light end column and finally 95% by volume ethanol water azeotrope is separated.

Ethanol Production from Catalytic Recycle Vapour Phase Hydration of Ethylene is shown below in figure.

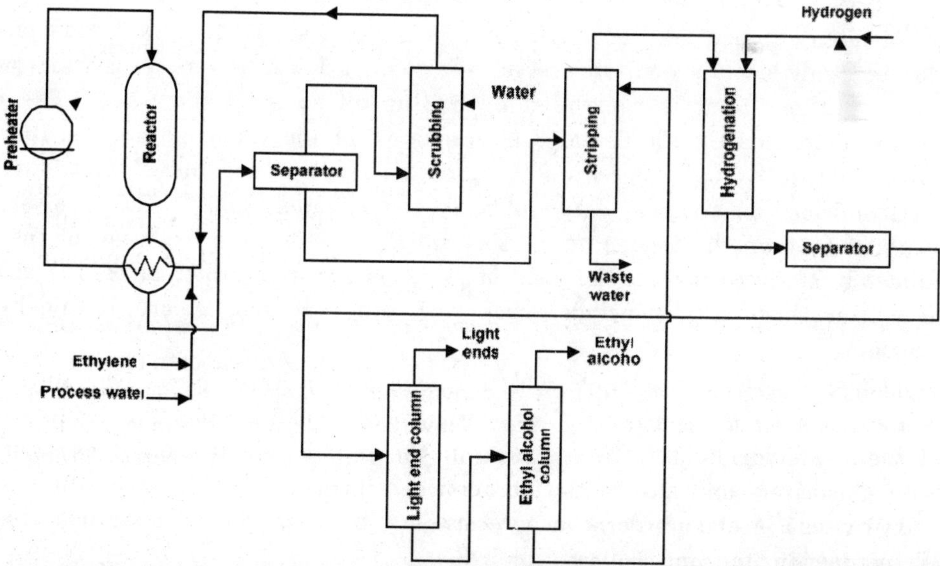

Figure: Ethanol Production from Catalytic Recycie Vapour Phase Hydration of Ethylene

Uses of Ethanol. Ethanol apart from its major use as a beverage is one of the most versatile chemicals and is one of the basic building blocks of the organic chemical industry.

Ethanol's many uses can be divided into solvent and chemical uses. As a solvent, ethanol dissolves many organic based materials such as fats, oils, and hydrocarbons. As a chemical intermediate, ethanol is a precursor for acetaldehyde, acetic acid, & diethyl ether, and it is used in the manufacture of glycol ethyl ethers, ethylamines, & many ethyl esters.

Q.100. ***Describe the manufacturing processes, properties and uses of Acetaldehyde (CH_3CHO).***

Ans. There are many ways to produce acetaldehyde. Historically, it was produced either by the silver-catalyzed oxidation or by the chromium activated copper-catalyzed dehydrogenation of ethanol. Currently, acetaldehyde is obtained from ethylene by using a homogeneous catalyst (Wacker catalyst). The catalyst allows the reaction to occur at much lower temperatures (typically 130°C) than those used for the oxidation or the dehydrogenation of ethanol (approximately 500°C for the oxidation and 250°C for the dehydrogenation).

Ethylene oxidation is carried out through oxidation-reduction (redox). The overall reaction is the oxidation of ethylene by oxygen as represented by:

$$CH_2 = CH_2 + 1/2 \rightarrow CH_3{-}\overset{\overset{\displaystyle O}{\|}}{C}{-}H \qquad \Delta H° = -218.6 \text{ KJ/mol}$$

The Wacker process uses an aqueous solution of palladium (II) chloride, copper (II) chloride catalyst system.

In the course of the reaction, the Pd^{2+} ions are reduced to Pd metal, and ethylene is oxidized to acetaldehyde:

$$CH_2 = CH_2 + PdCl_2 + H_2O \longrightarrow CH_3CHO + 2HCl + Pd°$$

The oxidation reaction may be carried out in a single-stage or a two- stage process. In the single-stage, ethylene, oxygen, and recycled gas are fed into a vertical reactor containing the catalyst solution. Heat is controlled by boiling off some of the water. The reaction conditions are approximately 130°C and 3 atmospheres. In the two-stage process, the reaction occurs under relatively higher pressure (approximately 8 atmospheres) to ensure higher ethylene conversion. The reaction temperature is approximately 130°C. The catalyst solution is then withdrawn from the reactor to a tube-oxidizer to effect the oxidation of the catalyst at approximately 10 atmospheres. The yield of acetaldehyde from either process is about 95%. By-products from this reaction include acetic acid, ethyl chloride, chloro acetaldehyde, and carbon dioxide.

Acetaldehyde is a colorless liquid with a pungent odor. It is a reactive compound with no direct use except for the synthesis of other compounds. For example, it is oxidized to acetic acid and acetic anhydride. It is a reactant in the production of 2-ethyl hexanol for the synthesis of plasticizers and also in the production of pentaerithritol, a polyhydric compound used in alkyd resins. Acetaldehyde is an intermediate for many chemicals such as acetic acid, n-butanol, pentaerithritol, and polyacetaldehyde.

Products made from Acetaldehyde and their uses are tabulated below.

Table: Product Profile of Acetaldehyde

Product	Uses
Pyridine, Picoline	Solvent, Drugs, Dyes, Agricultural chemicals
Chlor-aldehydes	Insecticides, Fungicides, Disinfectants
Acetaldol	1,3-Butylene Glycol (Polyesters),Urethane coating, Humcetant, Printing ink, Crotonaldehyde, n-butyl alcohol, n-butyric acid anhydride, 2-ethyl hexanol, Rubber accelerator, Sorbic acid
Paraldehyde	Rubber accelerator, Antioxidant dye,stuff
Per-acetic acid	Epoxidation reaction, Reagent in caprolactam, Synthetic glycerols
Penta-erythritol	Alkyl resin, Stabilizer, Plasticizers, Chlorinated polyether resin, Intumescents
Acetic, anhydride	Acetyl salicylic acid, Cellulose acetate, Esters
Acetic acid	Cellulose acetate, Vinyl acetate Chloro-acetic acid Ammonium acetate
Lactic acid	Food and, beverages, lactates, adhesives, Leather processing

Q.101. *Write a short note on Ethanol amines.*

Ans. A mixture of mono-, di-, and tri ethanol amines is obtained by the reaction between ethylene oxide (EO) and aqueous ammonia. The reaction conditions are approximately 30–40°C and atmospheric pressure:

$$6\ \overset{O}{CH_2\!-\!CH_2} + 3\ NH_3 \longrightarrow H_2NCH_2CH_2OH + HN(CH_2CH_2OH)_2 + N(CH_2CH_2OH)_3$$

The relative ratios of the ethanolamines produced depend principally on the ethylene oxide/ammonia ratio. A low EO/NH_3 ratio increases mono ethanolamine yield. Increasing this ratio increases the yield of di-and tri ethanol amines.

Table below shows the weight ratios of ethanolamines as a function of the mole ratios of the reactants.

Table: Weight ratios of ethanol amines as a function of the mole ratios of the reactants

	Moles of ethylene oxide/moles of ammonia		
	0.1	**0.5**	**1.0**
Monoethanolamine	75–61	25–31	12–15
Diethanolamine	21–27	28–32	23–26
Triethanolamine	4–12	37	65–59

Ethanolamines are important absorbents of acid gases in natural gas treatment processes. Another major use of ethanolamines is the production of surfactants. The reaction between ethanolamines and fatty acids produces ethanolamides .Lauric acid is the main fatty acid used for producing ethanolamides. Monoethanolamides are used primarily in heavy-duty powder detergents as foam stabilizers and rinse improvers.

Q.102. *Describe the various manufacturing processes of propylene.*

Ans. Propylene is second to ethylene as the largest volume hydrocarbon intermediate for the production of chemicals. Propylene is a byproduct of steam crackers and varying amount of olefins is produced from steam crackers depending on the type of feedstock. Other sources of propylene may be recovery of propylene from Fluid Catalytic Cracker (FCC) light ends, Propane dehydrogenation, Metathesis.

Some of the major processes for production of propylene is given below in Table.

Table: Propylene Production Technologies

Technology	Process description	
Olefin conversion technology	This process involves production of propylene from ethylene and 2-butenes in a fixed bed metathesis reactor containing proprietary catalyst, which promotes reaction of ethylene and 2-butene to form propylene and simultaneously isomerises 1-butene to 2-butene.	
Superflex Process	The process uses a fluidised bed catalytic reactor system using proprietary catalyst which converts low value feedstock to predominantly propylene and ethylene products. Low value light hydrocarbon streams from ethylene plant and refineries can be used, e.g. C4 and C5 olefin rich stream from ethylene plants, FCC naphtha, C4 stream, thermally cracked naphtha from visbreakers or cokers.	
Propylur Process	This process produces propylene beside ethylene from low value rich feeds ranging from C4-C8 from ethylene plant and refineries in a fixed bed reactor using proprietary catalyst. The process offers high selectivity towards propylene.	
UOP Oleflex Process	This process produces polymer grade propylene from propane and the process consist of a reactor, catalyst regeneration section and product separation and fractionation section. The process uses platinum catalyst.	
UOP/Hydro MTO Process	This process converts crude methanol (produced from synthesis gas using natural gas) to ethylene and propylene and can be operated either a maximum ethylene or a maximum propylene production mode using MTO-100 silico alumino phosphate synthetic molecular sieve based catalyst. The process utilizes fluidised bed reactor and regenerator.	
Methanol to propylene (MTP) Technology	This process produces propylene through methanol route using natural gas. In this process propylene is produced in two steps. First methanol is converted to dimethyl ether in reactor followed by reaction of methanol/DME in second reactor. Methanol can be produced from methane from conventional method.	

C4 hydrogenation and Meta-4 Process	This process involves production of polymer grade propylene plus an isobutylene rich stream or MTBE by upgrading low value C4 stream pyrolysis C4 cuts or butene rich cut. The process steps involve – butadiene and C4 acetylenes selective hydrogenation and butadiene hydroisomerisation, isobutylene removal or MTBE production and metathesis step for conversion of butene and ethylene to propylene. The two main equilibrium reactions taking place are metathesis and isomerisation.	
Olefin Ultra TM	A new ultra high activity ZSM-5 additive that provides the highest activity has been developed by Davision catalysts.	
KBR's MAXOFIN-3 TECHNOLOG Y	KBR's MAXOFIN process is based on fluidised bed cracking of gas oils and residue feeds using ZSM catalyst and proprietary MAXOFIN-3 catalyst additive. The process gives 15% or higher propylene yield from gas oil.	

A Typical Composition of FCC Gas Stream is tabulated below.

Table: Typical Composition of FCC Gas Stream

Products	Yield , weight percent
Dry gas (including ethylene)	12.7
Propane	6.5
Propylene	21.0
Butene	35.8

Propylene production by Catalytic Dehydrogenation of propane:

Catalytic dehydrogenation of light paraffins is of increasing importance because of the growing demand of olefins such as propylene and isobutene & n- butenes. Propane dehydrogenation accounts for 2% of the total world propylene production.

Catalytic dehydrogenation takes place at high temperature (650 0C) using platinum based or chromium-alumina or Fe, Cr/Al_2O_3 as catalyst. Reactor effluent treatment for the separation of hydrogen, propylene, and propane is not simple and total investment is high. These production units can be installed only in areas where field propane is available at low costs.

Methanol to Propylene:

This process produces propylene from natural gas via methanol by converting methanol to dimethyl ether in adiabatic reactor using high activity, high selectivity catalyst. The methanol, water, DME stream is then feed to series of MTP reactor where steam is added. The product stream is first processed for removal of traces of water, CO_2 and DME, followed by further processing for yielding polymer grade propylene.

Q.103. ***Describe the properties and uses of propylene.***

Ans. Propylene is a reactive compound that can react with many common reagents used with ethylene such as water, chlorine, and oxygen. However, structural differences between these two olefins result in different reactivities toward these reagents. For example, direct oxidation of propylene using oxygen does not produce propylene oxide as in the case of ethylene. Instead, an unsaturated aldehyde, acrolein, is obtained. This could be attributed to the ease of oxidation of allylic hydrogens in propylene. Similar to the oxidation reaction, the direct catalyzed chlorination of propylene produces allyl chloride through substitution of allylic hydrogens by chlorine. Substitution of vinyl hydrogens in ethylene by chlorine, however, does not occur under normal conditions. Many important• chemicals are produced from propylene.

Propylene is used in many of the world's largest and fastest growing synthetic materials & thermoplastics. The demand of propylene has increased rapidly during the last twenty years and primarily driven by polypropylene demand. Product profile of propylene is given below in table.

Table: Product Profile of Propylene

Product	Uses
Miscellaneous chemicals	1 butanol, 2-ethyl hexanol, Allyl chloride, Epichlorohydrin
Polymer	Polypropylene, Polyacrylamide, nylon 66, acrylic sheets
Propylene oxide	Polyether-polyols, glycol ethers, isopropyl amines, propylene carbamate, surfactants
Propylene glycol	Unsaturated polyester resins, food additives, cellophane, paints and coating, plasticisers, functional fluids, antifreeze, tobacco treatment
Acrylonitrile	Acrylic fiber, acrylic acid, acrylates, methyl methacrylates, adiponitrile
Isopropanol	Acetone, cosmetics, solvents, pharmaceuticals, isopropyl acetate
Polyols	Polyurethane and Polyester

Consumption of refinery grade propylene made up 9% of total consumption in 2009, chemical grade 23% and polymer grade 68%. Refinery grade propylene is consumed mainly for production of cumene and isopropyl alcohol. Chemical grade propylene mostly goes into oxo alcohol, propylene oxide and acrylonitrile production.

Figure below shows the important chemicals based on propylene.

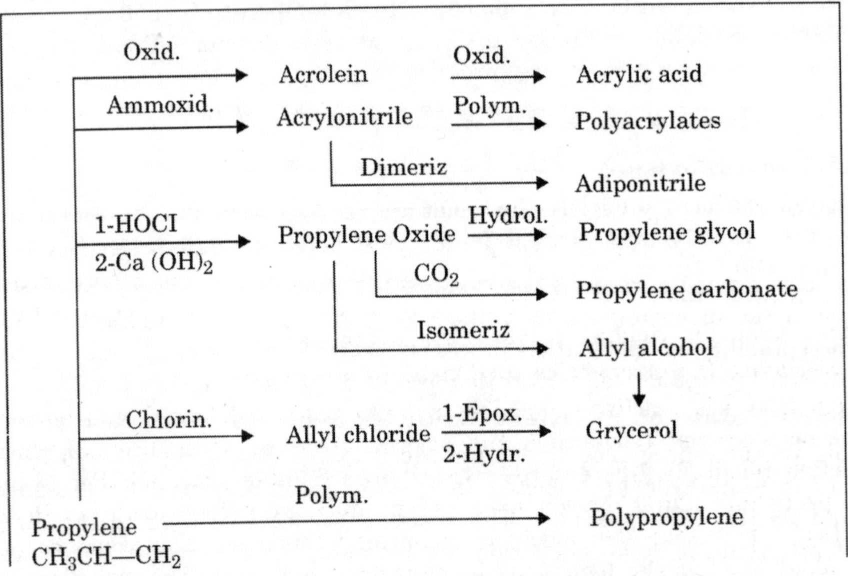

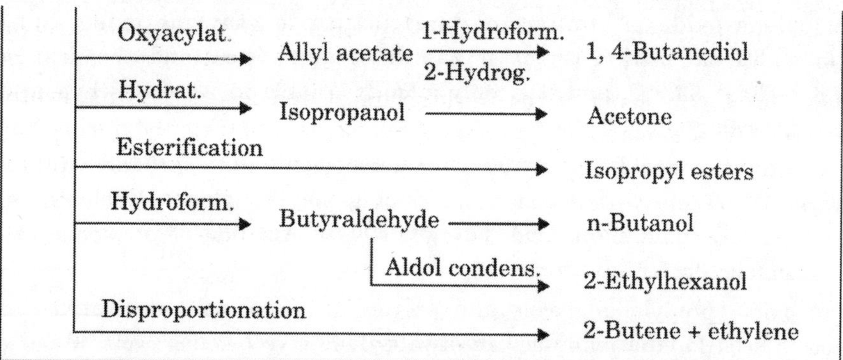

Figure: Important chemicals based on propylene.

Q.104. ***Describe the manufacturing process of Propylene Oxide.***

Ans. There are two major processes for the manufacture of propylene oxide: Propylene chlorohydrin process and propylene oxidation process using peroxides.

Propylene Chlorohydrin Process Route:

The chlorohydrination process consists of formation of propylene chlorohydrin by the reaction between hypochlorous acid and propylene. The propylene chlorohydrin is epoxidised to propylene oxide by a 10% solution of milk of lime or NaOH. Various steps involved are

- ***Propylene hypochlorination:*** Propylene is reacted with aquous chlorine resulting in the formation of propylene chlorohydrins. Unreacted propylene is recycled.
- ***Neutralisation:*** Neutrialisation of propylene chlorohydrins containing hydrochloric acid which is formed during the process.
- ***Dehydrochlorination:*** Reaction of propylene chlorohydrin with milk of lime or caustic soda to produce propyleneoxide.
- ***Purification:*** Distillation of crude propylene oxide for separation heavy ends.

Reactions involved are:

$$Cl_2 + H_2O = HOCl + HCl$$

$$\text{H–CH} = \text{CH} + \text{HOCl} \xrightarrow{350°C} \underset{\text{Propylene chlorohydrin}}{\text{CHCHOH–CHCl}}$$

$$2\,CH_3CHOH–CH_2Cl + Ca(OH)_2 \text{ or } NaOH \longrightarrow 2CH_3\underbrace{CH–CH_2}_{O} + CaCl_2 \text{ or } NaCl + 2H_2O$$

Byproducts formed during the reaction are 1,2-dichloropropane and chlorinated di-isopropyl ether. Some of the disadvantages and major economic drawbacks of the process which led to the wide acceptability of epoxidation processes are use of costly chlorine, production of weak calcium chloride byproduct, and corrosion problem due to chlorine handling.

Oxidation Route of Propylene Oxide manufacture using peroxide Compounds:

In this process, propylene and peracetic acid (in ethyl acetate) which is produced by oxidation of acetaldehyde are reacted in a series of three specially designed reactors at 50-80°C and 90-120 MPa pressure. The reaction products are fed to the stripper where a mixture of propylene & propylene oxide are obtained as top product while mixture of ethyl acetate & acetic acid is obtained as bottom product. Both mixtures are fed to two separated columns where separation of propylene oxide, ethyl acetate, acetic acid, and heavy end takes place.

Q.105. *Describe the properties and uses of Propylene oxide.*

Ans. Propylene oxide is similar in its structure to ethylene oxide, but due to the presence of an additional methyl group, it has different physical and chemical properties. It is a liquid that boils at 33.9°C, and it is only slightly soluble in water. (Ethylene oxide, a gas, is very soluble in water).

Similar to ethylene oxide, the hydration of propylene oxide produces propylene glycol. Propylene oxide also reacts with alcohols, producing polypropylene glycol ethers, which are used to produce polyurethane foams and detergents. Isomerization of propylene oxide produces allyl alcohol, a precursor for glycerol.

Propylene oxide, propylene glycols and polyols are important derivatives of propylene. propylene oxide is used for the manufacture of propylene glycol and polyols. Major consumption of propylene oxide is manufacture of polyurethane and polyester resins. Propylene glycol find major application in the manufacture of unsaturated polyester resins, food additives, pharmaceuticals and personal care, tobacco humectants, cellophane, paints and coatings. Polyols major use is in the manufacture of polyurethane.

Q.106 *Describe the manufacturing process of Propylene Glycol [$CH_3CH(OH)CH_2OH$]* .

Ans. Propylene glycol is made by hydrolysis of propylene oxide. The process steps involved are:

Reaction: Hydrolysis of propylene oxide resulting in formation of mono propylene glycols(MPG).

$$CH_3CH\overset{O}{\frown}CH_2 + H_2O \longrightarrow \underset{\text{Propylene glycol}}{CH_3CHOHCH_2OH}$$

Small amount of di propylene glycol (DPG) and tri propylene glycol (TPG)s are also formed

Concentration: Concentration of glycol solution in multiple effect evaporator.

Distillation: Separation of MPG, DIPG and TPG separated from MPG column in series of distillation column where MPG is separated in first column.

Propylene glycol (1,2-propanediol) is produced by the hydration of propylene oxide in a manner similar to that used for ethylene oxide. Depending on the propylene oxide/water ratio, di-, tri- and polypropylene glycols can be made the main products.

$$n\,CH_3CH\overset{O}{\frown}CH_2 + H_2O \xrightarrow{H^+} \underset{\text{Polypropylene glycol}}{HO[CH(CH_3)CH_2O]_nH}$$

Peroxides actually or potentially used to epoxidizepropylene

Peroxide feedstock	Epoxidation coproduct	Coproduct derivative
Acetaldehyde	Acetic acid	—
Isobutane	tert-Butyl alcohol	Isobutylene
Ethylbenzene	a-Phenylethyl alcohol	Styrene
Isopentane	Isopentanol	Isopentene and isoprene
Isopropanol	Acetone	Isopropanol

Q.107. ***Describe the manufacture, properties and uses of isopropanol (iso propyl alcohol).***

Ans. Isopropanol (2-propanol) is an important alcohol of great synthetic utility. The production of isopropanol from propylene occurs by either a direct hydration reaction (the newer method) or by the older sulfation reaction followed by hydrolysis.

In the direct hydration method, the slightly exothermic reaction could be effected either in a liquid or in a vapor-phase process.

$$CH_3CH = CH_2 + H_2O \longrightarrow CH_3CHOHCH_3$$

In the liquid-phase process, high pressures in the range of 80–100 atmospheres are used. A sulfonated polystyrene cation exchange resin is the catalyst commonly used at about 150°C. An isopropanol yield of 93.5% can be realized at 75% propylene conversion. The only important by-product is di-isopropyl ether (about 5%).

Gas phase hydration, on the other hand, is carried out at temperatures above 200°C and approximately 25 atmospheres. The ICI process employs WO_3 on a silica carrier as catalyst.

Older processes still use the sulfation route. The process is similar to that used for ethylene in the presence of H_2SO_4, but the selectivity is a little lower than the modern vapor-phase processes. The reaction conditions are milder than those used for ethylene.

Isopropanol is a colorless liquid having a pleasant odour; it is soluble in water. It is more soluble in hydrocarbon liquids than methanol or ethanol. For this reason, small amounts of isopropanol may be mixed with methanol-gasoline blends used as motor fuels to reduce phase-separation problems.

Isopropyl alcohol finds wide use as a solvent and raw material for other chemical products like acetone, isopropyl acetate, glycerol, isopropyl & disopropyl amines, corrosion inhibitor di-isoprpopyl ammonium nitrate, floatation agent isopropyl xanthate, isopropyl myristates etc.

About 50% of isopropanol use is to produce acetone. Other important synthetic uses are to produce esters of many acids, such as acetic (iso-propyl acetate, solvent for cellulose nitrate), myristic, & oleic acids (used in lipsticks and lubricants). Isopropyl palmitate is used as an emulsifier for cosmetic materials. Isopropyl alcohol is a solvent for alkaloids, essential oils, and cellulose derivatives.

Q.108. ***How is Acetone Produced from isopropanol.***

Ans. Acetone (2-propanone), is produced from isopropanol by a dehydrogenation, oxidation, or a combined oxidation dehydrogenation route. The dehydrogenation reaction is carried out using either copper or zinc oxide catalyst at approximately 450–550°C. A 95% yield is obtained:

$$CH_3CHOHCH_3 \longrightarrow CH_3\overset{\overset{\displaystyle O}{\|}}{C}CH_3 + H_2$$

The direct oxidation of propylene with oxygen is a non catalytic reaction occurring at approximately 90–140°C and 15–20 atmospheres. In this reaction hydrogen peroxide is co-produced with acetone. At 15% iso-propanol conversion, the approximate yield of acetone is 93% and that for H_2O_2 is 87%:

$$2CH_3CHOHCH_3 + O_2 \longrightarrow 2CH_3\overset{\overset{\displaystyle O}{\|}}{C}CH_3 + H_2O_2$$

The oxidation process uses air as the oxidant over a silver or copper catalyst. The conditions are similar to those used for the dehydrogenation reaction.

Acetone can also be co-produced with allyl alcohol in the reaction of acrolein with isopropanol. The reaction is catalyzed with an MgO and ZnO catalyst combination at approximately 400°C and one atmosphere. Hydrogen produced from the dehydrogenation of iso-propanol and adsorbed on the catalyst surface selectively hydrogenates the carbonyl group of acrolein:

$$CH_3CHOHCH_3 + CH_2{=}CH\overset{O}{\overset{\|}{C}}H \longrightarrow CH_3\overset{O}{\overset{\|}{C}}CH_3 + CH_2{=}CHCH_2OH$$

A direct route for acetone from propylene was developed using a homogeneous catalyst similar to Wacker system ($PdCl_2/CuCl_2$). The reaction conditions are similar to those used for ethylene oxidation to acetaldehyde. Today, most acetone is obtained via a cumene hydroperoxide process where it is co-produced with phenol.

Propertles and Uses of Acetone : Acetone is a volatile liquid with a sweet odor. It is miscible with water, alcohols, and many hydrocarbons. For this reason, it is a highly desirable solvent for paints, lacquers, and cellulose acetate.

As a symmetrical ketone, acetone is a reactive compound with many synthetic uses. Among the important chemicals based on acetone are methylisobutyl ketone, methyl methacrylate, ketene, and diacetone alcohol.

Q.109. ***Describe the method of producing cumene (iso propyl benzene).***

Ans. Cumene is made by alkylkating benzene with propylene using a catalyst.

$$C_6H_6 + CH_2{=}CHCH_3 \longrightarrow C_6H_5CH(CH_3)_2 \qquad \Delta H^{o}_{298} = -113 \text{ KJ/mol}$$

Either a liquid or a gas-phase process is used for the alkylation reaction. In the liquid-phase process, low temperatures & pressures (approximately 50°C & 5 atmospheres) are used with sulfuric acid as acatalyst.

In the vapor-phase process, the reaction temperature & pressure are 250°C & 40 atmospheres. Phosphoricacidon Kieselguhr is a commonly used catalyst. To limit polyalkylation, a mixture of propylene-propane feed is used. Propylene can be a slow as 40% of the feed mixture. A high benzene/propylene ratio is also used to decrease polyalkylation. A selectivity of about 97% based on benzene can be obtained.

Cumene (isopropylbenzene), a liquid, is soluble in many organic solvents but not in water. It is present in low concentrations in light refinery streams (such as reformates) and coal liquids. It may be obtained by distilling (cumene's B.P. is 152.7°C) these fractions.

Cumene is mainly used to produce phenol and acetone. A small amount of cumene is used to make a-methylstyrene by dehydrogenation.a-Methylstyrene is used as a monomer for polymer manufacture and as a solvent.

Q.110. *Describe the production of Acrylonitrile [CH_2=CHCN] by ammoxidation of propylene . List the major uses of acrylonitrile.*

Ans. Ammoxidation refers to a reaction in which a methyl group with allyl hydrogens is converted to a nitrile group using ammonia and oxygen in the presence of a mixed oxides ased catalyst. An application of this reaction produces acrylonitrile from propylene as per the following reaction :

$$CH_2 = CHCH_3 + NH_3 + 1.5\ O_2 \longrightarrow CH_2 = CHCN + 3H_2O$$

As with other oxidation reactions, ammoxidation of propylene is highly exothermic, so an efficient heat removal system is essential.

Acetonitrile and hydrogen cyanide are by-products that may be recovered for sale. Acetonitrile (CH_3CN) is a high polarity aprotic solvent used in DNA synthesizers, high performance liquid chromatography (HPLC), and electrochemistry. It is an important solvent for extracting butadiene from C_4 streams.

Both fixed and fluid-bed reactors are used to produce acrylonitrile, but most modern processes use fluid-bed systems. The Montedison-UOP process uses a highly active catalyst that gives 95.6% propylene conversion and a selectivity above 80% for acrylonitrile. The catalysts used in ammoxidation are similar to those used in propylene oxidation to acrolein. Oxidation of propylene occurs readily at 322°C over Bi-Mo catalysts. However, in the presence of ammonia, the conversion of propylene to acrylonitrile does not occur until about 402°C.

Uses of Acrylonitrile : Acrylonitrile (ACN) is used mainly as a monomer or co-monomer in the production of synthetic fibres, plastics and elastomers. The largest outlet is in the manufacture of acrylic fibres while the main growth sector for ACN is the acrylonitrile-butadiene-styrene (ABS)/styrene- acrylonitrile (SAN) market.

Copolymers of acrylonitrile with butadiene & styrene are the ABS resins and those with styrene are the styrene-acrylonitrile resins (SAN) that are important plastics. Most of the acrylonitrile is used for ABS resins and acrylic & modacry licfibers. Acrylonitrileis also a precursorforacrylic acid (by hydrolysis) and for adiponitrile (by anelectrodimerization).

Acrylonitrile is the key ingredient in the acrylic fiber used to make clothing & carpeting; in acrylonitrile-butadiene-styrene (ABS), a durable material used in automobile components, telephone & computer casings, sports equipment; and in nitrile rubber, which is used in the manufacture of hoses for pumping fuel.

Q.111. ***How is isoprene [$H_2C = C—CH_3 — CH = CH_2$] manufactured from propylene. State its uses.***

Ans. There are several routes for producing isoprene. The choice of one process over the other depends on the availability of the raw materials and the economics of the selected process.

While most isoprene produced today comes from the dehydrogenation of C_5 olefin fractions from cracking processes, several schemes are used for its manufacture via synthetic routes. One of the routes for its production is from Propylene (Goodyear Process) which is described below.

Isoprene is produced by the dimerization of propylene first to 2-methyl-1-pentene. The reaction occurs at 200°C and about 200 atmospheres in the presence of a tripropyl aluminum catalyst combined with nickel or platinum.

$$2\,CH_3CH{=}CH_2 \longrightarrow CH_3CH_2CH_2-\overset{\displaystyle CH_3}{\overset{|}{C}}{=}CH_2$$

The next step is the isomerization of 2-methyl-1-pentene to 2-methyl-2-pentene using an acid catalyst.

$$CH_3CH_2CH_2\overset{\displaystyle CH_3}{\overset{|}{C}}{=}CH_2 \longrightarrow CH_3CH_2CH{=}\overset{\displaystyle CH_3}{\overset{|}{C}}-CH_3$$

2-Methyl-2-pentene is finally pyrolyzed to isoprene.

$$CH_3CH_2CH{=}\underset{}{\overset{CH_3}{\overset{|}{C}}}{-}CH_3 \longrightarrow CH_2{=}\overset{CH_3}{\overset{|}{C}}{-}CH{=}CH_2 + CH_4$$

Isoprene (2-methyl 1,3-butadiene) is the second most important conjugated diolefin after butadiene. Most isoprene production is used for the manufacture of cis-polyisoprene, which has a similar structure to natural rubber. It is also used as a copolymer in butyl rubber formulations.

Q.112. ***Describe oxo processing of olefins for production of aldehydes and alcohols by Hydroformylation of olefins (Oxo-Reaction) .***

Ans. Hydroformylation of olefins (Oxo reaction) produces aldehydes with one more carbon than the reacting olefin. For example, when ethylene is used, propionaldehyde is produced. This reaction is important for the production of higher aldehydes that are further hydrogenated to the corresponding alcohols. The reaction is catalyzed with cobalt or rhodium complexes. Olefins with terminal double bonds are more reactive and produce aldehydes which are hydrogenated to the corresponding primary alcohols. With olefins other than ethylene, the hydroformylation reaction mainly produces a straight chain aldehyde with variable amounts of branched chain aldehydes.

The reaction is generally represented as:

$$2RCH{=}CH_2 + 2H_2 + 2CO \longrightarrow RCH_2CH_2CHO + R\overset{CH_3}{\overset{|}{C}}HCHO$$

The largest commercial process is the hydroformylation of propene (propylene), which yields n-butyraldehyde and isobutyraldehyde. n-Butyraldehyde (n-butanal) is either hydrogenated to n-butanol or transformed to 2- ethyl-hexanol via aldol condensation and subsequent hydrogenation. 2- Ethylhexanol is an important plasticizer for polyvinyl chloride.

Other olefins applied in the hydroformylation process with subsequent hydrogenation are propylene trimer & tetramer for the production of decyl & tridecyl alcohols, respectively, and C_7 olefins (from copolymers of C_3 and C_4 olefins) for isodecyl alcohol production.

Several commercial oxo processes are currently operative. Some use arhodium catalyst complex incorporating phosphine ligands $HRhCO(PPh_3)_2$ at relatively lower temperatures & pressures and produce less branched aldehydes. Older processes use a cobalt carbonyl complex $HCo(CO)_4$ at higher pressures & temperatures and produce a higher ratio of the branched aldehydes. The hydroformylation reaction using phosphine ligands occurs in an aqueous medium. A higher catalyst activity is anticipated in aqueous media than in hydrocarbons. Selectivity is also higher. Having more than one phase allows for complete separation of the catalyst and the products.

Table below shows the hydroformylation conditions of some commercial processes.

Table : Salient features of some commercial oxo processes for propylene hydroformylation

Process	Catalyst	Conditions	% Normal
Ruhrchemie	Co^{2+}, CoO	150°C, 300 atm.	70
BASF	$HCO(CO)_4$	150°C, 30 MPa	70
ICI	Co^{2+}	high pressure	70
Shell	CO/PR_3	180, 50 atm	88
UCC	$HRh(CO)(PPh_3)_3$	100, 30 atm	94

Note: PPh_3 is triphenyl phosphine.

Q.113. ***Describe the manufacturing processes of butadiene. What are its properties & uses?***

Ans. Steam cracker and catalytic dehydrogenation of butenees are the two major sources of butadiene. Butenes can be recovered from C_4 stream or produced by dehydrogenation of butanes. Butadiene (molecular formula $CH_2 = CH - CH = CH_2$) is obtained as the by-product from an naphtha cracker unit.

Butadiene is obtained mainly as a coproduct with other light olefins from steam cracking units for ethylene production. Other sources of butadiene are the catalytic dehydrogenation of butanes & butenes, and dehydration of 1, 4-butanediol. In India butadiene is manufactured from C_4 available from Naphtha & Gas Cracking units.

There are four major routes for production of butadiene:

- Steam cracking of naphtha
- Catalytic dehydrogenation of butenes
- Catalytic dehydrogenation of butanes
- Dehydrogenation-dehydration of ethanol (molassesroute)

Butadiene from C_4 stream of naphtha steam Cracker Plant:

C_4 cut from the steam cracker is first sent for butadiene recovery, which includes selective hydrogenation of acetylenics in the presence of palladium catalyst, then separation of butadiene. Extractive distillation process steps involved are:

- Extractive distillation in which acetylnic compound and butadiene are extracted in one or two stages.
- Recovery of solvent.
- Super fraction of butadiene stream for removal of acetylnic impurities.
- Water scrubbing butadiene depleted cut to recover the solvent.

Various solvents used for separation of butadiene are furfural, dimethyl formamide (DMF), n-methyl pyrrolidone (NMP), and dimethyl acetamide. Selective hydrogenation results in overall improvement in the economy with higher butadiene yield.

Catalytic dehydrogenation of Butanes

Catalytic dehydrogenation of butane is a two stage process involving :

1. catalytic dehydrogenation of butanes to butenes and
2. Catalytic dehydrogenation of butenes to butadiene

Properties and uses of butadiene : Butadiene is a colorless gas with a mild aromatic odor. Its specific gravity is 0.6211 at 20°C and its boiling temperature is–4.4°C.

Butadiene is easily liquefiable flammable gas used in the production of a wide variety of synthetic rubbers (SBR, BR, and NBR etc.), Acrylonitrile-butadiene-styrene (ABS), polymer resins and few chemical intermediates. The unique role of butadiene among other conjugated diolefins lies in its high reactivity as well as its low cost.

Butadiene is by far the most important monomer for synthetic rubber production. It can be polymerized to polybutadiene or copolymerized with styrene to styrene butadiene rubber (SBR).

Uses Butadiene is important raw material for production of a larger number of synthetic rubber and polymers such as styrene butadiene rubber (SBR), poly butadiene, chloroprene rubber, nitrile rubber, acrylonitrile butadiene styrene plastic. Other fastest growing use is in the manufacture of adiponitrile used in the manufacture of Nylon 66.

Butadiene is an important intermediate for the synthesis of many chemicals such as hexa-methylenediamine and adipic acid. Both are monomers for producing nylon. Chloroprene is another butadiene derivative for the synthesis of neoprene rubber.

Butadiene end use Consumption in India during 2010–11 is shown below in figure.

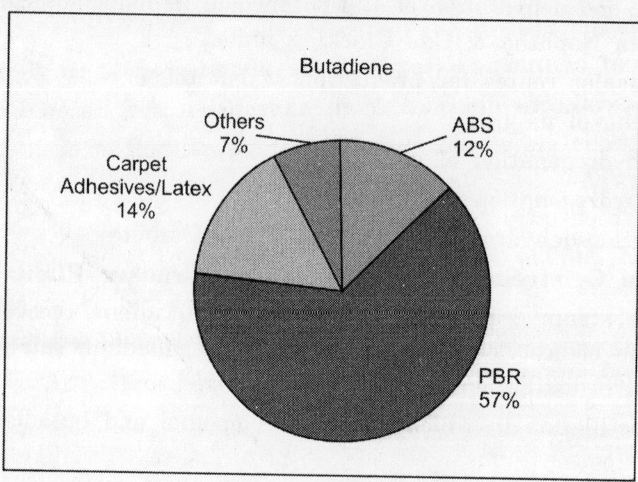

Figure : End use Consumption of Butadienein India during 2010–11

Q.114. *What are aromatics ? What are their uses ?*

Ans. Under the umbrella of 'aromatics', Benzene, Toluene & Xylene (BTX) and naphathalene are the major products. The reactivity of C_6, C_7, C_8 aromatics is mainly associated with the benzene ring. Aromatic compounds in general are liable for electrophilic substitution. Most of the chemicals produced directly from benzene are obtained from its reactions with electrophilic reagents. Benzene can be alkylated, nitrated, or chlorinated to important chemicals that are precursors for many commercial products.

Toluene and xylenes (methylbenzenes) are substituted benzenes. Al though the presence of methyl substituents activates the benzene ring for electrophilic attack, the chemistry of methylbenzenes for producing commercial products is more related to reactions with them ethyl than with the phenyl group. As an electron-withdrawing substituent (of methane), the phenyl group influences the methyl hydrogens and makes the more available for chemical attack. The methyl group could be easily oxidized or chlorinated as a result of the presence of the phenyl substituent.

Aromatic hydrocarbons especially benzene, toluene, xylene (BTX), and ethyl benzene are major feedstock for large number of intermediates which are used in the production of synthetic fibers, resins, synthetic rubber, explosives, pesticides, detergent, dyes, intermediates, etc. Styrene, linear alkyl benzene, and cumene are the major consumer of benzene.

The diverse end-use sectors of Aromatics products range from nylon, unsaturated polyester resins, phenolic resins, drugs, dyes, pesticides, polymers, detergents, solvents- inks, paints, others Polyurethanes, Carrier solvent for wide range of chemical specialties, Oil Field Chemicals to Food Acids.

Products profile of aromatics (ie, BTX) is shown below in figure.

Various uses of products derived from aromatics are listed below:

Naphthalene: Phthalic anhydride, insecticides, b-naphthol, moth balls, surface active agent, synthetic tanning

agents, dyes, rubber chemicals, solvents, agricultural uses.

Phenol: Adipic acid, aniline, Bisphenol-A, caprolactam, phenolic resin, pesticides, dyes, rubber chemical

Acetone: Methyl isobutyl ketone, methyl methacrylate, solvent methyl isobutyl carbinol.

LABs (Linear Alkyl Benzene): Detergent

Cyclohexane: Nylon 6, nylon 66, adipic acid.

Styrene: Polystyrene, SBR, ABS and SAN resins, polyester, acrylonitrile butadienes, styrene plastic protective coatings.

Maleic anhydride: Acids (fumaric, malic); lubricating oil additives; copolymers; agricultural chemicals, unsaturated polyester resin; 1,4-butanediol (Resins, polyurethane solvent, pharmaceuticals).

Chlorobenzene: Aniline, phenol, DDT, chloronitrobenzene

Aniline: Dyes and intermediates, rubber chemicals, drugs & pharmaceuticals, photographic chemicals, isocyanate.

BHC(Benzene Hexachloride): Pesticides.

Toluene di-isocynate (TDI): Polyurethane (Rigid foam, flexible foam, surface coatings).

Trinitrotoluene (TNT) : Explosives.

Benzoic acid: Caprolactam, pharmaceuticals and flavors, phthalates, terephthalic acid, Phenol, sodium benzoate - food preservatives.

p-cresol: Di-tert-butyl-p-cresol (antioxidants)

Isophathalic acid: Resins, unsaturated polyesters, plasticizers, other esters.

Phathalic anhydride: Plasticizers, polyester resins, alkyl resins, dyes &pigments, herbicides, isatoic anhydride, polyester polyols, phthalimide-fungicides.

TPA, DMT: Polyester fibers, films, polyethylene, terephthalate

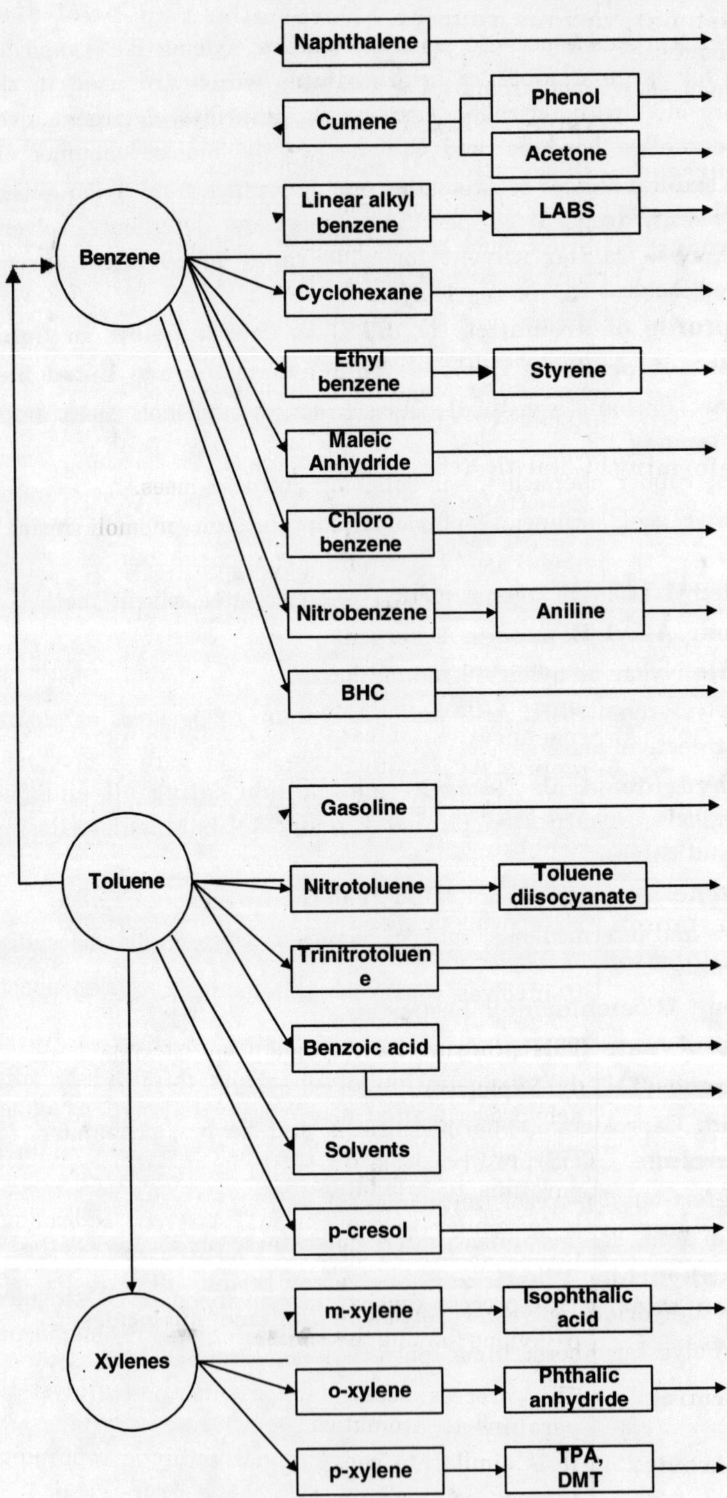

Figure : Product Profile of Aromatics

Q.115. List out various sources of aromatics and brief description of the production processes.

Ans. Various sources of aromatics are tabulated below.

Table : Various Sources of Aromatics

Processes/sources	Description of the processes
Coal Carbonisation (Coke oven plant)	From coke oven plant during carbonisation, benzene is obtained as by product which contains about 75,15 &10 percent of benzene, toluene & xylene respectively (obtained by distillation of benzol)
Steam cracking of hydrocarbons	Steam cracking of naphtha and light hydrocarbons like ethane & propane produce liquid product (pyrolysis gasoline) rich in aromatics containing about 65% aromatics about 50% of which is benzene. About 30% of benzene produced worldwide is from pyrolysis gasoline.
Catalytic Reforming of naphtha	Catalytic reforming is a major conversion process, which converts low octane naphtha to high-octane gasoline and produce aromatics rich in BTX. Major reactions involved are dehydrogenation of naphthalenes to aromatics, isomerisation of paraffins & naphthenes, dehydrocyclisation of paraffins to aromatics, and hydrocracking of paraffins.
BP-UOP Cyclar Process	In this process, BTX is produced by dearomatisation of propane and butane. The process consists of reaction system, continuous regeneration of catalyst, and product recovery. Catalyst is a proprietary zeolite incorporated with a non noble metalpromoter.
Dearomatisation of naphtha	Process consists of extraction of aromatics from high aromatic naphtha feed without prior reforming. The process is useful for naphtha having high aromatics.
Hydro dealkylation & disproportionation of toluene	Hydrodealkylation: It involves production of benzene by dealkylation of toluene either by catalytic or thermal process. Disproportionation: It involves conversion of toluene into benzene and xylenes.
Mitsubishi's Zforming Process	This process uses metallosilicate zeolite catalyst to promote dehydrogenation of paraffins followed by oligomerisation and dehydrocyclisation of paraffins followed by oligomerisation.
KTI Pyroforming	This process uses a shape selective catalyst to convert C_2 and C_3 paraffins to aromatics.
Cheveron'sAromax process	It is similar to conventional catalytic reforming processes and L-type zeolite catalyst.
Isomerisation and Isomerprocess	This process consists of conversion of C_8 stream into valuable o- & p- xylene having isomerisation and isomer separation stage.
KTI Pyroforming	This process uses a shape selective catalyst to convert C_2 & C_3 paraffins to aromatics.
Cheveron'sAromax process	It is similar to conventional catalytic reforming processes and L-type zeolite catalyst.

Q.116. *Describe the production processes and uses of benzene.*

Ans. Benzene is one of the primary building blocks for the petrochemical industry. It is found naturally in crude oil but it is also commercially produced via synthesis from other petroleum basedchemicals.

Benzene was originally produced by separation of benzol from coke oven gas followed by its rectification to produce benzene,toluene & xylene as a byproduct of coke production for the steel industry. However today, benzene is primarily produced as a by-product of refinery and steam cracker operations. There are also a number of other processes - toluene disproportionation (TDP), selective TDP (STDP) etc. - that produce benzene as a co-product. There is only one process, the hydrodealkylation (HDA) process that produces benzene on purpose.

World Benzene Production by Source is shown below in figure.

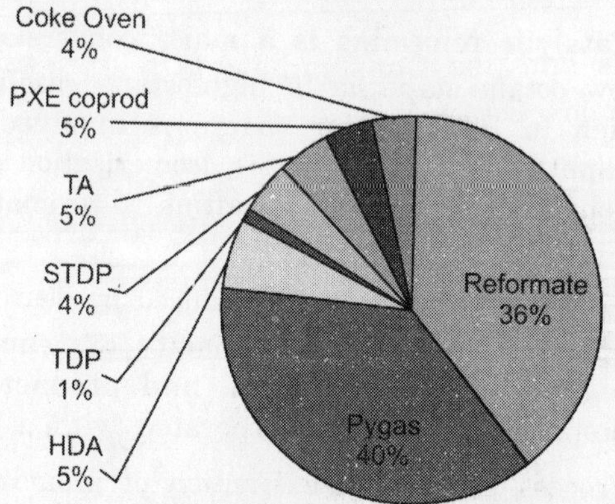

Figure : World Benzene Production by Source

Uses: Benzene (C_6H_6) is the most important aromatic hydrocarbon. It is the precursor for many chemicals that may be used as end products or intermediates. Almost all compounds derived directly from benzene are converted to other chemicals and polymers. For example, hydrogenation of benzene produces cyclohexane. Oxidation of cyclohexane produces cyclohexanone, which is used to make caprolactam for nylon manufacture. Due to the resonance stabilization of the benzene ring, it is not easily polymerized. However, products derived from benzene such as styrene, phenol, and maleic anhydride can polymerize to important commercial products due to the presence of reactive functional groups. Benzene could be alkylated by different alkylating agents, hydrogenated to cyclohexane, nitrated, or chlorinated.

Benzene is principally used as a primary building block for the production of Ethyl Benzene/Styrene, Cumene/phenol, cyclohexane/Caprolactum, Chlorobenzene/derivatives LAB and other industrialchemicals.

Ethylbenzene/styrene, cumene/phenol and cyclohexane consume over 80% of global benzene demand.

Benzene end use Consumption in india during 2010-11 is shown below in figure.

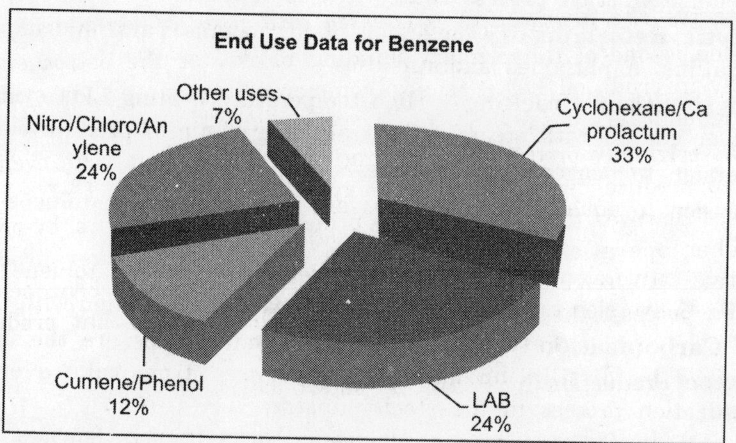

Figure : Benzene end use Consumption in india during 2010-11

Global Benzene Consumption End use wise is shown below in figure.

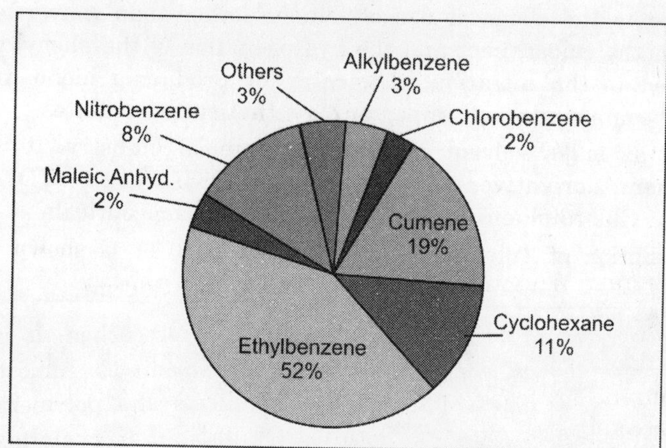

Figure : Global Benzene Consumption End use-wise

Q.117. ***Describe various production processes of toluene and its uses.***

Ans. Toluene is an aromatic hydrocarbon and it plays an essential role in the petrochemical industry. Many of the products we use in our everyday life stem from this ringed aromatic compound.

Production Processes for Toluene

It is commercially produced via synthesis from petroleum based chemicals. The Petrochemical industry's predominant sources of Toluene are refinery reformers and hycarbon steam crackers, which together account for over 98% of the world supply.

The toluene production process routes are as follows:

- **Catalytic Reforming** – Conversion of low octane paraffinic naphtha to hydrogen, isoparaffins, naphthenes andaromatics.
- **Pyrolysis Steam Cracking** – High temperature heating LPG's, condensate, naphtha, diesel or gas oil with steam, in the absence of air to produce olefins andaromatics.
- Conversion to benzene via hydrodealkylation, (HDA).
- Conversion to equal portions of benzene and xylenes via toluene disproportionation (TDP).
- Selective conversion to benzene and xylene via selective toluene disproportionation, (STDP). Conversion to benzene and xylene via transalkylation with C9 & C10aromatics.
- **Coal Carbonization** – Benzene, toluene and xylenes are the low purity products recovered by distillation of benzol recovered from coke oven gas in the coal carbonization process in the steel industry.

Uses of toluene : Toluene (methylbenzene) is similar to benzene as a mononuclear aromatic, but it is more active due to presence of the electron donating methyl group. However, toluene is much less useful than benzene because it produces more polysubstituted products. Most of the toluene extracted for chemical use is converted to benzene via dealkylation or disproportionation. The rest is used to produce a limited number of petrochemicals. The main reactions related to the chemical use of toluene (other than conversion to benzene) are the oxidation of the methyl substituent and the hydrogenation of the phenyl group. Electrophilic substitution is limited to the nitration of toluene for producing mono nitrotoluene and di nitrotoluenes. These compounds are important syntheticintermediates.

Toluene is used as major solvent in the petrochemical industry. It is also used in the manufacture of Toluene derivatives such as Toluene Di-Isocynate (TDI), Para-Cresol, Iso Butyl benzene (IBB), Chlorotoluene, Nitrotoluene andPharmaceuticals

End use Consumption of Toluene in India during 2010-11 is shown below in figure.

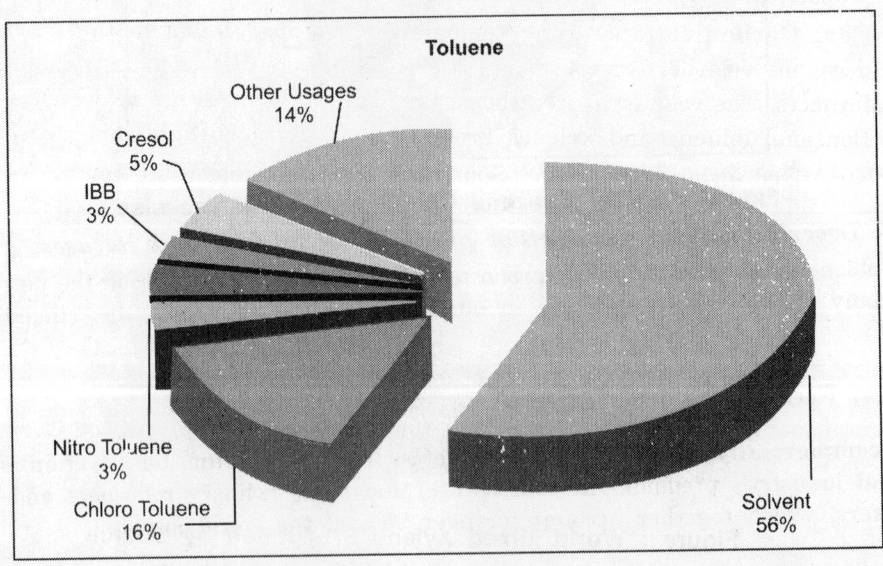

Figure : End use Consumption of Toluene in India during 2010-11

Mixed Xylenes: Mixed Xylenes (dimethylbenzenes) are an aromatic mixture composed of three isomers (o-, m-, and p-xylene). They are normally obtained from catalytic reforming and cracking units with other C_6, C_7, & C_8 aromatics. Separating the aromatic mixture from the reformate is done by extraction-distillation and isomerization processes.

Mixed Xylenesis aclear, colorless, sweet-smelling liquid that is very flammable. It is commercially produced via synthesis from other petroleum based chemicals.*para*-Xylene is the most important of the three isomers for producing terephthalic acid to manufacture polyesters. *m*-Xylene is the least used of the three isomers, but the equilibrium mixture obtained from catalytic reformers has a higher ratio of the meta isomer. *m*-Xylene is usually isomerized to the more valuable *p*-xylene.

Production Process for mixed xylenes : Mixed Xylenesare basic aromatic petrochemicals and are produced through ollowing severalroutes :

- **Catalytic Reforming**:– Conversion of low octane paraffinic naphtha to hydrogen, isoparaffins, naphthenes andaromatics.
- **Pyrolysis Steam Cracking**:– High temperature heating LPG's, condensate, naphtha, diesel or gas oil with steam, in the absence of air to produce olefins andaromatics.
- **Cyclar® (dehydrocyclodimerization):–** Catalytic conversion of propane/butane toaromatics.
- **Toluene disproportionation (TDP):–** Catalytic conversion of toluene to benzene andxylene.
- **Transalkylation:–** Catalytic conversion of toluene and C_9/C_{10} aromatics to benzene & xylene.
- The predominant supply sources for xylenes are refinery reformers and toluenedis-proportionation units which together account for over 93% of the world's supply.
- **Coal Carbonization:–** High temperature heating of coal in the absence of air to drive off volatile hydrocarbons. The resulting coke is used in steel industry blast furnaces. The volatile hydrocarbons, tar, light oil and gas are recovered as byproducts. Benzene, toluene and xylenes are recovered from the light oil.

Global Mixed Xylene Production Sourcewise is shown below in figure.

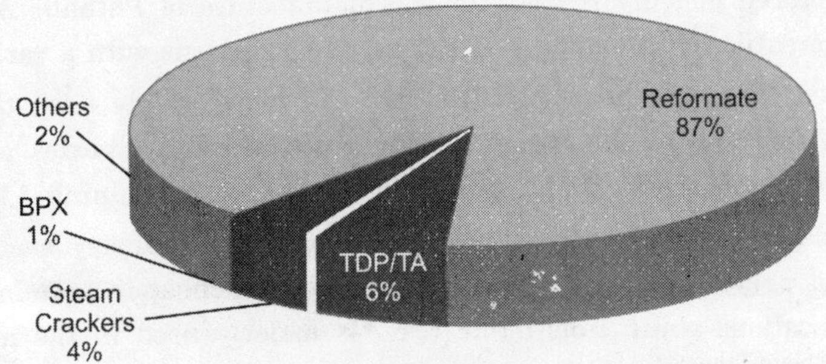

Figure : World Mixed Xylene Production by Source

Uses of mixed xylene (MX): Mixed Xylene are used as solvents in paints, printing inks, rubber, and leather industries.

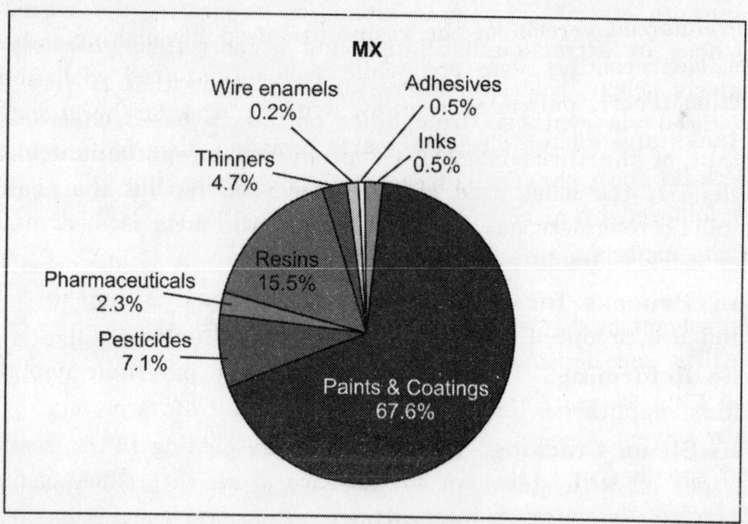

Figure : Mixed Xylene End Use Consumption in India during 2010-11

Approximately 65% of the isolated xylenes are used to make chemicals. The rest are either used as solvents or blended with gasolines.

Q.118. *Write a short note on OrthoXylene(OX).*

Ans. Ortho Xylene (OX) is one of the three isomers of Xylene. It has the second largest demand amongst three isomers. It is a colourless and tasteless liquid with a peculiar odour. It reacts explosively with strong oxidants.

Production Process: Ortho-xylene(OX) can be produced by catalytic reforming of naphtha as well as from pyrolysis gasoline.

Uses of ortho-xylene(OX) :

- Ortho Xylene is used for a variety of applications. A majority (92%) of all the Ortho-Xylene produced globally is used for the manufacture of Phthalic Anhydride (PA). PA is a very versatile intermediate used for organic synthesis with a variety of applications.
- Unsaturated Polyester Resins (UPRs) are the most widely used thermosetting resin accounting for 66% of the thermosetting resin market. Epoxies account for 23% of the market share while the others account for the remaining 11%.
- The two major types of UPRs used in the composite industry are Orthophthalic and Isophthalic polyester resin. Orthophthalic is the cheaper standard resin used for many applications while Isophthalic resin is majorly used in the marine construction industry due to its superior water resistance.

- The major useful characteristics of polyester resins are its corrosion resistance, low weight, modifiable mechanical properties, dimensional stability, low cost, easy processing andhandling.
- The un-reinforced version of the resins are used for making buttons, body fillers, coatings, clear casting resin etc. while resins reinforced with fibers are used for construction (doors, pultruded profiles, window, modular buildings), sanitary items (pipes, tanks and ducts), electrical parts (housing, fuse box etc.), automobile parts (hood, deck lid), boat pars (hull, decks etc.). Most of the UPR is used in the construction industry followed by marine, transportation and E&E.

Other than the major uses in PA and UPR, Ortho Xylene is also used for the following applications.

- Preferred solvent in the aromatics range due to its narrow distillation range, relatively low volatility and highsolvency.
- It is also used as a carrier solvent in many pesticide formulations.
- Used as a solvent inpaints.
- The good solvency of Ortho Xylene provides consistent tint development and tint stability in the ink formulations.
- It is also used as a solvent in the manufacture of Oil Field Chemicals like pour point depressants due to its relatively high flash point and high solvency.
- Manufacture of 3,4 Xylidine used for the manufacture of vitamin B_2.

End use Consumption of ortho-xylene (OX) in India during 2010-11 is shown below in figure.

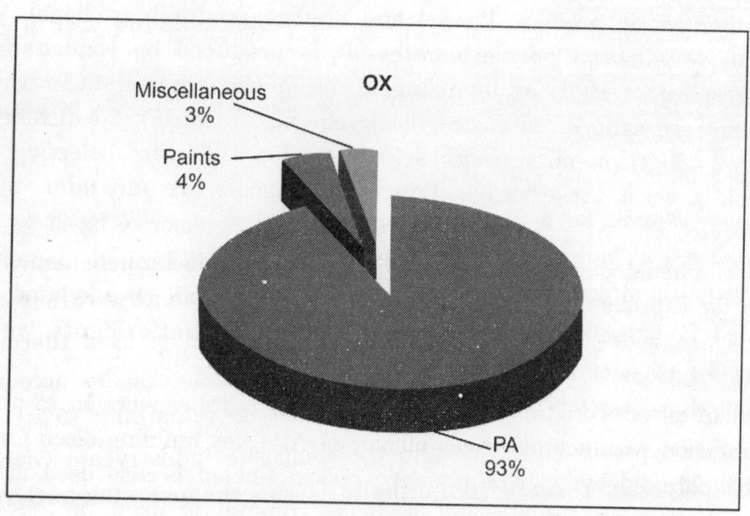

Figure : End use Consumption of ortho-xylene (OX) in India during 2010-11

Q.119. *Write a short note on HYDRODEALKYLATION OF TOLUENE to produce benzene*

Ans. Toluene is hydro dealkylated to benzene over a hydrogenation-dehydrogenation catalyst such as nickel. The hydrodealkylation of toluene with hydrogen is a hydro cracking reaction favored at higher temperatures and pressures. The reaction occurs at approximately 700°C and 40 atmospheres. A high benzene yield of about 96% or more can be achieved.

$$C_6H_5CH_3 + H_2 \longrightarrow C_6H_6 + CH_4$$

Dealkylation also can be effected by steam. The reaction occurs at 600–800°C over Y, La, Ce, Pr, Nd, Sm, or Th compounds, Ni-Cr_2O_3 catalysts, and Ni-Al_2O_3 catalysts at temperatures between 320–630°C. Yields of about 90% are obtained. This process has the advantage of producing, rather than using hydrogen .

Q.120. *Describe the manufacture, properties and applications of PHENOL.*

Ans. Phenol, also known as carbolic acid, is an organic compound derived from petrochemical sources. It is a white crystalline solid. Phenol is appreciably soluble in water and slightly acidic in nature. Various production routes for Phenol are :

- Phenol from Cumene
- Phenol from Benzoicacid
- Phenol from chlorobenzene
- Benzene Sulphonation

With the availability of propylene now phenol is made by cumene route with added advantage of acetone as by product. Phenol and acetone produceb is-phenol A, an important monomer for epoxy resinsand polycarbonates. It is produced by condensing acetone and phenol in the presence of HCI, or by using a cation exchange resin.Important chemicals derived from phenol are salicylic acid; acetylsalicyclic acid (aspirin); 2,4-dichlorophenoxy acetic acid (2,4-D), and 2,4,5- triphenoxy acetic acid (2,4,5-T), which are selective herbicides; and pentachlorophenol, a wood preservative. Other halophenols are miticides, bactericides, and leather preservatives.

About 12% of phenol demand is used to produce caprolactam, a monomer for nylon 6. The main source for caprolactam, however, is toluene.Phenol can be alkylated to alkylphenols. These compounds are widely used as nonionic surfactants, antioxidants, and monomers in resin polymer applications:

Major Uses of phenol : The major uses of phenol is its conversion to plastics or related materials. Condensation with acetone gives bisphenol–A, a key building block for polycarbonates. Condensation with formaldehyde gives phenolic resins. Phenol is also used in the preparation of cosmetics viz. sunscreen. In clinical medicine, phenol is used in the management of spasticity and refractory cancer pain.Globally Phenol was consumed mainly for production of bisphenol A & phenolic resins which accounted for 42 & 28% respectively of total phenol consumption in 2009. Major end uses of Phenol are: Bisphenol-A(BPA), Polycarbonate/Liquid Epoxy Resins, PF Resins, Caprolactum, Nylon fibers, resins & films.

End use Consumption of Phenol in India during 2010-11 is shown below in figure.

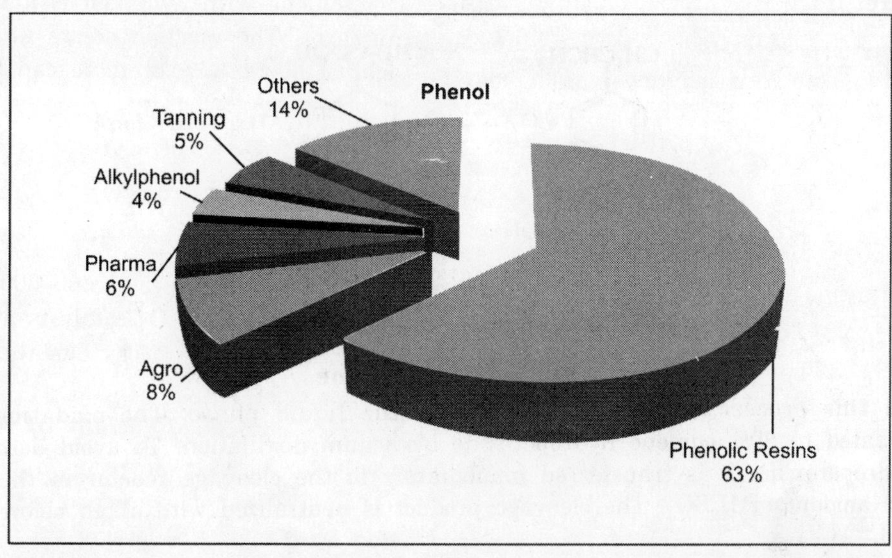

Figure : End use Consumption of Phenol in India during 2010-11

Phenol derivatives and their applications in India are tabulated below .

Table : Phenol derivatives and their applications in India

S.No.	Phenol Derivatives	Applications
1.	Phenol Formaldehyde resin	Automobile components, household electrical and domestic appliances Laminates and plywood industry Abrasive grinding wheel industry Foundry Industry
2.	Alkylated derivatives of phenol	Detergents, wetting agents, oil emulsifiers, textiles, lube oil additives, wire enamels, cutting fluid additives and rubber chemicals
3.	Chlorophenols	Fungicides, fumigants, Intermediates for Insecticides and herbicides
4.	Dyes	Textile Industry
5.	LeatherChemicals	Used for tanning & post tanning operations of raw hides & skins

Q.120. ***Describe the manufacturing process of Phenol & Acetone from Cumene.***

Ans. Phenol, C_6H_5OH (hydroxybenzene), is produced from cumene by a two-step process. In the first step, cumene is oxidized with air to cumene hydroperoxide. The reaction conditions are approximately 100–130°C and 2–3 atmospheres in the presence of a metal salt catalyst.

In the second step, the hydroperoxide is decomposed in the presence of an acid to phenol and acetone. The reaction conditions are approximately 80°C and slightly below atmospheric pressure.

$$C_6H_5CH(CH_3)_2 + O_2 \longrightarrow C_6H_5C(CH_3)_2OOH \qquad \Delta H = -116 \text{ KJ/mol}$$

Cumene   cumene hydroperoxide

$$C_6H_5C(CH_3)_2OOH \xrightarrow{H^+} C_6H_5OH + CH_3COCH_3$$

Phenol   **Acetone**

In this process, cumene is oxidized in the liquid phase. The oxidation product is concentrated to 80% cumene hydroperoxide byvacuum distillation. To avoid decomposition of the hydroperoxide, it is transferred immediately to the cleavage reactor in the presence of a small amount of H_2SO_4. The cleavage product is neutralized with alkali before it is finally purified.

After an initial distillation to split the coproducts phenol & acetone, each is purified in separate distillation and treating trains. An acetone finishing column distills product acetone from an acetone/water/oil mixture. The oil, which is mostly unreacted cumene, is sent to cumene recovery. Acidic impurities, such as acetic acid and phenol, are neutralized by caustic injection.

Cumene processes are currently the major source for phenol & coproduct acetone. Previously, phenol was produced from benzene by sulfonation followed by caustic fusion to sodium phenate. Phenol is released from the sodium salt of phenol by the action of carbon dioxide or sulfur dioxide.

Figure below shows the Mitsui Petrochemical Industries process for producing phenol & acetone from cumene.

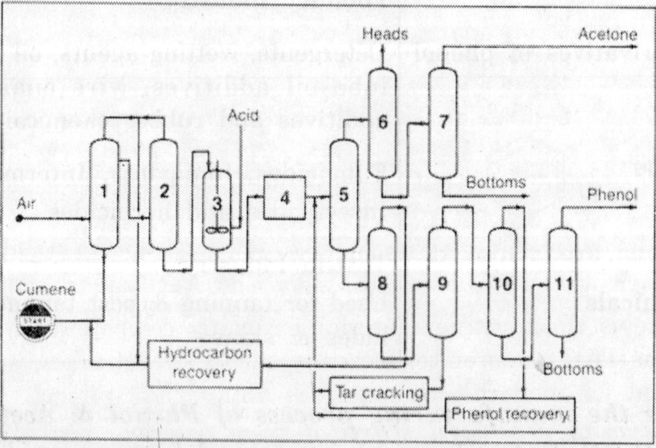

Figure. The Mitsui Petrochemical Industries process for producing phenol & acetone from cumene (1) autooxidation reactor, (2) vacuum tower, (3) cleavage reactor, (4) neutralizer, (5–11) purification train.

Q.121. *Describe the process of manufacturing phenol from toluene oxidation.*

Ans. Oxidizing toluene in the liquid phase over a cobalt acetate catalyst produces benzoic acid. The reaction occurs at about 165°C and 10 atmospheres. The yield is over 90%:

$$C_6H_5CH_3 + 1\frac{1}{2}O_2 \longrightarrow C_6H_5COOH + H_2O$$

The action of a copper salt converts benzoic acid to phenol. The copper, reoxidized by air, functions as a real catalyst. The Lummus process operates in the vapor phase at approximately 250°C. Phenol yield of 90% is possible:In the Lummus process, the reaction occurs in the liquid phase at approximately 220–240°C over Mg^{2+} + Cu^{2+} benzoate.Magnesium benzoate is an initiator, with the Cu^{2+} reduced to Cu^{1+}. The copper (1) ions are reoxidized to copper (II) ions.

Q.122. *Write a short note on chlorination of benzene to produce chlorobenzene.*

Ans. Chlorination of benzene is an electrophilic substitution reaction in which Cl^+ serves as the electrophile. The reaction occurs in the presence of a Lewis acid catalyst such as $FeCl_3$. The products are a mixture of mono-and dichlorobenzenes. The *ortho*-and the *para*-dichlorobenzenes are more common than meta-dichlorobenzene. The ratio of the monochloro to dichloro products essentially depends on the benzene/chlorine ratio and the residence time. The ratio of the dichloro-isomers (*o*- to *p*- to *m*-dichlorobenzenes) mainly depends on the reaction temperature and residence time:

Typical liquid-phase reaction conditions for the chlorination of benzene using $FeCl_3$ catalyst are 80–100°C and atmospheric pressure. When a high benzene/Cl_2 ratio is used, the product mixture is approximately 80% monochlorobenzene, 15% *p*-dichlorobenzene and 5%-dichlorobenzene.

Continuous chlorination processes permit the removal of mono-chlorobenzene as it is formed, resulting in lower yields of higher chlorinated benzene.

Monochlorobenzene is also produced in a vapor-phase process at approximately 300°C. The by-product HCl goes into a regenerative oxychlorination reactor. The catalyst is a promoted copper oxide on a silica carrier:

$$4\,HCl + O_2 \longrightarrow 2\,Cl_2 + 2\,H_2O$$

$$2\,C_6H_6 + 2Cl_2 \longrightarrow 2\,C_6H_5Cl + 2HCl$$

Higher conversions are achieved when temperatures of 234–315°C & pressures of 2.8-5.5 kg/cm^2 are used.

Monochlorobenzene is the starting material for many compounds, including phenol and aniline. Others, such as DDT, chloronitrobenzenes, polychlorobenzenes, and biphenyl, do not have as high a demand for monochlorobenzene as aniline and phenol.

Q.123. *How is styrene produced starting from benzene via ethyl benzene(EB)?*

Ans. The main process for producing EB is the catalyzed alkylation of benzene with ethylene:

$$C_6H_6 + CH_2{=}CH_2 \longrightarrow C_6H_5{-}CH_2{-}CH_3$$

benzene ethylene ethyl benzene

$AlCl_3$-HCl is commonly usedcatalysts for this reaction. Ethyl chloride may be substituted for HCI in a mole- for-mole basis. Typical reaction conditions for the liquid-phase $AlCl_3$ catalyzed process are 40–100°C and 2–8 atmospheres. Diethylbenzene and higher alkylated benzenes also form. They are recycled and dealkylated to EB.

The vapor-phase Badger process, can accept dilute ethylene streams such as those produced from FCC off gas. A zeolite type heterogeneous catalyst is used in a fixed bed process. The reaction conditions are 420°C & 15-20 kg/cm^2. Over 98% yield is obtained at 90% conversion. Poly ethylbenzene (polyalkylated) and unreacted benzene are recycled and join the fresh feed to the reactor. The reactor effluent is fed to the benzene fractionation system to recover unreacted benzene. The bottomscontaining ethylbenzene and heavier polyalkylates are fractionated in two columns. The first column separates the ethylbenzene product, and the other separates polyethylbenzene for recycling.

Ethylbenzene is mainly used to produce styrene. Styrene (vinylbenzene) is a liquid (b.p. 145.2°C) that polymerizes easily when initiated by a free radical or when exposed to light.

Dehydrogenation of ethylbenzene to styrene occurs over a wide variety of metal oxide catalysts. Oxides of Fe, Cr, Si, Co, Zn, or their mixtures can be used for the dehydrogenation reaction. Typical reaction conditions for the vapor-phase process are 600–700°C, at or below atmospheric pressure. Approximately 90% styrene yield is obtained at 30–40% conversion.

$$C_6H_5CH_2CH_3 \longrightarrow C_6H_5CH{=}CH_2 + H_2$$

Ethylbenzene styrene

In the Monsanto/Lummus Crest process, fresh ethyl benzene with recycled unconverted ethyl benzene are mixed with superheated steam. The steam acts as a heating medium and as a diluent. The endothermic reaction is carried out in multiple radial bed reactors filled with catalysts.

Q.124. ***Describe the production process of phthalic anhydride by catalyzed oxidation of o-xylene.***

Ans. Phthalic anhydride is mainly produced through catalyzed oxidation of *o*-xylene. A variety of metal oxides are used as catalysts. A typical one is V_2O_5 + TiO_2/Sb_2O_3. Approximate conditions for the vapor-phase oxidation are 375–435°C and 0.7 atmosphere. The yield of phthalic anhydride is about 85%:

$$C_6H_4(CH_3)_2 + 3O_2 \longrightarrow C_6H_4(CO)_2O + 3H_2O$$

o-xylene phathalic anhydride

Liquid-phase oxidation of *o*-xylene also works at approximately 150°C. Cobalt or manganese acetate in acetic acid medium serves as a catalyst. The major by-products of this process are maleic anhydride, benzoic acid, and citraconic anhydride (methylmaleic anhydride.

Phthalic anhydride's main use is for producing plasticizers by reactions with C_4–C_{10} alcohols. The most important polyvinyl chloride plasticizer is formed by the reaction of 2-ethylhexanol (produced via butyraldehyde) and phthalic anhydride.

Phthalic anhydride is also used to make polyester and alkyd resins. It is a precursor for phthalonitrile by an ammoxidation route used to produce phthalamide and phathilimide.

Q.125. ***How is maleic anhydride produced from benzene or n-butane/n-butene.***

Ans. Benzene oxidation is the oldest method to produce maleic anhydride. The reaction occurs at approximately 380°C and atmospheric pressure. A mixture of V_2O_5/MO_3 is the usual catalyst. Benzene conversion reaches 90%, but selectivity to maleic anhydride is only 50–60%; the other 40–50% is completely oxidized to CO_2:H_2O

$$\text{Benzene} + \tfrac{9}{2}\,O_2 \longrightarrow \text{maleic anhydride} + 2\,CO_2 + 2H_2O$$

Benzene maleic anhydride

Currently, the major route to maleic anhydride is the oxidation of n-butane . Maleic anhydride also comes from oxidation of n-butenes.

The production of maleic anhydride from n-butenes is a catalyzed reaction occurring at approximately 400–440°C and 2–4 atmospheres. A special catalyst, constituted of an oxide mixture of molybdenum, vanadium, and phosphorous, may be used.

$$CH_3CH{=}CHCH_3 + 3\,O_2 \longrightarrow \text{maleic anhydride} + 3\,H_2O$$

n-butenemaleic anhydride

Other routes to maleic anhydride are the oxidation of n-butane.

Maleic anhydride, a solid compound that melts at 53°C, is soluble in water, alcohol, and acetone, but insoluble in hydrocarbon solvents.

Maleic anhydride is important as a chemical because it polymerizes with other monomers while retaining the double bond, as in unsaturated polyester resins. These resins, which represent the largest end use of maleic anhydride, are employed primarily in fiber-reinforced plastics for the construction, marine, and transportation industries. Maleic anhydride can also modify drying oils such as linseed and sunflower.

As an intermediate, maleic anhydride is used to produce malathion, an important insecticide, and maleic hydrazide, a plant growth regulator.

Maleic anhydride is also a precursor for 1,4-butanediol through an esterification route followed by hydrogenation.

Q.126. ***How is Terephthalic Acid ($HOOCC_6H_4COOH$) produced?***

Ans. The main route for obtaining terephathalic acid (TPA) is the catalyzed oxidation of paraxylene. Cobalt acetate promoted with either NaBr or HBr is used as a catalyst in an acetic acid medium. Reaction conditions are approximately 200°C and 15 atmospheres. The yield is about 95%:

$$C_6H_4(CH_3)_2 + \frac{3}{2}O_2 \longrightarrow CH_3C_6H_4COOH + H_2O$$

p-Toluic acid

$$C_6H_4(CH_3)_2 + 3\,O_2 + 2CH_3OH \longrightarrow C_6H_4(COOCH_3)_2 + 4\,H_2O$$

Dimethylterephthalate

Special precautions are taken so that the reaction does not stop at the *p*-toluic acid stage. One approach is to esterify toluic acid as it is formed with methanol. This facilitates the oxidation of the second methyl group. The resulting dimethyl terephthalate (DMT) may be hydrolyzed to terephthalic acid.Another approach is to use an easily oxidized substance such as acetaldehyde or methylethyl ketone, which, under the reaction conditions, forms a hydroperoxide. These will accelerate the oxidation of the second methyl group. The DMT process encompasses four major processing steps: oxidation, esterification, distillation, and crystallization.

TPA can also be produced from benzoic acid by a disproportionation reaction of potassium benzoate in the presence of carbon dioxide. Benzene is the coproduct:

Potassium benzoate Terephthalic Acid benzene

$$2\,C_6H_5COOK \xrightarrow{CO_2} C_6H_4(COOK)_2 + C_6H_6$$

The reaction occurs in a liquid-phase process at approximately 400°C using ZnO or CdO catalysts. Terephthalic acid is obtained from an acid treatment; the potassium salt is recycled.

Terephthalic acid is an important monomer for producing polyesters. The main use of TPA and DMT is to produce polyesters for synthetic fiber and film.

Q. 127. *How is toluene diisocyanates (TDI) produced ?*

Ans. The dinitrotoluenes are important precursors for toluene diisocyanates (TDI), monomers used to produce polyurethanes.

Dinitrotoluenes are produced by nitration of toluene with a mixture of concentrated nitric and sulfuric acid at approximately 80°C. The main products are 2,4- and 2,6-dinitrotoluenes.

$$C_6H_5CH_3 + H_2SO_4 + HNO_3 \longrightarrow CH_3C_6H_3(NO_2)_2 + CH_3C_6H_3(NO_2)_2$$

2,4-Dinitrotoluene 80% 2,6-Dinitrotoluene 20%

The TDI mixture is synthesized from dinitrotoluenes by a first step hydrogenation to the corresponding diamines.The diamines are then treated with phosgene to form TDI. The yield from toluene is approximately 85%:

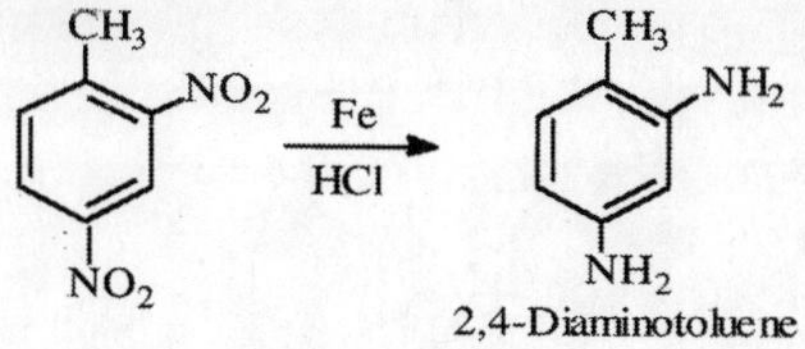

2,4-Diaminotoluene

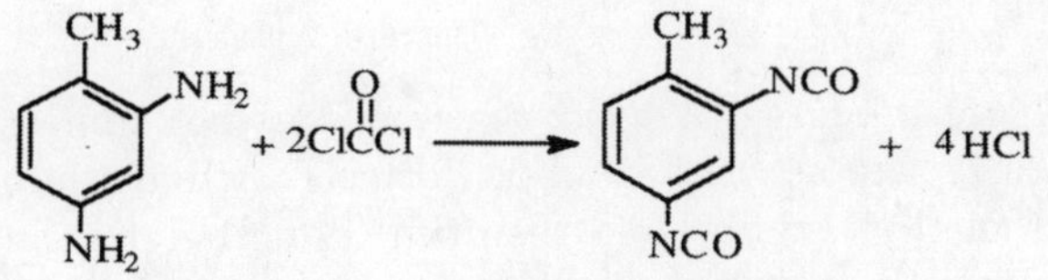

2,4-Toluene diisocyanate

An alternative route for TDI is through a liquid-phase carbonylation of dinitrotoluene in presence of $PdCl_2$ catalyst at approximately 250°C and 200 atmospheres.

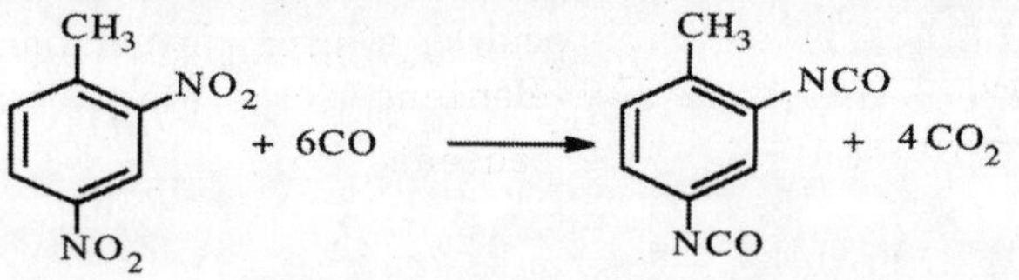

Trinitrotoluene TNT is a well-known explosive obtained by further nitration of the dinitrotoluenes.